The Colorado Plateau III

The Colorado Plateau III

INTEGRATING RESEARCH AND RESOURCES MANAGEMENT
FOR EFFECTIVE CONSERVATION

Edited by Charles van Riper III and Mark K. Sogge

The University of Arizona Press
Tucson

This volume is based on research presented at the Eighth Biennial Conference of Research on the Colorado Plateau. This conference was held at Northern Arizona University, Flagstaff, Arizona, and hosted by the U.S. Geological Survey Southwest Biological Science Center's Colorado Plateau Research Station, Merriam-Powell Center for Environmental Research, Bureau of Land Management, National Park Service, Diablo Trust, and the Center for Sustainable Environments at Northern Arizona University.

The University of Arizona Press
© 2008 The Arizona Board of Regents
All rights reserved

ISBN 978-0-8165-2738-0
Library of Congress Control Number: 2007936624

Manufactured in the United States of America on acid-free, archival-quality paper containing a minimum of 30% post-consumer waste and processed chlorine free.

13 12 11 10 09 08   6 5 4 3 2 1

This book was published from formatted electronic copy that was edited and typeset by the volume editors.

CONTENTS

FOREWORD

The Colorado Plateau is one of the most storied regions of the United States. The vistas range from the arches in southern Utah to the Vermillion Cliffs and the Grand Canyon in Arizona, to El Malpais in New Mexico. Many of the protected areas on the plateau are significant for their historical and cultural legacies. Americans have always had a place in their hearts for the romance of the Old West. No area embodies the allure and magic of those times and fantasies more than the Colorado Plateau. It is no wonder then that we seek to understand the unique relationships found on the Colorado Plateau.

This is the Eighth Proceedings that has been produced from papers presented at the series of biennial conferences over the past 16 years, highlighting research on the Colorado Plateau. The 21 chapters in this eighth volume are contributions from federal, state, and private sector researchers, who come together at Northern Arizona University every other year to share scientific information with land managers on the Colorado Plateau. This Colorado Plateau Biennial Conference series, supported by Northern Arizona University and the U.S. Geological Survey's Southwest Biological Science Center, focuses on providing information to U.S. Geological Survey partners, particularly land managers on the Colorado Plateau. The chapters of this book contribute to the ever-growing pool of scientific data that provides baseline information pertaining to the physical, cultural, and biological resources of the Colorado Plateau. Support for many of these studies has come from a spectrum of federal, state, and private partners concerned about the well-being of the plateau's resources. I applaud the efforts of the contributors. With modest funding and a broad base of public and institutional support, these authors have pursued important lines of work in the four states (Arizona, Utah, New Mexico, and Colorado) that comprise the Colorado Plateau biogeographic region. There is little question that the Colorado Plateau Biennial Conference has become the premier forum where scientists and land managers come together in public discussion about current research and resource management issues.

The Colorado Plateau remains one of the richest ecoregions in North America in terms of its high rates of plant endemism and species richness of invertebrates and vertebrates. It is also a region that has come under heavy pressure from human use, in terms of the diversion of its water resources and dramatic increases in eco-tourist activities. Combine this natural richness with new stresses posed by increasing human populations and changing climate patterns on plant and animal communities, and there is a clear need for more investment in science-based management of species and their habitats.

As a people, we face the prospect of extensive local and global environmental changes that continue to perturb the physical, cultural, and biological resources on lands of the Colorado Plateau. As the research branch for the Department of the Interior, we in the U.S. Geological Survey are committed to identify, in a scientific manner, information that can be used by land managers to protect our resources from detrimental change due to modern human influences. We must develop the information necessary to alert our managers, elected officials, and the public to the importance of their natural surroundings as elements of those basic resources that sustain us, inspire us, and represent our natural biological and environmental heritage.

ROBERT C. SZARO
U.S. Geological Survey
Chief Scientist for Biology
National Center, Reston, VA

DEDICATION

We are pleased to dedicate these Proceedings of the Eighth Biennial Conference of Research on the Colorado Plateau to Dr. "Sue" Haseltine, Anne Kinsinger, and Louella Holter. Sue and Anne have both provided excellent leadership for the USGS during recent times of change and challenge for the agency. Their efforts, both individually and collectively, have laid the foundation for the continuing success of the USGS Southwest Biological Science Center and its associated research stations. Sue Haseltine has been a strong national advocate for the effective integration of the Biology discipline into the broader USGS structure, and for a continued focus of USGS on meeting Department of Interior partner research needs. She is a champion of taking USGS science to new levels of technical excellence and societal relevancy. Anne Kinsinger has spearheaded an array of initiatives and programs designed to bring the entire range of USGS expertise and efforts to bear on addressing complex and large-scale issues, on systems ranging from Puget Sound to the Klamath watershed and the Sagebrush biome. She was also personally responsible for the formation of the Southwest Biological Science Center, which has provided a strong USGS biological science presence and focus on the arid lands of the Colorado Plateau and greater Southwest. Louella Holter's energy, dedication, and keen eye for editing have been instrumental in the actual production and publication of the Proceedings from the sixth, seventh, and now eighth Colorado Plateau Biennial Conferences. This will be the last biennial conference Proceedings that Louella edits, and we felt it appropriate that she be one of the people to whom we dedicate this volume.

Sue Haseltine was born and raised in Pittsfield, a small town in the middle of the spruce bog region of Maine. She grew up hunting and fishing in local haunts with her father, who was a wildlife biologist and biology teacher. After graduating from high school, Sue enrolled as a pre-medical student at Clark University in Worcester, Massachusetts. After two years, she transferred to the University of Maine and received her undergraduate degree (BS in Wildlife Science) in 1971. She went on to graduate work at Ohio State University, where she obtained a Masters degree (1973) and PhD (1976) for work on mechanisms of eggshell thinning by DDT/DDE in wild birds. Sue has been an active and well-known scientist and science manager for over 30 years. From 1976 to 1989, she was with the U.S. Fish and Wildlife Service Patuxent Wildlife Research Center in Laurel, Maryland; there, Sue served as Research Biologist, Section Leader and Branch Chief. Her other positions have included Chief of the Office of Quality Assurance (1987–1989) at the U.S. Fish and Wildlife Service Research and Development Branch, Director of

Sue Haseltine

the Northern Prairie Wildlife Research Center in Jamestown, North Dakota (1989–1992), and Assistant Regional Director of Refuges and Wildlife for the U.S. Fish and Wildlife Service Great Lakes–Big Rivers Region (1992–1995). Following the formation of the National Biological Service (NBS), Sue was appointed as Eastern Regional Biologist of NBS in 1995; she stayed in this position following the merger of NBS into the U.S. Geological Survey until 1998. Most recently, she has served the USGS as Chief Scientist for the Biological Resources Discipline (1998–2003), and is currently the Associate Director of Biology. Sue's research and management interests continue to focus on the integration of fish and wildlife needs over landscape scales, biological community restoration techniques, science-based approaches to conservation, and genetic and genomic techniques in ecological research. She is an active member of an array of professional science organizations, and regularly serves on research and conservation committees, boards, and teams. In addition to her love of science, Sue is an avid history buff, with particular interest in American and Russian history.

Anne Kinsinger was raised in Pasadena, California, where at an early age she developed a keen love of science and biology. She pursued these interests at the University of California, Davis, and in 1984 received a B.S. in Resource Science from the School of Agricultural and Environmental Science. Anne continued her education at Yale University, where she was awarded a Master of Forest Science in Natural Resource Economics and Policy from the School of Forestry and Environmental Studies in 1990. Initially attracted to the environmental policy arena, she served as Director of Fisheries and Wildlife Assessment for the National Fish and Wildlife Foundation (1990–1992). However, Anne found herself drawn back toward the field of science and took a job with the U.S. Fish and Wildlife Service as Chief of the Field Research Division of the Columbia Environmental Research Center (1992–1994), where she oversaw ecotoxico-

logical research at eight field stations located throughout the United States; during this time she was also involved in the formation of the National Biological Survey. Anne became Director of the Western Ecological Research Center in Sacramento, California in 1994, and helped lead her Center through the transition from NBS to the U.S. Geological Survey. From 1997 to 1999, she was Chief of Staff to the USGS Associate Director for Biology, and later became Chief of Strategic Planning and Analysis in the USGS Director's Office (1999–2001). More recently, as Western Regional Biologist (2001–2005), Anne was responsible for science and management administration of the six USGS Biological Science Centers in the West. In January 2006, Anne volunteered to serve as acting Regional Director of the USGS Western Region, then was formally selected as Regional Director in June 2007. In that role she has oversight of a wide range of scientific programs, including streamgaging; flood, fire and earthquake monitoring and research; fish and wildlife biology; ecosystem science; geographic information; coastal and marine studies; minerals assessments; and much more. Anne serves as lead and member of a variety of teams, boards, and committees that deal with science-based conservation of natural resources; her focus and successes have been on efforts that build collaborative partnerships to address large-scale and high-profile issues. Outside of work, Anne enjoys being outdoors in her garden, kayaking, hiking, and supporting her daughter's numerous athletic pursuits.

Anne Kinsinger

Louella Holter hails from Fox River Grove, a small town north of Chicago. After graduating from high school, she headed west to pursue post-graduation freedom in her Volkswagen minibus. Shortly thereafter in California, Louella traded her minibus for her first road bike and, in her words, "never looked back." This introduction to the world of bike riding spawned a life-long passion and dedication to biking and other outdoor sports. Louella has embarked on numerous bike tours across the United States and much of Europe, including several solo bike rides of a thousand miles or more. In 1978 she moved to Flagstaff, where she set down roots for the first time in her adult life. Louella had always loved reading and writing, and with her 100 word per minute typing skills she was in high demand and eventually was hired as a Technical Editor at

Northern Arizona University. By 1985, Louella took on the role of Associate Editor with the NAU Bilby Research Center. It was in this capacity that she began the technical editing and production of the Colorado Plateau Biennial Conference series. She enjoys working on such large, multi-author compilations because "it brings in every skill that you have to make a coherent whole." Louella will be retiring from NAU in October 2007, and we will miss the skills, energy, and commitment that she has brought to the last three conference proceedings. Based on her recent competitive performances, which include state, national, and international championships in velodrome (track) cycling and cross-country skiing, it is doubtful that Louella will have time to be bored in retirement. Between sporting events, she will undoubtedly continue to run the trails of the Grand Canyon, nurture her backyard lily pond, and spend numerous hours gardening and landscaping at her home in Flagstaff.

Louella Holter

It is because of their commitment to providing quality science as a basis for the conservation and management of our nation's resources, and their continued support of the USGS Southwest Biological Science Center, that we dedicate the Proceedings of the Eighth Biennial Conference of Research on the Colorado Plateau to Sue Haseltine, Anne Kinsinger, and Louella Holter.

INTRODUCTION AND ACKNOWLEDGMENTS

This is the eighth in a series of books that focus on research and resource management issues across the Colorado Plateau. The last three of these eight volumes have been published by the University of Arizona Press. These books highlight the integration of research into resource management efforts, as related to cultural, natural, and physical resources within the biogeographic province of the Colorado Plateau. This particular volume highlights aspects of cultural, vegetation and wildlife research, combined with a series of chapters explaining collaborative tools and decision-making processes that can be used to better manage resources on a larger scale over the Colorado Plateau. The mix of chapters, which covers many diverse research and resource management arenas, also addresses how scientists and land managers can better interface when dealing with management issues along the ever-widening wildland-urban interface within all regions of the Colorado Plateau.

The 21 chapters that constitute the core of this book were selected from 208 research papers, panel sessions, and posters presented at the Eighth Biennial Conference of Research on the Colorado Plateau. Held at Northern Arizona University on 7–10 November 2005 in Flagstaff, Arizona, the conference was hosted by the U.S. Geological Survey Southwest Biological Science Center's Colorado Plateau Research Station (CPRS), the Colorado Plateau Cooperative Ecosystem Studies Unit (CESU), the Merriam-Powell Center for Environmental Research, the Bureau of Land Management, the National Park Service, the Diablo Trust, and the Center for Sustainable Environments at Northern Arizona University. The theme of this conference revolved around research, inventory, and monitoring of lands over the Colorado Plateau, with a focus on tools that can be useful in the integration of biophysical and socioeconomic research into land management actions.

Any scientific work is never a single effort, but a direct result of assistance by many individuals. This book is no exception. We would especially like to thank the following scientific peer reviewers: James K. Agee, Craig Allen, Jan Balsom, Jayne Belnap, Julio Betancourt, Carl Bock, Michael Bogan, Ken Boykin, David Breshears, Sun Joseph Chang, Leslie Chow, Kenneth Cole, Anne Cully, Frank D'Erchia, Jeri DeYoung, Charles Drost, Thomas Edwards, Jeffrey Eidenshink, Laura Ellison, Helen Fairley, Sarah Falzarano, Peter Fflolliott, Karen Firehock, Peter Fulé, Bill Gannon, Randy Gimblett, Steve Gloss, Patty Guertin, Joan Hagar, William Halvorson, Richard Hereford, Calder Hibbard, Ron Hiebert, Collin Homer, Francis Howarth, Richard Hutto, Roy Jemison, Philip Jenkins, Matt Kaplinski, Ron Kerbo, Michael Kilgore, Robert Klaver, Julie Korb, Paul Krausman, A. Kristopher Lappin, Signa Larralde, Tom Loveland, Mary Ann Madej, Serena Mankiller, David Mattson, Dave Mikesic, Mark Miller, John Mosesso, Christopher O'Brien, Richard Ockenfels, Thomas O'Dell, Thomas O'Shea, Stewart Peck, David Peterson, Brian Powell, Julie Prior-Magee, David Propst, James Quinn, Peter Reinthal, Catherine Roberts, George Robinson, John Sauer, Roger Sayre, Jack Schmidt, J. Michael Scott, Harley Shaw, John Spence, Robert Szaro, Tad Theimer, Kathryn Thomas, Jan van Wagtendonk, Dave Wagner, and John Wehausen, all of whom unselfishly devoted their time and effort to improving each chapter that they reviewed.

This series of books has received continued financial support from the U.S. Geological Survey's Southwest Biological Science Center. Dennis B. Fenn, Laura Huenneke, Andrea Alpine, and Ron Hiebert all provided encouragement and/or financial assistance for this publication. Louella Holter and Emily Sherbrooke helped in many ways with editorial duties, and without their attention to detail this book would have never been a reality or finished on time for the 9th Biennial Conference. This will be the last Biennial Conference that Louella Holter will be involved with as Technical Editor, thus we felt it appropriate that she should be one of the people to whom this volume is dedicated (see Dedication section). We would also particularly like to thank Kim van Riper who contributed the line drawings that are found throughout this book. The dedicated USGS Colorado Plateau Research Station staff (S. Adson, T. Arundel, K. Cole, C. Drost, S. Durst, J. Hart, J. Holmes, M. Johnson, D. Mattson, E. Nowak, E. Paxton, K. Paxton, M. Saul, and especially R. Stevens) and volunteers and staff from the conference co-sponsors (especially Judy Buzard and Julye Evans of the Center for Sustainable Environments) provided much-needed assistance during the 8th Biennial Conference. We appreciate the assistance of Neil Cobb, who organized numerous special sessions at the conference. Finally, we express deep appreciation to our wives (Sandra Guest van Riper and Linda Sogge) and to our children for their support and understanding during the time that this book was in production.

This work, like other research compilations that are centered on a particular theme, should help to focus attention on investigations presently being conducted over lands of the Colorado Plateau. In particular, we hope that the public and private land stewards in Arizona, Utah, Colorado, and New Mexico, and in particular managers of our National Parks, U.S. Forest Service, Fish and Wildlife Service, Bureau of Reclamation, tribal lands, and the many new BLM national monuments, will be able to utilize the ideas and concepts presented within this book, to launch efforts toward enhanced management and stewardship of their lands. Finally, if the material in this volume can act as a stimulus of future research support for management of cultural, natural, and physical resources over the Colorado Plateau, it will make the organizational and editorial work of the past 2 years a worthwhile and productive effort.

CHARLES VAN RIPER III
USGS Southwest Biological Science Center
The Sonoran Desert Research Station
University of Arizona, Tucson

MARK K. SOGGE
USGS Southwest Biological Science Center
The Colorado Plateau Research Center
Northern Arizona University, Flagstaff

COLLABORATING TO
ACHIEVE CONSERVATION

CREATING SUCCESSFUL COLLABORATIONS IN THE WEST: LESSONS FROM THE FIELD

Whitney Tilt, Craig Conley, Michele James, Janet Lynn, Tischa Muñoz-Erickson, and Peter Warren

Two women approach one another on Main Street in a rural western town. On becoming aware of the other's presence, they cross to opposite sides of the street. Though of similar age and interests, each avoids the other because of their perceived differences—one supports timber cutting and believes the local timber industry's long tenure to be a central pillar in the community. The other woman is a relative newcomer whose anti-timber-harvest stance and other "outsider" views are equally strongly held and defended. In a town suffering economic depression, each woman views the other as the cause of her distress.

The town could be any one of the hundreds throughout the West where a richness of natural resources first attracted miners, loggers, and ranchers. More recently, such towns have attracted a growing immigration of newcomers whose livelihoods and sensitivities are often tied to economies and cultures outside the region. Amid traditional concerns about economic survival and resource utilization arises a growing concern with resource preservation. With these differences comes conflict.

A BATTLEFIELD OF INTERESTS

Not able to organize itself in ways that would build sustainable prosperity, the West bloodies itself in endless fights over whatever can momentarily pass as "economic development."
—Kemmis (1998)

The West has a rich history of fighting over its natural resources. In the beginning, log-gers, miners, and ranchers largely controlled the allocation decisions for water, timber, and range. Since the 1970s, however, other interests have enjoyed a greater and greater say in natural resources management. Flush with a feeling of empowerment or stung with a sense of lost opportunity, these factions have proved again and again their commitment to fight rather than settle. Peter Drucker (1994) describes the situation as "battlefields between groups, each of them fighting for absolute victory and not content with anything but total surrender of the enemy." However, victory in the natural resource arena has become increasingly difficult to declare. Instead, the legacy is one of procedural stalemate, lawsuits, and the zero-sum game of lobbying (Chrislip 2000; Snow 2001). Lost in this swirl of heat and smoke is a sense of community and the associated principle of neighboring.

Settlers to the West faced many hardships. While nature's challenges were met with individual hard work and personal courage, most settlers discovered that long-term tenure on the land required a little assistance from one's neighbors. Ranchers helped one another round up cattle off the open range and farmers helped neighbors harvest wheat before the locusts did.

In recent years, with a growing population of people "from away," the cohesiveness represented by "neighboring" has fractured. A growing population believes it doesn't need, nor is it indebted to, the larger community. Concerned about a society that "bowls alone," Robert Putnam (2000) warns that the nation's stock of social capital (the fabric of our connection with one another) has plummeted, impoverishing both communities and their citizens. The results are plain to see in the West, as "No Trespassing" signs proliferate, disputes are settled at the courthouse instead of the kitchen table, and stewardship of the land has become someone else's responsibility. The resulting loss of trust and sense of community from years of acrimony over natural resource management has led to a lack of civility or sense of community. To rediscover civility, restore community, and achieve improved conservation of natural resources, a new approach is needed.

AGE OF COLLABORATION?

A style of management that emphasizes people getting together to cooperatively solve shared problems seems almost like common sense. Yet most observers of the protracted conflicts over natural resource management in recent years agree that common sense is not so common.

—Wondolleck and Yaffee (2000)

In the recent past, a growing number of citizens and local governments across the West have been trying a different approach. Frustrated with divisiveness, they are creating processes that seek common ground, gain influence through inclusiveness, build social capital, and create a constituency for change (Chrislip 2000). Instead of a winner-take-all approach, warring parties discover reasons to work together, if only from simple exhaustion. "The ranchers know that if they are to continue to use the public's land, they need public support. The environmentalists recognize that if they want open space and habitat and a healthy watershed, the ranchers have to stay in business" (Marston 2001). This realization that existing approaches are not working and that a new approach is needed lies at the root of "collaboration."

To date, collaborative efforts have focused on a wide array of issues including water allocations, timber management, wildlife conflicts, range improvement, and rural community development. The concept of collaboration has begun to be codified into policy and law. The Healthy Forest Restoration Act of 2003 (P.L. 108-148), for example, calls for the development of community wildfire protection plans that must be "collaboratively developed" by local and state government representatives in consultation with the Forest Service, Bureau of Land Management, and other interested parties. The challenges of policy and law dictating "thou shalt collaborate" to agencies unaccustomed and untrained to undertake such activities will be a recurring theme in this paper.

This paper is based on the premise that a more collaborative approach to resource management provides the West's best chance for resolving conflict and restoring civility and dignified democratic discourse. If appropriate people are brought together to work constructively with good information, they will create effective visions and strategies for addressing the shared concerns of the community (Chrislip 2002).

We also address the need for community-based collaborations to address the stated concerns that local groups wield undue influence, that urban constituencies are increasingly disenfranchised, and that participants may possess dubious political and financial motivations (Cestero 1999; Coggins 2001; Dukes and Firehock 2001). Finally, we focus specifically on community-based collaborations—that is, processes undertaken at the local level by a range of citizen and government stakeholders—but the lessons articulated can be applied more broadly to collaborative endeavors in general.

METHODOLOGY

Our neighbor advised us to get together. Although we couldn't influence the rains, we could work together to change the other problems. We could be effective as a group. We could enlist the help of the very people who misunderstood us.

—Malpai Borderlands Group

As stated earlier, this paper draws on the experience of more than 125 collaborative projects supported by the Resources for Community Collaboration (RCC) program of the Sonoran Institute during the period

1998–2004, as well as dozens of others supported by the National Fish and Wildlife Foundation (NFWF) over the last 10 years. Launched in 1998 with a founding grant from the William and Flora Hewlett Foundation, the RCC works to provide financial and technical support to organizations undertaking collaborative efforts across western North America to resolve natural resource issues. The NFWF is a nonprofit organization, established by Congress in 1984, that develops and funds conservation partnerships benefiting fish, wildlife, and plants, and the habitat on which they depend.[1]

The lessons and learnings presented here are the result of project reports, conference proceedings, surveys, and personal communications produced by the projects listed in the nearby box. The primary sources of information and insight are the individual practitioners who shared their firsthand experience with the authors; the primary information sought was "lessons learned"—what worked and what did not. Where observations were common to more than one project, they were recorded and a typology was developed in which to frame similar learnings. To provide a sense of the breadth of projects, specific organizations are cited throughout the discussion. Often, many other projects reported similar learnings.

To further illustrate the potential of community-based collaborations, particularly the overlapping ingredients of success that arise within collaboratives addressing rangelands, we highlight three collaborative efforts in the southwestern United States: Malpai Borderlands, Rowe Mesa, and the Diablo Trust. These case summaries were prepared by Peter Warren, Craig Conley, and Tischa Muñoz-Erickson, respectively, as part of a panel presentation at the 2005 Eighth Biennial Conference of Research on the Colorado Plateau. The panel discussions were developed into a concise set of learnings by Janet Lynn and Michele James,

---

[1] Whitney Tilt served as the director of conservation projects for the National Fish and Wildlife Foundation from 1988 to 2002.

Sample Ground Rules

- Participants will attend all meetings.
- Personal attacks will not be tolerated.
- The motivations and intentions of participants will not be questioned.
- The personal integrity and values of participants will be respected.
- Stereotyping will be avoided.
- Commitments will not be made lightly and will be kept.
- Delay will not be employed as a tactic to avoid undesirable results.
- Disagreements will be regarded as problems to be solved rather than as battles to be won.

Legitimacy and Respect. All parties recognize the legitimacy of the interests and concerns of others, and expect that their interests will be represented as well.

Active Listening and Involvement. Participants commit to listen carefully to one another, ask questions for clarification, and make statements that attempt to educate or explain.

Responsibility. Each of us takes responsibility for getting our individual needs met, and for getting the needs of other participants met. Participants commit to keeping their colleagues or constituents informed about the progress of these discussions.

Honesty and Openness. Participants commit to stating needs, problems, and opportunities, not positions.

Creativity. Participants commit to search for opportunities and alternatives. A creative group can often find the best solution.

Consensus. Participants agree that any decision will be reached by consensus.

Separability. This process is in no way meant to detract from or interfere with current or other efforts, but to potentially arrive at a consensus-driven alternative.

Media. Participants agree that a climate that encourages candid and open discussion should be created. In order to create this climate, participants agree not to attribute suggestions, comments, or ideas of another participant to the news media or nonparticipants.

Freedom to Disagree. Participants agree to disagree.

Rumors. Participants agree to verify rumors at the meeting before accepting them as fact.

Freedom to Leave. Anyone may leave this process but only after telling the entire group why and seeing if the problem(s) can be addressed.

Dispute Resolution. Participants agree that in the event this effort is unsuccessful, all are free to pursue their interests in other forums without prejudice.

of the Ecological Monitoring & Assessment Program at Northern Arizona University.

This paper also benefits from the learnings of four other organizations committed to furthering collaborative approaches as a tool for conservation: the Community-Based Collaborations Research Consortium, the Ecosystem Management Initiative at the University of Michigan, the National Forest Foundation, and the Red Lodge Clearinghouse. The set of 11 lessons presented below reflect the authors' sense of the most important ingredients for success. As with any such anthology, the authors acknowledge the risks of omission and oversimplification.

LEARNINGS FROM THE FIELD

It takes an incredible amount of intestinal fortitude to stay there and be active and not leave the table. You stay there because it's important to tell people what you are for, not what you're against. That's the basis for true collaboration.

—Lynn Sherwood
Colorado Cattlemen's
Agricultural Land Trust
(Red Lodge Workshop 2001)

Drawing on the collective experience of RCC/NFWF-supported projects and other collaborative organizations, a number of lessons become clear. While not presented as an exhaustive or exclusive list, 11 lessons are critical for collaboration to succeed and for community-based collaboratives (CBCs) to function:

1. Understand what collaboration is and is not.

2. Recognize challenge and time involved.

3. Exhaust traditional approaches (ripeness).

4. Build a common vision (passion for place, a community of purpose).

5. Create an open, inclusive, and transparent process.

6. Ensure stakeholders are representative of the community.

7. Provide facilitation and process.

8. Develop a common factual base.

9. Secure operational funding.

10. Achieve and communicate results.

11. Meet or exceed applicable laws and be accountable.

1. Understand What Collaboration Is and Is Not

This stuff is really hard.
—Idaho Conservation League

Collaboration has become the process of choice for many elected officials, federal and state agencies, and community members faced with concerns about natural and social resources. Yet community-based collaboration remains a relatively new and uncalibrated tool for addressing and resolving resource conflicts.

To engage in collaboration, one needs to understand what collaboration is (and what it is not). For our purposes, collaboration is the process by which perceived adversaries enter into civil dialogue to collectively consider possible solutions. As such, collaboration represents a growing obligation to public participation that builds from the act of informing, the willingness to consult, and the invitation to cooperate and partner (IAP2 2004). Collaboration is stronger than cooperation and partnership because it requires the consideration of shared power and may be defined as a "shared responsibility for achieving results" (Chrislip 2002).

Under the above definition, collaboration raises the specter of shared power. Power relations are critical to initiating and successfully implementing collaborative efforts. Who has what decision-making authority, who has control of public opinion, and who aligns with whom are all elements of power that will come into play as community-based collaboration evolves. Who initiates the process, what parties are invited to the table, and who is excluded are further expressions of power relationships that must be recognized and addressed. Since the very conflicts to be addressed by a collaborative effort are likely the result of power inequities (real or perceived), many parties come to the collaborative table seeking some realignment of power, while other parties come to that same table to protect the status quo. Often, some authority or control is a critical incentive for participation; it is often a necessary companion to making collaborative groups responsible and accountable.

Putting these power considerations into a real context, a group of diverse stakeholders labors hard to reach agreement and collaboratively drafts a set of recommendations to a federal land management agency. The land management agency lauds the group for its efforts and then either ignores the group in its decision making or watches, powerless, as someone further up the chain of command renders a decision completely apart from the collaborative recommendations.

Conversely, an agency finds that while members of the collaborative reached agreement internally, the broader community was not adequately engaged and does not support the collaborative's decisions. Both situations illustrate the challenge of engaging in a collaborative effort where the powers and authorities, vis-à-vis a community-based collaboration, may be poorly defined. A majority of collaborative groups identified this as a major issue; these groups stressed the need for participants to engage in frank and continuing discussions on expected outcomes and the process of decision making by government agencies, and to agree on legal sideboards early on (Bureau of Land Management and Sonoran Institute 2000).

Most community-based collaborations in the western United States involve one or more federal land management agencies. To many stakeholders interested in working with federal agencies, agency representatives often appear more concerned with process than outcomes. On the agency side, many agency managers polled in various internal studies believe that collaboration violates one or more laws regarding their decision-making responsibilities. Coupled with a general aversion to risk taking and armed with a multitude of regulations, managers find it easy to identify rules and policies that obstruct their ability to collaborate (Tilt 2005).

To help a collaborative approach succeed, federal land management agencies can actively support field staff in their efforts at collaboration through training, developing improved performance measures that reward greater cooperation and enhanced public participation, and improving transition management so collaborative efforts are not derailed by personnel transfers.

2. Recognize Challenge and Time Involved

Collaboration is a long and exhausting process. Some people become burned out and disinterested while other relationships are indelibly forged for the long term.
—Utah Open Lands

The fastest way to move a cow is slow.
—Klamath Basin Ecosystem Foundation

In a world where everything is meant to be easier and faster, collaboration takes time—to explore and identify areas of potential common ground, to develop the necessary trust, to experiment with possible ways to address shared problems, to build the coalitions necessary for effecting policy changes, and to conduct the necessary project work, monitoring, and evaluation. A reading of eighteenth-century American history reminds us of the time and effort required to form a participatory democracy. Since each collaborative effort is formed and functions within its own context, few if any simple templates for success exist. Certain lessons and principles are applicable to collaborations as a whole, as captured here. Many other considerations, however, depend on an individual CBC's ability to adapt to the time- and space-specific context and content of their circumstances.

Utah Open Lands echoes another common experience: the continual need for steadfast nurturing of participants to ensure long-range maintenance of vision, goals, and enthusiastic participation of members.

Although it is tempting to find shortcuts, these tasks enable a group—especially one in which members do not trust each other—to work together and pull in the same direction. At the same time, a CBC must remember what many practitioners have learned the hard way: it takes weeks (if not months) to build trust and develop relationships; it takes only seconds to destroy them.

Another outcome of the long and potentially exhausting collaboration process is the reality that some participants burn out and others simply lose interest. Single-interest "whiners" will come and go, but effectively dealing with them can still take a long time and a lot of patience. As one practitioner dryly observed, "Don't start unless you are thick-skinned."

The Madison Valley Ranchlands executive director noted another challenge faced by CBCs: "Face it—nobody has the time or energy to go to meetings just for the sake of going." With the understanding that participants must remain motivated, CBCs should constantly look for ways to keep the process energized with an ongoing sense of accomplishment. CBCs have successfully used field trips, special events to celebrate milestones, and potluck dinners to involve members at the ground level. More than one CBC member mentioned how food and drink seem to bring a community together. Observers also commented on the need to have fun and maintain a sense of humor. These informal get-togethers help build respect and understanding among group members and throughout the community.

The majority of organizations polled noted that the social capital of working together to forge common goals extended far beyond individual project outcomes. While difficult to quantify, collaboration's impact on social capital cannot be ignored, especially since many practitioners believe it is the most significant outcome of their efforts. CBC practitioners routinely noted that some indelibly forged relationships emerge as the result of working together through countless meetings in search of common ground. Returning to the two women in the paper's introduction, in real life they became involved in a collaborative process; while the effort's outcome was undecided, they no longer crossed the street to avoid each other because they were no longer strangers.

3. Exhaust Traditional Approaches

We'd gotten awfully good at knowing what we were against, and decided it was time to figure out what we were for.

—Bill McDonald
Malpai Borderlands
(Cash 2001)

While working collaboratively seems like the obvious choice, it should be viewed as the method of "latter" resort, not the first.

Much as an apprentice is expected to spend years learning a trade before he is considered a master craftsman, a key ingredient for CBC success is the realization that traditional forums for redress have fallen short. To be successful, all parties involved in a collaborative effort must be motivated to work together. They must be willing to consider sharing power in the search to develop alternatives to the status quo. It is not enough to be told that a collaborative approach makes sense; it must become the collective desire of the group undertaking the effort.

A collaborative effort is initiated by a complex alchemy of factors (a more detailed discussion of these factors is beyond the scope of this paper). Practitioners engaged in collaborative efforts, however, commonly identified the element of having exhausted other approaches to resolution. Because conflict initially influences most collaborative efforts, the landscape is often marked by divergent interests entrenched in their own camps. They have explored a range of traditional approaches, such as lobbying, administrative appeals, and litigation, to resolve the conflict. When these approaches fail to reduce conflict, interest may grow in trying something different.

In 1998, a maze of regulations, paralyzing litigation regarding endangered species, and a loss of community due to economic instability brought ranchers, the Forest Service, and the Sonoran Institute together to form the Eagle Creek Watershed Group. The group's goal was to restore their namesake to a perennial stream. In electing to pursue formation of a watershed group, participants noted that a key ingredient was the exhaustion of other approaches to resolution. Grazing regulations had been hotly contested for years and unknown or unwanted animal species were granted protection with little or no local support for their conservation.

In Safford and other communities in east-central Arizona, residents saw economic prosperity ebbing from their communities; yet the traditional methods of appeal had brought little to no relief for ranchers, rangelands, or waterways. In this one corner of

Arizona, a small group of stakeholders who were veterans of failed processes were willing to try something new.

Given its focus on public lands, a second critical aspect found in Eagle Creek was the willingness of the district ranger to engage as a participant rather than a hesitant bystander. RCC-supported CBCs consistently noted the federal land managers' "willingness to take a chance" as a necessary ingredient to the CBC process.

Organizations intent on embracing collaborative approaches to conservation need to ask, "What would we be doing if we were not engaged in a CBC?" If the answer is "taking legal action," "maintaining our role as an outside agency expert," or "seeking a public referendum," the issue and participants are likely not ripe for engaging in a collaborative approach. If the answer is some variation on the theme of "we have tried everything short of breaking the law," the ground may be ripe for collaboration.

4. Build a Common Vision (Passion for Place, a Community of Purpose)

If you can get all of the stakeholders at the table and let them express their concerns, grievances, and needs, then trust begins to enter into the discussions.
— Madison Valley Ranchlands Group

Leave your mission at the door. While the individual capacities of each group lend strength to the whole, we have to occasionally re-focus on the issue at hand and subdue our own organizational interests for the greater good.
— Coalition for the Valle Vidal

The foundation for uniting a collaborative effort lies in forging a single vision built on a passion for place or a community of purpose. Passion may arise from a variety of sources, but most often it is the love of land and community that arises from tenure on it and in it. In practice, many efforts fail to ensure that a vision is developed common to all at the collaborative table. While many potential ingredients exist in development of a vision, surveyed practitioners recognize a number of consistent attributes:

1. Individuals must be passionate and committed. They may represent one or more agencies or organizations, but they draw on a personal desire to make the collaboration work.

2. The group must shape its own vision rather than adopt one already fashioned. Work to jointly develop a set of goal statements and purposes, develop a common vocabulary, and ensure that all stakeholders (including new members) receive an orientation to place them on equal footing with their peers.

3. A good vision focuses on what the group shares rather than on areas of disagreement. Success is glimpsed when individuals with different views are willing, at least on a trial basis, to put past antagonisms aside and work to build trust and solve problems.

4. A good vision statement acts as a touchstone for all members, serving as a milepost for where the group has been, where it is at the moment, and where it is going. It becomes the benchmark for defining success.

As the Rincon Institute and others have learned firsthand, most collaborative efforts form in the face of real or perceived crisis. Faced with this sense of urgency, it is difficult not to focus on short-term outcomes rather than focusing on the broader vision. But the long-term vision unites the greatest number of stakeholders and engenders the greatest sense of community. It is the "what" that continually helps define the "how."

A core of like-minded people often forms the nucleus of an emerging collaborative. It is tempting for this committed core to assume that others will share their vision and eagerness to participate, but they must commit to building a working vision that will resonate with the larger community. Experience shows that the collaborative effort must budget adequate time and effort for building a groundswell of interest, conducting outreach, and initiating project planning with the larger community. The core group must also work to constantly bring the currently unengaged into the process and be willing to allow the project's vision to evolve accordingly. Before approaching opinion leaders and other vital stakeholders, however, the emerging CBC must develop a compelling case for the tangible benefits that will accrue to the community from the project.

In southwestern Montana, the Big Hole River Foundation found its origins in a number of challenges—it arose as a response to drought, water allocation politics, and other social conflicts. It also arose from the shared values and concerns of the region's citizens and communities who collectively forged a vision "to understand, preserve, and enhance the free-flowing character of the Big Hole River, and to protect its watershed, culture, community, and excellent wild trout fishery." They bet their time and energy that a voluntary, collaborative approach would have a more profound and widespread impact than a litigious approach that would serve only to divide the stakeholders into pro and con camps and create a win-lose situation.

In southern British Columbia, the Columbia River Successful Communities Forum (SCF) decided to "dream big and see what happens." Before SCF's efforts, local governments and citizens likely would not have willingly embraced the notion of creating a citizen's guide to planning. But during 5 years of effort, local governments and citizens began to support and encourage the idea and actively participated in its development. Asked to measure their impact, SCF notes that

> focused public dialogue about the future is now not only possible, it's expected. That dialogue very clearly includes ecological, economic, and social factors. The notion that we need to protect functioning green spaces for ecological and economic reasons is taking hold, and creating controversy—this is not an issue that will go away any time soon. Official community planning has become the norm in this region. While the SCF is not responsible for this shift, it has played a significant role in the public's desire to be involved, and to be certain that those plans will reflect their values and hopes for the future, as opposed to just mitigating the impacts of growth.

Finally, as collaborative groups work to shape a common vision, some stakeholders may choose not to participate for ideological or other reasons. It is important to keep stakeholders who are not at the table in mind as a vision is fashioned, and to continually challenge the group to work to gain the participation of these individuals.

5. Create an Open, Inclusive, and Transparent Process

The earlier ALL stakeholders are involved in planning that will affect them, and the more transparent the decision-making process, the better the outcome.
—Friends of the Santa Cruz

As a basic tenet of representative government, the need for community-based collaboratives to be "open and transparent" is, at first glance, a statement of the obvious. To actually conduct a collaborative effort in this manner, however, presents more of a challenge. CBC practitioners and researchers provide some guidance.

A collaborative group operating as a self-appointed set of stakeholders might claim to represent the broader community but actually represent only a subset of special interests. In addition, the ability to exclude people from the collaborative table without accountability to the larger community appears to be more of a cabal than a collaborative.

Critics of the Quincy Library Group, for example, argued that the group's "community driven consensus" did not represent the full range of stakeholders, and asked to whom was the library group accountable (Cestero 1999). Resolution of these issues lies largely in a CBC's ability to involve the public "early, often, and ongoing" (Wondolleck and Yaffee 2000). Practitioners stress the need for collaborative groups to continually work to ensure that their process includes all stakeholders regardless of their views or opinions. CBCs must make sure that each participant understands his or her role in the collaborative and work to create a climate where all participants believe their opinion is important. It is also essential to glean input from everyone involved in the process so nobody at the table is surprised.

Teresa Jordan (1998), member of the Toiyabe Watershed and Wildlands Management Team, notes that while Wendell Berry entreats us to think locally and act locally, the dark side of local control is the potential for local tyranny. The collaborative process can escape the taint of localized tyranny only if it remains open and the "optics" of its actions are transparent. Two key aspects of an open process are (1) incorporating the attitudes and viewpoints of people who are not at the collaborative table and (2) insisting on including local experience-based knowledge in the collaborative project.

Idaho's Clearwater Elk Initiative resisted the impulse to jump right into solving the problem without first establishing rules and guidelines. They agreed upon operational guidelines that ensured a process open to all interested parties regardless of views, forged ground rules for meetings and discussions, and then worked to adhere to them so no one thought the project had more than one standard of conduct.

Additional themes emerge from the collective wisdom of Utah Open Lands and other practitioners. The need for open communication (internal and external) is noted as essential to maintaining trust. Leadership should be shared so that it is everyone's responsibility to keep the project moving rather than relying on one person in the group to be the "vision keeper" or "traffic cop." Meeting roles can be rotated so leadership and workload are shared rather than consolidated in a few individuals. In turn, this shared workload helps prevent burnout and enables smoother transitions of leadership should individual members of the group leave the process.

One final pragmatic observation from the field is that a written record of CBC process and actions is necessary. An open and transparent process is reflected in a comprehensive set of meeting minutes that includes such obvious items as attendance and decisions made.

6. Ensure Stakeholders Are Representative of the Community

A broad-based coalition is more believable, tangible, can reach a more diverse constituency, and has a more complete skill set to tackle major issues.
—Coalition for the Valle Vidal

We will send one representative to your first meeting. If he's comfortable with the process, he will
attend the second meeting; if he's not, we will send 50 to the next meeting.
—Northern Forest Pulpworkers

Dealing with people who are directly affected by grizzlies is more productive than dealing with formal elites, who may see an issue like grizzly conservation as an opportunity for grandstanding.
—Gravelly Range Grizzly Project

Building on lesson 5, stakeholders at the collaborative table must reflect the interests of the whole community—representative representation. RCC's experience is that the success of a CBC is directly linked to the effort's success in identifying stakeholders and opinion leaders in the community. Failure to address the issues of inclusiveness and diversity at the stakeholder table can render the collaborative process into little more than a replication of the power imbalances that already surround a set of issues.

A common criticism of community-based collaborations is that they are used as a way of avoiding established public processes (Dukes and Firehock 2001). Side-stepping the issue of whether established public processes serve either the public interest or natural resource stewardship, this criticism is easy to understand when legitimate interests are intentionally excluded from the process or elect not to participate. In addition, who represents whom—who has the proper portfolio to represent the environmental interests or those of industry? For the critic turned cynic, the stakeholder table often appears set by Capt. Renault's memorable line in the film "Casablanca" to "round up the usual suspects." The collaborative table needs to go beyond the "usual suspects" to provide a place for new voices and for the CBC to establish accountability to the larger community.

Recognizing the need for inclusiveness and diversity is a necessary step. Creating it at the collaborative table is the hard part. Half of the stakeholders surveyed in a random sample of 76 watershed-based stakeholder efforts in California and Washington noted that some critical interests were not effectively represented at the table (Leech 2004). Leech also noted that ordinary citizens often face a lack of motivation or other obstacles to participation, unlike agency, industry, and environmental representatives who can often participate as part of their jobs. As raised in lesson 5, other collaborations have observed similar challenges. A group is seen as either "self-selected"—choosing to define who gets to sit at the table from within a narrow view of stakeholders—or representing those who are willing to sit at the table regardless of the necessity of involving certain other interests for successful resolution of the issues at hand.

Although one or more disputes may have brought people to the table, it is people, not issues, who make the collaboration succeed or fail. With that in mind, participants will likely spend much more time on people issues than on natural resource issues. The personality factor is distracting, and there is a continuing need to focus on the areas of mutual interest and not on whom to blame. The experiences of the Calapooia Watershed Council, the Walla Walla Basin Watershed Council, and others offer additional insights:

1. Do not confuse constituents or partners with stakeholders. It is akin to the difference between eggs and ham—the chicken is interested, but the pig is committed.

2. Learn about and appreciate the various missions of your fellow collaborators even as you work to have them represent their knowledge and experience rather than their ideology or organizational mantra.

3. Protect ALL stakeholders' interests and avoid alienating one or more participants who may turn into spoilers.

4. Agency participants need to work on connecting with collaborative efforts, rather than directing them.

5. Failure to actively work to involve a diverse and representative range of stakeholders will likely result in failure of the CBC to accomplish its goals.

Practitioners consistently listed strong leadership as an essential ingredient for an extended life of a collaborative effort. Credible stakeholders in the collaborative who help convene, catalyze, and sustain the process are critical to the effort's success.

When viewed from the outside, a CBC drawn from diverse sectors of the community demonstrates the group's commitment to inclusiveness and provides a forceful statement to outside observers on all sides of the issue.

A diverse and representative stakeholder group is also the best defense against the potential problem of key players sitting at the table but not being "honest brokers." In the absence of leadership from key players, individuals may retain their individual rather than collective alliances and work to subvert the group's progress (Calapooia Watershed Council).

Government officials, industry representatives, and environmental organizations participate in a collaborative project as part of their jobs and typically receive some form of compensation for their investment of time. By contrast, many private citizens and individuals working for advocacy groups are not paid to participate and need to spend precious free time to do so. Asymmetries in available time and compensation can, de facto, lead to bias in representation and participation, often to the detriment of those who lack power under the status quo.

7. Provide Facilitation and Process

People want to work collaboratively and they are curious about the work of various conservation groups, but they also want to know that their time and energy have been invested in real progress. It is important to keep the planning work tied to results on the ground.

—Methow Conservancy

The facilitator must not presuppose to know the outcome of any collaboration. Collaborations are about listening, reflecting, sharing resources, and exploring potential approaches to the issue at hand. It is a process of group exploration and problem solving, and is driven by individuals' desire to improve on the status quo.

—Murie Center

Having set the collaborative table with a diverse and representative group of stakeholders, many of whom are likely leaders in the community, it is now time to "herd the cats." Heeding the advice of more than one seasoned practitioner to "never attempt to facilitate and lead at the same time," CBCs should consider engaging outside facilitators to help the group obtain its collective goals. In the experience of the CBC groups polled, strong facilitation experience was rare in emerging collaboratives, which required them to acquire skilled facilitators from the outside.

In selecting a facilitator, the most important attribute is that all participants in the collaborative process perceive the facilitator as legitimate and fair. The facilitator's purpose is to build a process, work with the group to establish sideboards, and then strive to make sure the sideboards are observed. A facilitator also makes sure that the quieter voices in the process don't get run over. As observed by one collaborative, the facilitator helped build mutual respect where environmentalists who have never ranched didn't tell ranchers how to ranch, and ranchers didn't run roughshod over the naturalistic interests of environmentalists.

Another part of a facilitated process is to keep the group focused on being proactive, not reactive—to focus on the vision, not on the past. A collaborative effort must work to make progress happen rather than sit back and see what happens. The primary role of effective facilitation is to establish and enforce ground rules for fairness and respectful behavior. The sample ground rules presented here (see box, next page) are adapted from those developed by the Saguache County Study Group in Colorado.

It is also important to continually foster a respectful and benevolent environment. The Sonoran Institute's publication "Beyond the Hundredth Meeting" makes clear the need for productive meetings right in its title (Cestero 1999). Collaboratives need to outline time commitments in advance so people can attend without fear that their lives will be swept away in meetings. Once in meetings, conveners look for ways to ensure that all members are heard and feel useful by utilizing smaller group meetings and delegating specific work assignments to subcommittees. Lastly, facilitators respect people's time by starting and ending meetings punctually.

As with many processes, success lies in the details. Something as simple as schedul-ing meetings becomes quite important. For example, meetings need to be convenient for all participants, not just a few. If a single set of convenient times proves elusive, then meeting schedules should rotate to accommodate the widest possible range of schedules. While staffs of many agencies and advocacy organizations are veterans of "attending meetings," many other citizens and stakeholders will not be comfortable with this particular form of social discourse (Bureau of Land Management 2003).

If a facilitator doesn't work well with the group, or a subset of the group, it is time to find a new facilitator. Idaho's Clearwater Elk Initiative had to change facilitators after five meetings: "We were hesitant to make the change, but it made a tremendous difference," one participant noted.

These are just a few examples of how a facilitated process works to establish an atmosphere where folks are willing to try something new—that is what community-based collaboration is all about.

8. Develop a Common Factual Base

Science that does not incorporate people who are involved in the subject of study is imperfect.

—Living Oceans Society

A major obstacle confronting resolution of many natural resource issues is their apparent complexity. Creating a common factual basis is critical in order to "bound the problem with credible information," in the words of Wondolleck and Yaffee (2000). Many CBCs note that ideological conflicts (Republican vs. Democrat, meat-eater vs. vegan, agnostic vs. Catholic) are surmountable barriers to progress, but conflict over issues of fact can incapacitate any collaborative process.

The first step is to recognize the need for a common basis of scientific information. The next is to recognize that the process for collecting that information must be a shared effort, not merely a stockpiling of data by one or more "experts." Since federal and state land management agencies are often repositories for natural resource information, their involvement in CBCs must go

Collaborative Organizations and Projects
Included in the Study

Cultural and Community Organizing

Big Island Resource Conservation Council (Hilo, HI); Kealakehe Ahupua 2020
Center for a Vital Community (Sheridan, WY); Teambuilding Retreat for Stewardship Workshops
Friends of Pronatura (Tucson, AZ); Community Training Workshops
Indigenous Community Enterprises (Flagstaff, AZ); Navajo Hogan Affordable Housing
Island Institute (Sitka, AK); Civic Collaboration Initiative
Living Oceans Society (Sointula, BC); Traditional Knowledge for Marine Planning
Mexicano Land Education & Conservation Trust (Espanola, NM); Land Grant Environmental Justice Project
Montana Preservation Alliance (Helena, MT); Tongue River Valley Natural and Cultural Preservation
Murie Center (Moose, WY); Teton Sustainability Project
Saguache County Environment and Economic Development (Saguache, CO); "Valley Wide" Summit
Tree New Mexico (Albuquerque, NM); Bluewater Ranch Restoration "Listening & Training"

Forest Use and Management

Backcountry Snowsports Alliance (Eldorado Springs, CO); Wolf Creek Winter Recreation Task Force
East Kootenay Environmental Society (Kimberly, BC); EKES Pulp Mill Project
Flathead Economic Policy Center (Columbia Falls, MT); Flathead Forestry Project
Gifford Pinchot Task Force (Vancouver, WA); Forests and Communities Collaborative Program
Grand Canyon Trust/Forest Foundation (Flagstaff, AZ); Restorative Forest Management
Idaho Conservation League (Boise, ID); Boulder-White Cloud Mountains
Jefferson Center for Education and Research (Wolf Creek, OR); Harvest of Alternate Forest Products
Quincy Library Group (Quincy, CA); Community Stability Proposal
San Miguel Watershed Coalition (Montrose, CO); GMUG National Forests Stakeholders
Siskiyou Regional Education Project (Cave Junction, OR); Community Involvement in RACs
San Isabel Foundation (Westcliffe, CO); Wet Mountain Collaborative Mapping Project
Tongass Conservation Society (Ketchikan, AK); Ketchikan Community Forest Planning
Western Colorado Congress (Montrose, CO); Red Mountain Pass Stakeholder's Meeting
Yaak Valley Forest Council (Troy, MT); Yaak Valley Forest Stewardship

Land Use

Beaverhead County Community Forum (Dillon, MT); Beaverhead County Housing
Big Hole River Foundation (Butte, MT); Big Hole River Conservation Corridor
California Oak Foundation (Oakland, CA); Salinas River Easements
Calapooia Watershed Council (Albany, OR); Management plan for Thompson's Mills
Capitol Land Trust (Olympia, WA); Springer Lake Community Planning
Columbia River Greenways Alliance (Invermere, BC); Community Guide to Citizen Involvement
Conservation Land Network (Bozeman, MT); Conservation Land Network
Copper River Watershed Project (Cordova, AK); Copper River Tourism Plan
Diablo Trust (Flagstaff, AZ); Colorado Plateau of Rangelands Planning
Earthlaw (Denver, CO); Front Range Riparian Protection
Friends of the Santa Clara River (Newbury Park, CA); Santa Clara River Enhancement
Gallatin County Open Lands Board (Bozeman, MT); Community Plan for Open Space
Georgia Strait Alliance (Nanaimo, BC); First Nations Involvement in Orca Pass
Gowgaia Institute (Queen Charlotte, BC); Haida Gwaii Ecosystem Planning
High Country Citizens Alliance (Crested Butte, CO); Upper Gunnison Valley Planning
Methow Conservancy (Winthrop, WA); Conservation Planning for the Methow Valley
Rincon Institute (Tucson, AZ); Cienega Corridor Conservation Council
Salmon River Mountains Working Group (Salmon, ID); Salmon River Mountains Working Group
Somenos Marsh Wildlife Society (Duncan, BC); Somenos Marsh Wildlife Refuge

Land Use (continued)

Swan Ecosystem Center (Condon, MT); Conservation Strategy for the Swan Valley of Montana

Utah Open Lands (Castle Valley, UT); Castle Valley Project

Mining and Energy DevelopmentCoalition for the Valle Vidal (Taos, NM); Valle Vidal

Northern Plains Resource Council (Billings, MT); Stillwater Mining "Good Neighbor" Agreement

Western Slope Environmental Resource Council (Paonia, CO); North Fork Coal Working Group

Ranching, Agriculture, Invasive Plants

Amigos Bravos/Taos County (Taos, NM); Taos County Weed Control

Catron County Citizens Group (Glenwood, NM); Gila NF Rangeland and Forest Management

Community Environmental Council (Santa Barbara, CA); Wine Industry Task Force

Eagle Creek Watershed Partnership (Safford, AZ); Working Rangeland Partnership

Hells Canyon Preservation Council (LaGrange, OR); Grazing Alternatives for Local Ranchers

Malpai Borderlands Group (Douglas, AZ); Malpai Stewardship

Northeastern Nevada Stewardship Group (Elko, NV); Elko Sagebrush Ecosystem Conservation Strategy

Quivira Coalition (Santa Fe, NM); Progressive Ranch Management Demonstration

Thunder Basin Grasslands Prairie Ecosystem Assn. (Douglas, WY); Thunder Basin Grasslands Project

Toiyabe Watershed and Wildlands Management Team (Austin, NV); Tipton Ranch Collaborative

Watershed & Water Use

1000 Friends of New Mexico (Santa Fe, NM); Acequia and Environmental Protection

Ecological Assn. of Hardy and Colorado Rivers (Mexicali, MX); Community Participation in Colorado River Delta Restoration

Amigos Bravos (Taos, NM); Somos Vecinos/We Are Neighbors

Applegate Partnership (Applegate, OR); Applegate Partnership

Community Foundation of Western Nevada (Reno, NV); Champions of the Truckee River

Friends of the Santa Cruz, Tubac, AZ; Viable Riparian Conservation Options in Santa Cruz County

Headwaters (Ashland, OR); Headwaters and Talent Irrigation Clean Water Collaboration

Henry's Fork Foundation (Aston, ID); Henry's Fork Watershed Council

Klamath Basin Ecosystem Foundation (Klamath Falls, OR); Klamath Basin Assessment Project

North Fork River Improvement Association (Hotchkiss, CO); North Fork Gunnison Restoration

Oregon Water Trust (Portland, OR); Enhancing Stream Flows in Rogue River Tributaries

Rio Grande Restoration (El Prado, NM); Acequia and the Santa Fe River

San Juan Citizens Alliance (Durango, CO); Dolores River Flows by Consensus

Santa Fe Watershed Association (Santa Fe, NM); Santa Fe/Rio Grande Stakeholders Meeting

Sierra Nevada Alliance (South Lake Tahoe, CA); Hydro Healing Project

South Yuba River Citizens League (Nevada City, CA); Yuba Watershed Council

Sun River Watershed Group (Great Falls, MT); Sun River Watershed Partners for Success

Truckee River Watershed Council (Truckee, CA); Truckee River CRM Plan

Walla Walla Basin Watershed Council (Milton-Freewater, OR); Walla Walla Habitat Conservation Collaboration

Wildlife Management

Clearwater Elk Collaborative (Lewiston, ID); Clearwater Basin Elk-Related Issues

Institute for Ecological Health (Davis, CA); Sacramento County HCP Collaborative

Northern Rockies Conservation Cooperative (Ennis, MT); Gravelly Range Grizzly Project

Madison Valley Ranchlands Group (Ennis, MT); Madison Valley Ranchlands Elk Management Program

Salmon River Mountains Working Group (Salmon, ID); Diamond Moose Grazing Project beyond the agency simply providing information. Regardless of the information's accuracy, stakeholders around the table must come to accept the science itself. The information cannot be force-fed to them by a group of self-proclaimed experts (who might already be viewed by many of the stakeholders as part of the problem). Case studies of CBCs involving the Forest Service or the Bureau of Land Management consistently point out the challenge of agency participants interacting with other participants as fellow community members rather than as authorities with command and control responsibilities (Dukes and Firehock 2001; Tilt 2005; Wondolleck and Yaffee 2000).

A look at water in the West is illustrative. Water allocation issues form a very complex web of laws, court decisions, operating decrees, and other forces. A collaborative effort focusing on water allocation issues needs access to both pertinent data and experienced professionals. Too often, outside experts simply dictate their findings to community groups rather than becoming part of the process. Further, they often overlook local or native knowledge, and the overall need for CBCs to achieve a collective comfort with the factual information provided. The Swan Ecosystem Center and other CBCs emphasize the need to build a process that recognizes the local or native knowledge of each community member, and to treat each as an expert in his or her own right.

To be effective, CBCs need to produce and present their own information from a community perspective (Gowgaia Institute). Clear information and an open forum to discuss how to use it are central to collaboration. The process of participating in informed discussions among diverse stakeholders (with equally diverse knowledge bases) also helps break down segregated silos of interest. The experience of Eagle Creek and others demonstrates that the collective development of solid information in a readily understood format helps to move participants from unyielding positions to respectful compromise. It also helps to move the overall group toward shared goals.

Finally, a factual basis does not reside in an inanimate assembly of data. Gaining a factual basis for resolving a set of issues is forged out on the land itself. Collaborative after collaborative noted the power of field trips and on-the-ground workshops to engender a growing sense of place and a greater understanding for how others view the same landscape. For the Quivira Coalition, field trips were the way they got to the "grassroots"—literally getting folks to look at plants and their roots as part of rangeland management. Each person in the group—the logger, the mushroom gatherer, the "tree-hugger"—has a unique view of themselves and unique perceptions of one another. Participating as a group helps us understand how love of the land can rightly manifest in a wide array of expressions.

9. Secure Operational Funding

Securing sufficient operational funding is a critical factor in launching, and maintaining a successful collaborative. Efforts to secure long-term, unrestricted, operational support are largely unsuccessful to date.
—Columbia River Greenways Alliance

The greatest threat to our project is a lack of dedicated staff time if key participants are not fully funded to engage in the collaborative process.
—Gifford Pinchot Task Force

The majority of organizations polled in this research face pressing and continuing challenges to identify sufficient funding to maintain their collaboratives. Although the majority of operational budgets are small, even by nonprofit organization standards, it remains difficult for these organizations to maintain stable budgets. Many collaboratives are successful in attracting sufficient funding for restoration projects, but the same sources are unwilling to provide funding for administration (South Yuba River Citizens League). It is a cold hard fact that an emerging collaborative effort must have some start-up resources to achieve some early success and interest. This success, in turn, is required to demonstrate the project potential that most funding sources want to see before they fund the project.

The Island Institute speaks for the vast majority of CBCs when it notes the sad lack of funders that support community-based collaboration at all, and the nearly total lack of funders who recognize that durable collaboration depends on extended effort. Funding for 1–3 years is generally insufficient to develop the local capacities needed to sustain healthy civic communities and their natural environment—what CBCs are offered translates into short-term speculation rather than essential long-term investment.

The Sonoran Institute's Resources for Community Collaboration program has faced these challenges firsthand. From 1998 to 2004, the program provided $640,000 to CBCs, with the program's funding consistently falling short of the demonstrated need. But the program's ability to consistently fund worthy projects year in and year out is limited due to financial constraints as well. Insights into the world of fundraising (Management Institute for Environment and Business 1993; Tilt 1996) include the following points:

1. Remember that people give to people. Develop relationships with the funding community. Unsolicited proposals seldom receive funding.

2. Develop a realistic budget for the project. Even volunteer organizations need more financial resources than anticipated to stay involved and vital.

3. Good deeds seldom attract funding on their own. Develop grant-writing skills as soon as possible within the collaborative, or find someone who can provide these skills.

4. Build institutional support (administrative overhead) into project funding.

5. Acknowledge your supporters. Say thank you, and then say thank you again.

10. Achieve and Communicate Results

We have a lot of technically competent people but they would have done something else for a career if they were interested in people. They are not the best communicators in many instances.

—Unnamed Forest Service Employee
(Wondolleck and Yaffee 2000)

"Nothing succeeds like success" is a common message from the field. Obviously CBC participants and those outside the process expect results in return for their time, effort, and patience. The following are some of the lessons put forward by the Swan Ecosystem Center, the Sun River Watershed Group, and others:

1. Identify specific actions that can be taken and then follow through to demonstrate some results.

2. Work on small do-able projects to gain skills and trust. Tackle controversial work later.

3. Accomplish tasks incrementally so you can continually acknowledge successes and reward your group and community with a celebration on each significant success story.

The need for good communication is also a constant theme heard from practitioners. While everyone acknowledges the need for it, few institutions are consistently good at it. In natural resource management, communications have too often been reduced to a robotic process of "public involvement" where public notice is provided, a requisite number of public hearings is conducted, and some agency makes a decision that appears totally divorced from any public input. This serves as a good model for what CBCs should not do. Some proactive lessons include the following:

1. Involve the public early and often.

2. Take full advantage of existing social networks in the community and target opinion leaders to involve them in the collaborative effort.

3. Work to familiarize the community with the project's goals and process.

4. Communicate by telephone, e-mail, and websites, but not at the expense of face-to-face interaction.

5. Keep accurate records of all events: participant lists, minutes, photos, articles, etc.

The experience of the Applegate Partnership in southwestern Oregon also cautions against seeking publicity before relationships and trust are fully developed. This early notoriety can cause damaging internal tension and conflict (KenCairn 1999).

11. Meet or Exceed Applicable Laws and Be Accountable

It is imperative for collaborative organizations to develop mechanisms for self-evaluation which allows for efficient use of funds, energy, and the planning of useful activities, as well as to transfer their story in the request of funding and support.
—Northeastern Nevada Stewardship Group

In today's world of competing interests and watchdogs, it is not enough to do "good work." CBCs must be capable of demonstrating their adherence to applicable federal and state laws and establishing sufficient monitoring and evaluation capacity to track and document project outcomes.

To be viewed as successful, both internally and externally, CBCs must demonstrate that their process meets or exceeds environmental law and policy. For example, many critics consider the Quincy Library Group collaboration to have represented a select group of special interests that successfully gained Congressional intervention to circumvent existing state and federal laws (Cestero 1999).

CBCs must also ensure that their monitoring and evaluation protocols are capable of assessing environmental and social progress. When monitoring, CBCs should remember that more measurement does not equal more understanding. There is a continual need for information triage because of the infinite amount of information available (Ecosystem Management Initiative 2004).

Finally, there is the importance of accountability for outcomes. Supporters and critics alike express the concern that CBCs do not pay enough attention to monitoring and evaluating outcomes. The environmental and social impacts of CBCs too often remain largely unknown. Although the body of thoughtful research on the subject of community-based collaborations is growing, most information remains anecdotal. One presenter at a gathering of researchers (the CBC/RCC 2003 Annual Meeting, Snowbird, Utah) noted the tendency to romanticize CBCs doing "on-the-ground conservation work" and urged the need for additional research to harden the benefits of utilizing a collaborative approach.

A CLOSER LOOK AT RANGELAND COLLABORATIONS

The 11 lessons presented above find field validation and additional depth in the first-hand experiences of the following three working collaborations. These rangeland efforts each demonstrate that collaborations can succeed and that adaptive learning can continue to flourish in difficult environments. Additionally, the overlap between these efforts demonstrates the potential for local collaborations to promote emerging social networks at a larger regional level.

The Malpai Borderlands Group, Arizona and New Mexico

The Malpai Borderlands Group began as informal discussions between a handful of concerned ranch families with ties to the land and ranching going back to the 1890s and early 1900s. Today, the group is a nonprofit organization led by local ranchers, with participation of state and federal agencies, scientists, the Nature Conservancy, and other stakeholders. The objective is to restore and maintain the natural processes that create and protect a healthy, unfragmented landscape that will support a diverse, flourishing community of human, plant, and animal life. Situated in the valleys of southeastern Arizona and southwestern New Mexico, the group aims to accomplish these goals by working to encourage profitable ranching and other traditional livelihoods that will sustain the open space nature of the land (Sayre 2005).

Two immediate concerns of both the ranchers and environmental interest groups involved with the Malpai Borderlands Group were the restoration of remaining native grasslands and their protection from further subdivision and development. Thus, the group immediately began to focus their energy on restoring fire to grasslands, implementing three major prescribed burns in conjunction with the U.S. Forest Service. In addition, they developed the first grassbank system, which allows forage on one ranch to be made available to another rancher's cattle in exchange for one or more conservation benefits and/or easements on neighboring or associated lands. The group now holds conservation easements on 12 ranches with a total of more than 75,000 acres of private land protected from development.

Several components have led to the sustainability and success of the Malpai Borderlands Group. One of the first steps to success was the creation of a mission statement that all stakeholders could agree upon. This statement has become an important reference point that continues to maintain focus and guide decisions about the group's future projects. Another critical ingredient was establishing trust among members such that any one member could speak comfortably in public on behalf of the collaboration as a whole. Although each member may have a different perspective on what they want the collaboration to accomplish, conveying their common goals to the public has been crucial in strengthening the collaborative effort. Monitoring has been another important component of the Malpai Borderlands Group's success. Two fundamental reasons to adopt a monitoring protocol for collaborative projects are the importance of educating land managers and decision makers on the progress and outcome of the project, and the added credibility and defensible base that monitoring provides for the group's actions. In addition, every year the Malpai Borderlands Group hosts several workshops to share ideas and experiences with people who are interested in similar locally organized efforts. Furthermore, their meetings have included visitors from Mexico, Canada, Brazil, Australia, Indonesia, and Kenya; there is thus a global awareness and desire for collaborations.

The Rowe Mesa Grassbank, New Mexico

Inspired by the work of the Malpai Borderlands Group, the collaborative Rowe Mesa Grassbank (RMG) was established in 1997 to demonstrate how grassbanking can serve as a practical tool for restoring national forest system lands in northern New Mexico. In 2004, the project was transferred from the Conservation Fund to the Quivira Coalition. At the core of the collaborative are five major partners: The Quivira Coalition, the U.S. Forest Service, the Northern New Mexico Stockman's Association, the New Mexico Cooperative Extension Service, and the current permittee participants. The three main goals of the RMG are to improve the ecological health of public grazing lands for the benefit of all, to strengthen the economic and environmental foundation of northern New Mexico's ranching tradition, and to demonstrate that ranchers, conservationists, and agency personnel can work together for the good of the land and the people.

The current political and social climate and the willingness of the various participants to collaborate during the initial stages of the RMG's formation were essential. Gaining the cooperation of the U.S. Forest Service and Stockman's Association was an important first step that enabled the RMG to gain the added trust of the current permittees and smaller landowners and ranchers from the surrounding communities. The collaboration has also found success through their ability to constantly adapt to the changing needs of the stakeholders as well as those expressed by outside interest groups. Thus, a new model is currently emerging on how to achieve the collaboration's goals and continue the success of the program while also being financially sustainable. This model depends on implementing a set of analytic and restoration tools as well as creating new relationships between people with an interest in public lands management. Another aspect vital to the success and management of the RMG has been the number of monitoring and restoration tools used to leverage forage in the grassbank for restoring land health on Rowe Mesa and in participating Forest Service grazing allotments. These tools include qualitative land health assessment, management-directed monitoring (including social and ecological), niche marketing, prescribed fire and post-fire grazing management, management of pinyon/juniper encroachment, and the use of professional herders. Furthermore, the RMG has also made continual efforts to improve education and outreach, strengthen ties to local communities, and focus on long-term management goals. This has brought the Rowe Mesa Grassbank national attention, and more important, acceptance and respect from local communities and interest groups.

The Diablo Trust, Arizona

Initially founded in 1993 by two ranches, Bar-T-Bar and Flying M, the Diablo Trust was created to link private and public values under one holistic goal: to create sustainable rangeland management that maintains the tradition of working ranches and provides economic viability while managing for ecosystem health. Situated east of Flagstaff, Arizona, collaborators of the Diablo Trust now include local ranchers, state and federal agencies, scientists, environmentalists, and other interested stakeholders.

One vital aspect that has led to the success of the Diablo Trust is the conviction that good land stewardship incorporates participatory research and monitoring projects. Introducing scientists into a collaborative environment has helped the Diablo Trust develop appropriate research questions that are relevant to the ranchers and that address perceived conflicts among stakeholders and the outside public (Sisk et al. 1999). In addition, integrated collaborative research and sound monitoring protocols can generate clear measures of effectiveness and progress in which to evaluate the success of the collaboration (Muñoz-Erickson and Aguilar-Gonzalez 2003).

Working with researchers at Northern Arizona University and Prescott College, the Diablo Trust incorporates both research and monitoring into rangeland conservation. This approach can foster collaboration by leveling the playing field among stakeholders by providing equal access of information to everyone, incorporating multiple sources of information and values, and engaging stakeholders in the data collection and generation of knowledge through multi-party monitoring projects (Sisk and Palumbo 2005; Muñoz-Erickson and Aguilar-Gonzalez 2003). Finally, science has also enhanced collaboration by bringing credibility to the process and by motivating the group to be accountable for their management actions.

The inclusion of research and monitoring in the collaborative process has brought several benefits to scientists as well. Collaborations provide scientists with the resources to "scale up" their studies from small plots to whole landscapes. In addition, the ability to collaborate with the people who manage the land results in more meaningful, insightful, and applicable science (Sisk and Palumbo 2005). In order to continue this fruitful relationship between stakeholders and scientists, the Northern Arizona University and Prescott College researchers have invested significant time into the collaborative process, anticipating a multi-decadal relationship. All stakeholders share the goal of sustaining research and monitoring over long periods to generate information that is relevant to an ecological system typified by slow responses interrupted by periodic bouts of drastic change.

By taking an active role and using this collaborative scientific approach, the Diablo Trust supports numerous successful projects, such as monitoring experimental vegetation plots over the long term, investigating the ecological effects of fire and grazing on grassland diversity and productivity, determining the effects of grazing on pronghorn habitat, discovering historical changes in grassland compositions, and developing the Integrated Monitoring for Sustainability project (Loeser et al. 2001; Muñoz-Erickson et al. 2004), which is a multi-party monitoring process that incorporates social and ecological well being and acknowledges their interrelationship. Through the IMfoS project, the Diablo Trust worked with the Northern Arizona University and Prescott College research team in developing the Holistic Ecosystem Health Indicator (HEHI) to assess and monitor the sustainability of the Diablo Trust's collaboratively managed rangelands. This monitoring tool measures ecological and social indicators of rangeland health and combines data from existing monitoring efforts, collected by different agencies, resource users, and volunteers, into a single data repository.

These research and monitoring efforts have brought national recognition to the

Diablo Trust, stimulating more collaboration with other groups such as the Malpai Borderlands Group and the Northwest Colorado Stewardship Council. It is the hope of the Diablo Trust that these efforts will enhance the group's adaptive management efforts by making information transparent, facilitating communication among stakeholders, and increasing learning efficiencies.

CONCLUSION

Don't concentrate on skeptics; concentrate on the eager learners.
—Quivira Coalition

Against the onslaught of sweeping change, Custer County [Colorado] offers us the rare glimpse of hope—that by setting aside our differences and focusing on our common love for the land, individual people can still make a difference.
—Todd Wilkinson (2004)

Community-based collaborations face many challenges. Many times the underlying problem and its solution are poorly understood, there is a paucity of data and little understanding of what the information means, and personnel and financial resources are small or nonexistent. In addition, conflicting values clutter the stage and innovation is often viewed as risky and expensive. Collaborations must bring together a diverse and representative group of stakeholders, and they must also embrace the amount of time, effort, and funding that is necessary to create and sustain a successful collaborative process. Resonating from each of our rangeland examples is the importance of gaining the trust of the stakeholders and outside interest groups by maintaining an open and transparent process that incorporates research and monitoring protocols in which to evaluate their goals.

Given these challenges, why would anyone elect to pursue community-based collaboration? The answer lies in the belief that collaboration represents an alternate approach that meets the national interest through local and place-based actions, and that the appropriate people brought together to work constructively with good information will create useful visions and strategies for addressing the shared concerns of the community. Collaborative groups have found that the process of collaboration is constantly changing and they are continuing to discover new methods of achieving landscape-scale conservation goals.

The power of community-based collaboration is its recognition that humans are part of the environment and a mandatory part of the solution. This paper assembled the field experience of dozens of practicing CBCs. Their experience confirms that community-based collaboration can be a fruitful road to long-term solutions. The three rangeland case studies also illustrate the potential that collaboration can magnify its impact beyond the community level through its connections with other similar groups, leading to emerging regional networks for resource management. But these solutions take time, determination, and strong people skills. Practicing CBCs have learned firsthand that good will, or at least a desire for it, is a fundamental prerequisite for collaboration. They point out the need to measure the benefits of CBCs in both social and biological terms, and to mark progress against a group's goals. They also point out many practical pieces of advice such as identifying an easily achievable first project to build trust and demonstrate the collaborative's worth. And practitioners stress over and over the importance of building relationships—CBCs are about working with people and building social capital.

Quoting historian Bernard DeVoto, Wallace Stegner dryly observed that the only true individualists in the West were usually found hanging from a rope, the other end of which was held by a group of cooperating citizens (Hahn 1998). In today's West, conflicts over natural resources are too important to be left to battles between individuals; they require involvement of the community, with its sense of place, its sense of economic foundation, and its collective capability to instill a sense of stewardship of natural resources. That is the lasting impact of community-based collaboration.

Successful communities and stewardship come from the ground up, originating within the community and involving citizens who make a conscious public commitment to a common vision that includes both a diversity of people and landscapes.
—Luther Propst, Sonoran Institute

LITERATURE CITED

Bureau of Land Management and Sonoran Institute. 2000. A Desktop Reference Guide to Collaborative, Community-Based Planning. Sonoran Institute, Tucson. 20 pages.

Bureau of Land Management. 2003. Leaving a 4 C's legacy: A framework for shared community stewardship. Report to the Assistant Secretary of Land and Minerals. Department of the Interior, Washington, D.C.

Cash, K. 2001. Malpai Borderlands: The searchers for common ground. In Across the Great Divide: Explorations in Collaborative Conservation and the American West, edited by P. Brick, D. Snow and S. Van de Wetering, pp. 112–121. Island Press, Washington, DC.

Cestero, B. 1999. Beyond the Hundredth Meeting: A Field Guide to Collaborative Conservation on the West's Public Lands. Sonoran Institute, Bozeman, Montana.

Chrislip, D. 2000. Transforming Civic Culture: Sitka, Alaska in the Year 2000. Assessment prepared for Island Institute, Sitka, Alaska.

Chrislip, D. 2002. Collaborative Leadership Fieldbook: A Guide for Citizens and Civic Leaders. Jossey-Bass, San Francisco.

Coggins, G. 2001. Of californicators, quislings, and crazies: Some perils of devolved collaboration. In Across the Great Divide: Explorations in Collaborative Conservation and the American West, edited by P. Brick, D. Snow and S. Van de Wetering, pp. 163–171. Island Press, Washington, DC.

Drucker, P. 1994. The age of social transformation. The Atlantic Monthly 274(5): 53–80.

Dukes, F.. and K. Firehock. 2001. Collaboration: A Guide for Environmental Advocates. University of Virginia, Institute for Environmental Negotiation, Charlottesville.

Ecosystem Management Initiative. 2004. Measuring Progress: An Evaluation Guide for Ecosystem and Community-Based Projects. School of Natural Resources and Environment, University of Michigan, Ann Arbor.

Hahn, M. 1998. Two rivers. In Reclaiming the Native Home of Hope, edited by Robert Keiter, pp. 34–42. University of Utah Press, Salt Lake City.

IAP2. 2004. Public participation spectrum. International Association of Public Participation. Available online at www.iap2.org/pracitionertools/.

Jordan, T. 1998. The truth of the land. In Reclaiming the Native Home of Hope, edited by Robert Keiter, pp. 43–49. University of Utah Press, Salt Lake City.

Kemmis, D. 1998. A democracy to match its landscape. In Reclaiming the Native Home of Hope, edited by Robert Keiter, pp. 2–14. University of Utah Press, Salt Lake City.

KenCairn, B. 1999. Applegate Partnership. In Beyond the Hundredth Meeting: A Field Guide to Collaborative Conservation on the West's Public Lands, edited by B. Cestero, pp. 41–45. Sonoran Institute, Tucson.

Leech, W.D. 2004. Is Devolution Democratic: Assessing Collaborative Environmental Management. California State University, Sacramento.

Loeser, M. R., T. D. Sisk, T. E. Crews, K. Olsen, C. Moran, and C. Hudenko. 2001. Reframing the grazing debate: Evaluating ecological sustainability and bioregional food production. In the Fifth Biennial Conference of Research on the Colorado Plateau, edited by C. W. van Riper III, K. A. Thomas, and M. A. Stuart, pp. 3–18. U.S. Geological Survey/FRESC Report Series, Flagstaff, Arizona.

Marston, E. 2001. Three Days in Estes Park. Collaborative Leadership in the West Conference, Estes Park, Colorado. Sonoran Institute, Bozeman, Montana.

Management Institute for Environment and Business. 1993. Conservation Partnerships: A Field Guide to Public-Private Partnering for Natural Resource Conservation. National Fish and Wildlife Foundation, Washington, D.C.

Muñoz-Erickson, T. A., and B. J. Aguilar-Gonzalez. 2003. The use of ecosystem health indicators for evaluating ecological and social outcomes of the collaborative approach to management: The case study of the Diablo Trust. Prepared for the National Workshop on Evaluating Methods and Environmental Outcomes of Community-based Collaborative Processes. Online Journal of the Community-based Collaborative Research Consortium, available at www.cbcrc.

Muñoz-Erickson, T. A., M. R. Loeser, and B. J. Aguilar-Gonzalez. 2004. Identifying indicators of ecosystem health for a semiarid ecosystem: A conceptual approach. In The Colorado Plateau: Cultural, Biological, and Physical Research, edited by C. W. van Riper III and K. L. Cole, pp. 139–152. University of Arizona Press, Tucson.

Putnam, R. 2000. Bowling Alone. Simon & Schuster, New York.

Red Lodge Clearinghouse. 2001. Next Year Country: A View from Red Lodge. Proceedings of a Workshop on Collaborative Resource Management in the Interior West. The Liz Claiborne/Art Ortenberg Foundation, New York.

Sayre, N. F. 2005. Working Wilderness: The Malpai Borderlands Group and the Future of the Western Range. Rio Nuevo, Tucson.

Sisk, T. D., and J. Palumbo. 2005. Collaborative science: Making research a participatory endeavor for solving environmental challenges. The Quivira Coalition 7(3).

Sisk, T. D., T. E. Crews, R. T. Eisfeldt, M. King, and E. Stanley. 1999. Assessing impacts of alternative livestock management practices: Raging debates and a role for science. In Proceedings of the Fourth Biennial Conference of Research on the Colorado Plateau, edited by C. W. van Riper III and M. A. Stuart, pp. 89–103. U.S. Geological Survey/FRESC Report Series USFSFRESC/COPL/1999/16.

Snow, D. 2001. Community-based collaborations in the West. Keynote address for the Collaborative Leadership in the West conference, Estes Park, Colorado. Sonoran Institute, Bozeman, Montana.

Tilt, W. 1996. Moving beyond the past: A grant-maker's vision for effective environmental education. Wildlife Society Bulletin 24(4): 621–626.

Tilt, W. 2005. Getting Federal Land Management Agencies to the Collaborative Table: Barriers and Remedies. Sonoran Institute, Bozeman, Montana.

Wilkinson, T. 2004. A Pilgrimage to Community: The Story of Custer County's Journey to Find Its Future. Sonoran Institute, Bozeman, MT.

Wondolleck, J. M., and S. L. Yaffee. 2000. Making Collaboration Work: Lessons from Innovation in Natural Resource Management. Island Press, Washington, D.C.

THE FRAME PROJECT—A COLLABORATIVE MODELING APPROACH TO NATURAL RESOURCE MANAGEMENT AT MESA VERDE NATIONAL PARK, COLORADO

Christine E. Turner, William H. Romme, Jim Chew, Mark E. Miller, George Leavesley, Lisa Floyd-Hanna, George San Miguel, Neil Cobb, Richard Zirbes, Roland Viger, and Kirsten Ironside

The rapid pace of social and environmental changes over the past few decades has presented significant challenges for the people who manage our public lands. Demographic shifts have placed large populations in close proximity to public lands and have resulted in increased public scrutiny of decision making on those lands. Natural resource managers are required to make informed decisions about multiple resources in complex natural systems in the face of competing and often conflicting objectives and values. The result is that natural resource managers are managing not only resources but also public expectations in the evolving nature of resource management (Tony Cheng, personal communication 2005). The challenge for today's natural resource manager is thus to optimize the management of multiple resources while minimizing the negative impacts of any given decision, and, at the same time, to engender trust and acceptance of the decision process. No small task.

In response to the increased expectations, two trends in natural resource management have emerged: a trend toward engaging stakeholders in participatory, collaborative processes, and a trend toward wider use of modeling to help manage the inherent complexity of natural systems. Collaborative engagement of stakeholders results in more inclusive and transparent decision making, which can engender greater acceptance of decisions and a wider sense of stewardship (Wondolleck and Yaffee 2000). The trend toward the use of numerical modeling in resource management addresses the need to accommodate the numerous and complex interactions of natural systems (Jakeman et al. 2006).

To adequately represent the inherent complexities of natural systems we need a way to fully address the interactions and feedback among individual components of the system. Although a large number of individual models are available to address individual components of natural systems, the coupling of these models is missing. The USGS's Modular Modeling System (MMS; http://www.brr.cr.usgs.gov/mms) offers an ideal framework to facilitate the integration and linking of process models and the execution of them in a coupled manner. The framework also facilitates adaptive management approaches where alternative scenarios and model combinations can be applied and refined iteratively with new scientific understanding and observations from monitoring results.

The principles of collaboration are helpful in situations in which knowledge is distributed among different parties. Although collaborative approaches to natural resource management often involve participation by the public and stakeholder groups (Wondolleck and Yaffee 2000), our collaborative process centers on a smaller set of participants—namely, resource managers, scientists, and modelers. We wanted to evaluate the dynamics of collaborative modeling by combining an integrated modeling approach (provided by MMS) with collaborative problem-solving approaches. We refer to this as a "collaborative modeling approach." The two major assumptions in the approach are that collaborative identification and framing of the science issues effectively links science to decision-making needs, and that integrated modeling approaches such as MMS provide the necessary framework to link information across the natural sciences, allowing for more integrated planning strategies. We postulated that the collabora- tive modeling approach would allow us to more effectively identify and link the perti- nent science to natural resource decision making. The value of collaborative modeling approaches is well recognized (Nicolson et al. 2002; van den Belt 2004). Now with MMS we can employ collaborative identification of the issue and selection of the appropriate models.

Our initial collaborative modeling efforts in Project FRAME (Framing Research in support of the Adaptive Management of Ecosystems) brought together resource man- agers, scientists, and modelers to address the management of pinyon-juniper (PJ) woodlands at Mesa Verde National Park. PJ woodlands are abundant on the Colorado Plateau (25 million ha), and PJ management involves a number of cross-cutting manage- ment issues, such as ecosystem restoration, fire management, grazing, off-highway vehicle use, and soil erosion. Moreover, a number of federal and university scientists and managers agreed to collaborate in this effort.

COLLABORATIVE MODELING APPROACH
Collaboration

The well-established principles of collabora- tion are articulated in a number of publica- tions, most notably for natural resource management in Wondolleck and Yaffee (2000) and McVicker and Bryan (2002). It is useful to stress what collaboration is, as well as what it is not. This is often the case with words that rapidly gain currency only to find their meaning diluted in the process. The term collaboration is sometimes inap- propriately used to describe interactions that do not meet the criteria for true collabora- tion. Key to true collaboration is a commit- ment to continuous engagement by all par- ties, which helps to create an environment of collective learning so that each party bene- fits from the perspectives and knowledge base that others bring to the table. Collabo- ration is not the same as seeking input, cooperating, hosting listening sessions, or engaging in outreach. As we shall see in the ensuing case study, a truly collaborative approach was key to success.

Our focus was on how to bring science into the collaborative problem-solving en- vironment. An important premise of our approach is that the knowledge to model a complex natural system is distributed, in our case among the resource managers, scientific experts, and physical-process modelers. The distributed knowledge can be elicited through the collective learning that is char- acteristic of collaborative problem-solving environments because of the continuous and active engagement of all participants. The scientists bring their diverse expertise to the discussion. Modelers provide the means to capture this expertise in numerical models. The resource manager contributes his knowledge of the ecosystem and also helps maintain the focus on the decision context so that pertinent science, not just "sound" science, is brought to bear on the resource management issues. In addition, framing the science issues collaboratively promotes in- terdisciplinary science approaches necessary to address most resource management needs. The collaborative process described here helps to frame the science needs em- bedded in resource management issues.

Modeling

Physical or numerical modeling of natural processes provides a way to analyze and assess the likely effects of alternative man- agement strategies on a variety of responses, such as hydrological and ecological re- sponses (e.g. Starfield 1997). To those not familiar with models, this can seem like a "black box" approach, but in essence, mod- eling is a systematic way to capture what people are already thinking. Models allow people to conceptualize the way they think things work while providing a framework for collaboration in which resource managers, scientific experts, and modelers can jointly develop the scientific input for the modeling effort (Starfield 1990). By involving all participants at all stages of the modeling it is possible to ensure that the important questions critical to resource management are addressed, that the scientific research that forms the basis of the modeling effort is as accurate as possible, and that the appropriate interrelationships in the natural system are captured. The results of the modeling simulations based on different scenarios allow resource managers to evaluate the possible range of outcomes. Collective learning occurs during the iterations required to refine the models because of increased understanding of the natural systems and the capabilities of the models. As a result, the resource managers and scientific experts learn how to interact with the models and to modify them to more accurately reflect the current state of knowledge of the ecosystem dynamics.

Constructing diagrammatic conceptual models—which visually depict interactions in ecosystems—provides the building blocks for quantitative numerical modeling. Conceptual models typically summarize existing knowledge and hypotheses about interrelationships among key system components and processes. These cause-and-effect relationships are illustrated diagrammatically with arrows that show how various parts of the system connect and interact. As a result, they can serve as important tools for communication among diverse audiences, aid in the identification of research needs, and inform the development of quantitative simulation models (Bestelmeyer et al. 2004). The communication function of conceptual models is often enhanced through their collaborative development (Heemskerk et al. 2003). Conceptual models provide abstractions of the ecosystems and the management issues that need to be quantified in order to provide the means to test, develop, and evaluate alternatives. Constructing diagrammatic conceptual models allows participating

scientists and resource managers to identify the structures of relationships for which physical modeling later provides the foundation. The conceptual models also help the modelers build a framework that can be flexible in testing knowledge.

THE ROLE OF MMS

Because natural systems are complex, the modeling approach used in ecosystem analysis must provide a means to express the interrelationships among components of the system. The models must therefore be able to accommodate integrated science approaches in order to accurately capture the feedback mechanisms in ecosystem dynamics. The USGS's Modular Modeling System (MMS) provides a modular framework in which to address these needs and is thus ideal for addressing cross-cutting resource management needs.

The MMS is an integrated system of computer software developed to provide the research and operational framework needed to enhance development, testing, and evaluation of physical-process modules and models; facilitate the coupling of models for application to complex, multidisciplinary problems; and provide a wide range of analysis and support tools for research and operational applications. MMS supports the integration of models and tools at a variety of levels of modular design. For process and single model applications, the MMS has a master library that contains compatible modules for simulating a variety of water, energy, and biogeochemical processes. A model for a specified application is created by coupling appropriate modules from the library. If existing modules cannot provide appropriate process algorithms, new modules can be developed and incorporated into the library. In addition to individual process models, the MMS also supports the development and application of tightly coupled models, loosely coupled models, and fully integrated decision support systems using both MMS and non-MMS modules, models, and analysis tools.

A geographic information system (GIS) interface is provided in MMS for applying

GIS tools to delineate, characterize, and parameterize topographical, hydrological, and biological features for use in a variety of lumped- and distributed-modeling approaches.

A set of tools is available for developing climate time series with which to run models. These include a climate generator, methods to downscale atmospheric model output, and methods to obtain and analyze historic climate data. Optimization and sensitivity analysis tools are also provided to analyze model parameters and evaluate the extent to which uncertainty in model parameters affects uncertainty in simulation results.

A major goal of the FRAME project (Framing Research in Support of Adaptive Management of Resources) is to link the vegetation dynamics model SIMPPLLE (SIMulating Patterns and Processes at Landscape Scales) with a variety of watershed, erosion, hydraulic, and ecosystem models in MMS to enable the assessment of the effects of alternative resource-management options on a variety of hydrologic and ecosystem processes. Output from SIMPPLLE is an ensemble of potential vegetation conditions projected years to decades into the future. Key components of the linked MMS models and SIMPPLLE are tools to estimate new parameters in MMS process-based models using vegetation and ecosystem attribute data from SIMPPLLE output, and a climate generator to provide time series of meteorological variables, such as precipitation and temperature, for use as input to the process-based models.

MMS is a key component of collaborative modeling approaches because, in addition to its adaptability, it serves as a framework for collaboration. The resource managers and scientific experts can work jointly with the modelers to choose the appropriate types of models needed to bring the appropriate science to bear on the issues. The natural resource managers and scientific experts also have the opportunity to provide their knowledge and scientific insights to tailor the models to fit the natural setting. The modular toolbox design also enables the immediate integration of advances in physical and biological sciences, GIS technology, computer technology, and data resources into the toolbox. Resource-management decision making thus benefits from the ability to constantly refine the models with state-of-the-art scientific information and technology.

MMS deals with complexity and the integrated science needed to address it because it allows the linkage of science information across the natural sciences. With the degree of integration that is permitted by the MMS, it is possible to evaluate the effects of potential management actions as they play out in the ecosystem. As a result, MMS is applicable to multi-objective resource management, allowing resource managers to develop more integrated planning strategies.

A FRAME PROJECT CASE STUDY

After designing the collaborative modeling approach to linking integrated science to natural resource management concerns, we needed an appropriate place and resource management issue to evaluate the approach. Project FRAME was proposed to test and refine the collaborative modeling approach by coupling collaborative approaches to framing science questions with modeling tools applicable to multi-objective resource management. Development of the strategy across a wide range of ecosystems will require a multi-year effort; however, we began with an initial focus on selected management issues in a single geographic area. We selected the Colorado Plateau region because it is an area dominated by federal lands, and because DOI and USDA agencies are currently reevaluating land management strategies because they face fundamental changes in ecosystems in this region (http://www.mpcer.nau.edu/direnet/). Drought provides the current focus for resource managers because many resource management plans were developed in the period 1978–1995, which were wet years in the region. Even in the absence of drought, however, resource management issues require a systems approach because choices made for each management objective have implications for other resources.

In partnership with the Colorado Plateau Cooperative Ecosystem Study Unit and the Merriam-Powell Center for Environmental Research, we selected the management of PJ woodlands at Mesa Verde National Park (MVNP) as our initial focus. PJ management involves a number of cross-cutting management issues—such as fire management, invasive species, insect infestation, and soil erosion and sediment deposition—that relate to ecosystem health, visitor safety, and cultural-site preservation. The natural resource manager at MVNP was enthusiastic about bringing science to bear on resource-management issues. MVNP was also an ideal location for the case study because of the numerous scientific datasets available for the modeling effort (e.g. Floyd et al. 2000 and 2004). Moreover, much of the park's remaining PJ woodlands lack the level of human disturbance that most other areas have experienced. A number of federal and university scientists, managers, and modelers agreed to participate in the effort (see http://www.mpcer .nau.edu/frame/ for more information).

A key resource-management objective in MVNP is to protect and maintain structurally and biologically the park's diverse old-growth PJ stands. Will it be possible to maintain some of the oldest stands as refugia, while at the same time allowing for the occurrence of natural disturbance and successional processes that may impinge on these valued stands? Fire and fire-management strategies have been dominant concerns in the park in recent decades. About 50 percent of PJ has burned since 1934, and two-thirds of the park has burned in less than a decade even with a policy of total fire suppression. The remaining old-growth PJ stands are 300–500 years old. Extensive tree-ring dating and mapping of past fires indicate a 400-year fire rotation (i.e., the time required for the cumulative area burned to equal the entire area of pinyon-juniper vegetation in the park). During the long time intervals between stand-replacing fires, small-scale disturbances (black stain, "normal" beetle kills, and lightning-ignited fires of small

extent) have led to development of dense old-growth stands that are susceptible to stand-replacing crown fires under extreme drought. The current drought and fire cycle has thus caused heightened concern among park management. Is continued fire suppression to protect PJ appropriate for a national park that is supposed to let nature take its course? Is nature even able to really take its course under current conditions (air pollution, exotic weeds, climate change)? The resource manager wanted to be able to evaluate the effects of various management choices in the park and to engage fire management in the process so that resource management and fire management plans could be complementary, resulting in more comprehensive planning and management strategies.

Framing the Question in the Decision Context

We established a collaborative modeling environment for the FRAME project at Mesa Verde National Park by convening a series of interactive workshops and field trips in the park, with all project participants present as often as possible. Workshop participants included the park natural resource manager, quantitative modelers, and other scientists with expertise in the dynamics and management of pinyon-juniper ecosystems. In a workshop setting, we began by framing the science issues for modeling. First, the natural resource manager of MVNP provided background information about the state of old-growth PJ in the park. With this focus we could begin to frame the science needs for ecosystem modeling in light of the resource manager's decision context. The natural resource manager established that his desired future condition (Table 1) was maintenance of healthy PJ woodland with preservation of the remaining old-growth PJ stands. Our strategy was to accomplish this goal in the context of integrated management of fire and of natural and cultural resources; in this early stage of the project, fire management, cultural resource management, and representatives from adjoining lands were not yet involved.

Table 1. Desired future conditions for pinyon-juniper ecosystems in Mesa Verde National Park described at the vegetative-patch level, the landscape level, and in terms of key maintenance processes (George San Miguel, personal communication).

Patch Level: All pinyon-juniper community patches in the park collectively consist of and are dominated by their entire range of structural and functional groups of native plant species as well as the full functioning of succession and other key ecological processes that are naturally characteristic of Mesa Verde pinyon-juniper landscapes, soil-geomorphic settings, climatic conditions, and successional stages.

Landscape Level: The Mesa Verde landscape is composed of a mosaic of native plant communities with a dynamic range of compositions and configurations (i.e. landscape structure) determined jointly by characteristic disturbance processes (e.g. fire, flood, drought, insect outbreak) and environmental constraints such as soil, topography, and climate. As a whole, the community patches in the landscape represent the full range of natural variability determined by the natural disturbance regime, including old-growth pinyon-juniper woodlands. For the NPS to retain the full range of successional stages and natural variability, spatial scales extending well beyond park boundaries and encompassing the minimum dynamic area may need to be considered.

Key Maintenance Processes: Natural disturbance processes (e.g. fire, climate), successional processes, and management treatments (e.g. efforts to control invasive exotic species) are tools to facilitate the maintenance of desired ecosystem conditions at patch and landscape scales.

Conceptualizing Ecosystem Dynamics

With the desired future condition determined, the scientists, modelers, and resource manager worked together to determine ecological factors and attributes related to PJ ecosystems that were pertinent to an integrated modeling effort. We engaged in collaborative dialogue to address the issue of ecological complexity. The process involved the challenge of not only identifying the complexity of interactions in the PJ ecosystem, but also reducing that complexity to facilitate the modeling task and the relevance of the model outputs for managers while also ensuring the scientific validity of outputs as well as user confidence. The goal is to maximize the utility of modeling for resource management. With the assistance

of a facilitator and a recorder, participants reviewed and discussed (1) drivers of ecosystem change and variability (natural disturbances, anthropogenic stressors, and management actions); and (2) ecosystem attributes that are affected by these drivers, amenable to quantitative modeling, and suitable for evaluating ecosystem conditions in relation to management objectives. In relation to this latter point, the desired future condition concept played a central role by focusing the dialogue among resource managers, scientists, and modelers. The workshop facilitator led the discussion while the recorder captured information in a table projected on a screen in front of the meeting room. Throughout the discussion, an explicit effort was made to identify modeling attributes that were directly related to NPS "vital signs"—environmental attributes selected for long-term monitoring by NPS staff for purposes of tracking status and trends in the condition or "health" of park ecosystems (Thomas et al. 2006).

Once the drivers of ecosystem change and variability (and the attendant response variables) were determined (Table 2), it was possible to use diagrammatic conceptual models that visually depict interactions among key components and processes of pinyon-juniper ecosystems (Figure 1). If a conceptual model had not already existed, we would have used the information in Table 2 to construct one. In this case, FRAME relied upon models previously developed by scientists and resource managers to support the identification of long-term monitoring needs in NPS units of the Colorado Plateau (Miller 2005; Vankat unpublished data). The key components and critical pathways identified in Table 2 as essential to modeling pinyon-juniper woodlands are captured in Vankat's conceptual characterization of the pinyon-juniper ecosystem (Figure 1).

During the course of the case study, ongoing field studies by FRAME project participants revealed that the rapid spread of cheatgrass (Bromus tectorum) in the park posed an increasing threat with respect to fire frequency and spread. In collaboration

Table 2. Drivers and measurable response variables included in the quantitative modeling effort as the result of collaborative dialogue between resource managers and scientists.

Drivers of Ecosystem Change and Variability				Response Variables: Ecosystem Condition Evaluation Criteria		
Natural Disturbances	Anthropogenic Stressors	Management Actions		Vegetative-Patch Attributes	Landscape Attributes	Watershed Attributes
Insect outbreaks	Landscape fragmentation	Fuel breaks; thinning	→	Fractional cover of native plant functional types (grasses, shrubs, trees)	Fire frequency and extent	Sediment production
Disease	Invasive exotic plants	Fire suppression	→		Total area and patch-size distributions of different types of vegetation	Runoff volume
Wildfire	Soil-surface disturbances	Management-ignited fire	→	Fractional cover of invasive exotic plants		Probability and magnitude of debris-flow events
Climatic episodes		Prescribed natural fire				

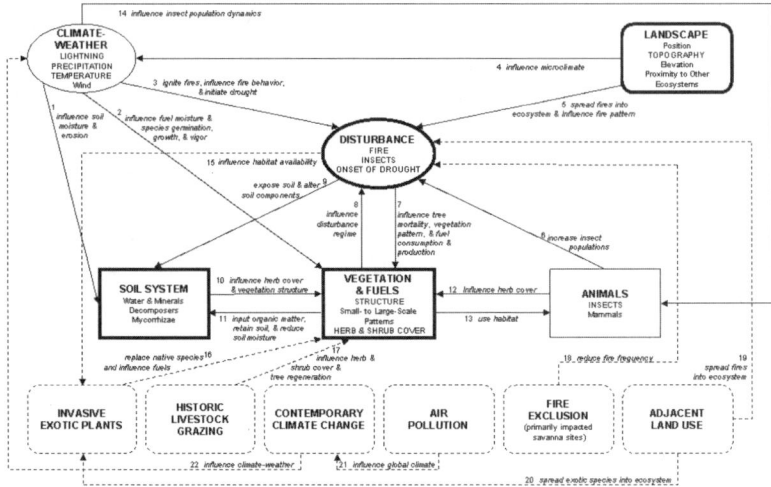

Figure 1. Pinyon-juniper ecosystem characterization model for the southern Colorado Plateau (Vankat, unpublished).

with the natural resource manager, a group consensus emerged: We decided that the spread of cheatgrass—and its implications for the PJ ecosystem—had become a priority. Vankat's diagram (Figure 1) identifies exotic species (including cheatgrass) and fire as two major components of the PJ ecosystem, shown schematically as two boxes. Another diagrammatic conceptual model (Figure 2), previously developed by one of the FRAME project members (Miller 2005), describes detailed interactions within the two boxes represented on Vankat's figure, and thus provided the necessary conceptual model for us to focus our efforts on the cheatgrass component of the PJ ecosystem (Figure 2). The decision of the group to focus our modeling efforts on the potential threat of cheatgrass to the ecosystem dynamics at MVNP illustrates the adaptive nature of a collaborative modeling approach. In resource management, it is often the case that priorities shift with time, and modeling efforts designed to meet the needs of resource management ideally should be able to accommodate these shifts. In our case study, the long-range management concern—preserving old-growth PJ—necessitated first addressing the role of cheatgrass as it per-

tained to fire regimes in the park, because cheatgrass posed the greatest immediate threat to the PJ ecosystem.

Cheatgrass is a non-native winter annual that germinates in the fall, grows slowly during the winter, and then grows rapidly in the early spring. By early summer it has set seed and died, creating a continuous fuel bed of quick-drying, flashy fine fuel that can readily carry fire, even without wind. In parts of the Great Basin and Colorado Plateau, cheatgrass has profoundly altered local fire regimes, from historically infrequent high-severity fires to frequent low-severity fires (Whisenant 1990; D'Antonio 2000); there is concern that this will happen in MVNP as well. The park's native flora exhibits a fascinating variety of adaptations for surviving fires that occur at long intervals, such as resprouting from roots and rhizomes or requiring a combination of heat and smoke to stimulate seed germination. However, some of these adaptations might be ineffective in the face of frequent fires. Indeed, in other areas where cheatgrass has altered the historical fire regime, some native plant species have been locally extirpated. In addition, an increase in the frequency of fast-spreading fires would pose a

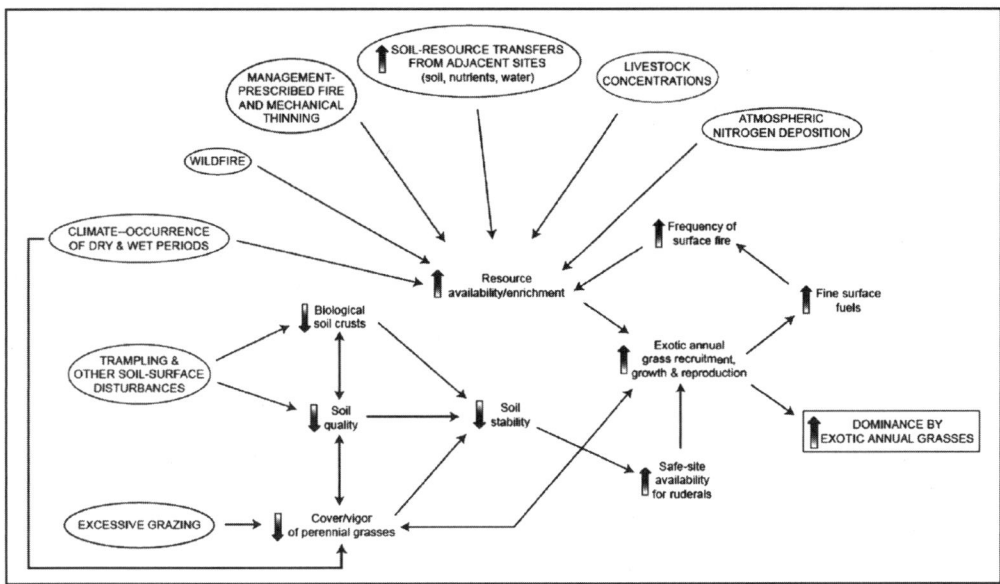

Figure 2. Ecological factors and processes that can interactively lead to increasing dominance of exotic annual grasses in pinyon-juniper ecosystems (from Miller 2005).

serious threat to MVNP's world-renowned cultural resources, last stands of old-growth woodlands, and visitor safety.

Cheatgrass has been present in MVNP for many years, especially in the deep canyon bottoms of the park's southern portion. However, it was never widespread until the last 5 years, when it began to expand its range across the mesa tops and into the highest elevations of the park. Apparently cheatgrass was previously limited by the relatively cool, moist conditions on the mesa tops, and was restricted therefore to the warmer, drier southern canyon bottoms. The unusually warm summers and winters of the past 5 years, coupled with heavy fall rains in all but one of those years—which is optimal timing of precipitation for cheatgrass germination—have allowed cheatgrass to rapidly expand its range, especially in places where fire or other disturbances have created bare ground. Cheatgrass is now a dominant species in much of the area that burned in the 2000 Bircher fire and elsewhere in MVNP.

Numerical Modeling

The construction and use of conceptual models comprise important first steps in ecosystem modeling efforts because they define the key relationships that capture ecosystem dynamics. In order to quantitatively evaluate the probability of interactions or consequences of decisions or events for resource management purposes, the relationships and interactions need to be modeled numerically, using the conceptual models as templates for selecting and constructing the numerical models. An underlying premise of the FRAME project is that quantitative simulation modeling can inform the decision-making process by enhancing managers' abilities to explore potential ecological consequences of different management alternatives (Starfield et al. 1995; van den Belt 2004).

The integration of all available modeling components using MMS is a long-term goal, whereas the results of our collaborative modeling process at MVNP identified cheatgrass invasion and the resulting fire regimes as urgent initial priorities for our numerical modeling efforts. A major concern of post-fire disturbances is the effect of erosion and debris flows on the park's cultural resources and visitor safety. The initial numerical modeling effort was therefore aimed at evaluating the potential impacts of cheatgrass invasion on fire frequency and fire-related processes (erosion and sedimentation). The case study was thus limited to the coupling of the landscape model SIMPPLLE with a debris-flow-generation model related to post-fire runoff and erosion.

Landscape Modeling

The SIMPPLLE model has a generic, object-oriented design that allows quantitative modeling of the interactions captured in the conceptual models. The model has been developed and used in a wide range of ecosystems (Chew et al. 2004; http://www.fs.fed.us/rm/missoula/4151/SIMPPLLE/). In our FRAME case study, SIMPPLLE was populated using the components and critical pathways represented by Figures 1 and 2. Selection and quantification of the numerical model parameters was made by the participants using scientific knowledge from the literature, field studies, and expert opinion. The existing SIMPPLLE design was compared to the diagrammatic conceptual models for the Colorado Plateau ecosystems to identify modifications (additions or changes) that had to be made to capture the specific knowledge within SIMPPLLE necessary to model the Colorado Plateau ecosystems for specific management issues. For example, the design of the categories in SIMPPLLE used to describe wildfire disturbance and species had to have additions made to capture the interaction between an invasive species' percent ground cover response to moisture conditions and a species-specific interaction with fire spread logic. Changes also had to be made in the category that represents land units to account for the level of soil information needed to predict the probability of the invasive species (Floyd et al. 2006).

SIMPPLLE computes a probability for the occurrence of a disturbance process for each plant community. These probabilities are determined by a combination of research results and expert opinion expressed as logic rules. These probabilities are determined not just by a plant community's attributes, but also by what exists around it, what processes are occurring around it, and what has occurred in the past. SIMPPLLE uses these process probabilities in a stochastic fashion, rather than using a transition matrix approach. There is no fixed transition rate of changes in the acres of vegetation states as a result of a disturbance process. Changes expressed for an entire landscape are the summation of changes at the plant community level. The range of possible combinations of outcomes for each plant community, as influenced by the interaction of the factors influencing disturbance probabilities in a simulation, results in a stochastic output.

Populating the model's structure with both numerical values and logic relationships, and validating its performance, involved a process of iterative interaction with the scientists and the resource managers. Selection and quantification of the modeling parameters were made by the participants using scientific knowledge from the literature, field studies, and expert opinion. Populating the structure is often the result of a consensus reached between scientists and resource managers. For example, decisions about how to identify vegetation species, size class, and density levels, and what level of soil survey information to use to describe land units, depends on what is available from inventories, what is needed to capture the dynamics, what is needed to predict probabilities of disturbance processes, and what is needed to address management issues. The system is designed with user-interface screens that facilitate the interaction with scientists and managers in making these choices. The model's behavior was validated at levels from individual plant communities to the entire landscape through a number of workshops with continuous and collaborative interaction among the modelers, scientists, and resource managers. The model had to display changes at an individual plant community level that were consistent with other research and experience. The disturbance processes at the landscape scale had to be consistent with past process history.

RESULTS

For the first modeling effort, which involved evaluating the potential impact of cheatgrass on the fuel conditions and fire potential in MVNP, we simulated fire and vegetative response to fire under two contrasting scenarios: (1) without any impact of cheatgrass on fire frequency or spread in the park and (2) with cheatgrass expanding at what we believe to be a maximum likely rate over the next 20 years and affecting fire frequency and fire spread. We recognize that neither of these scenarios might be exactly what will happen over the next 20 years in the real MVNP. However, our intent was not to forecast the future with precision, but rather to describe a range of potential futures within which the real future is likely to fall. In that sense, these two scenarios represent the best and worst outcomes likely to occur in the real park, if current trends continue. These scenarios do not include any management actions. The intent of this analysis is to help identify the potential need for scenarios that would include management actions. Both scenarios were quantified by making a set of 20 stochastic simulations for each one.

For both scenarios, we assumed a 20-year sequence that mimicked the weather patterns of the 1950s drought—that is, most years with below-average or average precipitation, but occasional years of above-average precipitation. We assumed a general drought condition in our simulations because atmospheric scientists predict that the current drought in the western United States will continue for at least another decade, and because we wanted to develop a worst-case scenario for the fire situation in MVNP. Because a single set of simulations cannot incorporate all of the many variables that could be of potential interest without becoming overwhelmingly complicated, neither of our scenarios included direct effects on fire behavior of the recent pinyon

mortality in the park that has resulted from drought, bark beetles, and black stain fungus. We also chose not to incorporate any effects of fire suppression. It would be possible to incorporate these effects in future simulations, if desired.

Once cheatgrass establishes in a plant community, the rate of change in canopy cover is identified through logic rules that include the level of moisture for the year and other disturbance processes. The change in a fire process, spreading from one plant community to another, was modified to identify what level of canopy cover of cheatgrass made a difference in the spread. The values used in these relationships were the result of an iterative process of making simulations with a range of values and evaluations by managers and scientists.

Adding cheatgrass changed the simulations by including logic that provided for the probability of cheatgrass occurring, its change in canopy cover once it is introduced, and its impact on the fire spread process. These logic rules depend on past disturbance processes, the other plant species present, the soil type, and the moisture for the yearly time step (Floyd et al. 2006). Combinations of these factors result in a specific probability of occurrence and a canopy cover level at which cheatgrass occurs. The highest probability of cheatgrass invasion and increase was in recently burned areas. In both burned and unburned areas the probability was higher in plant communities with a low number of species capable of prolific resprouting after fire (e.g. PJ woodlands), and on certain soil types known to be vulnerable to weed invasion (such as Mikim loam and Arabrab-Longburn soils). Probabilities of cheatgrass invasion and increase also were higher during wet years than during normal or dry years, in both burned and unburned areas.

The impact on fire spread across plant communities is simulated by adding a layer of fine fuels that is capable of spreading fire without being driven by a wind event. In SIMPPLLE, the fire spread logic was expanded to include the presence of cheatgrass and its canopy cover level. There was no change in the spread logic until it reached 45 percent cover. Above 45 percent cover the logic of fire spread was changed to be comparable to what is observed under conditions of dry fuels and high winds. We chose this 45 percent threshold along with the initial probability of occurrence and the change rates based on an iterative process of managers and scientists evaluating SIMPPLLE output across a range of input values. Professional judgment had to be used in selecting the final set of values because research has not yet clearly identified at what level cheatgrass initially occurs, changes on a yearly basis, or increases fire spread. In this second scenario, which incorporated an increase in cheatgrass distribution and abundance, at the end of each simulated year there was a probability of cheatgrass invading new portions of the landscape and of increasing in cover where it was already present.

20 Years of Simulated Fire Without Increasing Cheatgrass

We first ran SIMPPLLE under the assumptions (1) that cheatgrass would not have any impact on fire frequency or spread, and (2) that generally dry conditions would continue for 20 years. The result was very little fire in any of the 20 replicate simulations. The cumulative area burned over the entire 20 years was less than 600 acres in any of the 20 runs. Because we produced so little fire activity, we do not include any maps or further details of our results for this scenario.

We do note, however, that this is exactly the result we would expect given the assumptions that went into the simulations. Indeed, this scenario strongly resembles the actual fire regime that characterized MVNP for most of the twentieth century prior to 1996, that is, no significant fire spread in the great majority of years. Large fires occurred only in a few key years (1934, 1959, 1972, and 1989) when dry fuels and warm temperatures were accompanied by high winds—severe fire conditions that we did not incorporate in this scenario.

20 Years of Simulated Fire With Increasing Cheatgrass

Our second scenario was based on the assumptions of (1) progressive increases in the distribution and cover of cheatgrass, according to the probabilistic rules outlined above, and (2) generally dry conditions. Adding cheatgrass to the simulations resulted in a dramatic increase in total area burned. The average total area burned in each year, averaged across all 20 runs, ranged from less than 100 acres to 2400 acres. Smaller amounts of burned acreage were seen primarily in the first 5 years of the simulations, while cheatgrass is still expanding from its 2005 distribution. Once cheatgrass occupies its full potential extent across the park, the average area burned is more than 1000 acres in almost every year. Because the simulations are stochastic, there was much variability among the 20 runs. Every simulated year included at least one run with almost no area burned. (Note that this was a different run for each year; none of the individual runs produced a near-zero area burned in all or even many of the simulated years.) On the other hand, the maximum area burned in any of the 20 years ranged from 4000 to 8000 acres per year from year 6 to the end of the simulation. And at the end of 20 years, the median cumulative area burned was 22,880 acres. This represents a fire rotation of approximately 45 years for the park as a whole—a dramatic change from the historical fire rotation, which was measured in centuries.

To further explore the implications of such an increase in annual burning, we identified the individual simulation that produced the median cumulative total area burned over 20 years (22,880 acres) and mapped the locations of each year's fires as simulated by that particular run. Our focus was on locating the places that were burned more than once during the 20-year simulation; see Figure 3. A substantial amount of area was simulated to burn twice and several areas were simulated to reburn as many as five times in 20 years. For comparison see

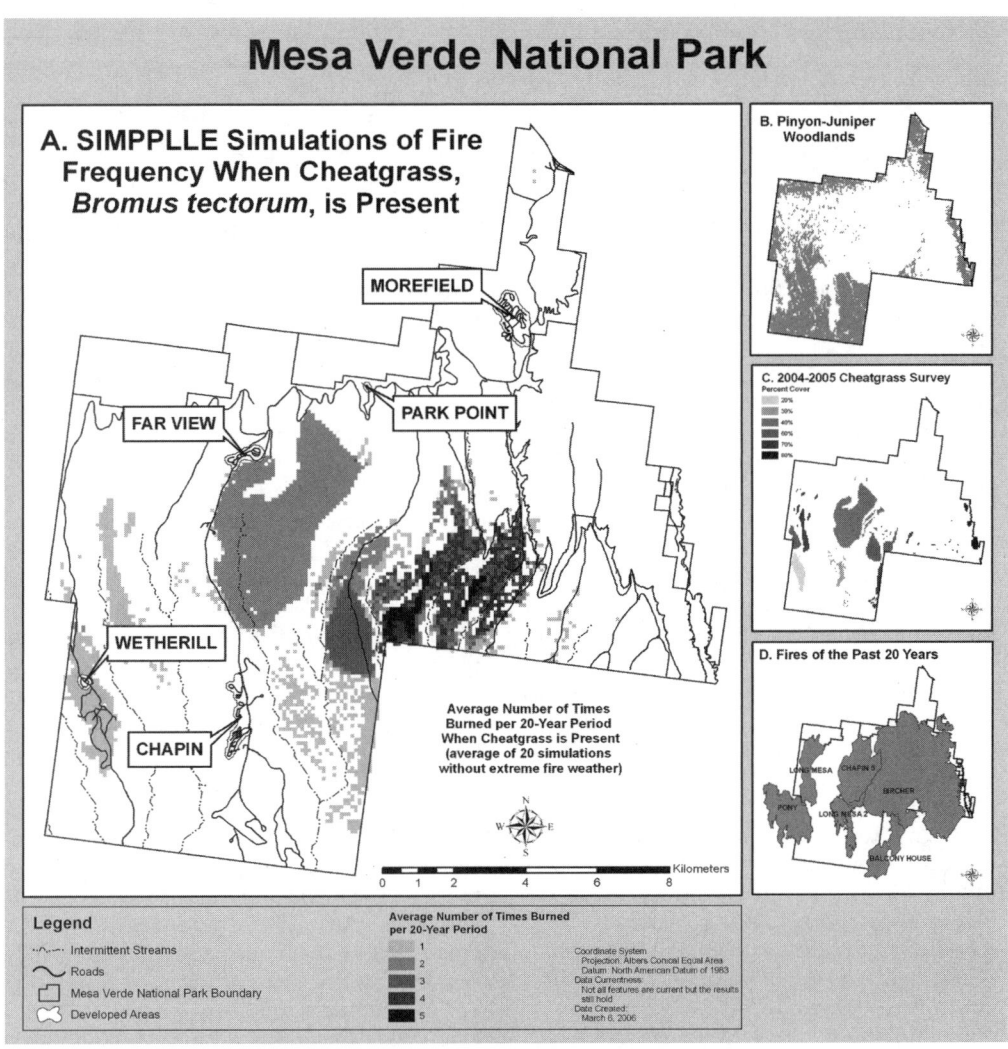

Figure 3. (a) Results of the SIMPPLLE simulations of fire frequency when cheatgrass, *Bromus tectorum*, is present at Mesa Verde National Park; (b) the remaining old-growth pinyon-juniper woodlands in the park; (c) mapped populations of cheatgrass during 2004 and 2005 surveys; (d) recent fires that have burned more than half of the park.

Figure 3b–d showing the PJ woodlands, the 2004–2005 cheatgrass survey, and fires of the past 20 years. The greatest concentration of the repeatedly burned areas in the simulations was in the south-central portion of the park, generally within the perimeter of the 2000 Bircher fire. Very little old-growth pinyon-juniper woodland was burned in the simulations.

Implications of Results for Cheatgrass and Fire

The results of these simulations raise serious concerns about cheatgrass invasion and its potential effect on fire frequency. Any substantial increase in the extent of annual burning might be of concern from the standpoint of conserving the park's native biota because most of the native fauna and flora are adapted to relatively infrequent fire. However, the most worrisome aspect of these projected changes in MVNP's fire regime is the demonstrated potential for frequent reburning, at intervals as short as a few years. Such a disturbance regime would be far outside the historical range of variability for this ecosystem, and would likely lead to substantial reductions and even local extirpation of many native plant species. For example, the park's pinyon and juniper need about 75 years after fire to become reestablished in burned areas. A 45-year fire rotation could thus prevent normal successional processes and adversely affect all of the native flora and fauna that depend on the woodland structure. At the same time, such a fire regime would create a nearly optimal environment for cheatgrass, musk thistle, and other non-native invasive species.

Debris-Flow-Potential Models

Debris flows are among the most hazardous consequences of rainfall on burned hillslopes. Because recently burned areas are vulnerable to debris flows during heavy precipitation events, the potential soil and hydrologic impacts of more frequent fires pose a serious management concern. The risk is greatest when locally intense precipitation falls within 1–3 years of a fire. We selected an empirical debris-flow model

(Cannon 2001; Cannon et al. 2004; Gartner 2005) to evaluate the potential for debris flows for the basins within the park following a cheatgrass-altered fire regime. This model has been used extensively throughout the intermountain West. The debris-flow model was incorporated into MMS to facilitate its linking with the output from SIMPPLLE.

The debris-flow model relates the probability and volume of a debris flow to a combination of geologic, soil, basin morphology, burn severity, and rainfall conditions. Basin, geologic, and soil characteristics can be obtained from databases that include digital elevation models (DEMs) and the USDA STATSGO soils database. Total rainfall and average rainfall intensity can be estimated using NOAA's *Precipitation-Frequency Atlas of the Western United States*. Using these digital databases and the fire-affected areas defined by SIMPPLLE, the GIS-based tools in MMS can be used to delineate and parameterize the debris-flow model. However, a major concern of MVNP was the potential for floods, erosion, and debris flows anywhere in the park and their impacts on cultural resources and visitor safety. Therefore, all basins in MVNP were evaluated for post-fire debris-flow potential.

The debris-flow model is limited to basins 25 km^2 or less in size. To focus on headwater areas in and adjacent to the park, and to avoid the lower canyon regions where the equations may not be appropriate for the steep sandstone walls and narrow valley bottoms, drainage basin size was limited to basins ranging from 2 to 10 km^2. All basins evaluated were assumed to burn completely at a moderate or high severity in order to provide a common basis for comparison among basins. The rainfall events selected for this application were the 2-year 1-hour and 100-year 1-hour storms. Total rainfall and average rainfall intensity for each storm were estimated using NOAA's atlas.

Debris-Flow Potential With Increasing Cheatgrass and Fire

The result of the debris-flow model application was that all of the basins showed an increase in the probability of a debris flow and in debris-flow volume when the 2-year and 100-year rainfall events were compared (Figures 4 and 5). Debris-flow probabilities and volumes were computed individually for each of 68 delineated basins. Results were then categorized into 10–20 percent probability classes and 20,000 m^3 volume classes (Figures 4 and 5) for the 2-year and 100-year rainfall events. The magnitude of the changes varied among simulated basins, reflecting underlying variation in topographic and soil characteristics.

These maps provide information that can be used to prioritize mitigation efforts, to aid in the design of mitigation structures, and to guide decisions for evacuation, shelter, and escape routes in the event that storms of similar magnitude to those evaluated here are forecast for the area. The potential for debris-flow activity after a fire decreases with time and the concurrent revegetation and stabilization of hillslopes. One can conservatively expect that the maps presented here may be applicable for approximately 3 years after the fires for the storm conditions considered here. Projected changes in the MVNP fire regime indicate the potential for frequent reburning, at intervals as short as a few years. This would bring an increased risk of significant debris-flow events, with the potential for substantial damage to water resources and cultural resources.

FRAME and PJ Ecosystem Management

The potential of the collaborative modeling approach developed in our FRAME case study goes beyond the boundaries of any particular department within a park or any land management unit. When we initially focused on PJ management as a natural resource issue in MVNP, we recognized the cross-cutting nature of the issue and that PJ management could best be addressed in the context of integrated management of fire and natural and cultural resources both within the park and on adjacent lands. It was discussed early in the FRAME project that Mesa Verde National Park may be too small a subset of an ecosystem to manage optimally without cooperation from neighboring lands. There are ecosystem drivers and stressors outside the park over which the NPS has no jurisdiction. For PJ management, for example, sufficient acreages of all PJ seral stages need to be maintained in a healthy state so that succession can lead back to a sound old-growth PJ community. To achieve that goal, we need to explore what size area ("minimum dynamic area") it takes to optimize the management of natural resources and to manage fire. Complementary management strategies across neighboring lands would be the optimum way to manage any part of the PJ ecosystem. As the modeling effort at MVNP progressed, fire management personnel from MVNP and land managers from adjacent BLM and Ute Mountain Ute tribal lands joined us in the workshops. They saw that we were developing a methodology that would address the full range of natural resource issues that they collectively face. Having all the adjoining land managers in the discussions increased the geographic area in which to view preservation of old-growth PJ woodland. This raised the possibility of preserving old-growth PJ on adjacent lands rather than exclusively within MVNP boundaries.

Collaboration across agency boundaries and across neighboring parcels of land is an emerging trend in land management. The collaborative modeling approach developed in the FRAME project provides a framework for collaborative decision making across agency boundaries. As a result, ecosystem-level land management can become a reality.

The FRAME project at Mesa Verde is a "proof-of-concept" case study that can be extended to other regions with PJ woodlands, the dominant vegetation type on the Colorado Plateau and the third largest vegetation type in the contiguous United States. Moreover, the FRAME collaborative modeling approach goes beyond any specific vegetation type. The approach can be incorporated into synthetic work where comparisons of managed, wildland, and urban landscapes would be possible. FRAME can also be used to incorporate information from regional drought studies, such as those promoted

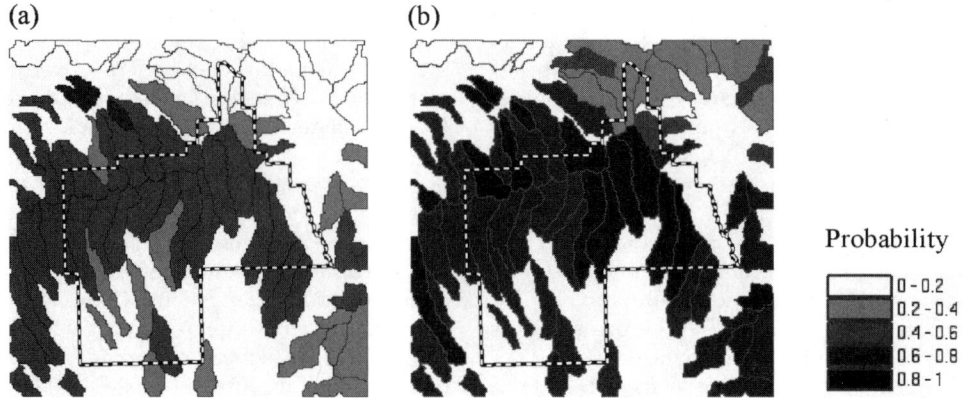

Figure 4. Probability of a debris flow for (a) 2-year 1-hour and (b) 100-year 1-hour rainfall events.

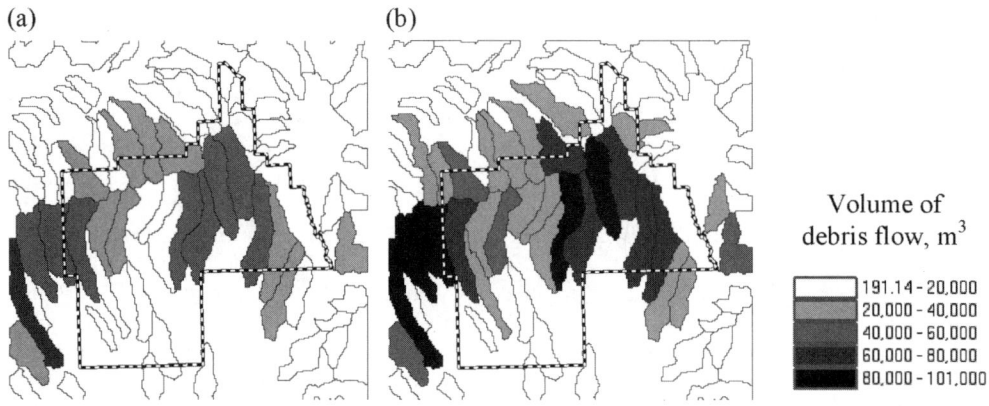

Figure 5. Volume of debris flow for (a) 2-year 1-hour and (b) 100-year 1-hour rainfall events.

by the Drought Impacts on Regional Ecosystems Network (DIREnet) to prioritize management decisions.

Ecosystem management is also expanding to include evaluation of large-scale drivers of system dynamics, such as climate. At Northern Arizona University, the landscape modeling effort under the auspices of FRAME has expanded to include a study whose goal is to make available a version of SIMPPLLE that incorporates output from global climate models (general circulation models). Species-specific responses across the landscape will be modeled through an interaction of changing susceptibility to disturbances, changing regeneration capabilities, and changing probabilities of disturbance processes. The expanded version of SIMPPLLE will therefore have the ability to track changes in species distribution as a response to climate change.

COLLABORATIVE MODELING AND THE FUTURE

The adaptive, collaborative modeling approach being developed in the FRAME project will provide land managers with the ability to evaluate the effects of alternative scenarios on multiple resources; the approach and the modular modeling system (MMS) are flexible enough to allow adjustments to changing conditions. Feedback from monitoring and assessment efforts can be used to refine the numerical models, creating an ideal adaptive management environment. Modeling can directly meet the needs of resource managers of various land management agencies because each agency can select the appropriate components of the model to apply. For example, NPS may not need to include grazing in their models, whereas BLM would.

An adaptive collaborative modeling approach also addresses a frequent concern expressed by both land managers and research scientists—the disparity between the scientists' desire to decrease the uncertainty of their understanding of complex natural systems through further research, and the resource manager's need to make the best possible decision in the near term based on the current state of knowledge. In the past, the conflict between the long-term and short-term perspectives has interfered with the ability to use science effectively in resource management decision making. The adaptive nature of MMS easily accommodates new research findings in addition to feedback from monitoring and assessment efforts. MMS provides an excellent way to both support decisions with current understanding and adapt to new scientific insights over time.

In addition to the specific models implemented thus far in the FRAME project, the modular modeling system provides the ability to link a variety of models, which offers particular promise in dealing with the complex natural resource issues that require incorporating knowledge from a broad range of scientific disciplines. The ability to evaluate the effects of alternative scenarios on multiple resources allows resource managers to optimize the management of multiple resources while minimizing the negative impacts of any one decision.

Our FRAME case study at Mesa Verde National Park was focused specifically on the collaborative modeling interactions of resource managers, scientists, and modelers. Engaging the public in collaborative modeling efforts, particularly those that are characterized by transparency and collective learning, can help build public trust in resource management decisions (Jakeman et al. 2006). The frameworks and models used in our study are open-source, allowing unrestricted use. In collaborative modeling efforts that include the public, facilitation of the process helps ensure that the principles of collaboration are honored, which is crucial to building and maintaining trust in the process. Key principles of collaboration that are crucial in these collaborative modeling settings include meaningful and continuous inclusion of interested parties, transparency, recognition of distributed knowledge, and fostering of a collective learning environment. The citizens on the Uncompahgre Plateau Project (http://www.upproject.org/) in western Colorado have run their own simulations of ecosystem dynamics, illustrating that modeling can be made readily accessible to all interested parties.

CONCLUSIONS

The overall strategy of the FRAME project is to combine the principles of collaboration with the adaptive capabilities of the USGS modular modeling system to develop a transportable, collaborative modeling approach to adaptive, multi-objective natural resource management. Although this will be a multi-year effort, the focus of our initial case study was management of pinyon-juniper woodlands at MVNP. The case study involved collaborative modeling efforts among resource managers, scientists, and modelers. The group collaboratively identified key system components, critical pathways, and associated conceptual models of pinyon-juniper ecosystem dynamics. The recent invasion and rapid spread of cheatgrass in the park has the potential to significantly alter the fire regime at MVNP by increasing fire frequency and impacting long-term vegetation successional patterns. This concern led us to focus on cheatgrass for the first modeling simulations. For the purposes of landscape modeling at MVNP, the SIM-PPLLE landscape model, a physical process model, was modified to capture the key ecosystem components and dynamics of the conceptual models. The SIMPPLLE model was further refined through an iterative process in which project scientific experts helped define probabilities.

Model results indicate the potential for frequent reburning, at intervals as short as a few years. These simulations suggest a projected fire rotation of approximately 45 years for the park as a whole—a dramatic change from the historic fire rotation, which was measured in centuries. Such a disturbance regime would be far outside the historical range of variability for the PJ ecosystem, and would likely lead to substantial reductions and even local extirpation of many native plant species. To evaluate the effects of frequent reburning on post-fire erosion and sedimentation, a debris-flow-potential model was incorporated in MMS to facilitate its linking with the output from SIMPPLLE. The results showed that the projected changes in MVNP's fire regime would bring an increased risk of significant debris-flow events, with the potential for substantial damage to water resources and cultural resources.

The FRAME case study at MVNP gave us an ideal opportunity to implement and refine the principles and components of a collaborative modeling approach. By coupling the principles of collaboration with integrated modeling approaches we are developing a collaborative modeling framework to facilitate adaptive, multi-objective resource management that is applicable across a wide range of ecosystems. Recent trends in natural resource management—toward integrated science approaches, co-management of public lands, adaptive management in the face of uncertainty, and public engagement in land-use decision making—are trends that developed in response to a greater appreciation of the inherent complexity, feedback mechanisms, and uncertainty in natural systems, plus increased public scrutiny of decisions on public lands. The FRAME collaborative modeling approach was developed to address the challenges faced by natural resource management, and provides a way to effectively link integrated science to natural resource management needs. The FRAME approach can also readily be adapted to engage the public in participatory natural resource management efforts.

ACKNOWLEDGMENTS

The FRAME (Framing Research in support of the Adaptive Management of Ecosystems) Project is a partnership among the USGS, NPS, BLM, USFS, BIA, the Ute Mountain Ute Tribe, universities, and research centers, including the Merriam-Powell Center for Environmental Research and the Colorado Plateau Cooperative Ecosystem Study Unit (CPCESU).

REFERENCES CITED

Bestelmeyer, B. T., J. E. Herrick, J. R. Brown, D. A. Trujillo, and K. M. Havstad. 2004. Land management in the American Southwest: A state-and-transition approach to ecosystem complexity. Environmental Management 34: 38–51.

Cannon, S. H. 2001. Debris-flow generation from recently burned watersheds: Environmental & Engineering Geoscience 7: 321–341.

Cannon, S. H., J.E. Gartner, M. G. Rupert, and J. A. Michael. 2004. Emergency Assessment of Debris-Flow Hazards from Basins Burned by the Padua Fire of 2003, Southern California. USGS Open-File Report 2004-1072.

Chew, J. D., C. Stalling, and K. Moeller. 2004. Integrating knowledge for simulating vegetation change at landscape scales. Western Journal of Applied Forestry 19(2): 102–108.

D'Antonio, C. M. 2000. Fire, plant invasions, and global changes. In Invasive Species in a Changing World, edited by H. A. Mooney and R. J. Hobbs, pp. 65–93. Island Press, Washington DC.

Floyd, M. L., W. H. Romme, and D. D. Hanna. 2000. Fire history and vegetation pattern in Mesa Verde National Park, Colorado, U.S.A. Ecological Applications 10: 1666–1680.

Floyd, M. L., D. D. Hanna, and W. H. Romme. 2004. Historical and recent fire regimes in piñon-juniper woodlands on Mesa Verde, Colorado, USA. Forest Ecology and Management 198: 269–289.

Floyd, M.L ., D. Hanna, W. H. Romme, and T. Crews. 2006. Predicting and mitigating weed invasions to restore natural post-fire succession in Mesa Verde National Park, Colorado, USA. International Journal of Wildland Fire 15: 247–259.

Gartner, J. E. 2005. Relations between wildfire-related debris-flow volumes and basin morphology, burn severity, material properties and triggering storm rainfall. Master's thesis, University of Colorado, Boulder.

Heemskerk, M., K. Wilson, and M. Pavao-Zuckerman. 2003. Conceptual models as tools for communication across disciplines. Conservation Ecology 7: 8. Available at http://www.ecologyandsociety.org/vol7/iss3/art8/.

Jakeman, A. J., R. A. Letcher, and J. P. Norton. 1006. Ten iterative steps in development and evaluation of environmental models. Environmental Modelling and Software 21:602–614.

McVicker, G., and T. Bryan. 2002. Community-based ecosystem stewardship. Unpublished paper posted to http://www.ksd.harvard.edu/sed (Science, Environment, and Development Group) on 9 April 2004.

Miller, M. E. 2005. The structure and functioning of dryland ecosystems—Conceptual models to inform long-term ecological monitoring. USGS Scientific Investigations Report 2005-5197. Available at http://pubs.usgs.gov/sir/2005/5197/.

Nicolson, C. R., A. M. Starfield, G. P. Kofinas, and J. A. Kruse. 2002. Ten heuristics for interdisciplinary modeling projects. Ecosystems 5: 376–384.

Starfield, A. M. 1990. Qualitative, rule-based modeling. Bioscience 40: 601–604.

Starfield, A. M. 1997. A pragmatic approach to modeling for wildlife management. Journal of Wildlife Management 61: 261–270.

Starfield, A. M., J. D. Roth, and K. Ralls. 1995. "Mobbing" in Hawaiian monk seals—The value of simulation modeling in the absence of apparently crucial data. Conservation Biology 9(1): 166–174.

Thomas, L. P., M. N. Hendrie (editor), C. L. Lauver, S. A. Monroe, N. J. Tancreto, S. L. Garman, and M. E. Miller. 2006. Vital signs monitoring plan for the Southern Colorado Plateau Network. Natural Resource Report NPS/SCPN/NRR-2006/002, National Park Service, Fort Collins, Colorado. Available at http://www.nature.nps.gov/im/units/scpn/index.htm.

van den Belt, M. 2004. Mediated Modeling—A System Dynamics Approach to Environmental Consensus Building. Island Press, Washington, D.C.

Whisenant, S. G. 1990. Changing fire frequencies on Idaho's Snake River Plain—Ecological and management implications. USDA Forest Service, General Technical Report INT-276.

Wondolleck, J. M, and S. L. Yaffee. 2000. Making Collaboration Work—Lessons from Innovation in Natural Resource Management. Island Press, Washington DC.

Assessing
Large-scale Land-use Issues

CONSERVATION STATUS OF THE COLORADO PLATEAU USING SOUTHWEST REGIONAL GAP ANALYSIS STEWARDSHIP DATA

Andrea E. Ernst and Julie S. Prior-Magee

The Colorado Plateau, as defined by the Nature Conservancy's (2005) Terrestrial Global Assessment Units, Ecoregions and Major Habitat Types, is perhaps one of the most biologically diverse ecoregions in North America. The conservation of such biodiversity is considered vitally important for the preservation of naturally functioning ecosystems (Ricketts et al. 1999). The Colorado Plateau ecoregion covers more than 196,500 km^2 across southeastern Utah, northern Arizona, northwestern New Mexico, and western Colorado (Figure 1). Unique geological formations and processes characterize this ecoregion, which is one of the most remote and unspoiled landscapes of the desert Southwest (Nie 1999; Tuhy et al. 2002). Its relative isolation, complex geology, and specialized landform features contribute to a highly diverse endemic flora and fauna, with approximately 290 species that are found nowhere else in the world (Kartesz and Farstad 1999).

The conservation of natural resources was brought to national attention in the late nineteenth century (Sellars 1997; Anderson 2000) as the result of concerted literary, scientific, and bureaucratic campaigns. Legislation was first introduced in 1882 to formally protect and preserve a portion of the Colorado Plateau under federal jurisdiction by setting aside the Grand Canyon as a public park (Anderson 2000). Then after a decade of failed attempts the Grand Canyon Forest Reserve was established, but grazing, mining, and logging were still permitted. The establishment of the Grand Canyon National Monument in 1906 served

to further limit development within the monument boundaries (Anderson 2000), and over time other national forests, parks, and monuments have been established (BLM 2000) that also provide areas specifically intended to preserve the region's natural and cultural diversity.

The Colorado Plateau is considered to be a remnant of the American frontier, with its extensive undeveloped landscapes and generally rural populations. Although it is mostly characterized by vast open spaces and a relative degree of isolation, there are signs of human alterations on more than 85 percent of the landscape (Ricketts et al. 1999; Tuhy et al. 2002); this includes dammed rivers and tributaries, suppression of natural fire patterns, and the invasion of exotic species. Improper grazing practices, mineral exploration and extraction, and the seasonal influx of tourists and recreationalists have also contributed to the loss and degradation of natural habitat.

One widely accepted strategy to help conserve biodiversity is to document the distribution of biologically rich areas, evaluate current levels of protection, and identify areas that may be in need of more adequate protection, with the goal of protecting representative samples of all major ecosystems with their full array of habitats and community dynamics (Noss 1992). The National Gap Analysis Program (GAP) addresses this need by creating digital maps that provide a visual representation of existing protected areas and of whatever elements of biodiversity may be present in the conservation network (Crist and Scott 1999). Gap analysis

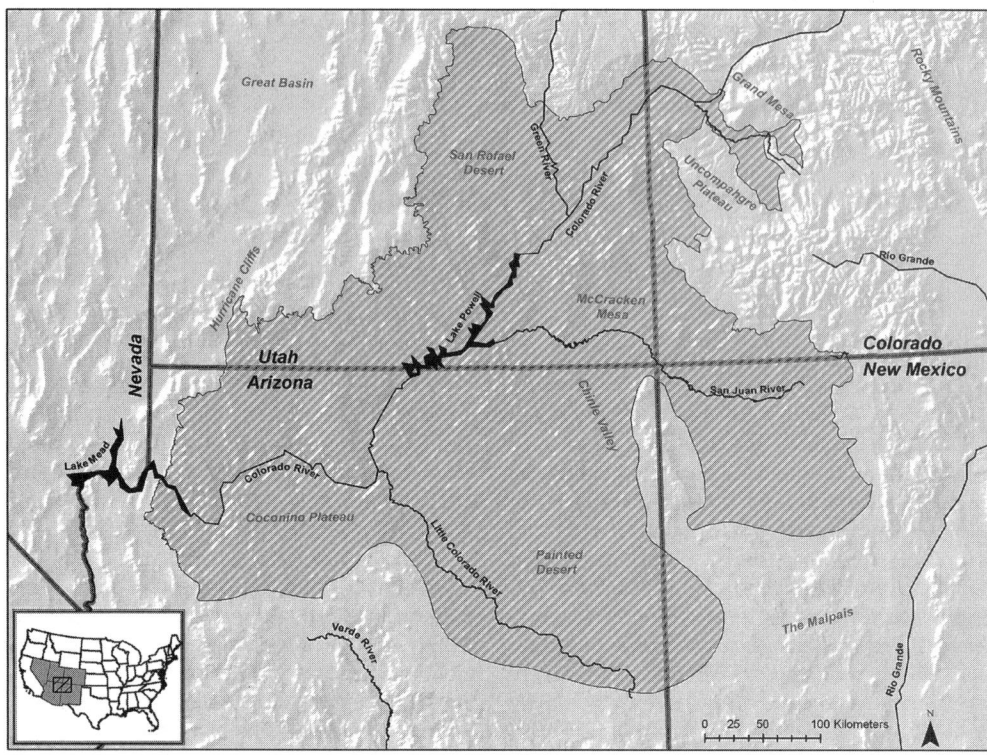

Figure 1. Map of the Colorado Plateau ecoregion and the Southwest Regional Gap Analysis Project study area.

seeks to identify "gaps" that might be filled by establishing reserves or by changing land management practices. Gap analysis uses the distribution of vegetation and vertebrate species as indicators of, or surrogates for, biodiversity. Digital GIS map overlays are used to identify individual species, species-rich areas, and vegetation types that are un-represented or underrepresented in existing biodiversity management areas (Scott et al. 1993), and these products are then used to develop conservation strategies and to pre-dict contributions of new management areas for biodiversity maintenance at landscape scales (Scott et al. 1991).

One unique and innovative aspect of GAP is the potential to analyze the manage-ment status of a particular biotic element throughout its entire range and across politi-cal boundaries (Crist and Jennings 1997). A key step in this process is the regionalization of GAP data, which provides the ability to analyze locally occurring elements in a watershed, or at ecoregion, national, or any other geographic scale (Crist and Jennings 1997). The Southwest Regional Gap Analysis Project (SWReGAP), the first such program in the region (Jacobs et al. 2001; Prior-Magee et al. 2007), was a highly interactive and cooperative endeavor that involved essen-tially all state and federal natural resource agencies, conservation organizations, tribal resource programs, and universities in Arizona, Colorado, New Mexico, Nevada, and Utah. Many private landowner groups and individuals also collected information and participated in data development. The cooperative nature of the project, combined with the involvement of natural resource agencies, was integral to its success.

The regional map of land stewardship, which comprises a fundamental data layer for the SWReGAP, combines attributes of land management, land ownership, and intent to manage for biodiversity. Stewardship has two main implications. First, legal ownership of the land does not always correlate to the actual land management practices applied to the property. Second, many land managers subdivide land tracts into various units to satisfy different management objectives. The spatial depiction of conservation lands provided by GAP facilitates visualization of the locations of managed and protected areas in relation to each other, which can have far-reaching applications in regional conservation planning. The Colorado Plateau provides a prime example of such planning; it is one of two ecoregions completely mapped into the SWReGAP database. The information provided by SWReGAP can thus very effectively assess general patterns in biodiversity and protection on the Colorado Plateau.

METHODS
Land Stewardship Mapping

Existing digital maps of land ownership were collected from the Bureau of Land Management (BLM) for each state in the SWReGAP area. These digital maps depict administrative land ownership with attributes that include federal, state, tribal, and private lands. These pre-existing maps are updated and maintained as a cooperative effort between the BLM and state land board offices. Geographic areas within a given state are often updated at different time periods and with different levels of detail. These data layers marked a reliable starting point for mapping land stewardship and provided important base information to which additional parcel information was aligned.

Various federal, state, tribal, and private sources were contacted directly to collect current boundary information, including identification of special or internal management areas such as private inholdings. If

data were not available in a digital format, paper maps were scanned and digitized manually. Data collected from the original source or actual parcel owner were considered more accurate than parcel information compiled from other generalized sources. A geographic database (geodatabase) was then created to serve as the common framework for all parcel information. Each boundary file was imported into the geodatabase and the GAP Management Coding System (USGS 2000) was applied. This hierarchical coding system was used to describe the managing agency and parcel type, and other attributes such as parcel name, alternative name, and source of the digital data were added when known. Information about parcel boundaries for special or internal management areas, non-governmental organizations, and local governments was merged into one continuous data layer.

One important function of the geodatabase is to define and preserve between-feature spatial relationships, which are maintained with a set of topological rules called "must not have gaps" and "must not overlap." These rules were used to address the numerous discrepancies that occur when combining different-scale data, different map projections, different dates of data production, and conversions of data from paper to digital form (Beardsley and Stroms 1993). Rules that address boundary discrepancies were often applied on a case-by-case basis. After resolving these internal boundary discrepancies, an additional topology rule, "must cover each other" was used to ensure an exact match with the boundaries of the individual state stewardship databases. The individual state stewardship databases were then merged to complete the stewardship map for the five-state SWReGAP region.

Biodiversity Management Status Categories

The GAP status code (Crist et al. 2000) describes the relative degree to which each area is managed for biodiversity. Individual land units are coded according to the level of commitment towards biodiversity management it receives. GAP developed a standard dichotomous key to make the categorization process more meaningful by providing a method in which status categories could be consistently applied to all parcels and by providing a process that is objective, more uniform, and replicable (Crist et al. 2000). The SWReGAP also used this key, but the resulting categorization is not exact and should be considered more as a guide than a definitive process (Prior-Magee et al. 1998; Crist et al. 2000).

Four biodiversity management status categories can be defined for the Colorado Plateau (after Scott et al. 1993; Edwards et al. 1994; Crist et al. 1996):

Status 1: An area having permanent protection from conversion of its natural land cover and a mandated management plan in operation to maintain a natural state within which disturbance events (of natural type, frequency, intensity, and legacy) are allowed to proceed without interference or are mimicked through management.

Status 2: An area having permanent protection from conversion of its natural land cover and a mandated management plan in operation to maintain a primarily natural state, but which may experience uses or management practices that degrade the quality of existing natural communities, including suppression of natural disturbance.

Status 3: An area having permanent protection from conversion of its natural land cover for the majority of the area, but subject to extractive uses of either a broad, low-intensity type (e.g. logging) or localized type (e.g. mining). It also confers protection to federally listed endangered and threatened species throughout the area.

Status 4: An area having no known public or private institutional mandates or legally recognized easements or deed restrictions held by the managing entity to prevent conversion of its natural habitat types to anthropogenic habitat types. The area generally allows conversion to unnatural land cover throughout.

Subjectivity can make distinguishing between Status 2 and Status 3 parcels difficult. This decision can be particularly critical to the final analytical mission of GAP because the division between these status codes is considered a division between more protected versus less protected areas. Individuals categorizing parcels can thus introduce their own subjectivity and individual bias. The lack of detailed information, such as specific management plans or maps of more detailed management boundaries, can also lead to the parcel receiving a lower status category. Decisions between Status 1 and Status 2 lands (the highest protection categories) can also be difficult, but for the most part these decisions are based on detailed management prescriptions such as the amount and capacity of grazing allotments, public access, or fire management practices. The difficulty of differentiating between Status 1 and 2 lands is clearly reflected in wilderness areas. Many individuals assume that wilderness areas automatically receive GAP Status 1 based on their high level of legally mandated protection, but management strategies can vary within the same wilderness area, between wilderness areas managed by the same management entity, or among areas managed by different entities; status categorizations should therefore be considered individually.

Several criteria were used to determine biodiversity management status for Colorado Plateau land parcels (USGS 2000; Crist et al. 2000). The first, and perhaps most important, was the permanence of protection from converting natural land cover to an unnatural state. The second was the relative amount of the land being managed for natural land cover. A 5 percent threshold was set by GAP as the maximum amount of a land unit that could be managed in an unnatural state and yet still achieve Status 1. GAP assumes that a land unit managed to retain all of its elements will maintain biodiversity longer than a land unit managed for only a single species or biotic element. GAP also incorporates current legal and institutional conservation mandates for each parcel. For example, units managed to allow or mimic natural disturbance regimes, such as fire, will maintain biodiversity longer than units that entirely suppress natural disturbances (Christensen et al. 1996).

Individual land management agencies were contacted and management plans for each parcel were reviewed in an effort to determine the appropriate conservation status objectively. When no known management plan was available, telephone interviews were conducted and interviewees were asked a list of standardized questions. Unnecessary bias in assigning status codes across the five-state region was minimized by having one person assign all status codes.

To compare the mapped distribution of biodiversity elements with their representation in the different categories of land ownership and management status, the SWReGAP stewardship map, including land status data, was clipped to the Colorado Plateau boundary (Figures 2 and 3). The inter-section of the stewardship data, the land cover map (Langs et al., this volume), and the terrestrial vertebrate habitat distribution maps and species-richness maps (Boykin et al., this volume) was then used to assess overall regional biodiversity.

RESULTS

Land Stewards of the Colorado Plateau

Forty-two percent of the Colorado Plateau (82,167 km^2) comprises lands under federal management (Table 1). The BLM manages 31 percent, the National Park Service manages 7 percent, and the U.S. Forest Service manages 4 percent. These stewards control a diverse collection of lands with many different legislative mandates, including a variety of national parks and monuments, wilderness areas, wilderness study areas,

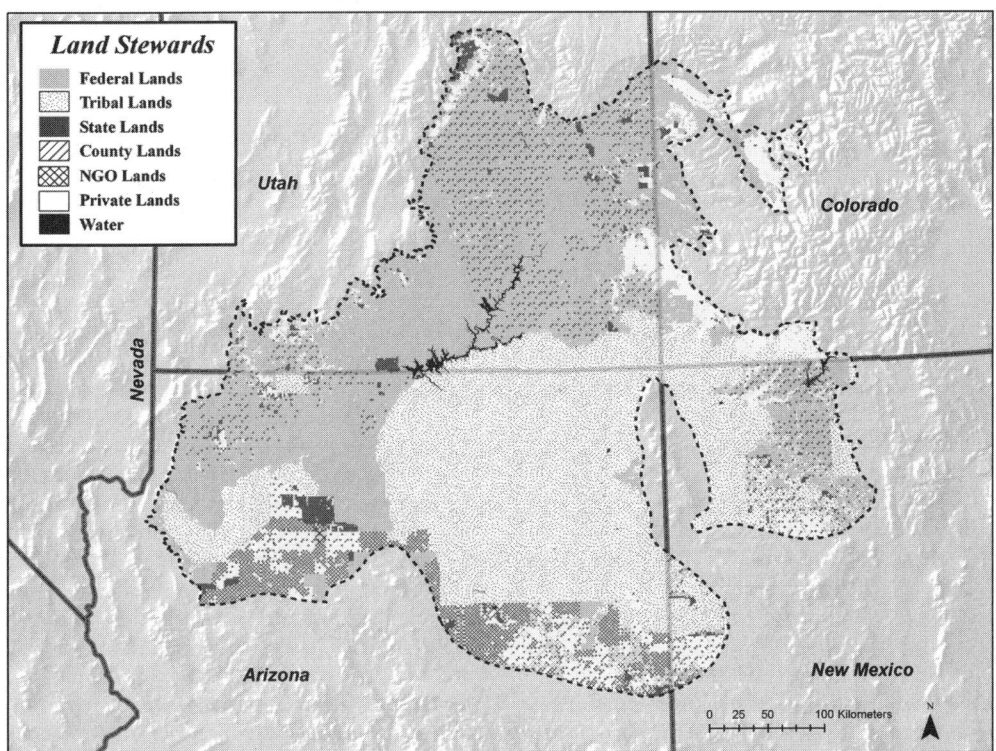

Figure 2. Map of the land stewards for the Colorado Plateau as depicted by the Southwest Regional Gap Analysis Project stewardship data.

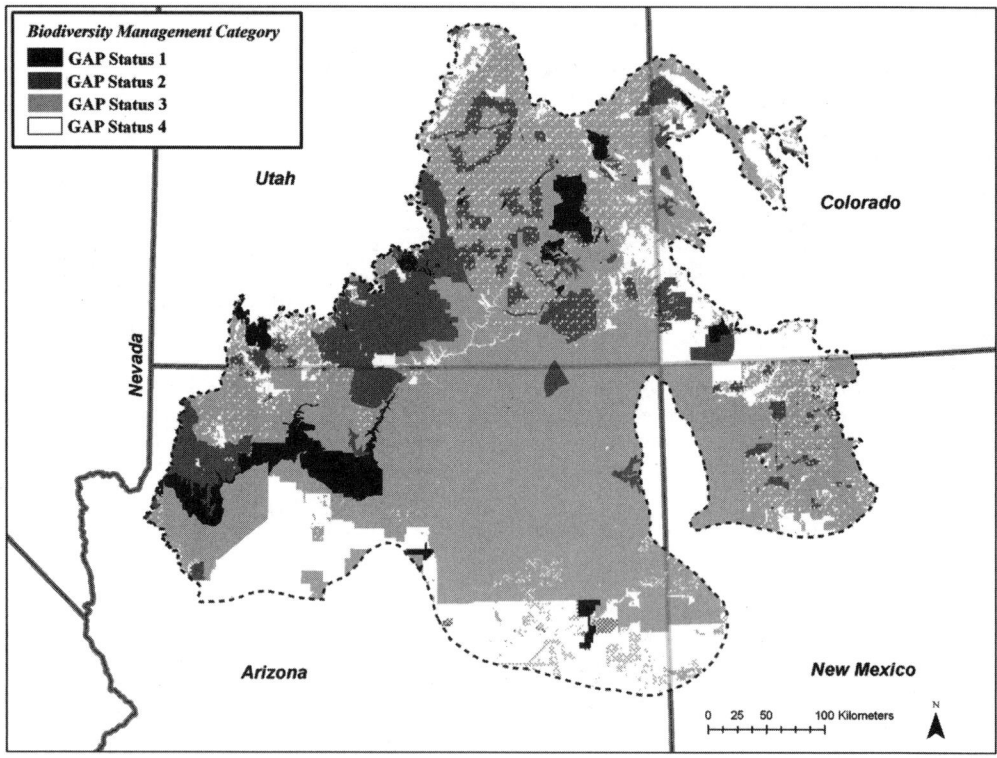

Figure 3. Map of the biodiversity management status categories for the Colorado Plateau as depicted by the Southwest Regional Gap Analysis Project stewardship data.

research natural areas, and areas of critical environmental concern. The largest federal parcels are the expanses of BLM public lands that are managed for multiple uses. Several national parks and monuments also contribute to the large federal land holdings in the ecoregion.

Tribal lands, the second-largest component of the Colorado Plateau ecoregion, comprise 36 percent (71,402 km²) of the landscape. The largest tribal land manager is the Navajo Nation, which has land holdings in Arizona, New Mexico, and Utah. The Navajo Nation manages the largest continuous land parcel on the Colorado Plateau. The Hopi and Hualapai Tribes also manage large landscapes in the ecoregion, along with many other tribes who manage reservations in the region. This area is thus one of

the top five most culturally diverse regions in North America (Nabhan et al. 2002).

Approximately 14 percent (27,470 km²) of the Colorado Plateau is managed privately. The combined state land departments, state wildlife agencies, and state park departments manage just 7 percent (13,935 km²) of the ecoregion. Most of the private lands and state lands are in Arizona, with almost twice the amount of state lands as in Utah. Utah and New Mexico have about the same amount of private land on the Colorado Plateau. The private land parcels tend to be mixed with the state trust lands and are typically small holdings that fragment landscape ownership. This is most obvious in Utah and Arizona where a distinct checkerboard land steward pattern is prevalent. This pattern was created historically with

Table 1. Land stewards and biodiversity management status categories of the Colorado Plateau ecoregion (land area in square kilometers).

Steward Category	Total Land Area		Status 1		Status 2		Status 3		Status 4	
	Area	%	Area	%	Area	%	Area	%	Area	%
Bureau of Land Management	60,234	30.65	657	7.42	20,677	85.41	38.900	32.86	—	—
Bureau of Reclamation	6	0.00	—	—	—	—	1	0.00	5	0.01
U.S. Fish & Wildlife Service	0.15	0.00	—	—	—	—	0.15	0.00	—	—
U.S. Forest Service	8188	4.17	278	3.14	525	2.17	7385	6.24	—	—
Department of Defense/ Department of Energy	7	0.00	—	—	—	—	7	0.01	—	—
National Park Service	13,732	6.99	7919	89.39	1418	5.86	4396	3.71	—	—
Tribal land	71,402	36.34	—	—	1026	4.24	67,222	56.79	3155	7.13
State parks and recreation	186	0.09	—	—	2	0.01	184	0.16	—	—
State Land Board	13,935	7.09	—	—	267	1.10	61	0.05	13,607	30.76
State wildlife reserves	275	0.14	2	0.02	271	1.12	1	0.00	0	0.00
Other state land	22	0.01	—	—	—	—	22	0.02	—	—
County land	7	0.00	—	—	—	—	—	—	7	0.01
The Nature Conservancy	212	0.11	2	0.03	24	0.10	185	0.16	—	—
Private land unrestricted for for development/ no known restriction	27,470	13.98	—	—	—	—	—	—	27,470	62.09
Water*	826	0.42	—	—	—	—	—	—	—	—
Total	196,500		8859	5	24,209	12	118,363	60	44,244	23

*The complexity of ownership, water rights, managing entities, and protection level categories of water resources are addressed in the aquatic component of GAP.

Table 2. Land stewards and biodiversity management status categories by state in the Colorado Plateau ecoregion.

Steward Category	Total Land Area (excluding water)		Status 1		Status 2		Status 3		Status 4	
	Area	%	Area	%	Area	%	Area	%	Area	%
Arizona	92,864	47.46	5657	63.86	5305	21.91	60,443	51.07	21,459	48.50
Colorado	16,124	8,24	303	3.42	2739	11.31	4821	4.07	8261	18.67
New Mexico	25,201	12.88	34	0.38	1103	4.56	18,534	15.66	5530	12.50
Utah	61,485	31.42	2865	32.34	15,062	62.22	34,565	29.20	8994	20.33
Total	195,674		8859	5	24,209	12	118,363	60	44,244	23

the establishment of trust lands which reserved a centrally located parcel in each surveyed township for support of public institutions, particularly public schools (Culp et al. 2005).

Biodiversity Management on the Colorado Plateau

Status 1 protection is afforded to approximately 5 percent (8859 km^2) of the Colorado Plateau ecoregion (Table 1). The National Park Service manages 90 percent of these Status 1 lands, including Grand Canyon, Canyonlands, Zion, and Arches National Parks. The BLM manages 7 percent of the Status 1 lands; most of these are small and isolated parcels in the form of administratively designated "Areas of Critical Environmental Concern" and "Outstanding Natural Areas." The Forest Service manages 3 percent of the Status 1 lands, of which the largest tract is the Kanab Creek Wilderness, and the Nature Conservancy manages just 1 percent as preserves. Altogether Arizona manages 64 percent of the Status 1 lands and Utah manages 32 percent (Table 2).

GAP Status 2 lands constitute 12 percent (24,409 km^2) of the entire Colorado Plateau ecoregion. Eighty-five percent of these lands are managed by the BLM; most are special management areas, particularly the newly designated national monuments. Grand Staircase–Escalante, Vermillion Cliffs, and Grand Canyon–Parashant National Monuments are some of the largest Status 2 parcels. The Park Service manages 6 percent and tribal land stewards manage 4 percent of Status 2 lands, respectively. Tribal parcels include Ute Mountain and Monument Valley tribal parks, and other Status 2 lands are defined as fossil areas, geological areas, wilderness study areas, and state wildlife areas. Most Status 2 lands are in Utah (62%) and Arizona (22%).

Of the approximately 60 percent (118,363 km^2) of the Colorado Plateau managed as Status 3 lands, 57 percent are tribal lands and 33 percent are managed by the BLM. These are primarily multiple-use lands. Approximately 51 percent of the Status 3 lands are located in Arizona.

Twenty-three percent (44,244 km^2) of the Colorado Plateau is managed as Status 4 lands. These parcels are primarily privately owned (62%) and have no known mandates to limit the alteration of natural land cover for anthropogenic uses. In addition to privately owned land, state land departments manage 31 percent of Status 4 lands. Most of the Status 4 lands are in Arizona (49%) with 19 percent occurring in Colorado.

DISCUSSION AND CONCLUSION

Based on SWReGAP stewardship mapping methods and conservation status definitions, 83 percent of the Colorado Plateau is currently managed as Status 3 and 4 lands, and 17 percent is being managed for long-term biodiversity with Status 1 and 2 protection. The conservation and management of the landscape surrounding the protected areas can be important in providing a buffer to enhance the ability of the protected areas to conserve long-term biodiversity. Human alterations such as damming of rivers and tributaries, suppression of natural fire patterns, the invasion of exotic species, improper grazing practices, mineral exploration and extraction, and the seasonal influx of tourists and recreationalists are continually affecting the landscape and biodiversity of the Colorado Plateau. As demands on natural resources continue to increase, human alterations could intensify, particularly on lands that currently enjoy no long-term protection against anthropogenic land conversion. To add to the complexity, it is unclear as to how fast these changes will occur and to what extent they will impact the long-term preservation of biodiversity on the Colorado Plateau.

The SWReGAP stewardship database provides a context for land managers to examine parcels using an ecoregional perspective (Crist et al. 2000), thus helping to identify the existing land conservation of the Colorado Plateau. This depiction of conservation lands also provides land managers with an idea of the proximity of similarly managed lands and the identity of adjacent land stewards and locations of lands that are unprotected or underrepresented in the network. Another important aspect in this depiction is the ability to recognize the potential that each land steward may provide as a partner for large-scale conservation efforts. Cooperation and collaboration among local, state, federal, and tribal governments, non-government organizations, universities, and private individuals could more effectively sustain the biodiversity of the Colorado Plateau, regardless of the status of land protection. Tribal entities are essential to the success of any collaboration effort, as together they manage 36 percent of the Colorado Plateau and thus constitute one of the largest land stewards in the region.

The planned Colorado Plateau conservation network would enable land managers to identify and better plan for specific conservation actions, and would also help managers to better assess the effects of isolation or loss of connectivity to similarly managed parcels. For example, Arches National Park is an isolated Status 1 parcel whereas others, such as Grand Canyon National Park and Grand Canyon–Parashant National Monument, are not as isolated and form a large cluster of Status 1 and Status 2 lands. Evaluating the connectivity between protected landscapes could also allow for newly planned conservation actions to establish corridors between protected area patches. Additional attributes such as parcel shape and persistence in the landscape are important in determining an area's ability to protect biodiversity. The spatial context afforded by GAP can help managers to identify new locations for future reserves that would provide the most effective approach to protecting biodiversity over the Colorado Plateau (Gergely et al. 2000; Scott et al. 2001).

LITERATURE CITED

Anderson, M. F. 2000. Polishing the Jewel: An administrative History of Grand Canyon National Park. Grand Canyon Association Monograph 11.

Beardsley, K., and D. Stroms. 1993. Compiling a digital map of areas managed for biodiversity in California. Natural Areas Journal 13: 177–190.

Bureau of Land Management. 2000. National Landscape Conservation System: Great American Landscapes—Healthy, Wild, and Open. Available at http://www.blm.gov/nlcs/index.html (accessed 27 January 2005).

Christensen, N. L., A. M. Bartuska, J. H. Brown, S. Carpenter, C. D'Antionio, R. Francis, J. F. Franklin, J. A. MacMahon, R. F. Noss, D. J. Parsons, C. H. Peterson, M. G. Turner, and R. G. Woodmansee. 1996. The report of the Ecological Society of America Committee on the scientific basis for ecosystem management. Ecological Applications. 6: 665–691.

Crist, P. J., and M. Jennings. 1997. Regionalizing state-level data. Gap Analysis Bulletin 6: 15–16.

Crist, P. J., and J. M. Scott. 1999. Identifying the gaps, locating the reserves: Some thoughts on getting gap analysis into conservation practice. Gap Analysis Bulletin 8: 14–16.

Crist, P. J., T. C. Edwards Jr., C. G. Homer, S. D. Bassett, and B. C. Thompson. 2000. Mapping and categorizing land stewardship. A Handbook for Gap Analysis, Version 2.1.0. USGS Gap Analysis Program, Moscow, Idaho.

Crist, P. J., B. Thompson, and J. Prior-Magee. 1996. Land management status categorization for Gap Analysis: A potential enhancement. Gap Analysis Bulletin 5, National Biological Survey, Moscow, Idaho.

Culp, P. W., D. B. Conradi, and C. C. Tuell. 2005. Trust Lands in the American West: A Legal overview and Policy Assessment. Available at http://www.trustland.org/about (accessed 2005).

Edwards, T. C., C. Homer, and S. Bassett. 1994. Land management categorization: A user's guide. A Handbook for Gap Analysis, Version 1. USGS Gap Analysis Program, Moscow, Idaho.

Gergely K., J. M. Scott, and D. Goble. 2000. A new direction for the National Wildlife Refuge System: National Wildlife Refuge System Improvement of 1977. Natural Areas Journal 20: 107–118.

Jacobs, S. R., K. A. Thomas, and C. A. Drost. 2001. Mapping land cover and animal species distributions for conservation planning: An overview of the Southwest Regional Gap Analysis Program in Arizona. In Proceedings of the Fourth Biennial Conference of Research on the Colorado Plateau, edited by C. van Riper III and M. A. Stuart, pp. 159–172. USGS/FRESC/COPL/1999/16. U.S. Department of the Interior.

Kartesz, J., and A. Farstad. 1999. Multi-scale analysis of endemism of vascular plant species. In Terrestrial Ecoregions of North America: A Conservation Assessment, edited by T. H. Ricketts, E. Dinerstein, D. M. Olsen, and C. Loucks, pp. 51–55. Island Press, Washington D.C.

Nabhan, G. P., P. Pynes, and T. Joe. 2002. Safeguarding species, languages, and cultures in the time of diversity loss: From the Colorado Plateau to global hotspots. Annals of the Missouri Botanical Garden 89: 164–175.

Nature Conservancy. 2005. TNC Terrestrial Global Assessment Units, Ecoregions and Major Habitat Types. Digital data. Arlington, Virginia.

Nie, M. A. 1999. Southern Utah wilderness and the meaning of the west. In Proceedings of the Fourth Biennial Conference of Research on the Colorado Plateau, edited by C. van Riper III and M. A. Stuart, pp. 195–210. USGS/FRESC/COPL/1999/16, U.S. Department of the Interior.

Noss, R. F. 1992. The Wildlands Project: A land conservation strategy. Wild Earth 2: 10–25.

Prior-Magee, J. S., B. C. Thompson, and D. Daniel. 1998. Evaluating consistency of categorizing biodiversity management status relative to land stewardship in the Gap Analysis Program. Journal of Environmental Planning and Management 41(2): 209–216.

Prior-Magee, J. S., K. G. Boykin, D. F. Bradford, W. G. Kepner, J. H. Lowry, D. L. Schrupp, K. A. Thomas, and B. C. Thompson, editors. 2007. Southwest Regional Gap Analysis Project Final Report. USGS Gap Analysis Program, Moscow, Idaho.

Ricketts, T. H., E. Dinerstein, D. M. Olsen, and C. Loucks. 1999. Terrestrial Ecoregions of North America: A Conservation Assessment. Island Press, Washington D.C.

Scott, J. M., B. Csuti, K. Smith, J. E. Estes, and S. Caicco. 1991. Gap analysis of species richness and vegetation cover: An integrated biodiversity conservation strategy. In Balancing on the Brink of Extinction: Endangered Species Act and Lessons for the Future, edited by K. Kohm, pp. 282–297. Island Press, Washington D.C.

Scott, J. M., F. Davis, B. Csuti, R. Noss, B. Butterfield, C. Groves, H. Anderson, S. Caicco, F. D'Erchia, T. C. Edwards Jr., J. Ulliman, and R. G. Wright. 1993. Gap Analysis: A geographic approach to protection of biological diversity. Wildlife Monographs 123: 1–41.

Scott, J. M., F. W. Davis, R. G. McGhie, R. G. Wright, C. Groves, and J. Estes. 2001. Nature reserves: Do they capture the full range of America's biological diversity? Ecological Applications 11: 999–1007.

Sellars, R. W. 1997. Preserving Nature in National Parks: A History. Yale University Press, New Haven, Connecticut.

Tuhy, J. S., P. Comer, D. Dorfman, M. Lammert, J. Humke, B. Cholvin, G. Bell, B. Neely, S. Silbert, L. Whitham, and B. Baker. 2002. A Conservation Assessment of the Colorado Plateau Ecoregion. The Nature Conservancy, Moab, Utah.

USGS Gap Analysis Program. 2000. A Handbook for Conducting Gap Analysis. Available at http://www.gap.uidaho.edu/handbook (accessed 24 February 2000).

Kim van Riper

A GAP ANALYSIS OF ECOLOGICAL SYSTEMS OF THE COLORADO PLATEAU ECOREGION USING SOUTHWEST REGIONAL GAP ANALYSIS LAND COVER

Lisa A. Langs, Kathryn A. Thomas, John H. Lowry, and Keith A. Schulz

Faced with competing and sometimes conflicting demands on our limited natural resources, managing for biodiversity is a significant challenge. The U.S. Geological Survey's Gap Analysis Program (GAP) takes a quantitative geographic approach to evaluating the representation of biota (i.e., vegetation and vertebrate species) in the context of current land management. Gap analysis determines the representation of biota in four biodiversity management status categories (Scott et al. 1993). Four primary GAP products are developed as inputs to the gap analysis: a land cover map showing ecological, disturbed, and land use categories, predicted habitat maps of vertebrate species, a land stewardship map that identifies land ownership boundaries, and a map of the GAP biodiversity management status categories associated with the land stewardship data. A fundamental assumption underlying gap analysis is that GAP biodiversity management status categories 1 and 2 provide adequate protection for maintaining the long-term viability of vegetation and vertebrate species occurring on those lands (Gap Analysis Program 2000).

The Southwest Regional Gap Analysis Project (SWReGAP) was a cooperative multi-institutional effort that included the states of Arizona, Colorado, Nevada, New Mexico, and Utah; this was the first regional gap analysis project (Prior-Magee et al. 2006). Activities were coordinated with institutions from each of the five states. Utah State University coordinated land cover mapping efforts and New Mexico State University coordinated animal habitat modeling and stewardship mapping (see Boykin et al., this volume). Other participating institutions included the USGS Southwest Biological Science Center in Flagstaff, Arizona; the U.S. Environmental Protection Agency, Landscape Ecology Branch, in Las Vegas, Nevada; the Colorado Division of Wildlife in Denver; and Colorado State University in Fort Collins. NatureServe ecologists supported the land cover mapping effort by developing a regional scale classification system of ecological systems to describe natural vegetation cover as part of the land cover legend and to ensure regionally consistent application of ecological system concepts.

Vegetation at the plant community or association and alliance levels, as described by the U.S. National Vegetation Classification System (USNVCS; Federal Geographic Data Committee 1997), is inherently difficult to map at regional scales. NatureServe addressed this problem by developing a new classification system—the Terrestrial Ecological Systems Classification framework for the coterminous United States (Comer et al. 2003). Ecological systems are groups of plant communities and sparsely vegetated habitats unified by similar ecological processes, substrates, and/or environmental gradients (Comer et al. 2003); these are linked to the USNVCS through characteristic associations and alliances. In naming ecological systems, biogeography and bioclimate were often used to define the predominant range of a type. For example, "Inter-Mountain Basins" refers to the Inter-Mountain Basins Division of Bailey (1997) and includes the Columbia Basin, Great Basin, Wyoming Basins, and Colorado Plateau Provinces. While the ecological systems concept includes existing vegetation as a component, the concept also incorporates environmental variables and ecological dynamics such as landform position, substrate, and climate, as well as hydrologic and fire regimes.

There are 15 distinct ecoregions within the SWReGAP area. These ecoregions are defined by the Nature Conservancy's (2005) Terrestrial Global Assessment Units, Ecoregions, and Major Habitat Types, which include modifications to original work done by Bailey (1995). The Colorado Plateau ecoregion, which spans four of the five states, is one of two ecoregions contained entirely within the five-state SWReGAP area (the other being the Utah High Plateaus). This ecoregional analysis would not have been possible using previous state-based GAP products without cross-walking and aggregating map units from each state's map legends. Here we use the products of the SWReGAP to provide an additional gap analysis of the Colorado Plateau ecoregion that focuses on natural land cover depicted using NatureServe's ecological systems. The results of our ecoregional gap analysis provide a perspective that addresses the uniqueness of the Colorado Plateau relative to the five-state, or regional, gap analysis. See Ernst and Prior-Magee (this volume) for a conservation assessment using SWReGAP stewardship data, and Boykin et al. (this volume) for a gap analysis of vertebrate species of the Colorado Plateau ecoregion.

THE COLORADO PLATEAU ECOREGION

The Colorado Plateau ecoregion, which covers portions of southeastern Utah, northern Arizona, western Colorado, and northwestern New Mexico, is approximately 196,500 sq km (48,535,500 acres) in size (see Figure 1 in Ernst and Prior-Magee, this volume). The ecoregion lies between the Great Basin to the west and the Southern Rocky Mountains to the east, with the Mojave and Sonoran Deserts bordering on the south and the Utah High Plateaus ecoregion to the north.

The Colorado Plateau has an arid to semiarid climate, where annual precipitation patterns vary widely depending on latitude and topographic effects. Long-term weather stations have recorded a median annual precipitation of less than 300 mm (12 inches) per year, with an overall range of 136–668 mm (5.4–26.3 inches) per year (Hereford et al. 2002). Northern regions of the Colorado Plateau have precipitation patterns resembling that of the Great Basin, with most precipitation occurring as snow during the winter months. In contrast, the southern Colorado Plateau receives most of its precipitation as high-intensity storm events during the summer "monsoon" season, in addition to the milder rainfalls and snows that come during winter.

The elevation ranges from about 350 m (1000 ft) in the lower reaches of the Grand Canyon to the 3870 m (12,721 ft) Mt. Peale in the La Sal Mountains of southeastern Utah. The majority of the Colorado Plateau lies below 1860 m (6100 ft). Many brightly colored, tilted, and erosional surfaces derived from extensive sedimentary rock materials are characteristic of the ecoregion. Igneous intrusions and volcanic activity also produced many of the mountain ranges, cinder and ash deposits, and basalt flows. Metamorphic rocks are generally restricted to deep canyons.

The myriad steep canyons, mesas, and badlands of the lower elevations provide a stark contrast to the rather isolated, laccolithic mountain ranges that rise above (McNab and Avers 1994). With such a large area containing diverse parent materials, topography, and climate, the soils are also quite diverse. Colorado Plateau soils are frequently aridic and poorly developed, except where they occur under more mesic moisture conditions on mountains and high plateaus. There are large areas of sandy soils derived from sandstone, as well as finer-textured and saline-sodic affected soils derived from marine shales. Aridisols—soils characteristic of dry climates—occur widely on plateaus, alluvial fans, and older stream terraces, with less-developed entisols in sand sheets, on rock outcrops, and along stream channels (Bailey 1995). Fragile microbiotic soil crusts are frequent and improve resistance to soil erosion by wind and water (St. Clair et al. 1993; Belnap et al. 2001). Although the Colorado Plateau is a relatively dry region, it is drained by several large rivers—the Green, Little Colorado, San Juan, and Virgin. All are tributaries to the Colorado River, which bisects the ecoregion.

Native vegetation consists of deep-rooted shrubs and grasslands in the high deserts and plateaus, rising into forested montane and subalpine zones. Alpine vegetation occurs only on the highest mountain ranges (La Sal and Henry Mountains in Utah and the San Francisco Peaks in Arizona). The unique combination of environmental factors and natural disturbance processes inherent to the Southwest and the Colorado Plateau makes this area one of the most floristically rich regions in the United States (Morin 1995). The Colorado Plateau is home to many rare plant communities (Grossman and Goodin 1995) and it is recognized for its relatively high level of endemism (Welsh 1978).

METHODS

We conducted a gap analysis for the Colorado Plateau ecoregion through a geospatial union of key environmental and management data layers using ESRI ArcGIS Desktop 9 software and Spatial Analyst extension. Input data consisted of three spatial databases developed by SWReGAP: land cover, land stewardship, and biodiversity management status categories. The analysis was constrained to the area within the boundary of the Colorado Plateau ecoregion (Nature Conservancy 2005).

Data Sets

The land cover map for the SWReGAP five-state region, completed in 2004, is available at http://earth.gis.usu.edu / swgap / index .html. SWReGAP mapped 125 land cover types at 1-acre minimum mapping units; of these, 109 are ecological systems, 12 are altered or disturbed types, 2 are developed types, 1 is an agricultural type, and 1 is an open water type. Within the Colorado Plateau ecoregion there are 77 land cover types; 62 of these are ecological systems, 11 are altered or disturbed types, 2 are developed types, 1 is an agricultural type, and 1 is an open water type.

The SWReGAP land cover data set was developed using a decision-tree (also known as a classification and regression tree, or CART) modeling approach (Breiman et al. 1984; Lawrence and Wright 2001; Falzarano et al. 2005). Primary inputs to the decision-tree were Landsat Enhanced Thematic Mapper Plus (ETM+) satellite images (three images per scene, taken between 1999 and 2001), derived Normalized Difference Vegetation Index (NDVI) values, derived tasseled cap (i.e. brightness, greenness, wetness) bands (Huang et al. 2002); elevation and associated data layers of slope, aspect, and landform (Manis et al. 2000); and approximately 93,000 ground reference points (of which 4304 were from the Colorado Plateau ecoregion). The CART tool, as developed by the USGS National Center for Earth Resources Observation and Science, integrated the decision-tree classifier software See5/ C5.0 (Quinlan 1993) with the spatially explicit modeling environment of ERDAS Imagine (version 8.6). SWReGAP evaluated 85 of the 125 land cover classes (91% of the total five-state area) using an internal validation approach (Shtatland et al. 2004). Overall, agreement for these 85 classes was 61 percent with a kappa statistic of 0.60. More information about the SWReGAP land cover map development can be found in Lowry et al. (2005).

The SWReGAP land stewardship data layer was completed in 2005 (http://fws-nmcfwru.nmsu.edu/swregap/Stewardship /Default.htm). The stewardship data layer depicts the boundaries of public and private (voluntarily provided) conservation lands at variable resolution (see Figure 2 in Ernst and Prior-Magee, this volume). Each land unit within the stewardship data layer is assigned to one of four biodiversity management status categories; these were then used to create a conservation status map (see Figure 3 in Ernst and Prior-Magee, this volume).

Status 1. An area having permanent protection from conversion of natural land cover and a mandated management plan in operation to maintain a natural state within which disturbance events (of natural type, frequency, intensity, and legacy) are allowed to proceed without interference or are mimicked through management.

Status 2. An area having permanent protection from conversion of natural land cover and a mandated management plan in operation to maintain a primarily natural state, but which may receive uses or management practices that degrade the quality of existing natural communities, including suppression of natural disturbance.

Status 3. An area having permanent protection from conversion of natural land cover for the majority of the area, but subject to extractive uses of either a broad, low-intensity type (e.g., logging) or localized intense type (e.g., mining). It also confers protection to federally listed endangered and threatened species throughout the area.

Status 4. There are no known public or private institutional mandates or legally recognized easements or deed restrictions held by the managing entity to prevent conversion of natural habitat types to anthropogenic habitat types. The area generally allows conversion to unnatural land cover throughout.

See Ernst and Prior-Magee (this volume) for a complete description of the development of the land stewardship data layer, patterns of ownership, and biodiversity management status within the Colorado Plateau ecoregion.

Gap Analysis

Gap analysis was conducted with a geospatial union of land cover, land stewardship, and biodiversity management status data layers using the tabulate area algorithm within the Spatial Analyst extension of ArcGIS 9. The tabulate area algorithm performs a cross-tabulation of the areas associated with two distinct data layers. We calculated the total area of each mapped land cover type by land steward and by biodiversity management status category, constrained within the Colorado Plateau ecoregion boundary.

Following GAP standards (Gap Analysis Program 2000), we evaluated the biodiversity management status of land cover on the Colorado Plateau by categorizing the results of the cross-tabulations using five thresholds of protection (0 to < 1%, 1 to < 10%, 10 to < 20%, 20 to < 50%, and ≥ 50%) for each land cover type's distribution within Status 1 and 2 lands. For example, a land cover type with less than 1 percent of its distribution in Status 1 and 2 lands has minimal legal mandate for conservation management. Thresholds of 10, 20, and 50 percent have been described in the literature for various conservation applications (Odum and Odum 1972; Specht et al. 1974; Miller 1984; Noss and Cooperrider 1994); however, other thresholds may be implemented depending on the application.

RESULTS

Ecological Systems

We found 77 land cover types within the Colorado Plateau ecoregion; 62 of these are ecological systems (Figure 1, Table 1). Only seven ecological systems have more than 5 percent of their mapped distribution within the ecoregion, which when combined represent approximately 75 percent of the total area; 48 ecological systems have 1 percent or less distribution within this ecoregion, of which 40 have 5 percent or less of their regional distribution within the Colorado Plateau ecoregion. These ecological systems have either naturally restricted ranges or are common but considered peripheral to the ecoregion.

The seven most abundant ecological systems on the plateau are Colorado Plateau Pinyon-Juniper Woodland (23% of the ecoregion), Inter-Mountain Basins Semi-Desert Shrub-Steppe (11%), Colorado Plateau Mixed Bedrock Canyon and Tableland (11%), Inter-Mountain Basins Semi-Desert Grassland (8%), Inter-Mountain Basins

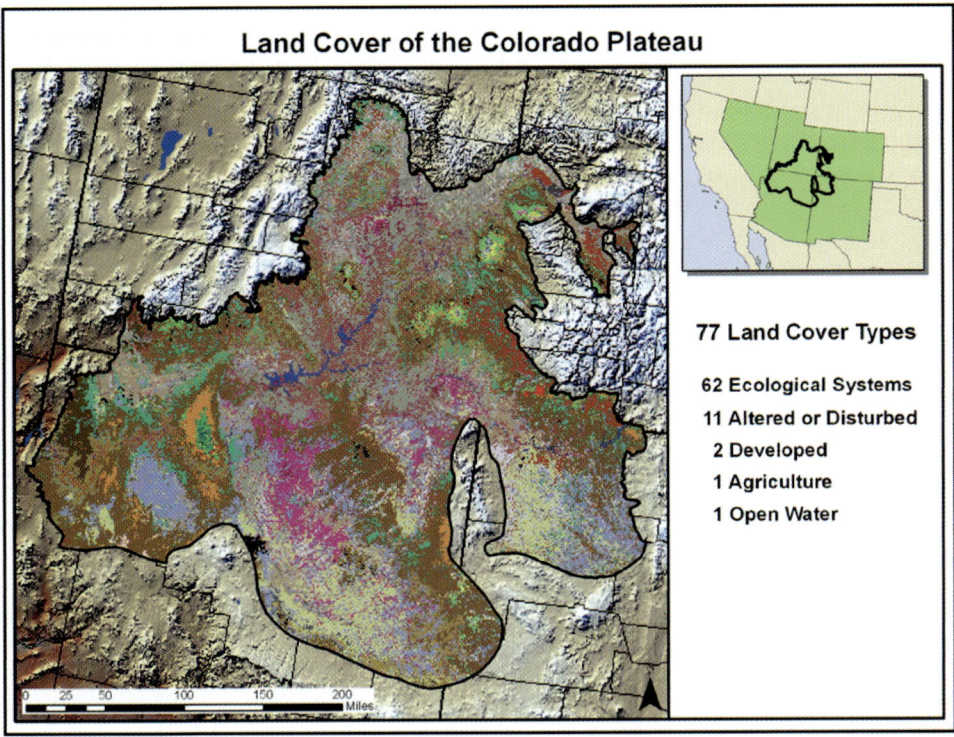

Land Cover of the Colorado Plateau

77 Land Cover Types

62 Ecological Systems

11 Altered or Disturbed

2 Developed

1 Agriculture

1 Open Water

Figure 1. Land cover of the Colorado Plateau ecoregion mapped by SWReGAP. See Table 1 for a list of mapped land cover types in the Colorado Plateau ecoregion.

Mixed Salt Desert Scrub (7%), Colorado Plateau Blackbrush–Mormon Tea Shrubland (7%), and Inter-Mountain Basins Big Sagebrush Shrubland (6%; Table 1).

Five ecological systems can be considered "nearly endemic" to the Colorado Plateau ecoregion; that is, more than 80 percent of the ecological system's total mapped distribution is within the ecoregion. These near endemics include Southern Colorado Plateau Sand Shrubland (99.7% of its mapped distribution falls within the ecoregion), Colorado Plateau Blackbrush–Mormon Tea Shrubland (99.6%), Colorado Plateau Mixed Bedrock Canyon and Tableland (86%), Inter-Mountain Basins Mat Saltbush Shrubland (82%), and Inter-Mountain Basins Shale Badland (82%; see Table 1).

Stewardship and Land Protection

Overall, 16.9 percent of the Colorado Plateau consists of Status 1 and 2 lands (4.5% and 12.4% respectively); these lands are assumed to be managed for long-term biodiversity. More than half the region is Status 3 (60.5%) and the remainder is Status 4 (22.6%). The holdings of 15 land stewards are mapped within the Colorado Plateau ecoregion: Native American (tribal) lands (36%), Bureau of Land Management (30%), private ownership (14%), National Park Service (7%), State Trust Lands (7%), and the U.S. Forest Service (4%). The U.S. Fish and Wildlife Service, Bureau of Recreation, Department of Defense and Energy, state parks, state wildlife reserves, county lands, private lands managed under conservation easements, and the

Table 1. Area and percent distribution of the land cover types within the Colorado Plateau ecoregion (total area = 196,500 sq km). Bold identifies the most abundant systems and italics identifies those that are "nearly endemic" (> 80% of regional distribution within the Colorado Plateau ecoregion).

Map Code	Land Cover Type	Area in Colorado Plateau Ecoregion (sq km)	Proportion of Colorado Plateau Ecoregion	Area in Ecoregion/ Area in SWReGAP Project
BARREN LANDS				
S010	*Colorado Plateau Mixed Bedrock Canyon & Tableland*	20,865	11%	86%
S012	Inter-Mountain Basins Active and Stabilized Dune	1,679	< 1%	54%
S015	Inter-Mountain Basins Playa	4	< 1%	< 1%
S011	*Inter-Mountain Basins Shale Badland*	2,694	1%	82%
S013	Inter-Mountain Basins Volcanic Rock & Cinder Land	441	< 1%	32%
S014	Inter-Mountain Basins Wash	7	< 1%	16%
S016	North American Warm Desert Bedrock Cliff & Outcrop	62	< 1%	2%
S022	North American Warm Desert Playa	< 1	< 1%	< 1%
S019	North American Warm Desert Volcanic Rockland	< 1	< 1%	< 1%
S002	Rocky Mountain Alpine Bedrock and Scree	55	< 1%	1%
S006	Rocky Mountain Cliff, Canyon and Massive Bedrock	304	< 1%	10%
DECIDUOUS FOREST				
S023	Rocky Mountain Aspen Forest and Woodland	274	< 1%	1%
EVERGREEN FOREST				
S039	**Colorado Plateau Pinyon-Juniper Woodland**	46,154	23%	47%
S040	Great Basin Pinyon-Juniper Woodland	2,691	1%	5%
S051	Madrean Encinal	< 1	< 1%	< 1%
S035	Madrean Pine-Oak Forest and Woodland	16	< 1%	< 1%
S112	Madrean Pinyon-Juniper Woodland	172	< 1%	1%
S032	Rocky Mtn Dry-Mesic Montane Mixed Conifer Forest & Woodland	451	< 1%	5%
S031	Rocky Mountain Lodgepole Pine Forest	< 1	< 1%	< 1%
S034	Rocky Mtn Mesic Montane Mixed Conifer Forest & Woodland	123	< 1%	2%
S028	Rocky Mtn Subalpine Dry-Mesic Spruce-Fir Forest & Woodland	161	< 1%	1%
S030	Rocky Mtn Subalpine Mesic Spruce-Fir Forest & Woodland	50	< 1%	< 1%
S038	Southern Rocky Mountain Pinyon-Juniper Woodland	21	< 1%	< 1%
S036	Southern Rocky Mountain Ponderosa Pine Woodland	3,703	2%	7%
MIXED FOREST				
S042	Inter-Mountain Basins Aspen-Mixed Conifer Forest & Woodland	134	< 1%	4%
SHRUB/SCRUB				
S058	Apacherian-Chihuahuan Mesquite Upland Scrub	1	< 1%	< 1%
S062	Chihuahuan Mixed Desert and Thorn Scrub	< 1	< 1%	< 1%
S059	*Colorado Plateau Blackbrush–Mormon Tea Shrubland*	13,250	7%	99.6%
S056	Colorado Plateau Mixed Low Sagebrush Shrubland	565	< 1%	24%
S052	Colorado Plateau Pinyon-Juniper Shrubland	7,514	4%	65%
S053	Great Basin Semi-Desert Chaparral	< 1	< 1%	< 1%
S054	**Inter-Mountain Basins Big Sagebrush Shrubland**	11,960	6%	11%
S045	*Inter-Mountain Basins Mat Saltbush Shrubland*	3,400	2%	82%
S065	**Inter-Mountain Basins Mixed Salt Desert Scrub**	13,293	7%	17%
S050	Inter-Mountain Basins Mtn Mahogany Woodland & Shrubland	2	< 1%	< 1%
S057	Mogollon Chaparral	247	< 1%	2%
S060	Mojave Mid-Elevation Mixed Desert Scrub	2,207	1%	13%
S046	Rocky Mountain Gambel Oak-Mixed Montane Shrubland	1,743	< 1%	9%
S047	Rocky Mountain Lower Montane-Foothill Shrubland	31	< 1%	1%
S069	Sonora-Mojave Creosotebush-White Bursage Desert Scrub	595	< 1%	1%
S070	Sonora-Mojave Mixed Salt Desert Scrub	< 1	< 1%	< 1%

Table 1 (continued)

Map Code	Land Cover Type	Area in Colorado Plateau Ecoregion (sq km)	Proportion of Colorado Plateau Ecoregion	Area in Ecoregion/ Area in SWReGAP Project
S129	Sonoran Mid-Elevation Desert Scrub	2	< 1%	< 1%
S063	Sonoran Paloverde-Mixed Cacti Desert Scrub	< 1	< 1%	< 1%
S136	*Southern Colorado Plateau Sand Shrubland*	6,998	4%	99.7%
S128	Wyoming Basins Low Sagebrush Shrubland	< 1	< 1%	< 1%
	GRASSLAND/HERBACEOUS			
S077	Apacherian-Chihuahuan Semi-Desert Grassland & Steppe	278	< 1%	1%
S075	Inter-Mountain Basins Juniper Savanna	3,591	2%	64%
S071	Inter-Mountain Basins Montane Sagebrush Steppe	371	< 1%	1%
S090	**Inter-Mountain Basins Semi-Desert Grassland**	15,973	8%	47%
S079	**Inter-Mountain Basins Semi-Desert Shrub-Steppe**	21,405	11%	45%
S115	Madrean Juniper Savanna	3	< 1%	< 1%
S083	Rocky Mountain Subalpine Mesic Meadow	53	< 1%	2%
S085	Southern Rocky Mtn Montane-Subalpine Grassland	93	< 1%	1%
	WOODY WETLAND			
S118	Great Basin Foothill and Lower Montane Riparian Woodland and Shrubland	< 1	< 1%	< 1%
S096	Inter-Mountain Basins Greasewood Flat	4,127	2%	17%
S094	North American Warm Desert Lower Montane Riparian Woodland and Shrubland	< 1	< 1%	< 1%
S097	North American Warm Desert Riparian Woodland & Shrubland	12	< 1%	3%
S020	North American Warm Desert Wash	11	< 1%	2%
S093	Rocky Mtn Lower Montane Riparian Woodland & Shrubland	554	< 1%	25%
S091	Rocky Mountain Subalpine-Montane Riparian Shrubland	6	< 1%	< 1%
	EMERGENT HERBACEOUS WETLAND			
S100	North American Arid West Emergent Marsh	6	< 1%	1%
S102	Rocky Mountain Alpine-Montane Wet Meadow	11	< 1%	1%
	ALTERED OR DISTURBED LAND COVER TYPES			
D01	Disturbed, Non-specific	3	< 1%	3%
D14	Disturbed, Oil well	7	< 1%	16%
D09	Invasive Annual and Biennial Forbland	326	< 1%	12%
D08	Invasive Annual Grassland	331	< 1%	4%
D06	Invasive Perennial Grassland	4	< 1%	< 1%
D04	Invasive Southwest Riparian Woodland & Shrubland	658	< 1%	41%
D02	Recently Burned	151	< 1%	7%
D11	Recently Chained Pinyon-Juniper Areas	358	< 1%	52%
D10	Recently Logged Areas	12	< 1%	1%
D03	Recently Mined or Quarried	131	< 1%	11%
	AGRICULTURE AND DEVELOPED TYPES			
N80	Agriculture	4,182	2%	6%
N22	Developed, Medium - High Intensity	391	< 1%	5%
N21	Developed, Open Space - Low Intensity	353	< 1%	5%
	OTHER COVER TYPES			
N31	Barren Lands, Non-specific	253	< 1%	18%
N11	Open Water	977	< 1%	9%

Nature Conservancy together account for less than 1 percent of the regional area. (See Figure 2 in Ernst and Prior-Magee, this volume, for the distributions of land ownership and GAP biodiversity management status categories within the Colorado Plateau ecoregion.)

Conservation Status of Ecological Systems

Sixteen ecological systems have less than 1 percent of their mapped distributions managed according to Status 1 or 2 criteria within the Colorado Plateau ecoregion (Table 2). It should be noted, however, that all of these ecological systems occur more commonly in other ecoregions and are peripheral to the Colorado Plateau. All contribute less than 1 percent to the total area of the Colorado Plateau ecoregion and less than 2 percent to the total regional area.

Eleven ecological systems have distributions of 1–10 percent within Status 1 or 2 lands (Table 2). Three of these have relatively low proportions of their total regional area represented within the Colorado Plateau ecoregion: Inter-Mountain Basins Greasewood Flat (17%), Inter-Mountain Basins Mixed Salt Desert Scrub (17%), and Inter-Mountain Basins Mountain Mahogany Woodland and Shrubland (< 1%). The two herbaceous wetland systems—North American Arid West Emergent Marsh and Rocky Mountain Alpine-Mountain Wet Meadow—are sparse within the ecoregion (1% each) and regionally (< 1% each). While the remaining six ecological systems are not abundant within the Colorado Plateau ecoregion, a substantial portion of each ecological system's regional area does occur within the ecoregion. These six ecological systems are Colorado Plateau Mixed Low Sagebrush Shrubland (< 1% ecoregional area, 24% of its regional area in the Colorado Plateau ecoregion); Inter-Mountain Basins Juniper Savanna (2% ecoregional area, 64% regional area); Inter-Mountain Basins Mat Saltbush Shrubland (2% ecoregional area, 82% regional area); Inter-Mountain Basins Semi-Desert Grassland (8% ecoregional area, 47% regional area); Inter-Mountain Basins Semi-Desert Shrub-Steppe (11% ecoregional area,

45% regional area); and Southern Colorado Plateau Sand Shrubland (4% ecoregional area, 99.7% regional area).

Sixteen ecological systems have mapped distributions of 10–20 percent in Status 1 or 2 lands (Table 2). These include four barren/sparsely vegetated lands, one deciduous and three evergreen forests, three shrubland/scrub types, two grassland/herbaceous types, and three woody wetlands. Several lower elevation shrubland and woodland ecological systems also occur within this conservation threshold; these are Colorado Plateau Pinyon-Juniper Woodland, Inter-Mountain Basins Big Sagebrush Shrubland, and Southern Rocky Mountain Ponderosa Pine Woodland. Other ecological systems include riparian, mesic meadow, aspen, and dune systems—all noteworthy for their relatively small patch size and high ecological importance. Although their total area contribution is low, 2 of the 16 ecological systems have more than 50 percent of their regional area in the Colorado Plateau ecoregion: Inter-Mountain Basins Active and Stabilized Dune (< 1% ecoregional area, 54% regional area) and Inter-Mountain Basins Shale Badland (1% ecoregional area, 82% regional area). Inter-Mountain Basins Volcanic Rock and Cinder Land and Inter-Mountain Basins Wash each represent less than 1 percent of the ecoregion, but they are notable because of their relatively high proportions of regional area within the Colorado Plateau ecoregion (32% and 16% respectively).

Twelve ecological systems have distributions of 20–50 percent within Status 1 or 2 lands (Table 2). These include three barren/sparsely vegetated lands, four evergreen forests, one mixed forest, three shrubland/scrub types, and one grassland/herbaceous type. Several of the barren and forested ecological systems within this threshold are afforded greater protection under U.S. Forest Service or Park Service jurisdiction (Colorado Plateau Mixed Bedrock Canyon and Tableland, Rocky Mountain Alpine Bedrock and Scree, Rocky Mountain Subalpine Dry-Mesic Spruce-Fir Forest and Woodland, and Rocky Mountain Subalpine

Table 2. Land cover types in the Colorado Plateau ecoregion, distribution in Status 1 and 2 lands, and conservation thresholds (total area = 33,067 sq km).

Map Code	Land Cover Type	Area in Status 1 & 2 Lands (sq km)	Conservation Thresholds (%)				
			< 1	1–< 10	10–< 20	20–< 50	≥ 50
S058	Apacherian-Chihuahuan Mesquite Upland Scrub	0	0%	–	–	–	–
S077	Apacherian-Chihuahuan Semi-Desert Grassland & Steppe	0	0%	–	–	–	–
S118	Great Basin Foothill and Lower Montane Riparian Woodland and Shrubland	0	0%	–	–	–	–
S015	Inter-Mountain Basins Playa	0	0%	–	–	–	–
S051	Madrean Encinal	0	0%	–	–	–	–
S115	Madrean Juniper Savanna	0	0%	–	–	–	–
S035	Madrean Pine-Oak Forest and Woodland	< 1	< 1%	–	–	–	–
S112	Madrean Pinyon-Juniper Woodland	0	0%	–	–	–	–
S016	North American Warm Desert Bedrock Cliff & Outcrop	0	0%	–	–	–	–
S094	North American Warm Desert Lower Montane Riparian Woodland and Shrubland	0	0%	–	–	–	–
S022	North American Warm Desert Playa	0	0%	–	–	–	–
S019	North American Warm Desert Volcanic Rockland	0	0%	–	–	–	–
S070	Sonora-Mojave Mixed Salt Desert Scrub	0	0%	–	–	–	–
S129	Sonoran Mid-Elevation Desert Scrub	0	0%	–	–	–	–
S063	Sonoran Paloverde-Mixed Cacti Desert Scrub	0	0%	–	–	–	–
S128	Wyoming Basins Low Sagebrush Shrubland	0 ·	0%	–	–	–	–
S056	Colorado Plateau Mixed Low Sagebrush Shrubland	17	–	3%	–	–	–
S096	Inter-Mountain Basins Greasewood Flat	261	–	6%	–	–	–
S075	Inter-Mountain Basins Juniper Savanna	54	–	2%	–	–	–
S045	Inter-Mountain Basins Mat Saltbush Shrubland	243	–	7%	–	–	–
S065	Inter-Mountain Basins Mixed Salt Desert Scrub	937	–	7%	–	–	–
S050	Inter-Mtn Basins Mtn Mahogany Woodland & Shrubland	< 1	–	8%	–	–	–
S090	Inter-Mountain Basins Semi-Desert Grassland	698	–	4%	–	–	–
S079	Inter-Mountain Basins Semi-Desert Shrub-Steppe	1095	–	5%	–	–	–
S100	North American Arid West Emergent Marsh	< 1	–	7%	–	–	–
S102	Rocky Mountain Alpine-Montane Wet Meadow	< 1	–	4%	–	–	–
S136	Southern Colorado Plateau Sand Shrubland	244	–	3%	–	–	–
S039	Colorado Plateau Pinyon-Juniper Woodland	8121	–	–	18%	–	–
S012	Inter-Mountain Basins Active & Stabilized Dune	171	–	–	10%	–	–
S054	Inter-Mountain Basins Big Sagebrush Shrubland	2056	–	–	17%	–	–
S011	Inter-Mountain Basins Shale Badland	345	–	–	13%	–	–
S013	Inter-Mtn Basins Volcanic Rock and Cinder Land	67	–	–	15%	–	–
S014	Inter-Mountain Basins Wash	1	–	–	12%	–	–
S097	North Amer. Warm Desert Riparian Woodland & Shrubland	1	–	–	11%	–	–
S023	Rocky Mountain Aspen Forest and Woodland	49	–	–	18%	–	–
S046	Rocky Mtn Gambel Oak-Mixed Montane Shrubland	199	–	–	11%	–	–
S093	Rocky Mtn Lower Montane Riparian Woodland & Shrubland	94	–	–	17%	–	–
S047	Rocky Mtn Lower Montane-Foothill Shrubland	3	–	–	10%	–	–
S083	Rocky Mountain Subalpine Mesic Meadow	9	–	–	18%	–	–
S091	Rocky Mtn Subalpine-Montane Riparian Shrubland	1	–	–	16%	–	–
S085	Southern Rocky Mountain Montane-Subalpine Grassland	14	–	–	15%	–	–
S038	Southern Rocky Mountain Pinyon-Juniper Woodland	3	–	–	15%	–	–
S036	Southern Rocky Mountain Ponderosa Pine Woodland	624	–	–	17%	–	–
S062	Chihuahuan Mixed Desert and Thorn Scrub	< 1	–	–	–	45%	–
S059	Colorado Plateau Blackbrush–Mormon Tea Shrubland	2659	–	–	–	20%	–
S010	Colorado Plateau Mixed Bedrock Canyon and Tableland	6714	–	–	–	32%	–
S052	Colorado Plateau Pinyon-Juniper Shrubland	3083	–	–	–	41%	–
S042	Inter-Mtn Basins Aspen-Mixed Conifer Forest & Woodland	35	–	–	–	26%	–
S071	Inter-Mountain Basins Montane Sagebrush Steppe	92	–	–	–	25%	–
S002	Rocky Mountain Alpine Bedrock and Scree	16	–	–	–	30%	–
S006	Rocky Mountain Cliff, Canyon and Massive Bedrock	107	–	–	–	35%	–

Table 2 (continued)

Map Code	Land Cover Type	Area in Status 1 & 2 Lands (sq km)	Conservation Thresholds (%)				
			< 1	1–< 10	10–< 20	20–< 50	≥ 50
S032	Rocky Mountain Dry-Mesic Montane Mixed Conifer Forest and Woodland	166	–	–	–	37%	–
S034	Rocky Mtn Mesic Montane Mixed Conifer Forest & Woodland	35	–	–	–	28%	–
S028	Rocky Mtn Subalpine Dry-Mesic Spruce-Fir Forest & Woodland	39	–	–	–	24%	–
S030	Rocky Mtn Subalpine Mesic Spruce-Fir Forest & Woodland	12	–	–	–	24%	–
S040	Great Basin Pinyon-Juniper Woodland	2167	–	–	–	–	81%
S053	Great Basin Semi-Desert Chaparral	< 1	–	–	–	–	100%
S057	Mogollon Chaparral	134	–	–	–	–	54%
S060	Mojave Mid-Elevation Mixed Desert Scrub	1493	–	–	–	–	68%
S020	North American Warm Desert Wash	8	–	–	–	–	79%
S031	Rocky Mountain Lodgepole Pine Forest	< 1	–	–	–	–	83%
S069	Sonora-Mojave Creosotebush-White Bursage Desert Scrub	365	–	–	–	–	61%
N31	Barren Lands, Non-specific	1	< 1%	–	–	–	–
D10	Recently Logged Areas	< 1	–	1%	–	–	–
N21	Developed, Open Space - Low Intensity	3	–	1%	–	–	–
D03	Recently Mined or Quarried	1	–	1%	–	–	–
N22	Developed, Medium - High Intensity	13	–	3%	–	–	–
N80	Agriculture	149	–	4%	–	–	–
D14	Disturbed, Oil well	< 1	–	6%	–	–	–
D06	Invasive Perennial Grassland	< 1	–	7%	–	–	–
D09	Invasive Annual and Biennial Forbland	29	–	9%	–	–	–
D11	Recently Chained Pinyon-Juniper Areas	43	–	–	12%	–	–
D08	Invasive Annual Grassland	41	–	–	12%	–	–
D04	Invasive Southwest Riparian Woodland & Shrubland	143	–	–	–	22%	–
N11	Open Water	69	–	–	–	24%	–
D01	Disturbed, Non-specific	1	–	–	–	43%	–
D02	Recently Burned	137	–	–	–	–	91%

Mesic Spruce-Fir Forest and Woodland). Colorado Plateau Blackbrush–Mormon Tea Shrubland and Colorado Plateau Pinyon-Juniper Shrubland are also notable in that they are each relatively abundant (composing 7% and 4% of the ecoregion, respectively) and have high levels of endemism within the Colorado Plateau ecoregion (99.6% and 65%, respectively; Table 2).

Seven ecological systems have 50 percent or more of their mapped distributions in Status 1 or 2 lands (Table 2). These include two evergreen forests, four shrubland/scrub types, and one woody wetland. Each of these ecological systems has fairly limited distributions and is somewhat peripheral to the Colorado Plateau ecoregion. They occur mostly within the national parks and monuments of southeastern and south-central Utah (e.g. Zion National Park, Grand Staircase–Escalante National Monument) and northwestern Arizona (e.g. Grand Canyon National Park, Grand Canyon–Parashant National Monument, and the BLM's Grand Wash Cliffs Wilderness Area), and are thus afforded higher conservation status.

DISCUSSION

Vegetation of the Colorado Plateau

The SWReGAP land cover map provides the first mapping of natural land cover of the Colorado Plateau using ecological systems. Descriptions of the most abundant ecological systems of the Colorado Plateau ecoregion as adapted from NatureServe's Explorer Database (http://www.natureserve.org/explorer/) are given in Table 3.

Table 3. Descriptions of the most abundant and nearly endemic* ecological systems of the Colorado Plateau ecoregion.

Code	Land Cover Type	Ecological Description**	Area (sq km) in Colorado Plateau Ecoregion	Area in CP Ecoregion / Area in SWReGAP	Relative Distribution
S039	Colorado Plateau Pinyon-Juniper Woodland	This woodland system occurs at lower elevations between 1500 and 2400 m and is often associated with warm dry sites on mountain slopes, mesas, plateaus, and ridges. *Pinus edulis* and/or *Juniperus osteosperma* may dominate the tree canopy in the southern part of the Colorado Plateau, occasionally with *J. monosperma* or *J. scopularum* as possible co-dominants at higher elevation. Shrubs commonly associated with this system include *Arctostaphylos patula, Artemisia tridentata, Cercocarpus intricatus, C. montanus, Purshia stansburiana, P. tridentata,* and *Quercus gambelii.* Herbaceous understory may include various warm season grasses and forbs, or may be absent depending on disturbance regime and land management practices.	46,142	47%	Centered on the Colorado Plateau
S059	Colorado Plateau Blackbrush–Mormon Tea Shrubland	This shrubland system is associated with bench lands, colluvial slopes, pediments, or alluvial fans, within 560–1650 m elevation. It often occurs in shallow, calcareous, non-saline, gravelly or sandy soils or in deeper soils on sandy plains where it may have invaded desert grasslands. The vegetation consists of extensive open shrublands dominated by *Coleogyne ramosissima* often with *Ephedra viridis, E. torreyana,* or *Grayia spinosa,* and sometimes *Artemisia filifolia* in sandy areas. The herbaceous layer is sparse with grass species such as *Achnatherum hymenoides, Pleuraphis jamesii,* or *Sporobolus cryptandrus.*	13,250	99.6%	Near endemic to the Colorado Plateau
S054	Inter-Mountain Basins Big Sagebrush Shrubland	This shrubland system occurs in broad basins between mountain ranges, plains, and foothills on deep, well-drained and non-saline soils between 1500 and 2300 m elevation. Sites are dominated by *Artemisia tridentata* ssp. *tridentata* and/or *A. tridentata* ssp. *wyomingensis* occasionally with scattered *Juniperus* sp., *Sarcobatus vermiculatus,* and *Atriplex* sp. *Purshia tridentata* or *Symphoricarpos oreophilus* may co-dominate in some stands, along with *Ericameria nauseosa* and *Chrysothamnus viscidiflorus* in disturbed areas. Perennial herbaceous components typically contribute less than 25% vegetative cover. Common grasses include *Achnatherum hymenoides, Bouteloua gracilis, Elymus lanceolatus, Festuca idahoensis, Hesperostipa comata, Leymus cinereus, Pleuraphis jamesii, Pascopyrum smithii, Poa secunda,* or *Pseudoroegneria spicata.*	11,960	11%	Occurs throughout western U.S.

Table 3 (continued)

Code	Land Cover Type	Ecological Description**	Area (sq km) in Colorado Plateau Ecoregion	Area in CP Ecoregion / Area in SWReGAP	Relative Distribution
S065	Inter-Mountain Basins Mixed Salt Desert Scrub	This open-canopied shrubland is associated with saline basins and alluvial slopes. Substrates are often saline with calcareous, medium- to fine-textured alkaline soils, but include some coarser-textured soils. One or more species of *Atriplex* typically dominate the woody layer including *A. confertifolia, A. canescens, A. polycarpa,* and *A. spinifera.* Other shrubs present to co-dominant may include *Artemisia tridentata* ssp. *wyomingensis, Chrysothamnus viscidiflorus, Ericameria nauseosa, Ephedra nevadensis, Grayia spinosa, Krascheninnikovia lanata, Lycium sp., Picrothamnus desertorum,* or *Tetradymia* sp. The herbaceous layer varies from sparse to moderately dense and is dominated by perennial grasses such as *Achnatherum hymenoides, Bouteloua gracilis, Elymus lanceolatus* ssp. *lanceolatus, Pascopyrum smithii, Pleuraphis jamesii, Pleuraphis rigida, Poa secunda,* or *Sporobolus airoides.* An assortment of forbs may also be present.	13,293	17%	Occurs throughout western U.S.
S079	Inter-Mountain Basins Semi-Desert Shrub-Steppe	This semi-arid shrub-steppe system typically has an open shrub layer where grasses represent more than 25% of the vegetation cover. It commonly occurs at lower elevations on alluvial fans and flats with moderate to deep soils. Grasses may include *Achnatherum hymenoides, Bouteloua gracilis, Distichlis spicata, Hesperostipa comata, Pleuraphis jamesii, Poa secunda,* and *Sporobolus airoides.* A variety of shrubs and dwarf-shrubs may co-dominate with the grass component including *Atriplex canescens, Artemisia tridentata, Chrysothamnus greenii, Chrysothamnus viscidiflorus, Ephedra* sp., *Ericameria nauseosa, Gutierrezia sarothrae,* and *Krascheninnikovia lanata. Artemisia tridentata* may be present but does not dominate.	15,973	47%	Occurs throughout western U.S.
S090	Inter-Mountain Basins Semi-Desert Grassland	This grassland system is found on dry plains and mesas, between 1450 and 2320 m elevation. It occurs in lowland and upland areas and may occupy swales, playas, mesa tops, plateau parks, alluvial flats, and plains, but sites are typically xeric. Substrates are variable including well-drained sandy or loamy-textured soils from igneous and metamorphic rocks as well as other fine-textured soils derived from sedimentary rock as well as other fine-textured soils. Dominant grasses may include *Achnatherum hymenoides, Aristida* sp., *Bouteloua gracilis, Hesperostipa comata, Muhlenbergia* sp., or *Pleuraphis jamesii.* Scattered shrubs may include species of *Artemisia, Atriplex, Coleogyne, Ephedra, Gutierrezia,* or *Krascheninnikovia lanata.*	21,405	45%	Occurs throughout western U.S.

Table 3 (continued)

Code	Land Cover Type	Ecological Description**	Area (sq km) in Colorado Plateau Ecoregion	Area in CP Ecoregion / Area in SWReGAP	Relative Distribution
S010	Colorado Plateau Mixed Bedrock Canyon and Tableland	This sparsely vegetated to barren system is characterized by steep cliff faces, narrow canyons, and open tablelands of exposed sedimentary rock such as sandstone, shale, and limestone. The vegetation may include scattered trees and shrubs with a sparse herbaceous layer. Common species include *Pinus edulis*, *P. ponderosa*, several species of *Juniperus*, *Cercocarpus intricatus*, and other short-shrub and herbaceous species, utilizing moisture from cracks and pockets where soil accumulates.	20,865	86%	Near endemic to the Colorado Plateau
S136	Southern Colorado Plateau Sand Shrubland	This shrubland system is a new ecological system that SWReGAP's land cover team described based on supporting field surveys and new evidence supporting its recognition. It occurs on windswept mesas, broad basins, and plains at low to moderate elevations (1300–1800 m). It is commonly associated with stabilized sandsheets or shallow to moderately deep sandy soils often with hummocks or small coppice dunes. Short shrubs with approximately 10–30% cover and a sparse graminoid layer dominate this open shrubland. *Ephedra cutleri*, *E. torreyana*, *E. viridis*, and *Artemisia filifolia* are characteristic species, and *Coleogyne ramosissima* is atypical. *Poliomintha incana*, *Parryella filifolia*, *Quercus havardii* var. *tuckeri*, or *Ericameria nauseosa* may be present to dominant locally. Characteristic grasses include *Achnatherum hymenoides*, *Bouteloua gracilis*, *Hesperostipa comata*, and *Pleuraphis jamesii*. Eolian processes are evident, such as pedicled plants, or small blowouts and dunes, but the generally higher vegetative cover and less prominent geomorphic features distinguish this system from Inter-Mountain Basins Active and Stabilized Dune.	6,998	99.7%	Near endemic to the Colorado Plateau (south-central portion)

* More than 80% of its regional distribution occurs within the Colorado Plateau ecoregion.
** Adapted from NatureServe.

Geographic Patterns of Protection

Gap analysis provides a means to assess the conservation of biodiversity using quantitative geographic criteria. The objective is to identify native species and ecological systems that are not sufficiently represented in existing conservation areas (Jennings 2000), as well as what proportion of their habitat or distribution falls within these conservation lands, which we refer to as "thresholds of conservation protection." The use of conservation thresholds allowed us to identify ecological systems with low representation in Status 1 and 2 lands (Table 2). It should be noted, however, that many ecological systems with low representation are rare within the ecoregion, are peripheral to the ecoregion, or are generally sparse throughout the five-state region. From our Colorado Plateau ecoregional analysis, we can assess the relative contribution this ecoregion specifically makes towards the conservation of these and other ecological systems. This is important because protecting systems at the edge of their range may capture important components of biodiversity (Channell and Lomolino 2000; Holt and Keitt 2005; Jaeger et al. 2005).

Using a conservative threshold for representation in protected lands (< 10% of an ecological system's area within Status 1 and 2 lands) and an assessment of the ecological system's total ecoregional distribution compared to its total regional distribution, we identified six ecological systems with minimal protection. Five more are identified if the next threshold level (10 to < 20% in Status 1 and 2 lands) is also considered. Furthermore, we noted three ecological systems with low levels of conservation protection ecoregionally, regional scarcity, and relative importance as water resources (North American Arid West Emergent Marsh, Rocky Mountain Alpine-Montane Wet Meadow, and Inter-Mountain Basins Wash). Management of these 14 ecological systems (Table 4) for conservation of biodiversity merits further evaluation.

A trend commonly seen in gap analysis, and which is reflected in the Colorado Pla-teau ecoregion, is that ecological systems that tend to occur at higher elevations have a higher percentage of representation in Status 1 and 2 lands. Land stewardship at higher elevations is often in the public domain. There are several high-elevation forested and exposed rock ecological systems in the Colorado Plateau with 24 percent or more representation in Status 1 and 2 lands because of their management by the U.S. Forest Service or Park Service. For example, Rocky Mountain Dry-Mesic Montane Mixed Conifer Forest and Woodland—a system characterized by the presence of *Pseudotsuga douglasii* (Douglas fir)—has 37 percent representation in Status 1 and 2 lands; it occurs mostly within Forest Service (59%) and Park Service (31%) lands. Conversely, several of the ecological systems within the 10 to < 20 percent conservation threshold (Colorado Plateau Pinyon-Juniper Woodland, Inter-Mountain Basins Big Sagebrush Shrubland, and Southern Rocky Mountain Ponderosa Pine Woodland) tend to occur at lower elevations on landscapes often managed for multiple human uses.

This spatial analysis of ecological systems in areas managed for conservation uses only one criterion for evaluating the level of protection afforded to ecological systems, but a number of threats can undermine even the most rigorous legal protection. For example, infestation by invasive non-native plants poses a threat to wetland, riparian, and upland ecological systems. The riparian areas in the region are highly susceptible to such invasion and in many places infestations of tamarisk (*Tamarix ramosissima*; Smith et al. 1998; Shafroth et al. 2005), Russian olive (*Elaeagnus angustifolia*; Katz and Shafroth 2003), and/or camelthorn (*Alhagi maurorum*; Kerr et al. 1965; Thomas 2006) have significantly degraded these systems. Likewise, cheatgrass (*Bromus tectorum*), Russian thistle (*Salsola* sp.), and halogeton (*Halogeton glomerata*) infestations have degraded grassland, steppe, savannah, and dune systems (West and Young 2000). Where mapped, the invasive (disturbed) land cover types (Invasive Southwest Riparian Woodland and Shrubland, Invasive Annual Grassland, Invasive

Table 4. Ecological systems with low levels of conservation protection (< 20% in Status 1 and 2 lands).

Map Code	Ecological System	Area in Colorado Plateau Ecoregion (sq km)	Area in Status 1 & 2 Lands (sq km)	Conservation Threshold		Area in CP Ecoregion/ Area in SWReGAP Project
				< 10%	10–20%	
S056	Colorado Plateau Mixed Low Sagebrush Shrubland	565	17	3%	–	24%
S075	Inter-Mountain Basins Juniper Savanna	3,591	54	2%	–	64%
S045	Inter-Mountain Basins Mat Saltbush Shrubland	3,400	243	7%	–	82%
S090	Inter-Mountain Basins Semi-Desert Grassland	15,973	698	4%	–	47%
S079	Inter-Mountain Basins Semi-Desert Shrub-Steppe	21,405	1,095	5%	–	45%
S136	Southern Colorado Plateau Sand Shrubland	6,998	244	3%	–	99.7%
S039	Colorado Plateau Pinyon-Juniper Woodland	46,154	8,121	–	18%	47%
S012	Inter-Mountain Basins Active and Stabilized Dune	1,679	171	–	10%	54%
S011	Inter-Mountain Basins Shale Badland	2,694	345	–	13%	82%
S013	Inter-Mountain Basins Volcanic Rock and Cinder Land	441	67	–	15%	32%
S093	Rocky Mountain Lower Montane Riparian Woodland & Shrubland	554	94	–	17%	25%
S102	Rocky Mountain Alpine-Montane Wet Meadow*	11	< 1	4%	–	1%
S100	North American Arid West Emergent Marsh*	6	< 1	7%	–	1%
S014	Inter-Mountain Basins Wash*	7	1	–	12%	16%

*Ecological systems noted for low levels of ecoregional conservation protection, regional scarcity, and relative importance as water resources.

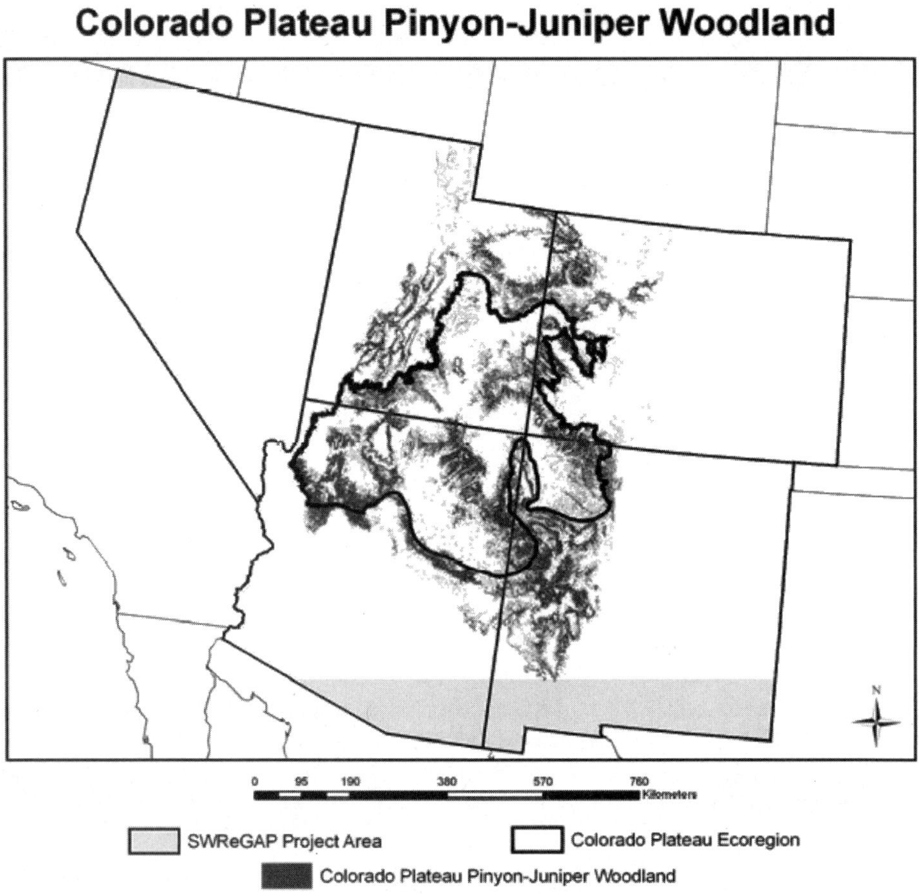

Colorado Plateau Pinyon-Juniper Woodland

Figure 2. Mapped distribution of Colorado Plateau Pinyon-Juniper Woodland within the five-state SWReGAP area.

Perennial Grassland, and Invasive Annual and Biennial Forbland) represent areas where the vegetation is dominated by these and other exotic weeds. Such species may also be present but to a lesser degree in natural ecological systems. Places where invasive land cover types are juxtaposed with natural systems pose areas of concern.

Many ecological systems on the Colorado Plateau are barren, or sparsely vegetated, or have open-canopied scrubby vegetation (West and Young 2000). These systems occur on soils that are easily erodible such as sand sheets, dunes, and shale badlands. Wind and water degradation of the soil leads to degradation of the vegetation supported in these substrates. Cryptogamic crusts play an important role in many of these systems by facilitating the infiltration of water, increasing fertility, and reducing erosion of the soil (Belnap et al. 2001).

Drought and increasing temperatures also pose a threat to the ecological systems of the Colorado Plateau. Many woody perennial plants died during the drought of 2002 either through direct desiccation or from the insect attacks that occurred over extensive areas of northern Arizona and

Southern Colorado Plateau Sand Shrubland

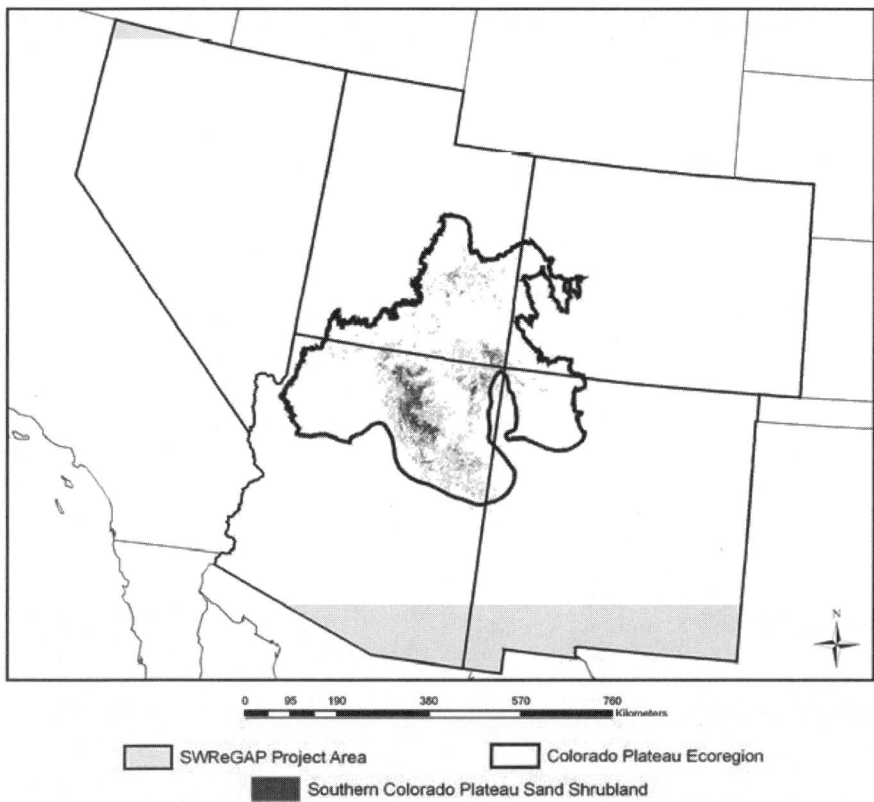

Figure 3. Mapped distribution of Southern Colorado Plateau Sand Shrubland within the five-state SWReGAP area.

southern Utah (Breshears et al. 2005). Additional mortality associated with insect attack and desiccation threatens Colorado Plateau Pinyon-Juniper Woodland—the most extensive ecological system within the Colorado Plateau ecoregion.

The mountain ranges within the ecoregion (La Sal, Henry, and Abajo Mountains of southeastern Utah and San Francisco Peaks of northern Arizona) are geographically isolated and are potentially threatened with higher temperatures from global climate change. Some species within their alpine and subalpine vegetation (e.g. Rocky Mountain Alpine Bedrock and Scree) may be at risk (Billings 1983). Somewhat lower in elevation, Rocky Mountain Aspen Forest and Woodland and Inter-Mountain Basins Aspen-Mixed Conifer Forest and Woodland are also experiencing ecological threats. Some upper-elevation aspen forests, typically maintained by mixed fire regimes, are being lost to the conifer forests that are migrating downslope into previously aspen-dominated areas because of effective fire-suppression and warmer temperatures (Bartos and Campbell 1998). At their lower boundary, aspen forests are showing poor regeneration and stunted growth patterns, related to heavy grazing by domestic livestock and increased browsing by elk (Johnson 1994; Bartos and Campbell 1998).

Conservation Opportunities

Ecological systems provide useful ecological units suited for "coarse filter" approaches to biodiversity conservation and natural resource management. The coarse filter approach assumes that species distributions are not random over the landscape, but rather occur in predictable assemblages, where the classification of ecological systems successfully partitions and describes the distribution of the component species (Anderson et al. 1999). A fine filter approach to biodiversity management is warranted for species or communities with highly specialized or restricted distributions that might otherwise be overlooked using a coarse approach. Fine filter assessments would be additional, yet complementary, to the coarse filter approach provided by SWReGAP's gap analysis. For example, narrow, linear riparian and wetland systems may require additional spatial analyses to refine their true position because of minimum mapping unit constraints. It is also likely that small, isolated patches of wetlands and/or ephemeral water bodies require higher-resolution mapping. For example, Colorado Plateau Hanging Gardens, an ecological system recognized by NatureServe, occurs in patches of less than 1 acre in size and thus did not meet the minimum unit for SWReGAP's land cover mapping.

In a previous conservation assessment, Tuhy et al. (2002) prepared an exhaustive portfolio of conservation targets within the Colorado Plateau ecoregion. Their approach incorporated both fine and coarse filter assessments, biophysical modeling of terrestrial and freshwater habitats, and occurrence, quality, and threat assessments of the conservation targets within the ecoregion. The data used in these analyses included previous state GAP products that needed to be joined and cross-walked into ecoregional layers, as well as fine-scale occurrence data of rare or imperiled plant and animal species or communities. Tuhy et al.'s (2002) results may be a useful reference in addition to the slightly coarser, yet more current results of SWReGAP.

Most of the Status 1 and 2 lands within the Colorado Plateau ecoregion are associated with national parks and monuments. While Status 1 and 2 lands represent a legal mandate to manage for conservation of biota, it is important to note that the actual management of Status 1 and 2 lands may not always achieve the legal mandate. Likewise, lack of a legal mandate for conservation management does not necessarily mean that Status 3 and 4 lands are not managed for protection of biodiversity. Status 3 lands may in fact receive protection where federally listed endangered and threatened species occur. Tribal entities (Havasupai, Hopi, Hualapai, Jicarilla-Apache, Kaibab-Paiute, Navajo, Ramah-Navajo, Southern Ute, Ute Mountain, and Zuni), State Trust Lands, and private land owners together comprise more than half (57%) of the land stewardship within the Colorado Plateau ecoregion. Continuing pressures from increased population growth, recreational use, and general development warrant ongoing evaluation of the land use and management goals of the various land stewards within the ecoregion. For example, existing land ownership patterns reflect the artifact of six sections for every Township being granted to State Trust Lands. Factoring in this pattern would be beneficial for identifying economic opportunities as well as contributions toward the conservation of biologically important areas.

Conservation at ecoregional scales requires the involvement of multiple partners and cooperative management among diverse land stewards. As suggested by Tuhy et al. (2002), partnerships with federal land management agencies, tribal entities, private land owners, academic institutions, and non-government organizations will play a vital role in ensuring successful long-term conservation on the Colorado Plateau. An example of a partnership that includes the Colorado Plateau is the Utah Partners for Conservation & Development, composed of state, federal, and natural resource agencies, universities, county and local governments, private land owners, conservation organizations, and other vested stakeholders, who are working cooperatively to manage and restore rangelands in Utah (Utah Division of Wildlife Resources 2005). Agreements such as these may help balance biodiversity management with development potential.

CONCLUSION

SWReGAP was the first formal GAP project designed at a regional, multi-state scale. From these data, we conducted a gap analysis of land cover for the Colorado Plateau ecoregion. Despite the abundance of national parks, national monuments, and wilderness areas within the Colorado Plateau, the extent of protected lands is limited for certain ecological systems. Using an area-based approach, we identified 14 ecological systems with "gaps" in their conservation management (Table 4). Six of the 14 ecological systems have 20 percent or more of their regional distribution within the Colorado Plateau but less than 10 percent of their ecoregional distribution has Status 1 or 2 protection (Colorado Plateau Mixed Low Sagebrush Shrubland, Inter-Mountain Basins Juniper Savanna, Inter-Mountain Basins Mat Saltbush Shrubland, Inter-Mountain Basins Semi-Desert Grassland, Inter-Mountain Basins Semi-Desert Shrub-Steppe, and Southern Colorado Plateau Sand Shrubland). Of particular note is the Southern Colorado Plateau Sand Shrubland, which is virtually endemic to the Colorado Plateau ecoregion (99.7% of its regional distribution occurs in the Colorado Plateau ecoregion), but only 3 percent of its ecoregional distribution is within Status 1 and 2 lands. Five ecological systems have at least 25 percent of their southwestern distribution in the Colorado Plateau ecoregion but less than 20 percent of their ecoregional distribution within Status 1 and Status 2 lands (Colorado Plateau Pinyon-Juniper Woodland, Inter-Mountain Basins Active and Stabilized Dune, Inter-Mountain Basins Shale Badland, Inter-Mountain Basins Volcanic Rock and Cinder Land, and Rocky Mountain Lower Montane Riparian Woodland and Shrubland).

Additionally, we point out three ecological systems that, despite their limited distribution within the ecoregion, are of partic-ular importance as water resources in this semi-arid landscape (North American Arid West Emergent Marsh, Rocky Mountain Alpine-Montane Wet Meadow, and Inter-Mountain Basins Wash). Not only do these ecological systems have minimal protection, but they also face an array of environmental threats, such as invasion by exotic plants, insect outbreaks, erosion, and drought, which pose distinct management challenges.

Gap analysis provides a standardized, geographical approach to assessing the status of conservation efforts currently in place. Gap data can serve as a basis for determining the persistence of existing land cover types and conservation areas over time using suitability or risk assessments (Groves et al. 2002). It also can provide the framework for testing alternative scenarios to promote a meaningful adaptive management process. Ultimately, the utility of gap products and gap analysis is to provide information to land managers and to keep common species common on the Colorado Plateau or any other place of interest. We encourage land managers to work collaboratively and to recognize opportunities where thoughtful land management can help balance the effects of the inevitable pressure for development for human use with a need to sustain our natural ecosystems and quality of life.

ACKNOWLEDGMENTS

The authors acknowledge the financial support provided by the USGS Biological Resources Discipline, the National Gap Analysis Program, and the efforts of the SWReGAP land cover mapping team, without which this project would not have been possible. The Intermountain Region Digital Image Archive Center at Utah State University and the USGS Colorado Plateau Research Station at Northern Arizona University provided additional financial support for the completion of this ecoregional gap analysis. We also thank the organizers of the Biennial Research Conference for providing the opportunity for us to share this work, and Andrea Ernst and Julie Prior-Magee for their thoughtful reviews of this paper.

LITERATURE CITED

Anderson, M., P. Comer, D. Grossman, C. Groves, K. Poiani, M. Reid, R. Schneider, B. Vickery, and A. Weakley. 1999. Guidelines for Representing Ecological Communities in Ecoregional Plans. The Nature Conservancy, Arlington, Virginia.

Bailey, R. G. 1995. *Descriptions of the Ecoregions of the United States*. U.S. Forest Service Miscellaneous Publication 1391 (revised), with separate 1:7,500,000 map. U.S. Department of the Interior, Forest Service, Washington D.C.

Bailey, R. G. 1997. Ecoregions Map of North America: Explanatory Note. Miscellaneous Publication 1548. U.S. Department of the Interior, Forest Service, Washington, D.C.

Bartos, D. L., and R. B. Campbell Jr. 1998. Decline of quaking aspen in the Interior West—Examples from Utah. Rangelands 20 (1): 17–24.

Belnap, J., J. H. Kaltenecker, R. Rosentreter, J. Williams, S. Leonard, and D. Eldridge. 2001. Biological soil crusts: Ecology and management. In BLM Technical Reference 1730-2, edited by P. Peterson, p. 110. U.S. Department of the Interior, Bureau of Land Management, U.S. Geological Survey Forest and Rangeland Ecosystem Science Center, Denver, Colorado.

Billings, W. D. 1983. Alpine vegetation. In North American Terrestrial Vegetation, edited by M. G. Barbour and W. D. Billings, pp. 537–572. 2nd edition. Cambridge University Press, New York.

Breiman, L., J. H. Friedman, R. A. Olshen, and C. J. Stone. 1984. Classification and Regression Trees. Wadsworth, Belmont, California.

Breshears, D. D., N. S. Cobb, P. M. Rich, K. P. Price, C. D. Allen, R. G. Balice, W. H. Romme, J. H. Kastens, M. L. Floyd, J. Belnap, J. J. Anderson, O. B. Myers, and C. W. Meyer. 2005. Regional vegetation die-off in response to global-change-type drought. In Proceedings of the National Academy of Sciences 102: 15144–15148.

Channell, R., and M. V. Lomolino. 2000. Dynamic biogeography and conservation of endangered species. Nature 403: 84–86.

Comer, P., D. Faber-Langendoen, R. Evans, S. Gawler, C. Josse, G. Kittel, S. Menard, S. Pyne, M. Reid, K. Schulz, K. Snow and, and J. Teague. 2003. Ecological Systems of the United States: A Working Classification of U.S. Terrestrial Systems. NatureServe, Arlington, Virginia.

Falzarano, S., K. Thomas, and J. Lowry. 2005. Using decision tree modeling in gap analysis land cover mapping: Preliminary results from northeastern Arizona. In The Colorado Plateau II: Biophysical, Socioeconomic, and Cultural Resources, edited by C. van Riper III and D. J. Mattson, pp. 87–100. University of Arizona Press, Tucson.

Federal Geographic Data Committee, Vegetation Subcommittee. 1997. FGDC Vegetation Classification and Information Standards—June 3, 1996 Draft. FGDC Secretariat, Reston, Virginia.

Gap Analysis Program. 2000. A handbook for conducting gap analysis. USGS Gap Analysis Program, Moscow, Idaho. Available at http://www.gap.uidaho.edu/handbook/default.htm. Accessed January 2006.

Grossman, D. H., and K. L. Goodin. 1995. Rare terrestrial ecological communities of the United States. In Our Living Resources: A Report to the Nation on the Distribution, Abundance, and Health of U.S. Plants, Animals, and Ecosystems, edited by E. T. LaRoe, G. S. Farris, C. E. Puckett, P. D. Doran, and M. J. Mac, pp. 218–221. U.S. Department of the Interior, National Biological Service, Washington, D.C.

Groves, C. R., D. B. Jensen, L. L. Valutis, K. H. Redford, M. L. Shaffer, J. M. Scott, J. V. Baumgartner, J. V. Higgins, M. W. Beck, and M. G. Anderson. 2002. Planning for biodiversity conservation: Putting conservation science into practice. BioScience 52 (6): 499–512.

Hereford, R., R. H. Webb, and S. Graham. 2002. Precipitation History of the Colorado Plateau Region, 1900–2000. U.S. Geological Survey Fact Sheet 119-02. Available at http://pubs.usgs.gov/fs/2002/fs119-02/.

Holt, R. D., and T. H. Keitt. 2005. Species' borders: A unifying theme in ecology. Oikos 108: 3–6.

Huang, C., B. Wylie, C. Homer, L. Yang, and G. Zylstra. 2002. Derivation of a tasseled cap transformation based on Landsat 7 at-satellite reflectance. International Journal of Remote Sensing 23: 1741–1748.

Jaeger, J. R., B. R. Riddle, and D. F. Bradford. 2005. Cryptic Neogene vicariance and Quaternary dispersal of the red-spotted toad (*Bufo punctatus*): Insights on the evolution of North American warm desert biotas. Molecular Ecology 14: 3033–3048.

Jennings, M. D. 2000. Gap analysis: Concepts, methods, and recent results. Landscape Ecology 15: 5–20.

Johnson, M. A. 1994. Changes in southwestern forests: Stewardship implications. Journal of Forestry 92: 16–19.

Katz, G. L., and P. B. Shafroth. 2003. Biology, ecology and management of *Elaeagnus angustifolia* L. (Russian olive) in western North America. Wetlands 23 (4): 763–777.

Kerr, H. D., W. C. Robacker, and T. J. Muzik. 1965. Characteristics and control of camelthorn. Weeds 13: 156–163.

Lawrence, R. L., and A. Wright. 2001. A rule-based classification systems using classification and regression trees (CART) analysis. Photogrammetric Engineering and Remote Sensing 67: 1137–1142.

Lowry, J. H, Jr., R. Ramsey, R. Boykin, D. Bradford, P. Comer, S. Falzarano, W. Kepner, J. Kirby, L. Langs, G. Manis, L. O'Brien, T. Sajwaj, K. Thomas, W. Rieth, S. Schrader, D. Schrupp, K. Schulz, B. Thompson, C. Velasquez, C. Wallace, E. Waller, and B. Wolk. 2005. Southwest Regional Gap Analysis Project: Final Report on Land Cover Mapping Methods. RS/GIS Laboratory, Utah State University, Logan. Available at http://earth.gis.usu.edu/swgap/. Accessed January 2006.

Manis, G., C. Homer, R. D. Ramsey, J. Lowry, T. Sajwaj, and S. Graves. 2000. The development of mapping zones to assist in land cover mapping over large geographic areas: A case study of the Southwest ReGAP project. In GAP Analysis Bulletin 9. U.S. Geological Survey, Biological Resources Division. Available at http://www.gap.uidaho.edu/Bulletins/9/bulletin9/default.html. Accessed January 2006.

McNab, W. H., and P. E. Avers. 1994. Ecological subregions of the United States: Section descriptions. USDA Forest Service, WO-WSA-5. Washington, D.C.

Miller, K. R. 1984. The Bali Action plan: A framework for the future of protected areas. In National Parks, Conservation, and Development, edited by J. A. McNeely and K. R. Miller, pp. 756–764. Smithsonian Institution Press, Washington D.C.

Morin, N. 1995. Vascular plants of the United States. In Our Living Resources: A Report to the Nation on the Distribution, Abundance, and Health of U.S. Plants, Animals, and Ecosystems, edited by E. T. LaRoe, G. S. Farris, C. E. Puckett, P. D. Doran, and M. J. Mac, pp. 200–205. U.S. Department of the Interior, National Biological Service, Washington, D.C.

Nature Conservancy. 2005. TNC Terrestrial Global Assessment Units, Ecoregions and Major Habitat Types—July 2005. Available at https://transfer.natureserve.org/download/longterm/Ecology/GIS_BASE_DATA or http://conserveonline.org/workspaces/ecoregional.shapefile/. Accessed January 2006.

Noss, R. F., and A. Y. Cooperrider. 1994. Saving Nature's Legacy. Island Press, Washington, D.C.

Odum, E. D., and H. T. Odum. 1972. Natural areas as necessary components of man's total environment. Proceeding of the North American Wildlife and Natural Resources Conference 39: 178–189.

Prior-Magee, J. S., K. G. Boykin, D. F. Bradford, W. G. Kepner, J. H. Lowry, D. L. Schrupp, K. A. Thomas, and B. C. Thompson, editors. 2006. Southwest Regional Gap Analysis Project Final Report. U.S. Geological Survey, Gap Analysis Program, Moscow, Idaho.

Quinlan, J. R. 1993. C4.5: Programs for Machine Learning. Morgan Kaufman, San Mateo, California.

Scott, J. M., F. Davis, B. Csuti, R. Noss, B. Butterfield, C. Groves, H. Anderson, S. Caicco, F. D'Erchia, T. C. Edwards Jr., J. Ulliman, and G. Wright. 1993. Gap analysis: A geographic approach to protection of biological diversity. Wildlife Monographs 123.

Shafroth, P. B., J. R. Cleverly, T. L. Dudley, J. P. Taylor, C. van Riper III, E. P. Weeks, and J. N. Stuart. 2005. Control of *Tamarix* in the western United States: Implications for water salvage, wildlife use, and riparian restoration. Environmental Management 35: 231–246.

Shtatland, E. S., K. Klienman, and E. M. Cain. 2004. A new strategy of model building in PROC LOGISTIC with automatic variable selection, validation, shrinkage and model averaging. SUGI 29 Proceedings, paper 191-29. Montreal, Canada. Available at http://www2.sas.com/proceedings/sugi29/191-29.pdf. Accessed October 2006.

Smith, S. D., D. A. Devitt, A. Sala, J. R. Cleverly, and D. E. Busch. 1998. Water relations of riparian plants from warm desert regions. Wetlands 18(4): 687–696.

Specht, R. L., E. M. Roe, and V. H. Boughlon. 1974. Supplement 7. Australian Journal of Botany Supplement Series.

St. Clair, L., J. R. Johansen, and S. R. Rushforth. 1993. Lichens of soil crust communities in the Intermountain area of the western United States. Great Basin Naturalist 53 (1): 5–12.

Thomas, K. A. 2006. Southwest Exotic Information Clearinghouse (SWEPIC) and Southwest Exotic Mapping Program (SWEMP). USGS Southwest Biological Science Center, Flagstaff, Arizona. Available at http://www.usgs.nau.edu/SWEPIC/index.asp. Accessed January 2006.

Tuhy, J. S., P. Comer, D. Dorfman, M. Lammert, J. Humke, B. Cholvin, G. Bell, B. Neely, S. Silbert, L. Whitham, and B. Baker. 2002. A Conservation Assessment of the Colorado Plateau Ecoregion. The Nature Conservancy, Moab Project Office, Utah.

Utah Division of Wildlife Resources. 2005. Utah's Watershed Restoration Initiative: Coming Together to Help Rangelands. Available at http://www.wildlife.utah.gov/watersheds/upcd.php. Accessed January 2006.

Welsh, S. L. 1978. Problems of plant endemism on the Colorado Plateau. Great Basin Naturalist Memoirs 2: 191–195.

West, N. E., and J. A. Young. 2000. Intermountain valleys and lower mountain slopes. In North American Terrestrial Vegetation, edited by M. G. Barbour and W. D. Billings. 2nd edition. Cambridge University Press, New York.

A GAP ANALYSIS OF TERRESTRIAL VERTEBRATE SPECIES OF THE COLORADO PLATEAU: ASSESSMENT FROM THE SOUTHWEST REGIONAL GAP ANALYSIS PROJECT

Kenneth G. Boykin, Charles Drost, and J. Judson Wynne

As part of the Southwest Regional Gap Analysis Project (GAP), we developed spatial habitat models of 819 vertebrate species for the region comprising Arizona, New Mexico, Colorado, Utah, and Nevada. Here we apply the results of the vertebrate habitat models to the Colorado Plateau region of northern Arizona, northwestern New Mexico, southwestern Colorado, and southern and eastern Utah. The Colorado Plateau boundaries encompass habitat for 581 vertebrate species from the original mapping effort. Total species richness is highest in areas of the Colorado and San Juan River drainages. We show, however, that patterns of richness vary among different vertebrate groups and subgroups (e.g. amphibians and bats). One important use of GAP data is to evaluate what proportion of the habitat of various species is managed for long-term conservation. These data can be expressed as "threshold" levels of species protection. We compare and contrast these GAP threshold species with lists developed by the state wildlife agencies of the southwestern states for "species of greatest conservation need" (SGCN). Our threshold lists differ from the SGCN lists because of their focus on longer-term protection of species that may still be quite common. In this way, the Southwest GAP data offer an alternative for land management and conservation planning in the region.

Conservation planning and assessments over large regions provide the ecological context necessary for landscape-scale man-agement decisions. Previous conservation assessments have been conducted for the Colorado Plateau (Tuhy et al. 2002) and portions of the plateau in Arizona (Arizona Game and Fish Department 2005a, 2005b), Colorado (Colorado Division of Wildlife 2005), New Mexico (New Mexico Department of Game and Fish 2005), and Utah (Utah Division of Wildlife Resources 2005). In addition, previous gap analysis projects have provided conservation information for the same area (Edwards et al. 1995; Thompson et al. 1996; Schrupp et al. 2001; Halvorson et al. 2002).

Gap analysis involves creating digital data sets of land cover, habitat models for terrestrial vertebrate species, and land stewardship, and analyzing the co-occurrence of these features on the landscape (See Ernst and Prior-Magee, this volume). Geographic Information System (GIS) maps and tables of these data sets allow for a wide variety of analyses for use in conservation planning. Over broad landscape scales, gap analysis provides a "coarse filter" approach to natural resources data and conservation assessments.

Gap analysis initially used a state-based approach (Scott et al. 1993), and now states in some areas have begun to collaborate on regional efforts (e.g., the Upper Midwest Gap Analysis Project; http://www.umesc.usgs.gov/umgaphome.html). The Southwest Regional Gap Analysis Project was the first formal multi-state gap analysis project, and additional regional projects are now in

progress in the Southeast (www.basic.ncsu
.edu/segap/) and Northwest (www.gap
.uidaho.edu/Northwest/2007_factsheet.doc).
This multi-state approach allows for gap
analyses to be completed for large regions
defined by natural biogeographic boundaries, rather than state or county boundaries.

Gap analysis relies on the concept of
wildlife-habitat relationships to model species habitats. Boykin et al. (2006) defined a
wildlife-habitat relationship as "a textual,
mathematical, graphical, or combination
statement that predicts abstractly or directly
what conditions are considered necessary
for a taxon's habitat to exist and where it
likely exists on a landscape." To be used in a
regional GIS-based model, habitat associations must be capable of being expressed as
a GIS theme, and data for that GIS theme
must be available for the region.

Drawing on data from the Southwest Regional Gap Analysis, we present preliminary
gap analyses for terrestrial vertebrates of the
Colorado Plateau. We also present sample
analyses for individual species and two
taxonomic groups (amphibians and bats) in
the region.

METHODS

The Southwest Regional Gap Analysis Project used a deductive modeling approach to
identify wildlife-habitat relationships for
each species (Boykin et al. 2006). We developed models for all species known to use
habitat in the five-state project area (Arizona, Colorado, Nevada, New Mexico, and
Utah) for significant portions of their life
cycles. Species were not included if they
were accidental, vagrant, or extirpated in the
project area, or were not considered to be
distinct species (see Boykin et al. 2006 for a
complete list of decision rules used in modeling). We developed habitat models from
literature reviews for each species using
specific associations of available GIS environmental variables. Variables used were
land cover, elevation, slope, aspect, distance
to hydrological features, landform (Manis et
al. 2001), soils, and mountains (see Boykin et
al. 2006). Models were constrained to the
current known range of the species (or

recent historic range, for species that have
experienced recent declines) using state,
regional, and national references. Range
data were converted to sub-basin watershed
units (8-digit hydrologic units, or "HUCs")
using the National Hydrography Data Set
(Boykin et al. 2006; see http://nhd.usgs
.gov/). These HUCs provided smaller base
units than currently described ecoregions for
the five-state study area.

One of the main objectives of vertebrate
species modeling within GAP is to intersect
habitat models with land stewardship data
to identify levels of long-term conservation
management. The resultant maps and tables
provide area and percent of area for species
habitat in each of four management status
categories (see Ernst and Prior-Magee, this
volume). We used the Colorado Plateau
Ecoregion as defined by Tuhy et al. (2002) to
define the spatial extent of our analyses, and
included habitat models for all species that
we considered to have habitat within the
boundaries of the Colorado Plateau.

Gap analysis uses a variety of thresholds
to provide conservation information to land
managers (Scott et al. 1993). Such thresholds
are useful in broad context, although specific
needs of individual species vary based on
their life history characteristics. For this assessment we used standard GAP reporting
thresholds for the percentage of each species' predicted area of occurrence that is on
lands managed for long-term protection
(Status 1 and 2 lands). These thresholds
provide a convenient reporting framework,
with the option to further analyze single
species, groups of species, or all species. In
the remainder of this paper, we refer to the
following thresholds of species habitat protection:

- Threshold 1: Less than 1% of predicted
  habitat is on Status 1 and 2 lands (least
  protected).
- Threshold 2: From 1% to less than 10% is
  on Status 1 and 2 lands.
- Threshold 3: Between 10% and 20% of
  predicted habitat on Status 1 and 2 lands.
- Threshold 4: Between 20% and 50% of
  predicted habitat on Status 1 and 2 lands.

- Threshold 5: More than 50% of predicted habitat is on Status 1 and 2 lands (most protected).

All of the states within the Southwest GAP area have recently identified "species of greatest conservation need" (SGCN) in accordance with their Comprehensive Wildlife Conservation Strategies (CWCS). The CWCS plans were submitted as a requirement for receiving State and Tribal Wildlife Grants from the U.S. Fish and Wildlife Service. State wildlife agencies identified these species based on criteria that were generally similar, but that differed from state to state. We reviewed these SGCN lists to provide a comparison and context for the Colorado Plateau species lists, based on perceived state concern for various species. Arizona identified 382 of the 819 species that Southwest GAP modeled as SGCN species (Table 1). Similarly, Colorado listed 114 species, New Mexico listed 85, and Utah listed 118.

We also evaluated species richness in the Colorado Plateau region using the GAP data. These species richness estimates were derived via two processes. We calculated the first species richness estimates using total species numbers (from the species habitat models) for each eight-digit hydrologic unit on the Colorado Plateau, to provide total richness of all species for the subregion. We also calculated species richness for all amphibians and for all bats using the predicted habitat models, to illustrate the application of smaller groupings of habitat models.

RESULTS

We identified 581 species that we predicted to have habitat within the Colorado Plateau Ecoregion (Tables 2 and 5). This was a majority (71%) of the species modeled in Southwest GAP for the entire five-state region (819 species). Analysis was conducted on 19 amphibians with predicted habitat on the Colorado Plateau (51% of all amphibians modeled in the region), 341 birds (78%), 143 mammals (67%), and 78 reptiles (60%). Full results for all models, including references, habitat data, modeling process, and textual and spatial models, can be found at http://fws-nmcfwru.nmsu.edu/swregap/.

The overall gap analysis (predicted habitat distribution by land management status) identified 43 species as "Threshold 1" species in the region, with less than 1 percent of the species' habitat on the Colorado Plateau managed for long-term conservation (Tables 2 and 3). There were 110 species in Threshold 2 (1–10% of habitat protected) with the majority of species in Thresholds 3 and 4. Only 20 species were Threshold 5 (Table 2). The following sections discuss patterns by major taxonomic group.

Amphibians

Of the four amphibian species identified in Threshold 1 (Table 3), two are closely associated with perennial aquatic systems—the Chiricahua leopard frog and lowland leopard frog. Within the Southwest, and specifically the Colorado Plateau, these aquatic and riparian systems are largely under private ownership. Couch's spadefoot has little or no association with permanent water sources. Two species (Great Plains toad and New Mexico spadefoot) were identified in Threshold 2 (see Table 5), and 13 species were placed in Thresholds 3 and 4. All but 3 of the 19 species in this analysis have been designated as SGCN by at least one state wildlife agency within Arizona, Colorado, New Mexico, and Utah.

Birds

Southwest GAP modeled 341 bird species on the Colorado Plateau. Of these, 286 were identified by one or more states as SGCN species. Of the 13 bird species identified within Threshold 1 (Table 3), 5 had less than 100 sq km of habitat identified on the Colorado Plateau (northern cardinal, gilded flicker, canvasback, Sprague's pipit, and bronzed cowbird). Those with more than 100 sq km of habitat were Mexican jay, curve-billed thrasher, dickcissel, white-throated sparrow, American pipit, brown thrasher, and whip-poor-will. All but three of these species (northern cardinal, gilded flicker, and brown thrasher) have been identified as SGCN.

There were 73 bird species identified in Threshold 2 (see Table 5). This covered a wide range of avian species, including 1

Table 1. Species of Greatest Conservation Need, by major taxonomic group, identified by the five states within the Southwest Regional Gap Analysis Project area.

Taxonomic Group	Arizona	Colorado	Nevada	New Mexico	Utah
Amphibians	11	4	2	7	8
Reptiles	32	8	13	8	32
Birds	268	81	60	48	42
Mammals	71	21	32	22	36
Total	382	114	107	85	118

Table 2. Total number of vertebrate species modeled by the Southwest Regional Gap Analysis Project on the Colorado Plateau, by gap analysis "threshold" levels. Thresholds represent proportion of habitat managed for long-term conservation, ranging from 1 (< 1% of habitat protected) to 5 (> 50% of habitat protected).

| | Threshold | | | | | |
	1 (< 1%)	2 (1–10%)	3 (10–20%)	4 (20–50%)	5 (> 50%)	Total
Amphibians	4	2	9	4	0	19
Birds	13	73	100	142	13	341
Mammals	15	22	63	38	5	143
Reptiles	11	13	23	29	2	78
Total	43	110	195	213	20	581

loon, 4 gallinaceous birds, 3 grebes, 1 egret, 4 owls, 38 passerines, 2 doves, 5 raptors, 7 shorebirds and gulls, 2 swifts, 1 hummingbird, 4 waterfowl, and 1 woodpecker. Of these, 15 species had less than 100 sq km of habitat mapped within the Colorado Plateau. Fifty-seven of the Threshold 2 species were identified as SGCN by one or more states within the region. There were 242 species of birds in Thresholds 3 and 4, and 13 species were in Threshold 5. Of these 255 species, 219 were identified as SGCN.

Mammals

Of the 143 mammal species modeled by the Southwest GAP project on the Colorado Plateau, 93 were identified as SGCN species. Fifteen species of mammals were in Threshold 1 (Table 3). Of these, eight species had less than 100 sq km of habitat mapped on the Colorado Plateau. Species with more than 100 sq km of mapped habitat and less than 1 percent of habitat in Status 1 and 2

lands included hooded skunk, southern plains woodrat, collared peccary, Osgood's mouse, cave myotis, southwestern myotis, and gray wolf (repatriated range). Eight of these 15 Threshold 1 mammals were identified as SGCN species.

There were a total of 22 mammal species in Threshold 2 (see Table 5). Three of these (desert kangaroo rat, Arizona gray squirrel, and hog-nosed skunk) had less than 100 sq km of habitat on the Colorado Plateau. The remaining 19 species included 3 carnivores (long-tailed weasel, the repatriated black-footed ferret, and wolverine), 3 hoofed mammals (white-tailed deer, the introduced Barbary sheep, and moose), and 13 rodents (spotted ground squirrel, house mouse, silky pocket mouse, Botta's pocket gopher, Gunnison's prairie dog, white-footed mouse, banner-tailed kangaroo rat, beaver, muskrat, thirteen-lined ground squirrel, Merriam's kangaroo rat, southern red-backed vole, and meadow vole). Sixteen of the 22 species

Table 3. Species with < 1 percent of predicted habitat on Status 1 and 2 lands on the Colorado Plateau. Status 1 and 2 lands are managed for long-term conservation purposes.

Taxa Group	Common Name	Scientific Name
Amphibian	Couch's spadefoot	*Scaphiopus couchii*
Amphibian	Colorado river toad	*Bufo alvarius*
Amphibian	Chiricahua leopard frog	*Rana chiricahuensis*
Amphibian	Lowland leopard frog	*Rana yavapaiensis*
Bird	Canvasback	*Aythya valisineria*
Bird	Whip-poor-will	*Caprimulgus vociferus*
Bird	Gilded flicker	*Colaptes chrysoides*
Bird	Black-billed cuckoo	*Coccyzus erythropthalmus*
Bird	Sprague's pipit	*Anthus spragueii*
Bird	Brown thrasher	*Toxostoma rufum*
Bird	Curve-billed thrasher	*Toxostoma curvirostre*
Bird	Bronzed cowbird	*Molothrus aeneus*
Bird	Northern cardinal	*Cardinalis cardinalis*
Bird	Dickcissel	*Spiza americana*
Bird	White-throated sparrow	*Zonotrichia albicollis*
Bird	Mexican jay	*Aphelocoma ultramarina*
Bird	American pipit	*Anthus rubescens*
Mammal	Preble's shrew	*Sorex preblei*
Mammal	Southwestern myotis	*Myotis auriculus*
Mammal	Cave myotis	*Myotis velifer*
Mammal	Pocketed free-tailed bat	*Nyctinomops femorosaccus*
Mammal	Snowshoe hare	*Lepus americanus*
Mammal	Hispid pocket mouse	*Chaetodipus hispidus*
Mammal	Southern plains woodrat	*Neotoma micropus*
Mammal	Mearns' grasshopper mouse	*Onychomys arenicola*
Mammal	Osgood's mouse	*Peromyscus gratus*
Mammal	Plains harvest mouse	*Reithrodontomys montanus*
Mammal	Hispid cotton rat	*Sigmodon hispidus*
Mammal	Hooded skunk	*Mephitis macroura*
Mammal	Lynx	*Lynx canadensis*
Mammal	Gray wolf	*Canis lupus*
Mammal	Collared peccary	*Pecari tajacu*
Reptile	Sonoran mud turtle	*Kinosternon sonoriense*
Reptile	Big Bend slider	*Trachemys gaigeae*
Reptile	Greater earless lizard	*Cophosaurus texanus*
Reptile	Regal horned lizard	*Phrynosoma solare*
Reptile	Crevice spiny lizard	*Sceloporus poinsettii*
Reptile	Checkered whiptail	*Cnemidophorus tesselatus*
Reptile	Rubber boa	*Charina bottae*
Reptile	Texas blind snake	*Leptotyphlops dulcis*
Reptile	Sonoran whipsnake	*Masticophis bilineatus*
Reptile	Checkered garter snake	*Thamnophis marcianus*
Reptile	Western coral snake	*Microides euryxanthus*

were identified as SGCN by at least one state (all of the above except house mouse, long-tailed weasel, white-footed mouse, Barbary sheep, moose, and meadow vole).

There were 63 species in Threshold 3 and 38 species in Threshold 4. Five species (i.e. round-tailed ground squirrel, desert pocket mouse, rock pocket mouse, marten, and western mastiff bat) were identified within Threshold 5.

Reptiles

Nine of 11 Colorado Plateau reptile species in Threshold 1 are on the periphery of their range, with less than 30 sq km of habitat mapped on the Colorado Plateau (Table 3). The two other species (greater earless lizard and crevice spiny lizard) have approximately 500 sq km of habitat, with less than 1 percent of that habitat in Status 1 and 2 lands. The crevice spiny lizard is not currently known to occur on the Colorado Plateau but its range extends near the southeast margin of the region, and suitable habitat was mapped on the plateau. Neither of the latter two species were identified by state agencies as SGCN.

Of the 13 reptile species in Threshold 2 (see Table 5), 7 are lizards and 6 are snakes. The six lizards with more than 100 sq km of habitat are the round-tailed horned lizard, Clark's spiny lizard, Great Plains skink, little striped whiptail, desert grassland whiptail, and lesser earless lizard. The Madrean alligator lizard (with < 100 sq km of habitat on the Colorado Plateau), round-tailed horned lizard, and a subspecies of the lesser earless lizard have been identified as SGCN. The six snake species we documented are the western diamondback rattlesnake, western hognosed snake, plains black-headed snake, Mojave rattlesnake, milk snake, and mountain patch-nosed snake. All but the western hog-nosed snake have been identified by at least one state as SGCN.

Fifty-two species fell within Thresholds 3 and 4 and only two species were identified in Threshold 5. Of the 78 reptile species mapped on the Colorado Plateau, 52 have been identified as SGCN species by one or more state agencies.

Species of Greatest Conservation Need

We compared the "threshold" habitat protection levels from the Southwest GAP data for the Colorado Plateau with the state SGCN lists (Table 4). In this comparison we omitted non-native species (e.g. European starling, house mouse), which are of little interest in a comparison of conservation priorities. As a predictor of conservation need, the GAP data diverge from the SGCN data. Of 43 species identified as Threshold 1 (little or no habitat managed for long-term protection) by the gap analysis, 28 (65%) were identified by one or more states as SGCN. As the threshold and level of protection increases, proportionately more species are identified by the states as SGCN. The proportion of species identified as SGCN reaches a maximum of 85 percent for Threshold 5, in which more than half of identified habitat is managed for long-term protection.

Table 4. Number of Colorado Plateau species in GAP habitat protection categories (thresholds), compared to numbers of those species identified as Species of Greatest Conservation Need (SGCN) by state wildlife agencies in the Southwest. Threshold 1 is least protected (< 1 % of modeled species range is managed for long-term conservation) and Threshold 5 is most protected (> 50 % of range is managed for long-term conservation).

	Threshold					
	1	2	3	4	5	Total
No. of species	43	106	194	213	20	576
No. of SGCN species	28	82	150	171	17	448
Percent SGCN	65.1	77.4	77.3	80.3	85.0	

Species Richness

Total species richness calculated from the Southwest GAP data for the Colorado Plateau averaged between 354 and 390 species per drainage subbasin (Figure 1). Species richness was higher in the eastern portion of the plateau, associated with the San Juan Mountains and the San Juan River, and on the western side of the plateau along the Colorado and Virgin Rivers. Compared to the entire Southwest GAP region, species richness on the Colorado Plateau was intermediate, with higher richness than more northern areas but lower richness than southern Arizona, much of New Mexico, and the Rocky Mountain front range in Colorado.

In addition to total richness of all vertebrate species, we also calculated species richness for two subgroups of vertebrates—amphibians and bats. Amphibian richness on the Colorado Plateau was highest in the northern and western areas of the plateau, associated with tributaries of the Colorado River (Figure 1). The southern and eastern extent of the plateau was relatively depauperate by comparison. The highest amphibian richness for the entire Southwest occurs in a broad band from southeastern Arizona across southern and central New Mexico to southeastern Colorado. High species richness in this area is a result of the convergence of several ecoregions, including the Great Plains, Rocky Mountains, Madrean, and Chihuahuan Desert. Most of Nevada and the northern deserts of Utah have relatively few amphibians, and most of the Rocky Mountain region is also poor in amphibian species.

Analysis of bat species richness on the Colorado Plateau identified areas of relatively high richness throughout much of the plateau (Figure 1). Areas of highest richness were identified around Zion National Park and the Music Mountains–Grand Wash Cliffs area. The relatively high numbers of bat species in the Zion area have been corroborated by other researchers conducting surveys at the park (M. Bogan, personal communication). Further analysis of bat species richness provides additional information on conservation opportunities on the Colorado Plateau. Analysis of potential gaps in bat habitat protection indicates that the most species-rich areas, where 18–19 species potentially occur, have 1 percent or less of the area effectively protected (Figure 2). Other areas that are relatively species-rich, supporting up to 11–17 species, range from 5 to 25 percent of the area effectively protected (Status 1 and 2 lands).

DISCUSSION

The goal of gap analysis is to provide data for conservation planning. It is one of many tools available and is intended to provide background information and ecological context for land managers. It provides broad landscape perspectives as well as providing data for land managers on conservation opportunities available to them. Analyses of GAP data often focus on Status 1 and 2 lands (lands managed for long-term conservation), but it is important to note that there are also many conservation opportunities on multiple-use lands (Status 3) and private lands (Status 4).

There are several caveats for the Southwest GAP models and the analyses presented here. The Southwest GAP vertebrate models are models of species habitat with no inferences as to differing levels of habitat quality or species abundance. The resulting maps show predicted distribution of habitat for each species, but do not necessarily indicate that the species occurs in all areas mapped as habitat. Further, model accuracy depends on available knowledge (literature sources and available regional data sets) and the resolution of the spatial data used in model development (Boykin et al. 2006). Mapped areas of predicted habitat for a species may or may not be occupied by that species, particularly at the periphery of the species' range. Hence, these habitat models tend to overestimate the extent of occupied range for a species on the landscape.

In reviewing these analyses it is important to recognize that species identified as

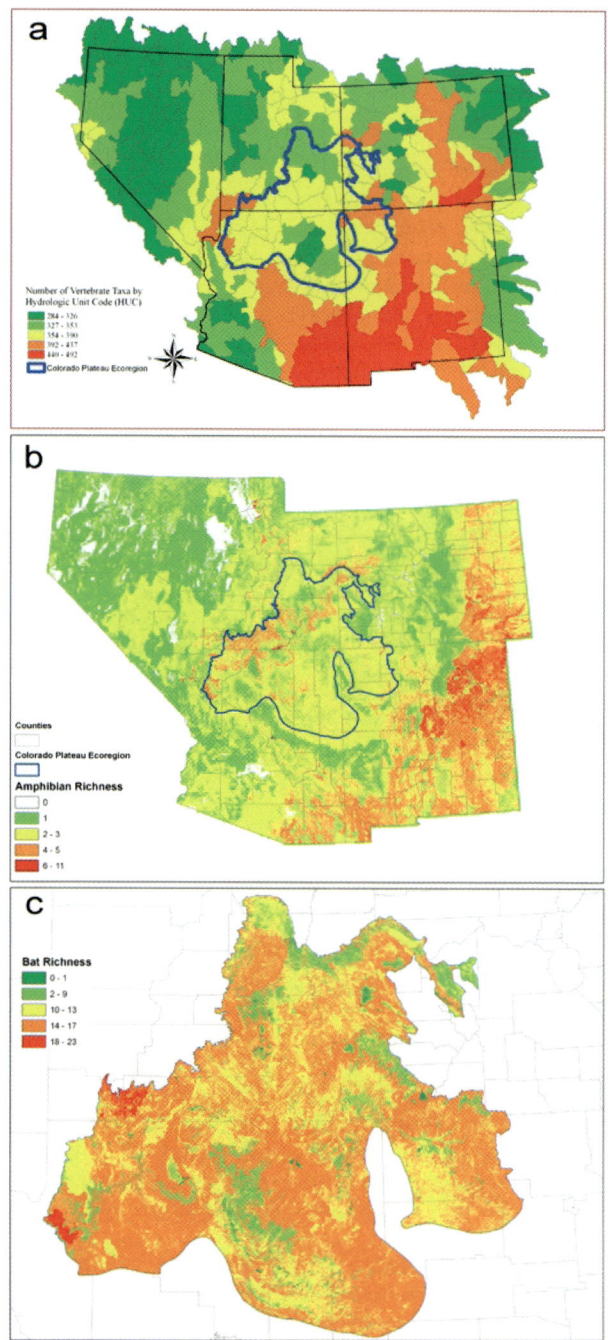

Figure 1. (a) Terrestrial vertebrate species richness by 8-digit hydrologic unit (HUC) in the south-western United States, as modeled by SReGAP, showing boundary of the Colorado Plateau. (b) Species richness of 37 amphibian species. (c) Bat species richness.

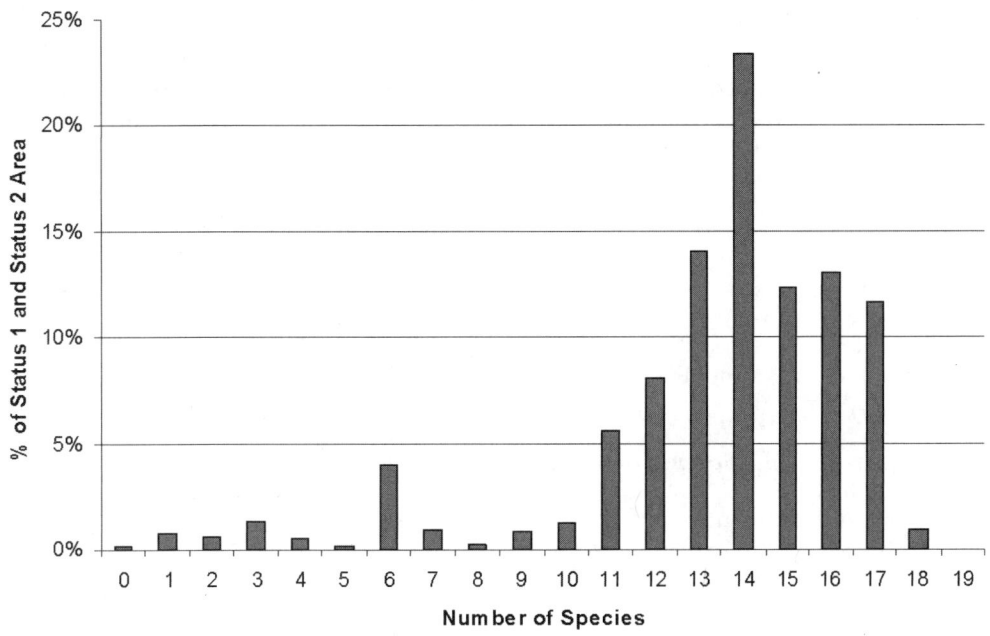

Figure 2. Level of long-term habitat protection, relative to species richness of bats on the Colorado Plateau. For each category of richness (number of bat species within an area), the vertical axis shows the proportion of that area that is within Status 1 and 2 lands (well protected over the long term).

GAP species may or may not need conservation attention; the species that are identified by the GAP filters should be evaluated case by case. For example, species that occur largely in "settled" habitats (e.g. the house finch, which is identified as a Threshold 2 species) will often be identified because such areas are not under conservation mandates. Wide-ranging species (e.g. the coyote) may also be identified as Threshold 1 or 2 in these analyses because such species are ubiquitous across the landscape and most of the landscape is within Status 3 or Status 4 lands. The threshold categories discussed here are somewhat arbitrary and different species doubtless need differing levels of habitat protection. The threshold level of 10 percent is simply a starting point for analysis. The analyses presented herein allow for preliminary generalizations, but further research is needed regarding the validity and specificity of these thresholds.

Species of Greatest Conservation Need

Our comparison of "Species of Greatest Conservation Need" and species identified as Threshold 1 and Threshold 2 by gap analysis highlights differences between the two lists. This is not surprising, because the two lists were prepared using very different approaches. The state SGCN lists take a bottom-up, species-centered approach, identifying species that are known or suspected to be threatened, or that are known to have experienced population declines. Species whose status is not well known, or that appear to be secure at the present time, do not typically receive attention. Gap analysis provides an alternative top-down approach, looking across the entire community of vertebrate species in the region and comparing them with a single currency—the percentage of each species' habitat that is considered well protected.

The GAP lists represent just a single filter; the species identified in this manner need to be examined further. The process produces some results that are clearly not of conservation concern (e.g. introduced species such as the house sparrow, identified as Threshold 2), and these can be easily elimi-nated. On the other hand, species identified by gap analysis as relatively well protected (e.g. Threshold 4) may still be threatened from causes other than habitat loss. One of the values of the GAP lists lies in this kind of careful examination, to identify species that may be at risk of decline or loss in the future.

This highlights another difference between the two approaches. Federal and state lists of threatened or at-risk species typically focus (as well they should) on those known to be declining or facing other apparent problems. Gap analysis provides the ability to identify species that may face declines in the future, because their habitat is subject to degradation or loss. These species may be common and widespread now, but with gradual loss of habitat they may experience significant declines in the future.

Both approaches have value for conservation planning, and they are best used in a complementary manner. Species that are currently threatened or in decline may need immediate attention. Over the longer term, however, conservation planning would benefit from efforts, like gap analysis, to protect overall diversity and to identify species that may become the threatened species concerns of the future.

Species Richness

Species richness is often used for conservation assessments and we present examples of such analyses here. Our analyses using all vertebrate species (by drainage subbasin) and two taxonomic groups (amphibians and bats, by habitat) demonstrate the utility of the overall data set. For most purposes, broad analyses (such as total vertebrate species richness) provide less insight than finer-scale explorations. Comparing areas based on total mammal richness, for example, can obscure areas of particular importance to specific groups of mammals. Areas of high total mammal richness may have low richness of bats, or no bats in the genus *Myotis*, or no nectar-feeding bats (this latter category includes two threatened species). In addition to analyses based on taxonomic groupings, a wide variety of other analyses are possible, using habitat or life history groupings, for example, depending on management or research interest. We encourage users of these data to more fully explore these possibilities.

Conservation and Management Considerations

Most of the Threshold 1 species identified within the boundaries of the Colorado Plateau are marginal species that are at or near the edge of their ranges. They have very limited mapped habitat (most < 100 sq km; see Table 5) and many are not currently known to occur on the plateau. In some cases, such species may be of particular interest precisely because they are present in limited habitat near the edge of their ranges. Some examples in this respect are the American pipit, which breeds in scattered alpine habitats on the Colorado Plateau, and the mountain plover, found in limited grassland habitats on the Colorado Plateau in Arizona and New Mexico. The southwestern myotis has a relatively limited total range, and occurs marginally along the southern edge of the Colorado Plateau.

Species identified as Threshold 2 (1–10% of mapped habitat managed for long-term protection) may be more instructive for understanding general patterns for conservation planning on the Colorado Plateau. Many of these species are grassland inhabitants. Examples in Threshold 2 include the Great Plains toad among amphibians, the Great Plains skink, little striped whiptail, lesser earless lizard, and milk snake among reptiles, the mountain plover, short-eared owl, savannah sparrow, prairie falcon, and others among birds, and the silky pocket mouse, spotted ground squirrel, Gunnison's prairie dog, and banner-tailed kangaroo rat among mammals. This pattern among vertebrate species adds further weight to other sources that have expressed concern over the decline of Southwest grasslands and the species they support (e.g. McClaran and Van Devender 1995).

CONCLUSIONS

The results presented here represent very preliminary analyses of the wealth of data produced by the Southwest GAP project. Even at a preliminary level, these analyses highlight some individual species and groups of species (e.g. those in grassland habitats) that may warrant additional consideration. Full data for all aspects of the Southwest GAP project (vegetation/land cover, animal habitat models, stewardship maps) are available at http://fws-nmcfwru.nmsu.edu/swregap/. The Southwest GAP project provides these data for use, review, and revision. The data provide baseline information for conservation on the Colorado Plateau. Particularly when combined with other current efforts these data provide researchers and managers with another tool for understanding the distribution of vertebrates on the Colorado Plateau, within the context of these species' habitats in the Southwest. An important objective of the Southwest GAP project has been to create models and processes that can be modified and updated using new data (or different perspectives on current data), such that the models remain current and continue to reflect an increased understanding of vertebrate species in the Southwest.

ACKNOWLEDGMENTS

The Southwest Regional Gap Analysis Project was the focus of more than 100 professionals during the life of the project. Nineteen habitat modelers spent long hours synthesizing data and creating the models. Without their efforts, this manuscript would not have been possible. Funding for the project was provided by the USGS Biological Resources Division. Senior author funding was provided by the New Mexico Cooperative Fish and Wildlife Research Unit during part of manuscript preparation.

Table 5. Colorado Plateau species in GAP Thresholds 1 (< 1% of habitat protected), 2 (1–10 %), and 3 (10–20 %), modeled as part of the Southwest Regional Gap Analysis Project. "Species of Greatest Conservation Need" are from draft lists available from the respective state natural resources agencies in Arizona, Colorado, New Mexico, and Utah, as of August 2005. In SGCN columns 1 = species, 2 = subspecies. Status columns list the modeled area of each species' habitat within lands of varying conservation status, from Status 1 lands (most protected) to Status 4 lands (least protected). Areas are in square km and percent of total predicted habitat for each species. Species are listed by major taxonomic groups, and within groups, by percent of habitat in Status 1 or 2 lands.

	SGCN				Total Area	Status 1		Status 2		Status 3		Status 4		Status 1 & 2		Threshold
	AZ	CO	NM	UT		Area	%	Area	%	Area	%	Area	%	Area	%	
AMPHIBIANS																
Couch's spadefoot (*Scaphiopus couchii*)	1	–	–	–	27	0	0	0	0	3	11.6	24	88.4	0	0.0	<1
Colorado River toad (*Bufo alvarius*)	–	–	1	–	61	0	0	0	0	1	1.1	60	98.9	0	0.0	<1
Lowland leopard frog (*Rana yavapaiensis*)	1	–	1	–	260	0	0.0	0	0.0	31	11.9	229	88.1	0	0.0	<1
Chiricahua leopard frog (*Rana chiricahuensis*)	1	–	1	–	93	0	0	1	0.8	31	32.8	62	66.4	1	0.8	<1
Great plains toad (*Bufo cognatus*)	–	–	–	1	12231	201	1.6	504	4.1	10194	83.3	1332	10.9	705	5.8	1-10
New Mexico spadefoot (*Spea multiplicata*)	–	–	–	1	9994	225	2.3	577	5.8	6195	62.0	2996	30.0	802	8.0	1-10
Mountain treefrog (*Hyla eximia*)	–	–	1	–	62	7	10.5	0	0	43	68.8	13	20.7	7	10.5	10-20
Boreal chorus frog (*Pseudacris maculata*)	–	–	–	–	48	1	2.3	5	9.9	16	32.6	27	55.2	6	12.2	10-20
Bullfrog (*Rana catesbeiana*)	–	–	–	–	644	37	5.8	54	8.3	153	23.7	400	62.1	91	14.1	10-20
Northern leopard frog (*Rana pipiens*)	1	–	–	1	351	26	7.4	28	7.9	143	40.8	154	43.8	54	15.3	10-20
Western toad (*Bufo boreas*)	–	–	2	1	4247	134	3.1	553	13.0	1576	37.1	1984	46.7	687	16.2	10-20
Tiger salamander (*Ambystoma tigrinum*)	2	–	1	–	194117	8416	4.3	23400	12.1	118055	60.8	44246	22.8	31816	16.4	10-20
Woodhouse's toad (*Bufo woodhousii*)	–	–	–	–	128924	4361	3.4	18412	14.3	78180	60.6	27971	21.7	22773	17.7	10-20
Canyon treefrog (*Hyla arenicolor*)	1	1	–	1	12675	786	6.2	1536	12.1	7481	59.0	2871	22.7	2322	18.3	10-20
Red-spotted toad (*Bufo punctatus*)	1	–	–	–	93920	4887	5.2	12991	13.8	54605	58.1	21437	22.8	17878	19.0	10-20

Table 5 (continued)

Species	SGCN AZ	SGCN CO	SGCN NM	SGCN UT	Total Area	Status 1 Area	Status 1 %	Status 2 Area	Status 2 %	Status 3 Area	Status 3 %	Status 4 Area	Status 4 %	Status 1 & 2 Area	Status 1 & 2 %	Threshold
REPTILES																
Rubber boa (*Charina bottae*)	–	–	–	1	6	0	0	0	0	5	77.9	1	22.1	0	0	<1
Checkered whiptail (*Cnemidophorus tesselatus*)	–	–	–	–	13	0	0	0	0	11	84.7	2	15.3	0	0	<1
Greater earless lizard (*Cophosaurus texanus*)	–	–	–	–	501	0	0.0	0	0.0	7	1.4	494	98.6	0	0.0	<1
Sonoran mud turtle (*Kinosternon sonoriense*)	2	–	–	–	0	0	0	0	0	0	0	0	100	0	0	<1
Texas blind snake (*Leptotyphlops dulcis*)	–	1	–	–	2	0	0	0	0	1	60	1	40	0	0	<1
Sonoran whipsnake (*Masticophis bilineatus*)	2	–	–	–	25	0	0	0	0	1	2.8	24	97.2	0	0	<1
Western coral snake (*Micruroides euryxanthus*)	2	–	–	–	19	0	0	0	0	0	0	19	100	0	0	<1
Regal horned lizard (*Phrynosoma solare*)	–	–	1	–	1	0	0	0	0	0	0	1	100	0	0	<1
Crevice spiny lizard (*Sceloporus poinsettii*)	–	–	–	–	598	0	0.0	0	0.0	135	22.6	463	77.4	0	0.0	<1
Checkered garter snake (*Thamnophis marcianus*)	–	–	–	0	0	0	0	0	0	91.7	0	8.3	0	0	<1	
Big Bend slider (*Trachemys gaigeae*)	–	–	–	–	0	0	0	0	0	0	83.8	0	16.2	0	0	<1
Madrean alligator lizard (*Elgaria kingii*)	–	–	2	–	1	0	0	0	1	0	9.1	1	90	0	1	1-10
Round-tailed horned lizard (*Phrynosoma modestum*)	1	–	–	–	821	1	0.1	10	1.2	568	69.2	242	29.5	11	1.3	1-10
Clark's spiny lizard (*Sceloporus clarkii*)	–	–	–	–	14944	20	0.1	219	1.5	3823	25.6	10883	72.8	239	1.6	1-10
Western diamondback rattlesnake (*Crotalus atrox*)	1	–	1	–	11606	1	0.0	197	1.7	4591	39.6	6816	58.7	199	1.7	1-10
Great Plains skink (*Eumeces obsoletus*)	–	–	–	–	29163	381	1.3	215	0.7	10543	36.2	18024	61.8	596	2.0	1-10
Little striped whiptail (*Cnemidophorus inornatus*)	–	–	–	–	13002	428	3.3	421	3.2	8285	63.7	3868	29.7	849	6.5	1-10
Western hog-nosed snake (*Heterodon nasicus*)	–	–	–	–	25759	97	0.4	1756	6.8	20332	78.9	3574	13.9	1853	7.2	1-10

Table 5 (continued)

Species	SGCN				Total Area	Status 1		Status 2		Status 3		Status 4		Status 1 & 2		Threshold
	AZ	CO	NM	UT		Area	%	Area	%	Area	%	Area	%	Area	%	
Plains black-headed snake (*Tantilla nigriceps*)	1	–	–	–	1434	2	0.1	102	7.1	1107	77.2	223	15.6	104	7.3	1-10
Desert grassland whiptail (*Cnemidophorus uniparens*)	–	–	–	–	317	24	7.6	3	0.8	26	8.2	264	83.3	27	8.4	1-10
Mojave rattlesnake (*Crotalus scutulatus*)	2	–	–	1	1166	48	4.1	52	4.4	332	28.4	735	63.0	100	8.5	1-10
Milk snake (*Lampropeltis triangulum*)	2	–	1	1	48193	599	1.2	3605	7.5	27528	57.1	16461	34.2	4204	8.7	1-10
Mountain patch-nosed snake (*Salvadora grahamiae*)	1	–	–	–	10936	63	0.6	952	8.7	6125	56.0	3796	34.7	1016	9.3	1-10
Lesser earless lizard (*Holbrookia maculata*)	2	–	–	–	61489	570	0.9	5396	8.8	47007	76.4	8516	13.9	5966	9.7	1-10
Gila spotted whiptail (*Cnemidophorus flagellicaudus*)	–	–	–	–	338	27	7.9	7	2.2	30	9.0	274	80.9	34	10.1	10-20
Black-necked garter snake (*Thamnophis cyrtopsis*)	2	1	–	1	47814	1148	2.4	4002	8.4	26534	55.5	16130	33.7	5150	10.8	10-20
Many-lined skink (*Eumeces multivirgatus*)	–	–	–	1	74676	1277	1.7	6914	9.3	47150	63.1	19336	25.9	8190	11.0	10-20
Painted turtle (*Chrysemys picta*)	2	–	–	–	3654	185	5.1	258	7.1	2076	56.8	1135	31.1	443	12.1	10-20
Western fence lizard (*Sceloporus occidentalis*)	–	–	–	–	3526	227	6.4	219	6.2	2318	65.7	763	21.6	446	12.6	10-20
Rosy boa (*Charina trivirgata*)	2	–	–	–	93	0	0	12	12.9	53	56.7	28	30.5	12	12.9	10-20
Collared lizard (*Crotaphytus collaris*)	–	1	–	–	51838	1971	3.8	5544	10.7	28230	54.5	16093	31.0	7515	14.5	10-20
Racer (*Coluber constrictor*)	2	–	–	–	35637	1410	4.0	4007	11.2	24931	70.0	5289	14.8	5417	15.2	10-20
Western rattlesnake (*Crotalus viridis*)	2	–	–	–	166195	7418	4.5	17953	10.8	102190	61.5	38634	23.2	25371	15.3	10-20
Glossy snake (*Arizona elegans*)	2	–	–	–	38566	1658	4.3	4369	11.3	25503	66.1	7036	18.2	6027	15.6	10-20
Ring-necked snake (*Diadophis punctatus*)	–	–	–	1	37272	1299	3.5	4575	12.3	16688	44.8	14710	39.5	5874	15.8	10-20
Long-tailed brush lizard (*Urosaurus graciosus*)	–	–	–	–	13	2	15.2	0	0.6	11	84.2	0	0	2	15.8	10-20
New Mexico whiptail (*Cnemidophorus neomexicanus*)	–	–	–	–	1369	147	10.8	71	5.2	526	38.4	625	45.6	219	16.0	10-20

Table 5 (continued)

Species	SGCN				Total Area	Status 1		Status 2		Status 3		Status 4		Status 1 & 2		Threshold
	AZ	CO	NM	UT		Area	%	Area	%	Area	%	Area	%	Area	%	
Smooth green snake (*Liochlorophis vernalis*)	–	–	–	1	15638	401	2.6	2121	13.6	8774	56.1	4342	27.8	2523	16.1	10-20
Night snake (*Hypsiglena torquata*)	2	1	–	1	141148	4080	2.9	19153	13.6	87312	61.9	30603	21.7	23233	16.5	10-20
Plateau striped whiptail (*Cnemidophorus velox*)	–	–	–	1	100943	3708	3.7	13140	13.0	59102	58.5	24993	24.8	16848	16.7	10-20
Spotted leaf-nosed snake (*Phyllorhynchus decurtatus*)	–	–	–	1	11	2	16.5	0	0.6	9	82.8	0	0	2	17.2	10-20
Eastern fence lizard (*Sceloporus undulatus*)	–	–	–	–	153549	6610	4.3	20347	13.3	95015	61.9	31577	20.6	26957	17.6	10-20
Sagebrush lizard (*Sceloporus graciosus*)	–	–	–	–	130672	4583	3.5	19179	14.7	83489	63.9	23421	17.9	23762	18.2	10-20
Long-nosed snake (*Rhinocheilus lecontei*)	–	1	–	1	8629	118	1.4	1488	17.2	6265	72.6	759	8.8	1605	18.6	10-20
Western whiptail (*Cnemidophorus tigris*)	–	–	–	–	48868	3329	6.8	5865	12.0	32746	67.0	6928	14.2	9194	18.8	10-20
Sidewinder (*Crotalus cerastes*)	2	–	–	1	31	0	0.3	6	19.4	14	46.4	11	33.9	6	19.7	10-20
Bullsnake (*Pituophis catenifer*)	–	–	–	–	145877	7697	5.3	21351	14.6	86071	59.0	30758	21.1	29048	19.9	10-20
BIRDS																
Sprague's pipit (*Anthus spragueii*)	1	–	1	–	34	0	0	0	0	34	100	0	0	0	0	<1
Northern cardinal (*Cardinalis cardinalis*)	–	–	–	–	0	0	0	0	0	0	100	0	0	0	0	<1
Gilded flicker (*Colaptes chrysoides*)	–	–	–	–	7	0	0	0	0	0	3.2	7	96.8	0	0	<1
Mexican jay (*Aphelocoma ultramarina*)	1	–	–	–	115	0	0.0	0	0.0	93	80.3	23	19.7	0	0.0	<1
Dickcissel (*Spiza americana*)	1	1	–	–	374	0	0.0	0	0.0	109	29.0	265	70.9	0	0.0	<1
Curve-billed thrasher (*Toxostoma curvirostre*)	–	1	–	–	882	0	0.0	0	0.0	27	3.0	855	97.0	0	0.0	<1
Bronzed cowbird (*Molothrus aeneus*)	1	–	–	–	39	0	0	0	0.1	3	7.4	36	92.6	0	0.1	<1
Black-billed cuckoo (*Coccyzus erythrophthalmus*)	3	–	–	1	20	0	0	0	0.2	0	1.8	20	98.1	0	0.2	<1

Table 5 (continued)

Species	SGCN				Total Area	Status 1		Status 2		Status 3		Status 4		Status 1 & 2		Threshold
	AZ	CO	NM	UT		Area	%	Area	%	Area	%	Area	%	Area	%	
White-throated sparrow (*Zonotrichia albicollis*)	1	–	–	–	171	1	0.4	0	0.1	56	32.4	115	67.2	1	0.4	<1
American pipit (*Anthus rubescens*)	1	–	–	–	1264	2	0.2	5	0.4	68	5.4	1189	94.1	7	0.5	<1
Canvasback (*Aythya valisineria*)	1	–	–	–	21	0	0.1	0	0.6	21	97.2	0	2.2	0	0.6	<1
Brown thrasher (*Toxostoma rufum*)	–	–	–	–	773	0	0.0	5	0.6	73	9.5	695	89.9	5	0.6	<1
Whip-poor-will (*Caprimulgus vociferus*)	1	–	–	–	718	0	0.0	5	0.7	113	15.8	599	83.5	5	0.7	<1
Common grackle (*Quiscalus quiscula*)	–	–	–	–	2239	0	0.0	22	1.0	186	8.3	2032	90.7	22	1.0	1-10
Eastern phoebe (*Sayornis phoebe*)	1	–	–	–	49	0	0	0	1	1	2	48	97	0	1	1-10
Eastern meadowlark (*Sturnella magna*)	1	–	–	–	6688	66	1.0	11	0.2	1337	20.0	5275	78.9	77	1.1	1-10
Red-eyed vireo (*Vireo olivaceus*)	1	–	–	–	14	0	0	0	1.2	1	9.7	13	89.1	0	1.2	1-10
Blue-throated hummingbird (*Lampornis clemenciae*)	1	–	–	–	12	0	0	0	1.6	0	0.7	12	97.7	0	1.6	1-10
Ovenbird (*Seiurus aurocapillus*)	–	–	–	–	547	1	0.2	8	1.5	269	49.2	268	49.1	9	1.7	1-10
Upland sandpiper (*Bartramia longicauda*)	–	1	–	–	303	0	0.1	5	1.7	14	4.7	284	93.6	5	1.8	1-10
Inca dove (*Columbina inca*)	–	–	–	–	193	3	1.4	1	0.4	7	3.7	182	94.5	3	1.8	1-10
Black-necked stilt (*Himantopus mexicanus*)	–	–	–	1	3	0	0	0	1.8	2	53.4	1	43.9	0	1.8	1-10
Wood thrush (*Hylocichla mustelina*)	–	–	–	–	251	0	0.1	5	1.8	11	4.3	235	93.8	5	1.9	1-10
Mountain plover (*Charadrius montanus*)	1	1	1	1	3337	40	1.2	35	1.0	2389	71.6	873	26.2	74	2.2	1-10
Hooded oriole (*Icterus cucullatus*)	1	–	1	–	18	0	2.1	0	0.1	3	18.4	15	79.4	0	2.2	1-10
Palm warbler (*Dendroica palmarum*)	–	–	–	–	85	1	1.3	1	1	5	5.8	78	91.8	2	2.3	1-10
Blackpoll warbler (*Dendroica striata*)	–	–	–	–	105	2	1.8	1	0.5	30	28.8	72	68.9	2	2.3	1-10

Table 5 (continued)

Species	SGCN				Total Area	Status 1		Status 2		Status 3		Status 4		Status 1 & 2		Threshold
	AZ	CO	NM	UT		Area	%	Area	%	Area	%	Area	%	Area	%	
Clark's grebe (Aechmophorus clarkii)	1	1	–	–	31	0	0.9	0	1.4	29	91.4	2	6.3	1	2.4	1-10
Franklin's gull (Larus pipixcan)	1	–	–	–	206	2	1.2	2	1.2	94	45.5	108	52.2	5	2.4	1-10
Chestnut-collared longspur (Calcarius ornatus)	1	1	–	–	14731	205	1.4	162	1.1	9454	64.2	4910	33.3	367	2.5	1-10
Cassin's sparrow (Aimophila cassinii)	1	1	–	–	23211	258	1.1	462	2.0	11259	48.5	11232	48.4	720	3.1	1-10
Redhead (Aythya americana)	1	–	–	–	41	1	1.6	1	1.7	33	81.2	6	15.6	1	3.3	1-10
Western grebe (Aechmophorus occidentalis)	1	1	–	–	35	1	1.8	1	1.7	31	89.4	2	7.2	1	3.4	1-10
Marsh wren (Cistothorus palustris)	1	–	–	–	4	0	3.3	0	0.1	1	15.5	3	81.1	0	3.4	1-10
House sparrow (Passer domesticus)	–	–	–	–	4921	5	0.1	161	3.3	481	9.8	4275	86.9	165	3.4	1-10
Mississippi kite (Ictinia mississippiensis)	1	–	–	–	605	0	0.0	24	4.0	161	26.6	420	69.4	24	4.0	1-10
Ring-necked pheasant (Phasianus colchicus)	–	–	–	–	2283	0	0.0	93	4.1	202	8.9	1988	87.1	93	4.1	1-10
Great-tailed grackle (Quiscalus mexicanus)	1	–	–	–	8512	79	0.9	270	3.2	2346	27.6	5817	68.3	349	4.1	1-10
European starling (Sturnus vulgaris)	–	–	–	–	21258	251	1.2	627	2.9	10344	48.7	10036	47.2	878	4.1	1-10
Snow bunting (Plectrophenax nivalis)	–	–	–	–	13526	58	0.4	549	4.1	8282	61.2	4636	34.3	607	4.5	1-10
McCown's longspur (Calcarius mccownii)	1	1	–	–	4236	135	3.2	61	1.5	1789	42.2	2250	53.1	197	4.6	1-10
Scaled quail (Callipepla squamata)	1	1	1	–	19452	31	0.2	872	4.5	14524	74.7	4025	20.7	903	4.6	1-10
Short-eared owl (Asio flammeus)	1	1	–	–	48015	597	1.2	1657	3.5	25940	54.0	19822	41.3	2253	4.7	1-10
Snow goose (Chen caerulescens)	1	–	–	–	4852	63	1.3	166	3.4	850	17.5	3773	77.8	229	4.7	1-10
Ring-billed gull (Larus delawarensis)	1	–	–	–	128	4	3.2	2	1.5	57	44.2	66	51.1	6	4.7	1-10
Black phoebe (Sayornis nigricans)	1	–	–	–	1084	36	3.3	15	1.4	68	6.3	965	89.1	51	4.7	1-10

Table 5 (continued)

Species	SGCN				Total Area	Status 1		Status 2		Status 3		Status 4		Status 1 & 2		Threshold
	AZ	CO	NM	UT		Area	%	Area	%	Area	%	Area	%	Area	%	
Bullock's oriole (Icterus bullockii)	1	–	–	–	5482	32	0.6	229	4.2	703	12.8	4519	82.4	260	4.8	1-10
Savannah sparrow (Passerculus sandwichensis)	1	–	–	–	42359	554	1.3	1506	3.6	23566	55.6	16734	39.5	2060	4.9	1-10
Rock dove (Columba livia)	–	–	–	–	5287	25	0.5	247	4.7	712	13.5	4302	81.4	273	5.2	1-10
Long-billed curlew (Numenius americanus)	1	1	1	1	11140	251	2.3	354	3.2	5223	46.9	5313	47.7	605	5.4	1-10
Black swift (Cypseloides niger)	–	1	1	1	5	0	0.1	0	5.4	1	27.6	4	66.9	0	5.5	1-10
Cattle egret (Bubulcus ibis)	1	–	–	–	10869	211	1.9	396	3.6	5816	53.5	4447	40.9	607	5.6	1-10
American tree sparrow (Spizella arborea)	–	–	–	–	30554	340	1.1	1367	4.5	19164	62.7	9682	31.7	1707	5.6	1-10
Downy woodpecker (Picoides pubescens)	1	–	–	–	5903	48	0.8	307	5.2	904	15.3	4644	78.7	354	6.0	1-10
House finch (Carpodacus mexicanus)	1	–	–	–	27504	538	2.0	1165	4.2	14637	53.2	11164	40.6	1703	6.2	1-10
Eastern kingbird (Tyrannus tyrannus)	1	–	–	–	185	0	0.1	11	6.1	25	13.8	148	80.0	11	6.2	1-10
White-throated swift (Aeronautes saxatalis)	1	1	–	–	18198	467	2.6	689	3.8	10913	60.0	6129	33.7	1156	6.4	1-10
Pacific loon (Gavia pacifica)	1	–	–	–	17	0	0	1	6.5	6	38.1	9	55.4	1	6.5	1-10
Surf scoter (Melanitta perspicillata)	1	–	–	1	17	0	0	1	6.5	6	38.1	9	55.4	1	6.5	1-10
Gunnison sage-grouse (Centrocercus minimus)	–	1	–	–	4074	13	0.3	263	6.5	1225	30.1	2574	63.2	276	6.8	1-10
Lark bunting (Calamospiza melanocorys)	1	1	–	–	4183	6	0.2	291	7.0	2188	52.3	1697	40.6	298	7.1	1-10
Prairie falcon (Falco mexicanus)	1	1	–	–	66087	1704	2.6	3044	4.6	39513	59.8	21826	33.0	4748	7.2	1-10
Boreal owl (Aegolius funereus)	–	1	–	1	52	0	0.2	4	7.1	22	41.6	27	51.1	4	7.3	1-10
Northern bobwhite (Colinus virginianus)	2	1	–	–	2	0	0	0	7.3	0	2.7	1	90	0	7.3	1-10
Indigo bunting (Passerina cyanea)	1	–	–	–	4328	58	1.3	256	5.9	1045	24.1	2969	68.6	314	7.3	1-10

Table 5 (continued)

Species	SGCN				Total Area	Status 1		Status 2		Status 3		Status 4		Status 1 & 2		Threshold
	AZ	CO	NM	UT		Area	%	Area	%	Area	%	Area	%	Area	%	
Harris's sparrow (*Zonotrichia querula*)	1	1	–	–	7074	73	1.0	443	6.3	2749	38.9	3809	53.8	516	7.3	1-10
Rough-legged hawk (*Buteo lagopus*)	1	–	–	–	55904	896	1.6	3308	5.9	31262	55.9	20438	36.6	4204	7.5	1-10
Horned grebe (*Podiceps auritus*)	1	–	–	–	193	8	4.3	6	3.1	147	76.2	32	16.4	14	7.5	1-10
Lapland longspur (*Calcarius lapponicus*)	1	–	–	–	11712	126	1.1	791	6.8	8839	75.5	1955	16.7	917	7.8	1-10
Common barn-owl (*Tyto alba*)	–	–	–	–	60198	2042	3.4	2717	4.5	34562	57.4	20877	34.7	4759	7.9	1-10
Northern rough-winged swallow (*Stelgidopteryx serripennis*)	1	–	–	–	5860	162	2.8	342	5.8	1174	20.0	4182	71.4	504	8.6	1-10
Canyon towhee (*Pipilo fuscus*)	–	–	–	–	46810	2068	4.4	2022	4.3	25067	53.6	17654	37.7	4089	8.7	1-10
Bank swallow (*Riparia riparia*)	1	–	1	–	5837	169	2.9	337	5.8	1167	20.0	4164	71.3	506	8.7	1-10
Pectoral sandpiper (*Calidris melanotos*)	1	–	–	–	867	1	0.2	75	8.7	322	37.2	468	54.0	77	8.8	1-10
Yellow-headed blackbird (*Xanthocephalus xanthocephalus*)	1	–	–	–	4909	140	2.8	294	6.0	1357	27.6	3119	63.5	434	8.8	1-10
Red-faced warbler (*Cardellina rubrifrons*)	1	–	1	–	686	61	8.9	0	0.0	403	58.8	222	32.3	61	8.9	1-10
American crow (*Corvus brachyrhynchos*)	1	–	–	–	8494	376	4.4	378	4.4	2581	30.4	5159	60.7	754	8.9	1-10
Snowy owl (*Nyctea scandiaca*)	–	–	–	–	756	3	0.3	65	8.6	206	27.3	483	63.8	67	8.9	1-10
American goldfinch (*Carduelis tristis*)	–	–	–	–	1328	38	2.9	82	6.2	303	22.8	905	68.2	120	9.0	1-10
Northern harrier (*Circus cyaneus*)	1	1	–	–	36664	657	1.8	2645	7.2	20365	55.5	12997	35.4	3302	9.0	1-10
Canada goose (*Branta canadensis*)	1	–	–	–	6625	229	3.5	408	6.2	1466	22.1	4523	68.3	637	9.6	1-10
Vesper sparrow (*Pooecetes gramineus*)	1	1	–	–	63259	2202	3.5	3886	6.1	34122	53.9	23049	36.4	6088	9.6	1-10
Horned lark (*Eremophila alpestris*)	1	–	–	–	74767	2658	3.6	4558	6.1	44478	59.5	23073	30.9	7216	9.7	1-10
Western meadowlark (*Sturnella neglecta*)	1	–	–	–	60932	2055	3.4	3847	6.3	34566	56.7	20464	33.6	5902	9.7	1-10

Table 5 (continued)

Species	SGCN				Total Area	Status 1		Status 2		Status 3		Status 4		Status 1 & 2		Threshold
	AZ	CO	NM	UT		Area	%	Area	%	Area	%	Area	%	Area	%	
Red-winged blackbird (Agelaius phoeniceus)	1	–	–	–	7412	267	3.6	470	6.3	1739	23.5	4936	66.6	737	9.9	1-10
Red-tailed hawk (Buteo jamaicensis)	–	–	–	–	75101	2778	3.7	4688	6.2	43967	58.5	23667	31.5	7467	9.9	1-10
Marbled godwit (Limosa fedoa)	1	1	–	–	6412	229	3.6	410	6.4	1505	23.5	4268	66.6	640	10.0	10-20
Summer tanager (Piranga rubra)	1	–	–	–	489	36	7.3	13	2.6	185	37.8	256	52.2	49	10.0	10-20
Gadwall (Anas strepera)	1	–	–	–	1368	73	5.3	66	4.8	433	31.7	796	58.2	139	10.1	10-20
Lewis's woodpecker (Melanerpes lewis)	1	1	1	1	9091	406	4.5	515	5.7	3298	36.3	4871	53.6	921	10.1	10-20
Stilt sandpiper (Calidris himantopus)	–	–	–	–	489	0	0.0	50	10.3	174	35.5	265	54.2	50	10.3	10-20
Mallard (Anas platyrhynchos)	1	–	–	–	1063	66	6.2	45	4.2	232	21.8	720	67.8	111	10.4	10-20
Common redpoll (Carduelis flammea)	–	–	–	–	10350	81	0.8	998	9.6	5337	51.6	3934	38.0	1079	10.4	10-20
Greater white-fronted goose (Anser albifrons)	1	–	–	–	5864	229	3.9	385	6.6	1415	24.1	3835	65.4	614	10.5	10-20
American coot (Fulica americana)	–	–	–	–	6728	248	3.7	456	6.8	1618	24.1	4405	65.5	705	10.5	10-20
Thayer's gull (Larus thayeri)	–	–	–	–	73	6	8.3	2	2.3	25	33.6	41	55.8	8	10.6	10-20
Baird's sandpiper (Calidris bairdii)	–	–	–	–	608	1	0.2	64	10.5	204	33.6	339	55.7	65	10.7	10-20
Calliope hummingbird (Stellula calliope)	1	–	–	–	968	9	0.9	95	9.8	840	86.8	24	2.4	104	10.7	10-20
Whimbrel (Numenius phaeopus)	–	–	–	–	447	24	5.5	25	5.6	45	10.1	352	78.9	49	11.0	10-20
Abert's towhee (Pipilo aberti)	1	–	1	1	171	13	7.6	6	3.4	39	22.9	113	66.1	19	11.0	10-20
Semipalmated sandpiper (Calidris pusilla)	1	–	–	–	82	0	0	9	11.2	29	35.7	43	53.1	9	11.2	10-20
Blue grosbeak (Guiraca caerulea)	–	–	–	–	25294	1857	7.3	973	3.8	12919	51.1	9546	37.7	2829	11.2	10-20
Sage thrasher (Oreoscoptes montanus)	1	–	–	1	140848	3745	2.7	12862	9.1	89412	63.5	34829	24.7	16607	11.8	10-20

Table 5 (continued)

Species	SGCN				Total Area	Status 1		Status 2		Status 3		Status 4		Status 1 & 2		Threshold
	AZ	CO	NM	UT		Area	%	Area	%	Area	%	Area	%	Area	%	
Bendire's thrasher (Toxostoma bendirei)	1	–	1	1	64443	1918	3.0	5734	8.9	41672	64.7	15119	23.5	7652	11.9	10-20
Lucy's warbler (Vermivora luciae)	1	–	1	1	920	43	4.6	67	7.3	301	32.7	510	55.4	109	11.9	10-20
Le Conte's sparrow (Ammodramus leconteii)	–	–	–	–	97	9	9.6	2	2.4	73	74.9	13	13.1	12	12	10-20
Greater sage-grouse (Centrocercus urophasianus)	–	1	–	1	2984	64	2.2	294	9.8	1844	61.8	782	26.2	358	12.0	10-20
Common snipe (Gallinago gallinago)	–	–	–	–	186	6	3.1	17	9.0	40	21.3	124	66.6	22	12.1	10-20
Red knot (Calidris canutus)	–	–	–	–	174	0	0.2	21	12.0	60	34.8	92	52.9	21	12.2	10-20
Killdeer (Charadrius vociferus)	–	–	–	–	8141	399	4.9	609	7.5	2524	31.0	4610	56.6	1007	12.4	10-20
California quail (Callipepla californica)	–	–	–	–	477	14	2.9	47	9.8	107	22.4	309	64.9	61	12.7	10-20
Burrowing owl (Athene cunicularia)	2	1	1	1	106285	3643	3.4	9916	9.3	66091	62.2	26635	25.1	13559	12.8	10-20
Yellow warbler (Dendroica petechia)	1	–	1	–	451	20	4.5	38	8.4	130	28.9	263	58.2	58	12.9	10-20
Sharp-tailed grouse-Columbian (Tympanuchus phasianellus columbianus)	–	1	–	1	110	0	0.0	14	12.9	65	58.9	31	28.2	14	12.9	10-20
Brewer's blackbird (Euphagus cyanocephalus)	–	–	–	–	42392	2134	5.0	3369	7.9	24147	57.0	12742	30.1	5503	13.0	10-20
White-winged crossbill (Loxia leucoptera)	–	–	–	–	265	3	1.2	31	11.8	163	61.6	67	25.5	34	13.0	10-20
Merlin (Falco columbarius)	1	–	–	–	76361	2603	3.4	7481	9.8	43110	56.5	23168	30.3	10083	13.2	10-20
Northern shrike (Lanius excubitor)	1	–	–	–	43879	1200	2.7	4593	10.5	23086	52.6	15001	34.2	5793	13.2	10-20
Fox sparrow (Passerella iliaca)	1	–	–	–	8282	503	6.1	591	7.1	3628	43.8	3560	43.0	1094	13.2	10-20
Northern saw-whet owl (Aegolius acadicus)	1	–	–	–	10573	712	6.7	709	6.7	4149	39.2	5004	47.3	1421	13.4	10-20
Blue grouse (Dendragapus obscurus)	1	1	1	–	5838	371	6.4	454	7.8	3370	57.7	1643	28.1	825	14.1	10-20

Table 5 (continued)

Species	SGCN				Total Area	Status 1		Status 2		Status 3		Status 4		Status 1 & 2		Threshold
	AZ	CO	NM	UT		Area	%	Area	%	Area	%	Area	%	Area	%	
Least tern (*Sterna antillarum*)	1	1	1	–	214	9	4.1	21	10.0	46	21.7	138	64.2	30	14.1	10-20
Ruddy turnstone (*Arenaria interpres*)	–	–	–	–	69	9	12.8	1	1.5	55	80.7	3	5	10	14.3	10-20
California gull (*Larus californicus*)	1	–	–	–	674	80	11.9	16	2.4	211	31.3	367	54.3	97	14.3	10-20
Sage sparrow (*Amphispiza belli*)	1	1	1	1	115710	3050	2.6	13565	11.7	73162	63.2	25933	22.4	16615	14.4	10-20
Pine siskin (*Carduelis pinus*)	–	–	–	–	22229	993	4.5	2205	9.9	13179	59.3	5852	26.3	3198	14.4	10-20
Flammulated owl (*Otus flammeolus*)	1	1	–	–	31349	1942	6.2	2571	8.2	18743	59.8	8093	25.8	4513	14.4	10-20
Wood duck (*Aix sponsa*)	1	–	–	–	1883	126	6.7	148	7.9	830	44.1	779	41.4	275	14.6	10-20
Green-winged teal (*Anas crecca*)	1	–	–	–	476	54	11.3	18	3.7	167	35.1	238	49.9	71	15.0	10-20
Barrow's goldeneye (*Bucephala islandica*)	1	1	–	–	329	9	2.7	40	12.3	83	25.4	196	59.7	49	15.0	10-20
Common raven (*Corvus corax*)	1	–	–	–	167706	6703	4.0	18614	11.1	100710	60.1	41679	24.9	25317	15.1	10-20
Dark-eyed junco (*Junco hyemalis*)	1	–	–	–	116348	4102	3.5	13641	11.7	69336	59.6	29269	25.2	17743	15.3	10-20
Rose-breasted grosbeak (*Pheucticus ludovicianus*)	1	–	–	–	579	20	3.4	69	11.9	234	40.4	256	44.3	89	15.3	10-20
American wigeon (*Anas americana*)	1	–	–	–	436	54	12.3	13	3.1	138	31.6	231	53.0	67	15.4	10-20
Northern waterthrush (*Seiurus noveboracensis*)	1	–	–	–	1647	69	4.2	185	11.2	755	45.9	637	38.7	254	15.4	10-20
Long-eared owl (*Asio otus*)	1	–	–	–	143458	5999	4.2	16185	11.3	83032	57.9	38242	26.7	22184	15.5	10-20
Swainson's hawk (*Buteo swainsoni*)	1	1	–	–	106427	4642	4.4	11850	11.1	62018	58.3	27917	26.2	16492	15.5	10-20
Lazuli bunting (*Passerina amoena*)	1	1	–	–	3661	196	5.4	377	10.3	1205	32.9	1883	51.4	573	15.6	10-20
Blue-winged teal (*Anas discors*)	1	–	–	–	425	54	12.7	13	3.1	140	32.9	218	51.3	67	15.7	10-20
Sandhill crane (*Grus canadensis*)	1	1	–	1	31573	1535	4.9	3417	10.8	16237	51.4	10384	32.9	4952	15.7	10-20

Table 5 (continued)

Species	SGCN AZ	CO	NM	UT	Total Area	Status 1 Area	%	Status 2 Area	%	Status 3 Area	%	Status 4 Area	%	Status 1 & 2 Area	%	Threshold
Northern pintail (*Anas acuta*)	1	1	1	–	421	53	12.7	13	3.1	137	32.6	217	51.6	67	15.8	10-20
Yellow-billed loon (*Gavia adamsii*)	–	–	–	–	396	59	14.9	4	0.9	297	74.9	37	9.3	63	15.8	10-20
Sanderling (*Calidris alba*)	1	–	–	–	766	39	5.1	85	11.2	403	52.7	238	31.1	124	16.2	10-20
Ross's goose (*Chen rossii*)	1	–	–	–	2398	207	8.6	184	7.7	852	35.5	1155	48.2	391	16.3	10-20
Little blue heron (*Egretta caerulea*)	1	–	–	–	210	24	11.6	10	4.8	123	58.5	53	25.1	34	16.4	10-20
Glaucous gull (*Larus hyperboreus*)	–	–	–	–	249	26	10.4	15	6.0	122	48.8	86	34.7	41	16.5	10-20
Great horned owl (*Bubo virginianus*)	1	–	–	–	195622	8849	4.5	24202	12.4	118330	60.5	44241	22.6	33051	16.9	10-20
Belted kingfisher (*Ceryle alcyon*)	1	–	–	–	567	28	5.0	68	12.0	223	39.4	247	43.6	96	16.9	10-20
Loggerhead shrike (*Lanius ludovicianus*)	1	1	1	–	179034	7558	4.2	22816	12.7	107390	60.0	41270	23.1	30374	17.0	10-20
Brown-headed cowbird (*Molothrus ater*)	–	–	1	–	134998	5266	3.9	17699	13.1	78059	57.8	33974	25.2	22965	17.0	10-20
Mourning dove (*Zenaida macroura*)	1	–	1	–	186824	8369	4.5	23449	12.6	112390	60.2	42616	22.8	31818	17.0	10-20
Williamson's sapsucker (*Sphyrapicus thyroideus*)	–	1	–	1	5138	377	7.3	499	9.7	3255	63.4	1006	19.6	876	17.1	10-20
American kestrel (*Falco sparverius*)	1	–	–	–	164097	7365	4.5	20995	12.8	94993	57.9	40744	24.8	28360	17.3	10-20
Western sandpiper (*Calidris mauri*)	1	–	–	–	27307	1500	5.5	3248	11.9	15778	57.8	6781	24.8	4748	17.4	10-20
Golden eagle (*Aquila chrysaetos*)	1	1	–	–	180979	8541	4.7	23215	12.8	109480	60.5	39743	22.0	31756	17.5	10-20
Hepatic tanager (*Piranga flava*)	1	–	–	–	27082	2247	8.3	2549	9.4	15695	58.0	6591	24.3	4796	17.7	10-20
Lesser scaup (*Aythya affinis*)	1	1	–	–	404	59	14.6	13	3.3	239	59.2	93	23.0	72	17.8	10-20
Black-and-white warbler (*Mniotilta varia*)	1	–	–	–	8556	749	8.8	774	9.0	4738	55.4	2295	26.8	1523	17.8	10-20
Chipping sparrow (*Spizella passerina*)	1	–	–	–	110035	5033	4.6	14560	13.2	61168	55.6	29274	26.6	19593	17.8	10-20

Table 5 (continued)

	SGCN				Total Area	Status 1		Status 2		Status 3		Status 4		Status 1 & 2		Threshold
	AZ	CO	NM	UT		Area	%	Area	%	Area	%	Area	%	Area	%	
Orange-crowned warbler (*Vermivora celata*)	1	–	–	–	105217	3929	3.7	14814	14.1	63145	60.0	23329	22.2	18743	17.8	10-20
Bohemian waxwing (*Bombycilla garrulus*)	–	–	–	–	12385	184	1.5	2034	16.4	4337	35.0	5830	47.1	2218	17.9	10-20
Gray flycatcher (*Empidonax wrightii*)	1	1	–	1	99798	3879	3.9	13957	14.0	57647	57.8	24315	24.4	17835	17.9	10-20
Golden-crowned kinglet (*Regulus satrapa*)	1	–	–	–	51756	2397	4.6	6850	13.2	31352	60.6	11157	21.6	9247	17.9	10-20
Northern mockingbird (*Mimus polyglottos*)	1	–	–	–	173057	8311	4.8	22809	13.2	103730	59.9	38207	22.1	31120	18.0	10-20
Zone-tailed hawk (*Buteo albonotatus*)	1	–	–	–	379	65	17.2	4	0.9	47	12.4	263	69.4	69	18.1	10-20
White-tailed ptarmigan (*Lagopus leucurus*)	3	1	1	–	10	2	18.1	0	0	6	62	2	19.8	2	18.1	10-20
Black-billed magpie (*Pica hudsonia*)	1	–	–	–	20239	719	3.6	2966	14.7	8760	43.3	7793	38.5	3685	18.2	10-20
Rock wren (*Salpinctes obsoletus*)	1	–	–	–	159034	7863	4.9	21125	13.3	95275	59.9	34771	21.9	28988	18.2	10-20
Wild turkey (*Meleagris gallopavo*)	2	–	2	–	136034	5616	4.1	19294	14.2	79293	58.3	31831	23.4	24910	18.3	10-20
Gray-crowned rosy-finch (*Leucosticte tephrocotis*)	–	–	–	–	80	0	0.3	15	18.2	55	69.3	10	12.2	15	18.5	10-20
American dipper (*Cinclus mexicanus*)	1	–	–	–	101	4	4.0	15	14.7	24	23.6	58	57.6	19	18.7	10-20
Say's phoebe (*Sayornis saya*)	1	–	–	–	24276	1868	7.7	2660	11.0	11934	49.2	7814	32.2	4528	18.7	10-20
Common nighthawk (*Chordeiles minor*)	1	–	–	–	150333	7116	4.7	21250	14.1	86904	57.8	35064	23.3	28365	18.9	10-20
Yellow-breasted chat (*Icteria virens*)	1	–	–	–	2475	239	9.7	227	9.2	1112	44.9	896	36.2	467	18.9	10-20
Cassin's vireo (*Vireo cassinii*)	1	–	–	–	675	28	4.1	100	14.8	272	40.2	276	40.8	128	18.9	10-20
Black-crowned night-heron (*Nycticorax nycticorax*)	1	–	–	–	155	8	5.1	22	13.9	50	31.9	76	49.1	30	19.0	10-20
Swainson's thrush (*Catharus ustulatus*)	1	–	–	–	270	1	0.5	50	18.7	78	29.0	140	51.8	52	19.2	10-20
Hammond's flycatcher (*Empidonax hammondii*)	1	–	–	–	3222	270	8.4	349	10.8	2005	62.2	597	18.5	620	19.2	10-20

Table 5 (continued)

Species	SGCN AZ	SGCN CO	SGCN NM	SGCN UT	Total Area	Status 1 Area	Status 1 %	Status 2 Area	Status 2 %	Status 3 Area	Status 3 %	Status 4 Area	Status 4 %	Status 1 & 2 Area	Status 1 & 2 %	Threshold
Western kingbird (*Tyrannus verticalis*)	1	–	–	–	134389	6881	5.1	18878	14.0	80181	59.7	28449	21.2	25759	19.2	10-20
Ferruginous hawk (*Buteo regalis*)	1	1	1	1	42624	2019	4.7	6187	14.5	25138	59.0	9279	21.8	8206	19.3	10-20
Bald eagle (*Haliaeetus leucocephalus*)	1	1	1	1	109147	5777	5.3	15335	14.0	61527	56.4	26508	24.3	21112	19.3	10-20
House wren (*Troglodytes aedon*)	1	–	–	–	107645	5092	4.7	15631	14.5	62290	57.9	24632	22.9	20723	19.3	10-20
Cinnamon teal (*Anas cyanoptera*)	1	–	–	–	319	51	15.9	12	3.8	124	38.7	133	41.5	63	19.8	10-20
Common black-hawk (*Buteogallus anthracinus*)	1	–	1	–	10	2	15.4	0	4.3	1	11.8	7	68.4	2	19.8	10-20
Yellow-rumped warbler (*Dendroica coronata*)	1	–	–	–	126836	7422	5.9	17652	13.9	72678	57.3	29083	22.9	25074	19.8	10-20
Bushtit (*Psaltriparus minimus*)	–	–	–	–	122433	5880	4.8	18464	15.1	71129	58.1	26960	22.0	24344	19.9	10-20
MAMMALS																
Gray wolf (*Canis lupus*)	2	1	2	1	1178	0	0.0	0	0.0	230	19.5	948	80.5	0	0.0	<1
Hispid pocket mouse (*Chaetodipus hispidus*)	1	–	–	–	49	0	0	0	0	31	63.7	18	36.3	0	0	<1
Snowshoe hare (*Lepus americanus*)	–	–	2	–	15	0	0	0	0	9	61.2	6	38.8	0	0	<1
Lynx (*Lynx canadensis*)	–	1	–	1	28	0	0	0	0	15	52.7	13	47.3	0	0	<1
Hooded skunk (*Mephitis macroura*)	1	–	–	–	801	0	0.0	0	0.0	120	15.0	681	85.0	0	0.0	<1
Southern plains woodrat (*Neotoma micropus*)	–	–	–	–	2773	0	0.0	0	0.0	1046	37.7	1726	62.3	0	0.0	<1
Pocketed free-tailed bat (*Nyctinomops femorosaccus*)	–	–	1	–	25	0	0	0	0	0	1	24	99	0	0	<1
Mearns' grasshopper mouse (*Onychomys arenicola*)	–	–	–	–	59	0	0	0	0	41	69.8	18	30.2	0	0	<1
Collared peccary (*Pecari tajacu*)	–	–	–	–	151	0	0.0	0	0.0	1	0.9	150	99.1	0	0.0	<1
Osgood's mouse (*Peromyscus gratus*)	–	–	–	–	320	0	0.0	0	0.0	85	26.5	235	73.5	0	0.0	<1

Table 5 (continued)

Species	SGCN AZ	CO	NM	UT	Total Area	Status 1 Area	%	Status 2 Area	%	Status 3 Area	%	Status 4 Area	%	Status 1 & 2 Area	%	Threshold
Plains harvest mouse (*Reithrodontomys montanus*)	–	–	–	–	0	0	0	0	0	0	100	0	0	0	0	<1
Hispid cotton rat (*Sigmodon hispidus*)	2	–	–	–	59	0	0	0	0	41	69.8	18	30.2	0	0	<1
Preble's shrew (*Sorex preblei*)	–	1	1	1	6	0	0	0	0	5	88.2	1	11.8	0	0	<1
Southwestern myotis (*Myotis auriculus*)	–	–	–	–	550	0	0.0	3	0.5	239	43.5	308	56.0	3	0.5	<1
Cave myotis (*Myotis velifer*)	–	–	–	–	308	0	0.0	2	0.5	23	7.5	283	92.0	2	0.5	<1
Arizona gray squirrel (*Sciurus arizonensis*)	1	–	2	–	1	0	0	0	1	0	9.1	1	90	0	1	1-10
Wolverine (*Gulo gulo*)	–	1	–	1	297	0	0.1	4	1.2	166	55.8	128	42.9	4	1.3	1-10
White-footed mouse (*Peromyscus leucopus*)	–	–	–	–	40228	538	1.3	146	0.4	22875	56.9	16669	41.4	684	1.7	1-10
Hog-nosed skunk (*Conepatus mesoleucus*)	–	1	–	–	65	0	0	1	2.1	4	5.5	60	92.4	1	2.1	1-10
White-tailed deer (*Odocoileus virginianus*)	2	–	2	–	21199	401	1.9	46	0.2	6931	32.7	13821	65.2	447	2.1	1-10
Meadow vole (*Microtus pennsylvanicus*)	–	–	–	–	252	0	0.0	8	3.0	94	37.1	151	59.8	8	3.0	1-10
Black-footed ferret (*Mustela nigripes*)	1	1	–	1	43519	386	0.9	1458	3.3	30126	69.2	11549	26.5	1843	4.2	1-10
Barbary sheep (*Ammotragus lervia*)	–	–	–	–	8228	16	0.2	383	4.7	7056	85.8	773	9.4	399	4.8	1-10
Desert kangaroo rat (*Dipodomys deserti*)	1	–	–	1	0	0	0	0	4.8	0	60.6	0	34.6	0	4.8	1-10
Moose (*Alces alces*)	–	–	–	–	404	1	0.2	24	6.0	98	24.3	281	69.5	25	6.2	1-10
Silky pocket mouse (*Perognathus flavus*)	2	–	–	1	88342	756	0.9	4826	5.5	56106	63.5	26655	30.2	5581	6.3	1-10
Spotted ground squirrel (*Spermophilus spilosoma*)	1	–	–	1	99128	1696	1.7	4616	4.7	66085	66.7	26731	27.0	6312	6.4	1-10
Thirteen-lined ground squirrel (*Spermophilus tridecemlineatus*)	1	–	–	1	6130	21	0.3	389	6.3	3472	56.6	2248	36.7	410	6.7	1-10
Southern red-backed vole (*Clethrionomys gapperi*)	1	–	–	–	685	17	2.5	38	5.6	263	38.4	367	53.5	55	8.1	1-10

Table 5 (continued)

Species	SGCN				Total Area	Status 1		Status 2		Status 3		Status 4		Status 1 & 2		Threshold
	AZ	CO	NM	UT		Area	%	Area	%	Area	%	Area	%	Area	%	
Gunnison's prairie dog (*Cynomys gunnisoni*)	1	1	1	1	80379	2526	3.1	4205	5.2	50710	63.1	22938	28.5	6731	8.4	1-10
Botta's pocket gopher (*Thomomys bottae*)	2	1	–	–	85290	1368	1.6	6101	7.2	53291	62.5	24530	28.8	7469	8.8	1-10
Muskrat (*Ondatra zibethicus*)	1	–	–	–	6988	189	2.7	463	6.6	1648	23.6	4688	67.1	652	9.3	1-10
Beaver (*Castor canadensis*)	1	–	1	–	9739	227	2.3	698	7.2	3187	32.7	5628	57.8	925	9.5	1-10
Long-tailed weasel (*Mustela frenata*)	–	–	–	–	59180	1445	2.4	4157	7.0	35955	60.8	17623	29.8	5602	9.5	1-10
House mouse (*Mus musculus*)	–	–	–	–	89815	2227	2.5	6489	7.2	55387	61.7	25713	28.6	8715	9.7	1-10
Merriam's kangaroo rat (*Dipodomys merriami*)	1	–	–	–	5239	207	3.9	307	5.9	3526	67.3	1199	22.9	514	9.8	1-10
Banner-tailed kangaroo rat (*Dipodomys spectabilis*)	1	–	–	–	27935	304	1.1	2431	8.7	20815	74.5	4385	15.7	2735	9.8	1-10
Heather vole (*Phenacomys intermedius*)	–	–	–	–	1040	21	2.0	87	8.3	512	49.2	420	40.4	108	10.4	10-20
Stephens' woodrat (*Neotoma stephensi*)	1	–	–	1	29861	883	3.0	2294	7.7	19241	64.4	7443	24.9	3177	10.6	10-20
Virginia opossum (*Didelphis virginiana*)	2	–	–	–	19	0	0.3	2	10.7	15	76	3	13	2	11	10-20
White-tailed prairie dog (*Cynomys leucurus*)	–	1	–	1	12117	525	4.3	848	7.0	8082	66.7	2662	22.0	1373	11.3	10-20
White-tailed antelope squirrel (*Ammospermophilus leucurus*)	2	–	–	–	112009	2447	2.2	10421	9.3	72996	65.2	26146	23.3	12868	11.5	10-20
Masked shrew (*Sorex cinereus*)	–	–	–	–	12075	266	2.2	1133	9.4	7044	58.3	3632	30.1	1399	11.6	10-20
Gray-collared chipmunk (*Tamias cinereicollis*)	1	–	–	–	525	61	11.7	0	0.0	366	69.7	98	18.6	61	11.7	10-20
Rock mouse (*Peromyscus nasutus*)	–	–	–	1	55920	452	0.8	6119	10.9	38178	68.3	11171	20.0	6571	11.8	10-20
Uinta ground squirrel (*Spermophilus armatus*)	–	–	–	–	1878	4	0.2	223	11.9	902	48.0	749	39.9	227	12.1	10-20
Brush mouse (*Peromyscus boylii*)	1	–	–	–	125093	4208	3.4	11104	8.9	80465	64.3	29316	23.4	15312	12.2	10-20
Eastern cottontail (*Sylvilagus floridanus*)	–	–	–	–	31229	2181	7.0	1621	5.2	12791	41.0	14636	46.9	3802	12.2	10-20

Table 5 (continued)

Species	SGCN				Total Area	Status 1		Status 2		Status 3		Status 4		Status 1 & 2		Threshold
	AZ	CO	NM	UT		Area	%	Area	%	Area	%	Area	%	Area	%	
Desert shrew (*Notiosorex crawfordi*)	1	–	–	1	124219	4608	3.7	10654	8.6	76814	61.8	32145	25.9	15261	12.3	10-20
Kit fox (*Vulpes macrotis*)	1	1	–	1	49743	1275	2.6	4889	9.8	32917	66.2	10662	21.4	6164	12.4	10-20
California myotis (*Myotis californicus*)	1	–	–	–	139928	4684	3.3	13157	9.4	87602	62.6	34485	24.6	17841	12.8	10-20
Brazilian free-tailed bat (*Tadarida brasiliensis*)	1	–	–	–	144272	5142	3.6	15005	10.4	87249	60.5	36876	25.6	20147	14.0	10-20
Montane vole (*Microtus montanus*)	2	–	2	1	4338	77	1.8	533	12.3	2564	59.1	1164	26.8	610	14.1	10-20
Mink (*Mustela vison*)	–	–	–	–	7833	322	4.1	796	10.2	3159	40.3	3556	45.4	1119	14.3	10-20
Western jumping mouse (*Zapus princeps*)	–	–	–	–	157	3	2.1	19	12.2	77	49.0	58	36.7	22	14.3	10-20
American pika (*Ochotona princeps*)	–	–	2	1	161	10	6.0	14	8.4	100	61.9	38	23.7	23	14.4	10-20
White-throated woodrat (*Neotoma albigula*)	1	–	–	–	148762	5965	4.0	15784	10.6	90140	60.6	36873	24.8	21749	14.6	10-20
Arizona myotis (*Myotis occultus*)	1	1	1	–	145820	5550	3.8	15925	10.9	86398	59.2	37947	26.0	21475	14.7	10-20
Cliff chipmunk (*Tamias dorsalis*)	–	–	–	–	84385	4273	5.1	8205	9.7	47763	56.6	24144	28.6	12477	14.8	10-20
River otter (*Lontra canadensis*)	–	–	1	1	5183	259	5.0	519	10.0	2125	41.0	2280	44.0	778	15.0	10-20
Western harvest mouse (*Reithrodontomys megalotis*)	1	–	–	–	137277	5060	3.7	15739	11.5	81997	59.7	34481	25.1	20799	15.2	10-20
Desert cottontail (*Sylvilagus audubonii*)	–	–	–	–	139054	4720	3.4	16361	11.8	81520	58.6	36453	26.2	21081	15.2	10-20
Deer mouse (*Peromyscus maniculatus*)	–	–	–	–	164447	6765	4.1	18381	11.2	99272	60.4	40029	24.3	25146	15.3	10-20
Ord's kangaroo rat (*Dipodomys ordii*)	1	–	–	–	106371	2790	2.6	13594	12.8	69093	65.0	20894	19.6	16384	15.4	10-20
Striped skunk (*Mephitis mephitis*)	1	–	–	–	132874	3584	2.7	17295	13.0	85671	64.5	26324	19.8	20879	15.7	10-20
Dwarf shrew (*Sorex nanus*)	1	1	–	1	5126	318	6.2	509	9.9	3138	61.2	1162	22.7	827	16.1	10-20
Badger (*Taxidea taxus*)	1	–	–	–	151295	6467	4.3	17868	11.8	90976	60.1	35984	23.8	24335	16.1	10-20

Table 5 (continued)

| Common name (Scientific name) | SGCN | | | | Total Area | Status 1 | | Status 2 | | Status 3 | | Status 4 | | Status 1 & 2 | | Threshold |
	AZ	CO	NM	UT		Area	%	Area	%	Area	%	Area	%	Area	%	
Mule deer (Odocoileus hemionus)	2	–	1	1	183129	8210	4.5	21708	11.9	110980	60.6	42231	23.1	29918	16.3	10-20
Bison (Bos bison)	1	1	–	–	831	27	3.2	112	13.4	564	67.9	129	15.5	138	16.6	10-20
Yuma myotis (Myotis yumanensis)	–	–	–	1	164528	6879	4.2	20716	12.6	100980	61.4	35952	21.9	27596	16.8	10-20
Mexican woodrat (Neotoma mexicana)	2	–	–	–	71422	2608	3.7	9407	13.2	38251	53.6	21155	29.6	12015	16.8	10-20
Coyote (Canis latrans)	1	–	–	–	195189	8790	4.5	24190	12.4	118060	60.5	44149	22.6	32980	16.9	10-20
Western small-footed myotis (Myotis ciliolabrum)	–	–	–	–	190783	8551	4.5	23727	12.4	115100	60.3	43405	22.8	32278	16.9	10-20
Western pipistrelle (Pipistrellus hesperus)	1	–	–	–	187254	8240	4.4	23415	12.5	113140	60.4	42459	22.7	31655	16.9	10-20
Big brown bat (Eptesicus fuscus)	–	–	–	–	164814	7075	4.3	21012	12.7	98221	59.6	38506	23.4	28087	17.0	10-20
Porcupine (Erethizon dorsatum)	1	–	–	–	186757	8616	4.6	23442	12.6	112729	60.4	41969	22.5	32059	17.2	10-20
Townsend's big-eared bat (Corynorhinus townsendii)	–	1	–	1	172907	7324	4.2	22640	13.1	102203	59.1	40741	23.6	29963	17.3	10-20
Long-legged myotis (Myotis volans)	–	–	–	–	145397	6363	4.4	18775	12.9	86846	59.7	33413	23.0	25138	17.3	10-20
Big free-tailed bat (Nyctinomops macrotis)	1	–	–	1	184628	8588	4.7	23298	12.6	111300	60.3	41442	22.4	31886	17.3	10-20
Pallid bat (Antrozous pallidus)	1	–	–	–	165220	6654	4.0	22029	13.3	97312	58.9	39225	23.7	28683	17.4	10-20
Fringed myotis (Myotis thysanodes)	–	1	–	1	185144	8656	4.7	23623	12.8	110126	59.5	42740	23.1	32278	17.4	10-20
Brown bear (Ursus arctos)	–	1	–	1	136727	6354	4.6	17401	12.7	85042	62.2	27930	20.4	23755	17.4	10-20
Black-tailed jack rabbit (Lepus californicus)	–	–	–	–	176039	8273	4.7	22780	12.9	104020	59.1	40966	23.3	31053	17.6	10-20
Mountain lion (Puma concolor)	1	–	–	–	179184	8456	4.7	23106	12.9	109640	61.2	37982	21.2	31562	17.6	10-20
Gray fox (Urocyon cinereoargenteus)	1	–	–	–	158431	6954	4.4	21147	13.3	97108	61.3	33222	21.0	28101	17.7	10-20
White-tailed jack rabbit (Lepus townsendii)	–	1	2	–	9263	411	4.4	1239	13.4	5697	61.5	1916	20.7	1650	17.8	10-20

Table 5 (continued)

Species	SGCN				Total Area	Status 1		Status 2		Status 3		Status 4		Status 1 & 2		Threshold
	AZ	CO	NM	UT		Area	%	Area	%	Area	%	Area	%	Area	%	
Pinyon mouse (*Peromyscus truei*)	–	–	–	–	127437	4950	3.9	17681	13.9	76475	60.0	28331	22.2	22631	17.8	10-20
Mountain cottontail (*Sylvilagus nuttallii*)	2	–	–	–	106867	3933	3.7	15097	14.1	64208	60.1	23629	22.1	19030	17.8	10-20
Montane shrew (*Sorex monticolus*)	–	–	–	–	8597	479	5.6	1061	12.3	4894	56.9	2163	25.2	1540	17.9	10-20
Wyoming ground squirrel (*Spermophilus elegans*)	–	–	–	1	704	6	0.8	121	17.2	362	51.4	215	30.6	127	18.0	10-20
Long-tailed vole (*Microtus longicaudus*)	1	–	–	–	14413	548	3.8	2054	14.3	7908	54.9	3903	27.1	2602	18.1	10-20
Wapiti (*Cervus elaphus*)	2	–	–	–	69561	3284	4.7	9436	13.6	41204	59.2	15637	22.5	12720	18.3	10-20
Hoary bat (*Lasiurus cinereus*)	–	–	–	–	147752	6869	4.6	20210	13.7	88538	59.9	32135	21.7	27079	18.3	10-20
Rock squirrel (*Spermophilus variegatus*)	–	–	–	–	99000	3719	3.8	14766	14.9	52622	53.2	27893	28.2	18485	18.7	10-20
Pygmy rabbit (*Brachylagus idahoensis*)	–	–	–	1	135	3	2.0	23	16.8	67	49.6	43	31.6	25	18.8	10-20
Bobcat (*Lynx rufus*)	1	–	–	–	163420	8304	5.1	22459	13.7	97639	59.7	35018	21.4	30763	18.8	10-20
Ermine (*Mustela erminea*)	–	–	–	–	3767	287	7.6	428	11.4	2268	60.2	785	20.8	715	19.0	10-20
Silver-haired bat (*Lasionycteris noctivagans*)	–	–	–	–	151401	7107	4.7	21878	14.5	90741	59.9	31676	20.9	28984	19.1	10-20
Arizona pocket mouse (*Perognathus amplus*)	1	–	–	–	2729	290	10.6	246	9.0	1515	55.5	678	24.9	536	19.6	10-20
Red squirrel (*Tamiasciurus hudsonicus*)	1	–	–	–	5249	582	11.1	454	8.6	3387	64.5	826	15.7	1036	19.7	10-20

LITERATURE CITED

Arizona Game and Fish Department. 2005a. Arizona's Comprehensive Wildlife Conservation Strategy: 2005–2015. Phoenix.

Arizona Game and Fish Department. 2005b. Arizona's Comprehensive Wildlife Conservation Strategy: State of the State (Companion Document B). Phoenix.

Boykin, K. G., B. C. Thompson, R. A. Deitner, D. Schrupp, D. Bradford, L. O'Brien, C. Drost, S. Propeck-Gray, W. Rieth, K. Thomas, W. Kepner, J. Lowry, C. Cross, B. Jones, T. Hamer, C. Mettenbrink, K. J. Oakes, J. Prior-Magee, K. Schulz, J. J. Wynne, C. King, J. Puttere, S. Schrader, and Z. Schwenke. 2006. Predicted Animal Habitat Distributions and Species Richness. In Southwest Regional Gap Analysis Final Report, edited by J. S. Prior-Magee. USGS Gap Analysis Program, Moscow, Idaho. Available at http://fws-nmcfwru.nmsu.edu/swregap/.

Colorado Division of Wildlife. 2005. Colorado's Comprehensive Wildlife Conservation Strategy. Denver.

Edwards, T. C., Jr., C. H. Homer, S. D. Bassett, A. Falconer, R. D. Ramsey, and D. W. Wright. 1995. Utah Gap Analysis: An Environmental Information System. Technical Report 95-1, Utah Cooperative Fish and Wildlife Research Unit, Utah State University, Logan.

Halvorson, W. L., K. Thomas, L. Graham, M. R. Kunzmann, P. S. Bennett, C. Van Riper, and C. Drost. 2002. The Arizona GAP Analysis Project Final Report. USGS Biological Resources Division, Western Ecological Research Center, University of Arizona, Tucson.

Manis, G., J. Lowry, and R. D. Ramsey. 2001. Pre-Classification: An Ecologically Predictive Landform Model. GAP Analysis Bulletin 10. USGS Biological Resources Division. Available at http://www.gap.uidaho.edu/Bulletins/10/preclassification.htm.

McClaran, M. P., and T. R. Van Devender. 1995. The Desert Grassland. University of Arizona Press, Tucson.

New Mexico Department of Game and Fish. 2005. Comprehensive Wildlife Conservation Strategy for New Mexico. Santa Fe.

Scott, J. M., F. Davis, B. Csuti, R. Noss, B. Butterfield, C. Groves, H. Anderson, S. Caicco, F. D'Erchia, T. C. Edwards Jr., J. Ulliman, and R. G. Wright. 1993. Gap analysis: A geographical approach to protection of biological diversity. Wildlife Monographs 123.

Schrupp, D. L., W. A. Reiners, T. G. Thompson, L. E. O'Brien, J. A. Kindler, M. B. Wunder, J. F. Lowsky, J. C. Buoy, L. Satcowitz, A. L. Cade, J. D. Stark, K. L. Driese, T. W. Owens, S. J. Russo, and F. D'Erchia. 2001. Colorado Gap Analysis Program: A Geographic Approach to Planning for Biological Diversity. Final Report. USGS/BRD Gap Analysis Program and Colorado Division of Wildlife, Denver.

Thompson, B. C., P. J. Crist, J. S. Prior-Magee, R. A. Deitner, D. L. Garber, and M. A. Hughes. 1996. Gap Analysis of Biological Diversity Conservation in New Mexico Using Geographic Information Systems. Research Completion Report. New Mexico Cooperative Fish and Wildlife Research Unit, Las Cruces.

Tuhy, J., P. Comer, D. Dorfman, M. Lammert, B. Neely, L. Whitham, S. Silbert, G. Bell, J. Humke, B. Baker, and B. Cholvin. 2002. An Ecoregional Assessment of the Colorado Plateau. The Nature Conservancy, Moab Project Office.

Utah Division of Wildlife Resources. 2005. Utah's Comprehensive Wildlife Conservation Strategy (Draft). Salt Lake City.

ADDRESSING WILDLIFE ISSUES

A HISTORICAL ASSESSMENT OF PRONGHORN MANAGEMENT ON ANDERSON MESA: MISCALCULATIONS AND REMEDIES

David E. Brown

No area in Arizona is more associated with pronghorn than Anderson Mesa in Game Management Unit 5. More than a fourth of all of the pronghorn in the "Millennium" edition of the Arizona Wildlife Trophy Book came from this unit or from areas restocked with animals from Anderson Mesa (Lewis 2000). Three of the top five pronghorn trophies in the Boone and Crockett Club's North American Record Book are from Coconino County, where Anderson Mesa is located (Byers and Bettas 1999). Anderson Mesa was the site of Arizona's first legal pronghorn hunt after statehood, and this stony plateau has been a focal point for pronghorn studies since the early 1930s when the Coconino National Forest estimated the population to be between 4000 and 5000 head (Nichol 1931; Knipe 1944a; Edwards 1950). Pronghorn studies on Anderson Mesa have ranged from developing survey and capture methodologies (Wilkins and Welles 1944; Edwards 1947; Wallmo 1951), determining seasonal food habits (Wallmo 1951; Gay 1984), and evaluating reproductive performance (Erling 1956a; 1956b), to evaluating the effects of coyote control and other factors on fawn recruitment (Arrington 1947; Arrington and Edwards 1951; Neff and Woolsey 1979, 1980; Neff et al. 1985).

Declines in pronghorn recruitment rates and population size in the 1970s resulted in an intensive study to determine if aerial gunning of coyotes could improve the pronghorn population on Anderson Mesa. Although aerial gunning was expensive and politically unpopular, these studies indicated that pronghorn fawn recruitment could be improved by applying such control practices (Neff and Woolsey 1979, 1980; Neff et al. 1985). When pronghorn recruitment and population numbers again declined in the 1990s, however, coyote reduction efforts were no longer deemed an effective solution, initiating a demand to determine the reasons why pronghorn recruitment on Anderson Mesa was chronically below herd maintenance levels (Yoakum 2003).

STUDY AREA

Anderson Mesa was named after Jim Anderson, an early settler (Barnes 1988). This 1950–2195 m (6400–7200 ft) high plateau south and east of Flagstaff encompasses about 1036 sq km (400 sq mi) of volcanic basalt, a substrate commonly referred to as malpai or "bad lands" (Figure 1). Annual precipitation averages 457 mm (18 in) and the primary vegetation is a combination of intermountain prairie and plains grassland interspersed with stands of ponderosa pine (*Pinus ponderosa*) forest and pinyon-juniper (*Pinus edulis–Juniperus* spp.) woodland (Brown 1994). Perennial streams are lacking, and natural water sources are limited to a few springs and a number of ephemeral lakes. In addition to pronghorn, Anderson Mesa has been an important area for mule deer (*Odocoileus hemionus*) and wild turkey (*Meleagris gallopavo*), and beginning in the 1970s, elk.

Anderson Mesa's pronghorn are only part-time residents. All of the pronghorn

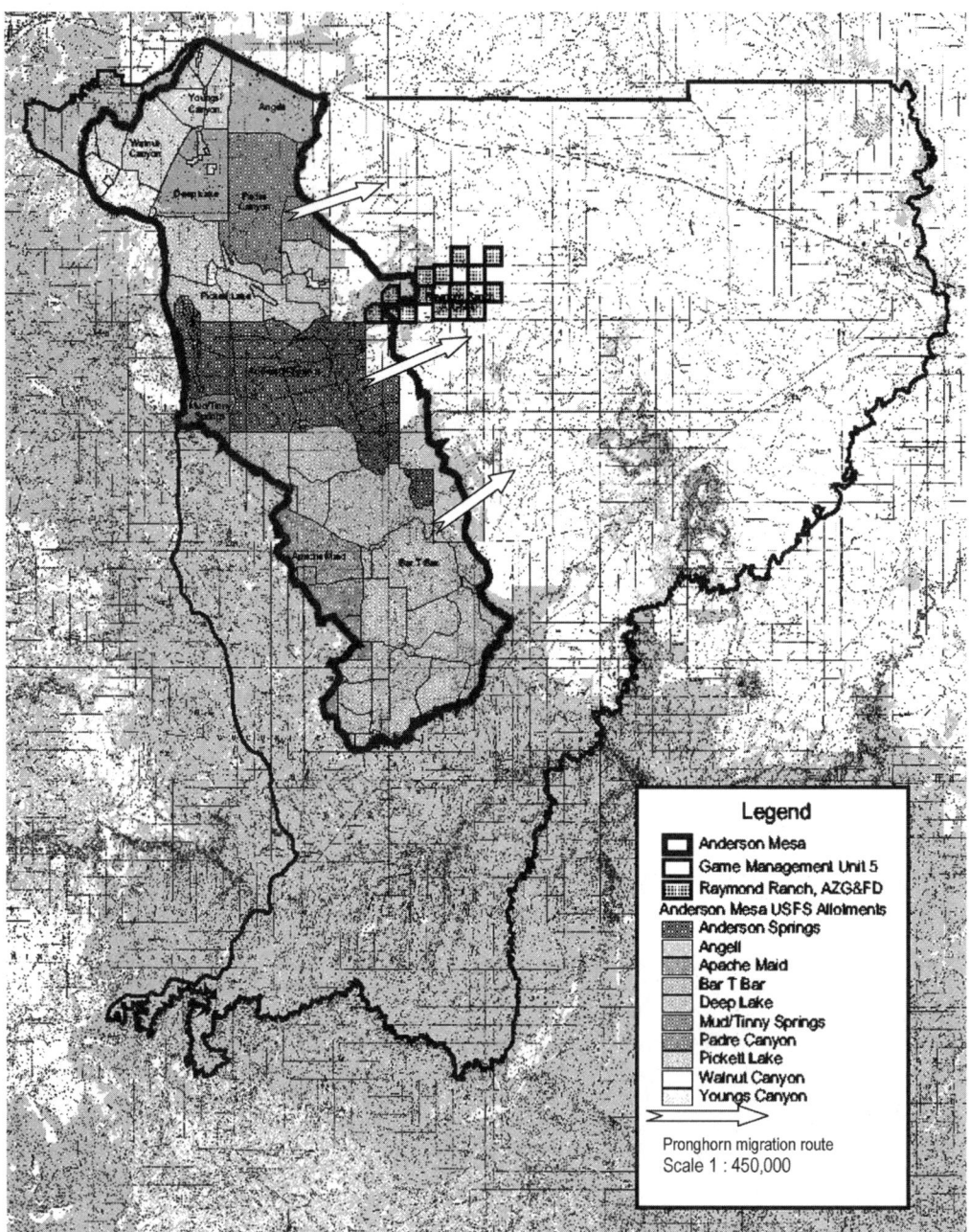

Figure 1. Map of Anderson Mesa and Game Management Unit 5 showing historic pronghorn migration routes.

leave the mesa in some winters, and some pronghorn leave the mesa during most winters. Hemmed in by pine forests on the north, west, and south, the pronghorn drift eastward in November or December to drop off of the plateau's rim in search of a more moderate climate (Figure 1). Following migratory routes down Grapevine, Anderson, Padre, and other canyons, the pronghorn pass through a belt of junipers (*Juniperus* spp.) to forage on winter fat (*Eurotia lanata*), sages (*Artemisia* spp.), and saltbush (*Atriplex canescens*), which extend northward toward the Little Colorado River. Here, the migratory pronghorn mix with other pronghorn, at least some of which appear to be permanent residents of the grassy lowlands.

All is not well with Anderson Mesa and its pronghorn. Fewer than 300 animals were surveyed in Game Management Unit 5 in 2002, and fewer than 10 males were harvested during the autumn hunt. The reason given for these low numbers is a chronic loss of fawns (Miller and Drake 2005). During the 1990s, fewer than 15 does in 100 raised a fawn to 6 months of age. Dissatisfied with this poor reproductive performance, and fearful that Anderson Mesa's pronghorn were headed for oblivion, the Arizona Wildlife Federation sued the Coconino National Forest over its management of Anderson Mesa's resources. Government agency personnel were equally concerned, and the Flagstaff regional office of the Arizona Game and Fish Department (AGFD) formed a coalition of biologists, U.S. Forest Service (USFS) resource managers, area ranchers, and sportsmen/conservationists to develop a cooperative management program to increase pronghorn numbers (Arizona Game and Fish Department and Coconino National Forest 2004).

HISTORICAL PERSPECTIVE

With the abandonment of Walnut Canyon and other Anasazi culture sites in the fourteenth century, Anderson Mesa was without human residents (Reid and Doyel 1986). The first non-Indian to visit Anderson Mesa was most likely a Hispanic sheepherder from the Rio Grande Valley, or an itinerant "mountain man" passing through in the 1840s. The first written descriptions of the country around Anderson Mesa were by Americans intent on surveying their newly acquired territories after the Mexican War. Writers included members of the 1851 Sitgreaves military reconnaissance (Davis 2001), personnel with the Whipple transportation route survey of 1853–1854 (Davis 2001), and the great Mormon pathfinder Jacob Hamblin (1881). On arriving in the vicinity of the San Francisco Peaks in December 1853, Whipple (1941:65) and his party found large herds of pronghorn "on a sweep of lava plains covered with an excellent growth of grama grass," and Whipple reported following the trail of at last "one hundred" pronghorn in the same area. Whipple's (1941:67) report went on to describe "forked-horned antelope" as being "everywhere," but there is no mention of Indians or elk—only antelope, deer, hares, and turkeys, all of which he described as abundant (Whipple 1941).

Four years later, Lieutenant Edward Fitzgerald Beale and his camel caravan pioneered a wagon route from Fort Defiance westward to the Colorado River. Writing of the country north of Anderson Mesa in the summer of 1857, Beale (1858:50, in Davis 2001) described the landscape glowingly:

> We traveled rapidly over a lovely country of open forest and mountain valley, which continually drew exclamations of delight and surprise from every member of the party. Even the stoicism and indifference to beauty of scenery so characteristic of the lower class of Spanish population was moved and as we passed successive vales and glades, filled with verdant grass knee high to our mules, dotted with flowers, and edges skirted by gigantic pines, they constantly gave vent to their delight in fervent ejaculations.

Anderson Mesa's verdure encouraged settlement, and an attempt by Jacob Hamblin to colonize the country east of the San Francisco Peaks was thwarted only by having to deal with the troubles that Mormon colonists were having with Navajos and Hopis at Moenkopi (Hamblin 1881). Other pioneers recognized the livestock-raising potential of Anderson Mesa, but because of Indian troubles and a lack of markets, stock raising would remain transitory until after the Civil War. It was not until the summer of 1876 that the Daggs brothers herded a large flock of sheep across Anderson Mesa preparatory to setting up headquarters at Chavez Pass, where they eventually pastured 50,000 sheep. Later that same summer, William Ashurst located a sheep ranch on Anderson Mesa proper, and other sheep ranches were established near what is now Upper Lake Mary. By this time, Mormon colonists along the Little Colorado River were using Grapevine Springs and Anderson Mesa as a summer pasture for sheep wintering around Canyon Diablo (Neff 1974a, 1974b).

Cattlemen were not far behind, and bovines were grazing on Anderson Mesa as early as 1877 (Neff 1974a). Thereafter, settlers steadily moved on to the mesa, building cabins and corrals near the few available springs, and running livestock on the open range. Animal numbers were not especially large, however, until after 1881 when the Atlantic and Pacific Railroad connected Canyon Diablo with Albuquerque and the East. With a market for meat and wool, livestock numbers multiplied, prompting an editorial in the Flagstaff newspaper (the *Arizona Champion*, 31 October 1885): "Many of our ranches which a few years ago had ample water and pasturage for the number of cattle maintained are now said to be overstocked and new pastures are being sought."

But newcomers kept coming, some to settle the land under the Homestead Act of 1862 and others to run livestock on the open range (Neff 1974c). That the homesteader could be hard on pronghorn is attested to in a 19 January 1884 article in the Flagstaff *Weekly Champion* that reported a homesteader named Kelly coming upon seven pronghorn, which had fallen on a sheet of ice near Chavez Pass. Having no firearm, Kelly trussed the helpless animals up with rope along with five more trapped on a small island that he secured with his lariat. Leaving the 12 bound pronghorn on their sides, he went home for an axe, which he used to butcher his captives. Later, on delivering the animals to a Flagstaff market, he showed off the bloody axe to a crowd of incredulous but admiring townsfolk.

With the completion of the railroad between Holbrook and Flagstaff, the government ceded every other section of land within 64 km (40 mi) of the right-of-way to the railroad, which in turn sold it to the Aztec Land and Cattle Company, a stock corporation commonly known for its "Hash Knife" brand. The Hash Knife promptly moved in 40,000 cattle on its patented sections, which were also the winter range for other ranchers as well as Anderson Mesa's pronghorn. There not then being any fences, the Hash Knife livestock mingled with the cattle and sheep already present, resulting in a number of sheep men having to trail their flocks to irrigated lands in the Salt and Gila River valleys as there was now insufficient winter forage to support them (Neff 1974a, 1974b).

Pronghorn, at least on Anderson Mesa proper, were still abundant, however. In the summer of 1885, J. R. W. Hitchcock and a friend embarked from the frontier town of Flagstaff to take in the sights. After hiring a local citizen as a guide and outfitter, the men left to inspect the Indian ruins at Walnut Canyon and to do a little deer hunting. Proceeding a short distance beyond the ruins, Hitchcock (1886:400) provided a first-hand description of Anderson Mesa: "We entered a vast natural park, probably ten miles long and five wide. It was covered with a luxuriant growth of grass nearly knee-high, but seared and yellow, which seemed to afford an excellent grazing ground."

Here, and in the adjacent "parks," the three men had "a succession of adventures with antelopes, the principal if not sole game of the region." After unsuccessfully stalking "a large herd of antelopes feeding in the open ground temptingly near the edge of the forest," the men climbed what may have been Pine Hill to observe the "great forest park, dotted with herds of antelopes and fringed with scattered pines, beyond which the forests stretched away in solid ranks to the southern horizon."[1]

---

[1]Don Neff used Pine Hill as an observation post to observe pronghorn fawning activity 90 years later.

More and more settlers were now coming to Anderson Mesa, and it is difficult today to visualize how many people and livestock would soon be living there—local areas like Hay Lake supported a population of 60 people by 1894 (Neff 1974c). Every spring and cienega soon served as headquarters for a sheep or cattle ranch, causing the *Champion* (1 December 1888) to editorialize that "many portions of the Territory are now overstocked to an alarming extent … all available ranges where a natural supply of water can be had are now located and settled upon and those seeking ranges are compelled to buy or intrude on other parties property."

Such demands on Anderson Mesa's forage could not of course last. The early 1890s were drought years and the numbers of livestock in Arizona thus peaked in 1891 (Mike Pallesen, Arizona Agricultural Statistics Service, personal communication). The winter of 1892–93 was bitterly cold, with January storms sweeping across the Mesa for days on end, and even the winter range near Winslow was covered with 18 inches of crusted snow and ice. Cattle starved and froze by the thousands, and range inventories were halved (Barnes 1913). This was compounded by 1893 being a year of financial crisis. Most corporate livestock owners went broke or sold out, as did the Hash Knife in 1901, leaving only ranching families who had no one to sell to and nowhere to go (Neff 1974b). The situation for the pronghorn was equally dire. Arizona rancher and historian Will Barnes later reported that by 1898 he could only count 250 antelope in the country he had formerly ranched between Holbrook and Winslow—an area that he estimated contained 50,000 pronghorn on his arrival in 1883 (Barnes files, Arizona Historical Society). On the basis of such reports, the Territorial Legislature closed the antelope season in 1905 for 5 years, or until March of 1911 (Brown 2007).

In 1898, President William McKinley signed an Executive Order withdrawing the San Francisco Mountains and Black Mesa forest reserves from homesteading, thus retaining most of the unsettled portions of Anderson Mesa in the public domain. After some boundary adjustments, both reserves were reorganized as the Coconino National Forest in 1908. The newly created Forest Service favored cattle over sheep due to the government's bias in favor of small family ranches over corporations; at one point, in 1902, President Theodore Roosevelt had almost signed an order banning sheep from the reserves altogether (Lauver 1938). Government management was minimal, however, and few limits were imposed on numbers of livestock pastured. Sheep herds were restricted to particular sites, but cattle numbers depended largely on how many animals the permittee thought he could run. Everyone with a legitimate claim or homestead was allowed to use the adjacent national forest, as long as a grazing fee of $1.00 was paid for each head of stock. Most stockmen used the Forest Service lands for cow-calf operations or as lambing grounds, as there was no charge for young animals. Generally unpopular with territorial politicians, the Forest Service concentrated on being a good neighbor rather than a government moneymaker, and actual livestock numbers could have been twice the permitted numbers (Neff 1974a, 1974b; Brown 2007).

The Coconino National Forest was largely open range and remained so until 1927, with only individual homesteads and corrals being fenced. Until then the only government fence was a drift fence built in 1915 that ran from General Springs on the Mogollon Rim to a few miles west of Flagstaff (Neff 1974b). Only the area east of the fence that included Anderson Mesa was open to grazing of both sheep and cattle. Although cattle were allowed to drift between the summer and winter range depending on forage conditions, sheep had to be herded from site to site. Heavy grazing was favored to reduce fires, and livestock numbers were not stringently regulated (Lauver 1938, cited in Brown 2007). In 1910, 5463 cattle were permitted on Anderson Mesa, or about 1 cow per 16.2 ha (40 acres). Sheep grazing was more intense and may have been as high as 1 sheep per 0.47 ha per month (Neff 1974a, 1974b). The Forest Service range staff was concerned that much of the Coconino National Forest had been "eaten or sheeped out" by 1904, and that the native bunchgrasses had been depleted prior to 1910 (Miller 2001). Nor was the status of the pronghorn thought to be improving. The new Arizona State Legislature opted not to reopen the season on antelope in the 1912 State Game Code (Brown 2007).

Good rains and good livestock prices characterized the years preceding America's entry into World War I, and a lenient government favored an increase in grazing permits (Lauver 1938). The result was that more permits were issued for both sheep and cattle in 1918 than at any other time in the history of the Coconino National Forest (Figure 2). Increased numbers necessitated distributing livestock to all parts of the National Forest; every available water source was developed and numerous stock tanks were dug with horse-drawn fresno scrapers. Grazing both sheep and cattle was the norm on Anderson Mesa, and animals were allowed on the summer range as early as possible (Neff 1974a, 1974b). Meanwhile, State Land Commissioner Obed Lassen selected the remaining government "checker-board" sections on the winter range as per the *in lieu selection* process that compensated the state for "state school sections" taken in by the National Forests (Brown 2007).

Some stability had been achieved by the permit system that gave grazing lease preferences to base property owners, but range management on the Coconino remained in a formative stage. To alleviate some of the political hard feelings arising from creating national forest reserves by Executive Order, Congress passed two more homestead acts—The Forest Homestead Act of 1906 and the Stock-Raising Homestead Act of 1916 (Brown 2007). The former allowed claimants to take up to 160 acres (65 ha) of meadow or other non-forested lands within the national forests for farming, and the latter allowed an individual to homestead 640 acres (259 ha) and graze 50 head of cattle on lands the Forest Service deemed appropriate for such use. Settlement continued accordingly, and the peak period of homestead filing on Anderson Mesa was between 1915 and 1925, with the last homestead patented in 1935 (Neff 1974b, 1974c). By the early 1920s, there were 21 cattle and 11 sheep permits on Anderson Mesa, coinciding with the peak

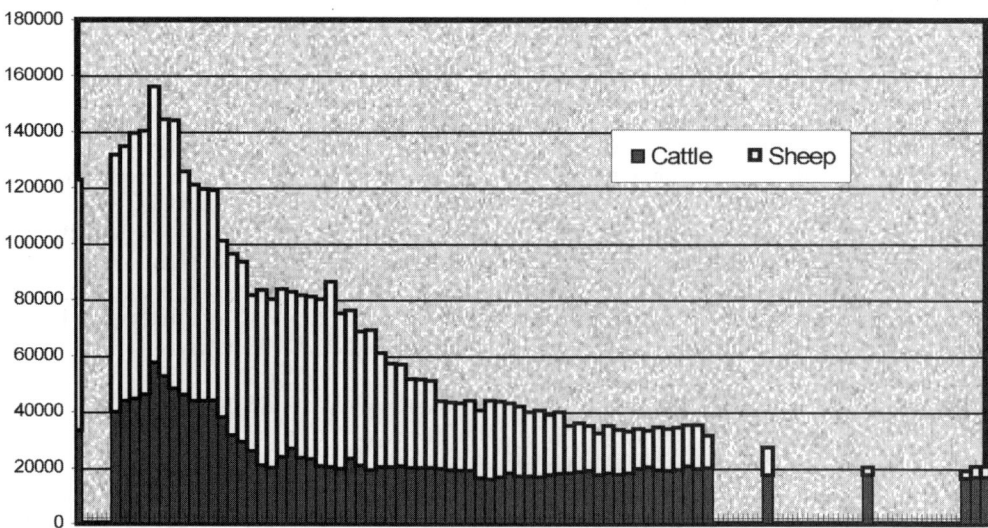

Figure 2. Estimated livestock numbers for Coconino National Forest, 1910-2000.

number of people living there, all of whom were attempting to eke out a living by ranching, dry-farming, wood-cutting, operating stills, and living off the land in general. Small settlements and even school districts were formed with at least 30 people living in the Anderson Pass District (Neff 1974b, c).

All of this activity had a great impact on Anderson Mesa and its wildlife. But unlike many other areas, Anderson Mesa never lost all of its pronghorn. Nor, according to Neff's (1974c) interviews with the remaining residents of Anderson Mesa, were these animals now being overtly harassed. Most homesteaders and stockmen stated that they actually liked "antelope." This was also the opinion of E. W. Nelson, a biologist with the U.S. Biological Survey, who while en route to the Arizona Strip in September, 1909, spent some time on the pronghorn's winter range south of Winslow. Nelson (cited in Brown 2007) concluded that pronghorn were "still rather generally distributed throughout the cedar and pinyon belt south of the Atchison Topeka and Santa Fe railroad, where these animals were said to be at least holding their own and are probably increasing in number." Nelson attributed their presence to protection by stockmen, and estimated the number of antelope to be between 60 and 70 southwest of Sunset Pass (where they ranged into ponderosa pine forest), with another 100 between Chevelon and Jack's Canyons.

Another U.S. Biological Survey biologist, E. A. Goldman, visited Mormon Lake on 1 May 1913, and found it full. He went on to describe the grassland of Anderson Mesa as "extending for 15 or 20 miles southward, and containing numerous prairie dogs." He remained at Mormon Lake until 29 July when he went to Anderson Mesa to collect pronghorn for the U.S. National Museum. Goldman's 1913 field notes describe the "stony mesa" as "rolling, mainly open country, and rather thickly covered with grass, but here and there scattered over the plain are belts and isolated trees and groups of junipers, pinyons, and a few ponderosa pines." Operating out of the Yaeger Ranch, he spent 31 July hunting on foot and unsuc-

cessfully stalking an individual pronghorn. On 3 August he saw two pronghorn near Pine Hill and killed one. The following day, he saw four more and shot two. Goldman's field notes suggest that these animals, while flighty and not especially numerous, were not especially rare either (Brown 2007).

The lax grazing regulations attendant with the war years had irreversibly altered the pronghorn's habitat. Many of the West's rangelands, including at least parts of Anderson Mesa, had been heavily overgrazed, and game numbers were thought to be at all-time lows. Pronghorn were now considered to be in danger of extinction, so much so that a national conference was scheduled in 1921 to develop a plan to prevent the animal's total disappearance (Nelson 1925). Attendant with the conference was an international census to be conducted between 1921 and 1923 using regional estimates made by foresters, game wardens, predator control agents, ranchers, and other knowledgeable people. Areas with enough public land and enough pronghorn would be considered potential sites for a national antelope refuge (Nelson 1925).

Results of the survey, although almost certainly too conservative, were not as discouraging as many had feared. More than 30,000 pronghorn were thought to still survive, 26,000 of them in the United States, mostly in Wyoming, Montana, and northern Nevada. The total population of pronghorn in Arizona, however, was estimated at only 651 animals in 18 locations. Only 113 antelope were thought to remain on Anderson Mesa—too few for a national antelope refuge (Nelson 1925).

What the people at the conference did not realize was that a recovery was already underway. In an attempt to rectify poor range conditions brought on by the war years, the U.S. Forest Service had begun reducing livestock permit numbers in 1918 and continued doing so for the next 10 years. The Coconino National Forest was no exception; bunch-grasses and such important herbs as deer-vetch (*Vicia* spp.) had virtually disappeared from large areas (Leiberg et al. 1904; Plummer 1904; Talbert and Hill 1922;

Lotfield 1924; Merrick 1932; Cooper 1960; Miller 2001). Fewer sheep and cattle permits were now being issued each year on Anderson Mesa. Some reductions were draconian; the Hennessey Sheep Co., ranging around Ashurst Lake, was cut from 3292 to 2185 head between 1924 and 1925—a one-year reduction of 34 percent (Neff 1974a, 1974b). By the late 1920s, sheep numbers on Anderson Mesa had been reduced by 40 percent and cattle numbers by 65 percent. Large numbers of feral horses and trespass stock were rounded up and removed. Allotments had begun to be fenced about 1927, and by 1930, dual use by both cattle and sheep had ended. Each pasture was now managed either for sheep or cattle, with only two sheep allotments remaining on Anderson Mesa. Perhaps equally or more important, the 8-month grazing season on Anderson Mesa that began the first half of April was reduced in 1922 to a 5-month grazing season (Neff 1974a, 1974b).

RECOVERY OF PRONGHORN ON ANDERSON MESA

Improved range conditions, coupled with increased protective measures, undoubtedly benefited pronghorn on Anderson Mesa, and on 23 September 1927 the *Coconino Sun* reported that "the antelope is one of the rarest animals in Arizona. Everyone here is inclined to protect the antelope and because of it the number is rapidly increasing. Possession of a dead antelope carries a minimum penalty of $100 fine."

When the Arizona Game Protective Association was formed in 1923, the pronghorn was adopted as a logo, and the animal was touted as Arizona's conservation success story (Kartchner 1931). Forest Supervisor Edward G. Miller proudly estimated a population of 1400 pronghorn on the Coconino National Forest in 1928. In 1929, the population estimate was 1500, and in 1930 it was 3300. Between 1500 and 2000 of these animals were reported to reside on Anderson Mesa (USFS 1932).

Going from 113 animals in 1923 to more than 3000 in 1930, although possible, is by any modeling standard a difficult proposi-

tion that requires an unexploited population with a male-to-female ratio of 54:100 to have an annual fawn recruitment rate near the maximum of 180 fawns per 100 does for several years. A more likely scenario is that the 1923 population estimate was too conservative, and Miller's estimates too optimistic. Nonetheless, it was clear that pronghorn on Anderson Mesa were making a comeback as stated by the AGFD biologist (Knipe 1944a: 33) charged with determining factors governing Anderson Mesa's pronghorn fluctuations: "The antelope herds in Arizona have built up from near extinction to their present numbers with little attention from agencies concerned with their status."

Knipe's statement notwithstanding, there were several reasons why the pronghorn population on Anderson Mesa recovered so rapidly. Subsistence hunting slackened after 1919, and range pressures decreased as the Forest Service gained more control over grazing permittees. Most important, however, was that every summer from 1914 through 1927 received above-average moisture—a situation that occurred again from 1929 through 1933, according to the Palmer Drought Severity Index. The PDSI is a monthly water balance index, standardized to local climates, that considers previous precipitation and temperature data to determine a region's relative dryness. Thus, the PDSI can be used as an index to plant growth and foraging conditions (Palmer 1965). An index value of 0 is considered normal, a –2 value is a moderate drought, –3 is a severe drought, and –4 is an extreme drought (Table 1; July PDSI for north-central Arizona, NOAA). Not only were the 1920s generally wet, resulting in the production of abundant "new" vegetation, but delayed entry of livestock on to the summer range and cessation of dual use also must have had a beneficial effect on the mesa's pronghorn forage.

Other factors may have also been at work. According to the *Coconino Sun*, hundreds of fires occurred on the Coconino National Forest between June of 1923 and the early summer of 1929, several of them on or near Anderson Mesa. Although all were

Table 1. July Palmer Drought Severity Index, survey, hunt, and other Unit 5 management information, 1950–2004 (data modified from Neff and Woolsey 1984).

Year	July PDSI	Prong-horn Seen	Fawns/ 100 Does	Prong-horn Harvest	Elk Seen	Calves/ 100 Cows	Elk Har-vest	Deer Seen	Fawns/ 100 Does	Deer Har-vest	Bison	Calves/ 100 Cows	Est. Cattle	Est. Sheep	Management Notes
2004	-3.68	450	39	10	1440	40	1863	139	53	102	51	58	6907	0	Change in survey procedures
2003	-4.04	397	29	11	1856	51	1844	279	34	102	50	52	5455	0	—
2002	-5.76	266	11	14	1814	26	1492	234	39	57	75	44	3676	0	Coop. management measures implemented
2001	1.18	503	23	5	1545	32	1448	249	19	85	104	63	5289	0	Extreme drought; some aerial gunning
2000	-5.41	280	19	9	1635	19	2178	377	43	100	100	53	7496	0	Public concern over pronghorn management
1999	1.5	546	10	16	2603	32	1868	386	20	120	90	51	7,729	0	AWF questions management of AM (Anderson Mesa)
1998	2.82	566	10	15	2691	44	2181	518	40	200	92	59	7668	0	—
1997	0.24	557	11	22	2775	49	1765	637	38	164	103	68	7608	0	Some aerial gunning of coyotes
1996	-5.05	421	1	22	1882	43	1287	589	41	233	92	59	7150	0	—
1995	0.28	636	6	24	1963	34	1195	443	32	262	71	56	7094	0	Very dry winter
1994	0.9	681	12	46	2828	42	2546	944	37	396	85	72	7428	0	—
1993	4.22	839	18	36	2382	40	3043	467	43	352	113	75	7367	0	Prop. 200 bans trapping on public land
1992	4.67	712	7	40	2481	56	1994	763	59	398	116	79	7306	0	—
1991	-0.25	964	21	57	2814	51	1990	404	49	321	119	83	7247	0	—
1990	0.21	1189	24	75	2288	52	1896	622	40	393	114	70	7389	0	—
1989	-3.52	1146	34	74	2203	42	1497	302	36	351	111	63	7,893	0	—
1988	4.27	1161	23	80	2152	49	1564	384	43	428	113	74	5995	0	Aerial gunning of coyotes
1987	3.46	948	25	44	1995	57	1247	335	51	477	116	75	6087	0	Aerial gunning of coyotes
1986	1.78	1215	14	44	1359	56	933	168	51	457	113	84	6186	0	Aerial gunning of coyotes
1985	4.88	1390	25	116	771	46	1019	546	41	576	109	85	6291	0	—
1984	1.62	1107	43	120	1011	64	1105	403	50	501	107	81	6392	0	—
1983	4.86	1096	57	103	846	43	927	169	27	467	107	74	6493	0	—
1982	2.06	863	62	73	717	56	723	224	49	553	93	74	6595	0	Aerial gunning of coyotes
1981	0.33	674	39	53	567	62	863	247	44	519	89	59	6697	0	Aerial gunning of coyotes
1980	5.01	616	37	21	678	56	694	181	54	362	91	59	6798	0	Trapping, aerial gunning of coyotes
1979	5.53	504	16	28	306	59	772	279	47	217	103	81	6,900	0	Coyotes trapped
1978	2.29	444	30	33	642	55	659	201	51	248	97	66	7332	0	Coyotes trapped

Table 1 (continued)

Year	July PDSI	Prong-horn Seen	Fawns/ 100 Does	Prong-horn Harvest	Elk Seen	Calves/ 100 Cows	Elk Har-vest	Deer Seen	Fawns/ 100 Does	Deer Har-vest	Bison	Calves/ 100 Cows	Est. Cattle	Est. Sheep	Management Notes
1977	-2.11	472	30	28	593	64	647	202	50	286	96	60	6988	0	Coyotes trapped
1976	1.86	357	20	30	356	59	557	135	56	332	120	80	7379	0	Coyotes trapped; last fire on AM (200 ac)
1975	1.96	408	28	36	578	62	575	163	65	369	139	61	7770	0	—
1974	-3.09	433	41	35	438	66	348	160	61	395	167	42	9067	0	—
1973	6.14	259	15	38	271	66	355	259	61	239	208	77	9000	0	Elk numbers increasing
1972	-3	389	45	33	340	54	323	198	64	296	214	85	9262	0	345 kV power-line built on AM
1971	-5.1	520	52	32	214	57	363	170	56	281	275	56	9167	0	1080 banned; Grapevine allotment open to grazing in 1972
1970	-1.9	414	80	56	847	56	483	207	53	252	215	77	8685	0	—
1969	0.37	328	86	49	186	54	288	290	78	371	239	70	8,044	0	—
1968	-0.24	153	38	0	419	60	307	187	83	302	227	—	9547	0	Closed season
1967	-1.8	1046	60	0	245	60	335	162	45	288	215	60	9262	0	Est. 80% loss from Jan storm; no hunt
1966	-0.51	672	67	112	280	61	381	245	61	244	220	71	9000	0	—
1965	3.17	581	26	56	261	63	351	248	57	189	223	56	8312	0	—
1964	-2.73	491	46	50	237	54	240	150	52	247	192	80	8550	0	—
1963	-5.11	681	43	67	148	49	215	169	36	219	115	71	7785	0	Includes 3 does
1962	-1.59	613	23	51	158	51	182	173	31	406	212	75	7785	1490	Unit 5 divided into 5A & 5B
1961	-2.61	868	28	41	90	47	194	183	47	633	253	54	8977	1600	Major p-j control project completed
1960	-2.12	1135	48	163	88	39	105	179	45	782	218	68	8325	1750	Includes 60 does; 34 pr translocated
1959	-4.23	1350	80	122	53	58	185	102	46	869	208	59	9,082	1547	—
1958	-0.78	888	34	69	129	45	72	120	43	517	206	69	9220	1593	83 pronghorn translocated
1957	0.78	963	32	27	110	57	46	129	26	429	212	72	9298	1640	—
1956	-3.42	1211	62	74	143	45	39	—	21	—	205	—	9372	1687	—
1955	-1.87	1389	66	76	99	28	26	—	52	—	192	—	9449	1773	—
1954	-1.64	1588	71	131	130	57	155	—	36	—	172	—	9525	2740	—
1953	0.85	1466	57	421	85	22	86	—	68	—	152	—	9595	2786	—
1952	3.67	1596	76	219	—	42	—	—	49	—	132	—	9675	2833	Increase in fences and stock tanks
1951	-4.26	1328	62	248	—	42	—	—	38	—	112	—	8546	3770	Severe winter storm in fall of 1952
1950	-1.85	1812	90	166	—	43	—	—	73	—	92	—	8761	2926	Drought; 167 pronghorn + 175 bison transl. in 1950–51; 4000 pronghorn est. on CNF; 166 bison to RR

extinguished, some "consumed several hundred acres," and at least one burned more than a thousand. Favorable moisture conditions, fire, and reduced livestock pressures can greatly enhance the growth of nutritious forbs (Brockway et al. 2002), and the effects of this increased nutritional level on pronghorn fawn recruitment, while unmeasured at the time, is the most rational explanation for increased pronghorn numbers between 1923 and 1930.

Many ranchers and game rangers nonetheless attributed the pronghorn's recovery during the 1920s to a decrease in mortality brought on by predator control and a reduction in subsistence hunting. There may be some merits to their argument, even though no government predator control program was in effect on Anderson Mesa in the 1920s (Dodson 1991). In those days, ranchers and farmers used strychnine to kill coyotes and other "varmints." So effective was this measure, coupled with the use of traps, that wolves (*Canis lupis*), the principal predator of adult pronghorn, had been largely eliminated from Coconino County by 1925 (Brown 1983). The take of coyotes and bobcats (*Lynx rufus*) on Anderson Mesa must have also been substantial due to the large numbers of settlers then present and the high price of fur. In 1927, the average price of a coyote pelt had increased to the equivalent of $26.87 in today's dollars, in 1928 it rose to $36.33, and in 1929 it was $38.00 (M. Zornes, AGFD, personal communication). Such figures represented a small fortune to a farm boy's family struggling to eke out a living on Anderson Mesa.

As in several other Western states that were also experiencing resurgences in their pronghorn population, the Arizona Game and Fish Commission had established a number of state game refuges in the 1910s and 1920s. Most of these were on lands administered by the National Forests, and though open to grazing and other land uses, they were closed to hunting with the understanding that surplus game numbers would repopulate depleted areas elsewhere. Game numbers had indeed increased, and by 1931, ranchmen were complaining that excess numbers of game animals were damaging the forage they were leasing. The Forest Service therefore commissioned a special committee to inspect five of the most contentious refuges, one of which was Hay Lake Antelope Range on Anderson Mesa, which had been specifically set aside for pronghorn. Among the committee's inspectors were several USFS rangers, a representative of the Arizona Wool Growers' Association, an Arizona Cattle Growers' Association representative, E. A. Goldman of the U.S. Biological Survey, and A. A. Nichol of the University of Arizona. Another committee member, State Game Warden Ken Kartchner, was not present for the Hay Lake inspection.

In their report on the Hay Lake Antelope Range, the Committee (U.S. Forest Service 1931:3) reported that their "examination of the area shows heavy browsing of what appears to be important antelope forage such as juniper, rabbit brush (*Chrysothamnus*), snakeweed (*Gutierrezia*), gooseberry, antelope weed (*Eriogonum*), and squawbush (*Rhus*). It was with some surprise that some members of the committee noted the obvious browsing of juniper by antelope which had been regarded mainly as grass feeding animals."

Forest Service personnel stated that antelope had greatly increased in recent years, and that the heavy browsing of junipers was confirmed by Nichol's collection of four pronghorn stomachs in 1930. The Committee recommended that some antelope be trapped and relocated in vacant areas elsewhere in the state.

In a supplemental report, Goldman (1931: 3–4) expanded on the Committee's account, stating that the antelope on Anderson Mesa had become "very numerous" since his previous visit in 1913, and that he had "considerable concern" regarding antelope forage conditions and the high-lining of so many junipers. He encouraged AGFD to complete an antelope trap then under construction, and recommended that some animals be removed even though such action would be unpopular at both state and national levels. Goldman had also expressed great concern with the number of elk on the Sitgreaves National Forest, and recommended an immediate reduction in numbers lest these animals increase and spread further.

Nichol (1931), who did not sign the Committee report, felt obliged to file his own report as a university representative. Nichol considered the state's involvement too tepid, and the Forest Service's report as lacking public support. He considered the antelope in danger of overusing their food supply as junipers and other unpalatable plants were being taken, and recommended a dietary study and a hunting season to curtail the antelope's rapidly increasing numbers. As for elk on the adjacent Sitgreaves National Forest, he considered these animals to already constitute a serious problem on the summer range. Unlike the others, Nichol included in his report a number of photos of Anderson Mesa that showed "high-lined" junipers and a seriously damaged wolfberry (*Lycium andersonii*). Other photos showed not all of the junipers to be high-lined, and although ground cover was short, there was only a moderate invasion of young juniper trees.

Indeed, Arizonans would have had an antelope season in the 1930s, but the recently appointed Arizona Game and Fish Commission was reluctant to open a season on an animal that had been so recently imperiled. Despite a continued rise in the Forest Service's estimate of the antelope herd on Anderson Mesa to 4600 head in 1933, the Commission continued to postpone a hunt in hopes that the animals would expand to other areas. Instead, many of the antelope succumbed to drought and a lack of forage during the summers of 1934 and 1939 (Knipe 1942b; Edwards 1950). Hundreds more died during the severe winter of 1936–1937 (O'Connor 1940; Knipe 1942c, 1944a).

One citizen who took a special interest in Anderson Mesa's pronghorn was a sportsman and English professor at Arizona State Teachers School in Flagstaff (now Northern Arizona University) named Jack O'Connor. Taking on the mesa's antelope as a personal conservation project, O'Connor began what

would be a lifetime outdoor writing career in the May 1934 issue of *Arizona Wildlife and Sportsman*, penning an impassioned editorial, "Arizona's Antelope Problem," that called for a hunt to reduce their numbers. When the Arizona Game and Fish Commission failed repeatedly to open a pronghorn season, he really let them have it in a February 1940 editorial:

> Bad winters killed the animals by the hundred and coyotes flocked in from all over northern Arizona to feast on the weakened and dead animals. They stayed to kill the fawns every spring. To my notion, the Anderson Mesa situation is as good a lesson in how not to manage game as the Kaibab once was. Nine years ago we had 5,000 antelope. With good management the herd should have increased 30 percent annually. The increase between 1930 and 1939 should have been 13,500 animals. Instead, we have today probably around 2,000 antelope.
>
> In other words we have wasted about 15,000 antelope! All you Arizonans did with your antelope, Dr. H. L. Shantz, head of the game management division of the U.S. Forest, told me recently, was to furnish a free lunch for the coyotes! (O'Connor 1940:5)

O'Connor was especially impressed with the trophy quality of the bucks on Anderson Mesa. After his friend and hunting partner J. C. McGregor collected a male specimen for the Museum of Northern Arizona in 1935, Jack obtained permits to take a male and female for the Arizona State Museum in Tucson, declaring Anderson Mesa to have the finest "bucks" to be found anywhere (O'Connor 1939a). But although he considered Anderson Mesa to be the "most thickly populated pronghorn range in North America," he could never decide just what factors were responsible for the turnaround in the animal's numbers (O'Connor 1939b). The general consensus at the time, stated by State Game Warden Ken Kartchner in 1931, and echoed by O'Connor through the 1930s, was that the initial increase in the 1920s had been due to a reduction in illegal killing, improved range conditions, and the removal of wolves, coyotes, and mountain lions (*Puma concolor*) by ranchers and government hunters. Years later, O'Connor (1961) would change his mind and opine that a decline in numbers and trophies after 1940 was due to too much land being closed to livestock grazing, and that overgrazing by cattle was good for pronghorn.

The first Federal-Aid to Wildlife Restoration Act project authorized to study Arizona's big game resulted in Ted Knipe being sent to Anderson Mesa to determine why the pronghorn population had declined and whether the population could sustain an open season. Knipe (1940) estimated that the population was down to 800 pronghorn, and the animals were suffering from poaching, predation, poor winter range, fences, and possibly disease and too many adult males; his survey showed a male-to-female-to-fawn ratio of 238:203:112, with only 8 animals unclassified. Although Knipe noted that the range was composed largely of snakeweed (*Gutierrezia sarothrae*) and pingue (*Actinea richardsonii*), and that competition was "acute," he made no mention of drought or livestock numbers as factors in the pronghorn's decline. Most of the grasses present were western wheatgrass (*Agropyron smithii*) and other intermountain cool-season grasses, with saltbush (*Atriplex canescens*) and wolfberry listed as important browse plants; the most-used browse on the poorer winter range was winterfat (*Eurotia lanata*). Few deer were noted, and no mention was made of grama grass (*Bouteloua gracilis*). Only one elk, an adult female, was seen, and

the observation was described as unusual. His recommendation was to have a hunt to remove 75 males.

Knipe's recommendation (1940) was considered too conservative and Nichol (1941) and McGuire (1941a, 1941b, 1941c), who estimated the population at 2200 animals, instead recommended that 300 males be removed to reduce intraspecific competition and spread the male-to-female ratio. Emphasizing the need for the latter, the AGFD recommended 400 "buck-only" permits at $5.00 each. The Commission complied, authorizing in 1941 a 20 September through 5 October season for Anderson Mesa—the first legal pronghorn hunt in Arizona since statehood.

To prevent crowding, the hunt was held in two segments (Knipe 1941). A check station was operated during both hunts and animals checked in were aged, weighed, and measured. In total, 287 males were harvested, for a hunter success rate of 73 percent. Biological data collected at the check station indicated that pronghorn on Anderson Mesa attained maximum size by age 2, and maximum weight by age 3–4 (Knipe 1941, 1942a; Table 2).

Not only was it becoming clear that pronghorn could not be "stock-piled," but a survey the following winter showed that the hunt had resulted in little change in the 1:1

Table 2. Age and sizes of buck pronghorn collected on and around Anderson Mesa in 1941, 1942, and 1943 (Knipe 1943).

Age class	1941	1942	1943	Total
Young (24 yrs)	36%	37.5%	55%	181 (39%)
Prime (58 years)	33%	47%	36%	248 (50%)
Aged (9+)	31%	15.5 %	9%	61 (11%)
Mean weight	36.7 kg (81 lbs)		34.1 kg (75.3 lbs)	
Mean horn length	33.0 cm (13")		31.0 cm (12.2")	
Mean horn circumference	14.2 cm (5.6")		13.7 cm (5.4")	
Mean horn spread	32.8 cm (12.9 ")		31.3 cm (12.3")	
Mean body length	142.3 cm (56")		141.3 cm (55.6)"	
Heaviest buck	45.8 kg (101 lbs)		40.8 kg (90 lbs)	
Longest horn	47.0 cm (18.5")		45.7 cm (18")	
Longest circumference	16.5 cm (6.5")		17.0 cm (6.7")	
Greatest spread	62.3 cm (24.5")		43.2 cm (17")	
Longest body length	165.2 cm (65")		155.0 cm (61")	

male-to-female ratio (Knipe 1942a, 1942b, 1942c). As a result, the AGFD initiated an extensive pronghorn survey of all of northern Arizona in 1942, and recommended that the number of permits be nearly doubled (Knipe 1942b, 1942d). This hunt generated a different kind of controversy. O'Connor (1943) thought that the season should be in April when the bucks' horn sheaths were newly erupted and would make better trophies. He argued that the antelope's pelage was better in the spring and that there would be less chance of shooting females as the sexes were then separated. These arguments, however, along with the complaint that pronghorn venison was less edible during the autumn rut, were successfully refuted (Knipe 1942c, 1942d).

Although hunters in 1942 harvested 487 male pronghorn for a 68 percent success rate, the 1942 hunt also failed to widen the male-to-female ratio to the degree anticipated. Nonetheless, the autumn season had proven itself to be highly popular after all, and the hunts on Anderson Mesa were the subject of favorable articles by Jack O'Connor (1945) and other writers in *Arizona Wildlife Sportsman*, the magazine of the Arizona Game Protection Association.

The next year, 1943, the Commission opened most of northern Arizona to antelope hunting, and again, to limit crowding, 1072 permits were spread over two hunting seasons—the first hunt was from 30 September through 4 October and the other was 7–11 October. More than 500 pronghorn were harvested, mostly on Anderson Mesa—a figure that would prove to be an all-time record (Knipe 1944b; Table 2).

The imposition of gas rationing and other restrictions due to World War II precluded authorizing hunts in 1944 and 1945. But Arizonans were now more optimistic about the pronghorn's future. The increase in populations in northern Arizona promised surplus bucks to hunt as well as animals to restock vacant habitats elsewhere (Niehuis 1943).

Indeed, biologists had already attempted to transplant pronghorn from Anderson Mesa but with difficulty, with many of the animals "dying of shock" either during capture or shortly afterwards (Niehuis 1943). Hunting pronghorn also remained controversial, as both sportsmen and wildlife agency personnel were more conservative in the 1940s than today. The AGFD, then under the direction of Tom Kimball, was reluctant to recommend another antelope hunt unless it could be convincingly demonstrated that surplus animals were available to harvest. Kimball believed that, in order to sustain hunting, pronghorn fawn recruitment had to be at least 50 fawns for 100 females. In 1944, the AGFD's surveys showed only 30 fawns per 100 females. In 1945, the fawn-to-female ratio was 39:100. While not unusual by today's standards, these ratios were then considered nothing short of disastrous. Even in 1946, when surveys showed a statewide fawn-to-doe ratio of 62:100, and a male-to-female ratio of 1:2, the AGFD recommendation was not to have a hunt.

Some AGFD biologists thought the reason for the pronghorn's low reproductive performance in the mid-1940s was a combination of inadequate summer precipitation and fawn predation by coyotes (Wilkins 1944; Frost 1945a; Arrington 1947). An intensive coyote-trapping program was therefore implemented on Anderson Mesa and in other fawning areas, with mixed results. Although AGFD personnel were convinced that trapping coyotes increased fawn recruitment, the fawn-to-female ratio in the summer of 1947 remained below the hoped-for 75:100. Again, for the fourth year in a row, no hunt was recommended (Table 2).

An intensive coyote reduction program, using the new, lethal predacide Compound-1080, was instituted throughout much of northern Arizona during the winter of 1947–48. Perhaps because of the use of 1080, or due to its use in combination with favorable weather conditions, the percentage of females observed with fawns in 1948 jumped to 74 percent. But still no pronghorn hunt was recommended! Game rangers patrolling the last hunt in 1943 had reported finding many wounded and dead animals. People both within and outside the AGFD questioned whether Arizona's pronghorn could sustain hunting regardless of how many fawns were produced (O'Connor 1944).

The fawn-to-female ratio on Anderson Mesa in 1949 was 79:100, and after much deliberation, and public persuasion, the AGFD recommended a pronghorn season if a stringently controlled hunt could prevent flock-shooting and protect the female component of the population. The Commission agreed, and approved 606 "antelope license-tags" in four stratified seasons, each hunt opening at noon. In addition to certain restrictions on the use of vehicles, a higher ethical behavior was encouraged by instituting a Sportsman's Award program. Points were given for taking males with 12-inch (30.5 cm) horns or better, with additional points awarded for each additional inch (2.54 cm). Killing a pronghorn with one or two bullets allowed a hunter to accrue 20–25 additional points. Points could also be earned by taking proper care of the carcass, and the hunter who earned 100 points achieved a perfect score.

The 1949 antelope hunt was highly successful (Table 1). Not only was hunter behavior exemplary, but hunter success was 76 percent. Thirty-six animals had horns of more than 40.6 cm (16 in), and seven exceeded 43.2 cm (17 in). Four measured 45.7 cm (18 in). One hundred and eighteen hunters earned Sportsman's Award certificates for having 80 points or more; 45 sportsmen had scores of more than 90 points and 4 achieved perfect scores of 100. All agreed that, if these standards could be maintained, and coyote populations controlled, a pronghorn season could be an annual event (Edwards 1950).

The only concern was the apparent reduction in pronghorn horn size since 1941 (Edwards 1950), a decrease seemingly explained by declining age means (Table 1). What were not then appreciated were the environmental effects of winter temperatures and forage conditions on horn growth (Brown and Mitchell 2005). Conditions for horn growth in 1941 were excellent, with a mean monthly minimum temperature of −7.1° C (19.3° F) during the previous October

through March horn growth period, and a July PDSI of 5.54 preceding the hunt (Brown et al. 2006a). In contrast, the winters preceding the 1942 and 1943 hunts had mean monthly minima of −9.5° C (14.9° F) and −7.6°C (18.4° F), respectively, with July PDSIs of −0.48 and −2.52. In 1949, when buck measurements were also less than in 1941, the preceding winter's monthly minimum was −11.3° C (11.8° F).

Pronghorn management was also progressing on another front. In 1942, the Arizona Game and Fish Commission accepted the AGFD's recommendation to acquire the 8223 ha (20,320 ac) Raymond Ranch below Anderson Mesa as a pronghorn wintering area. This purchase, the AGFD's first large habitat acquisition, was made possible by the 1937 Pittman-Robertson Act. It was hoped that enough animals could now be trapped en route to the wintering range below Anderson Mesa to restock southern Arizona despite Knipe's (1944a) suggestion that pronghorn from the Dugas area were more ecologically suited for that purpose. Drift traps were set up on the winter range to capture migrating antelope (Webb 2005).

After a difficult beginning, a permanent "antelope trap" was constructed at the mouth of Grapevine Canyon in 1943, and by 1945, more than 100 head of antelope had been trapped and released in four locations in southern Arizona (Wallmo 1951; Webb 2005). Even greater success was to follow, and in 1949, 72 pronghorn were shipped to the AGFD's new wildlife refuge at Fort Huachuca—the largest single release in Arizona to that time. Another 60 pronghorn were translocated to southern Arizona in 1951, and 57 pronghorn were released in 1952 in San Rafael Valley. By the time trapping and transplanting operations on Anderson Mesa ceased in 1959, nearly 650 pronghorn had been captured, of which 433 were released at 12 sites in Arizona (Munig 2004). Most of the remaining animals were used in a mark and release study (Webb 2005). While purchasing Raymond Ranch, the AGFD also acquired the Grapevine Allotment on Anderson Mesa and the pasture adjacent to the pronghorn trap (Webb 2005).

No sooner had Raymond Ranch been acquired than Arrington (1945) recommended a land exchange and the introduction of up to 75 bison. Arrington had noted that most of the pronghorn on Anderson Mesa normally wintered south of Raymond Ranch, and only passed through the ranch's pastures after heavy snows. In 1947, a truckload of 42 bison was shipped from the AGFD's overstocked buffalo ranch in Houserock Valley, which with four bulls from Oklahoma, served to form the nucleus of the Raymond Ranch herd. The impact of a new five-strand fence around the refuge's 8.9 sq ha (22 sq mi) perimeter, along with the partitioning of additional pastures, was not immediately appreciated, but with the eventual increase in the bison herd to more than 200 head, conditions for pronghorn deteriorated. But because pronghorn fawn recruitment in the late 1940s and early 1950s was usually above 60 fawns per 100 females, there was little immediate concern (Powell 1952a, 1952b, 1952c).

The 1950s and 1960s were years of management consolidation; annual surveys were conducted each summer using light aircraft and season-setting arrangements were standardized. Domestic sheep numbers continued to be reduced and the Grapevine Allotment was retired for wildlife use. The loss of a large number of pronghorn after a severe storm in January of 1949 was followed by exceptionally high fawn recruitment, suggesting that pronghorn might be density dependent, and a limited number of "antlerless" permits were authorized in the unit adjacent to Anderson Mesa (Powell 1955b). In an effort to open up the habitat and reverse a long-term decline in grassland area on Anderson Mesa (Figure 3), the Forest Service initiated a massive pinyon-juniper eradication program in 1962. These were the happy years, and no one expressed any concern over Anderson Mesa's pronghorn or the increased fencing and stock-tank building then taking place (Powell 1956, 1957b; Welsh 1958a, 1958c).

Figure 3. Juniper invasion along the head of Grapevine Canyon—a principal migration route for pronghorn wintering off Anderson Mesa. Photo by Rick Miller, AGFD.

These years also marked the beginnings of the coyote control debate. Frost (1945a) had noted that, even though fawn recruitment was too low in 1945 to recommend a hunt, fawn recruitment was 25 percent higher where predator control had been conducted, compared to an adjacent control area south of Diablo Canyon. J. T. Wright (1946), however, found that only 6 of 38 coyote stomachs collected between 27 April and 25 August 1945 contained pronghorn remains, and only one of these was a fawn. Wright (1946), like Frost (1945b), considered coyote control of doubtful value, and thought that low fawn recruitment was a function of vitamin deficiencies due to drought and poor forage conditions. Heddings (cited in Edwards 1950) agreed, noting that the fawn drop on Anderson Mesa was over by 10 June with no indication of predation by coyotes. Wallmo (1951), who was conducting a food habits study on Anderson Mesa, also thought that coyote control was ineffective in increasing fawn recruitment and that nutritious forage was the key to good fawn survival. Wallmo (1951) found the most important browse plants on Anderson Mesa to be brickelia (*Brickellia* spp.), silver sage (*Artemisia ludoviciana*), Wright's buckwheat (*Eriogonum wrightii*), sheep loco (*Astragulus nothoxys*), spiderling (*Boerhaavia coccinea*), wild lettuce (*Lactuca graminifolia*), and wild grape (*Vitus arizona*), with only moderate use of wild aster (*Aster* spp.), feather dalea (*Dalea formosa*), telegraph-plant (*Heterotheca subaxillaris*), morning glory (*Ipomoea* sp.), giant-hyssop (*Agastache rupestris*), deer vetch (*Lotus greenei*), and wild primrose (*Oenothera* sp.).

Others were of a different opinion (Arrington 1947). Edwards (1947, 1950) considered predator control to have mixed results, with control efforts having more effect in dry years than wet ones, summer rains being beneficial to the survival of fawns. The result was a series of pronghorn studies on Anderson Mesa that included tagging hundreds of fawns (Powell 1954a, 1954b, 1955b). The period of fawn drop on Anderson Mesa was brief—between 8 and 18 May—and fawn recruitment rates were generally good

during this time (Powell 1955b). Larry Powell (1952b, 1953), the AGFD's antelope biologist, concluded that recruitment rates were indeed influenced by predator control even though he noted that areas experiencing low rainfall had low fawn recruitment. Although Powell's conclusions became generally accepted by AGFD biologists, the role of coyotes in fawn recruitment remained controversial in that even though fawns had been observed being taken by golden eagles (*Aquila chrysaetos*), no direct predation by coyotes had been noted (Powell and Webb 1956). It was also noted that Anderson Mesa had the best range conditions of any antelope range in the state, and photographs taken during the fawn cap-ture project show a robust growth of bunch-grasses and grass cover (Peterson 1956). Erling (1956a, 1956b) noted that all of the 2-year-old does captured had bred as yearlings, and that even a few yearlings had bred as fawns.

Hunt data continued to be gathered, and although some "nice trophies" were taken, pronghorn on Anderson Mesa had lighter average dressed weights (34 kg, 75 lbs) than pronghorn in eastern Arizona and west of Flagstaff—38 kg (83.7 lbs) and 36 kg (79.3 lbs), respectively (Powell 1954a). Annual pronghorn hunts were standardized in 1954, and "any antelope" hunts were initiated in adjacent units (Powell 1955a, 1955b). Powell (1956, 1957a, 1957b) noted that few large trophies had been taken during the drought year of 1955, and a drought in 1956 resulted in lower fawn recruitment and fewer pronghorn, implying that some adult mortality had occurred. Although fawn-to-doe ratios in Unit 5 were relatively high in 1956, at 62:100, recruitment fell to what was presumed to be maintenance levels in 1957 and 1958 (Table 1).

Welsh (1958a, 1958b) reported pronghorn numbers to be generally down in 1958 due to drought, even though he thought that range conditions were improving generally. Fawn recruitment on Anderson Mesa remained below 50 fawns per 100 females, however, and for the fourth year in a row the number of animals observed declined (Welsh 1958c). Moreover, the pronghorn

Figure 4 (a and b). Thirty-three years of change on a vegetation transect on the Pickett Lake allotment of Anderson Mesa. The grasses, mostly blue grama, have increased in density since 1952 (4a) whereas forbs and weed numbers have declined in the 1985 photo (4b, next page). Note the moderate increase and frequency of the junipers and ponderosa pines.

harvest on Anderson Mesa in 1958 was the lowest since 1951 (Welsh 1959; Table 1). Welsh, now the AGFD's pronghorn biologist, reported seeing more coyotes in 1958 than in the past, and Compound-1080 was recommended as a pronghorn management prescription throughout much of northern Arizona. And, despite being in a period of general drought, pronghorn fawn recruitment jumped to 80 fawns per 100 females in 1959, with an improvement in observation and harvest levels, before again dropping off in the early 1960s (Table 1).

Other changes were also occurring, some of them subtle. U.S. Forest Service range transects showed that except in protected areas, bunch-grasses and herbs formerly present on the mesa had been replaced with blue grama and noxious plants such as snakeweed (USFS files; Figure 4). Although pronghorn feed primarily on weedy forbs, the quality of the forbs appeared to be declining, and high-quality pronghorn foods such as deer-vetch (*Vicia americana*), green and yellow pea (*Lotus* spp.), and penstemon (*Penstemon* spp.) were in short supply (Wallmo 1951). And, although cattle permit numbers remained about the same, elk numbers were on the rise; the harvest of elk in Unit 5 grew from less than 100 in 1959 to nearly 400 in 1967.

In 1958 the AGFD was reorganized, and Roger Bumstead, the AGFD's elk biologist resigned. Levi Packard, the new Region II Supervisor in Flagstaff, then adopted a policy that would let elk numbers increase.

b 1985

Figure 4b. Photos courtesy of Mormon Lake Ranger District, Coconino National Forest.

In contrast to Region I and the White Mountains, no "antlerless" or "any elk" permits were issued in Unit 5 for more than 20 years, and when antlerless hunts resumed, the number of female elk taken would prove to be lower than needed to keep the elk population in check (Table 1).

The weather also had an unpleasant surprise for pronghorn on Anderson Mesa, when on 13 December 1967 a combination winter storm and northern cold front moved into Arizona from the west and north (White 1969). Sometime before dawn it began to snow, and it kept snowing until 19 December. During that time, 211 cm (83 in) of snow were recorded at Flagstaff, and 109 cm (40 in) of snow fell at Winslow, located 610 m (2000 ft) lower in elevation. No sooner had it stopped snowing than a dense fog descended on the Little Colorado River and the sun was not seen for several more days. A massive temperature inversion now settled in, and temperatures did not rise above freezing for more than a week. Deep snow not only prevented pronghorn from finding food, but the snow drove the animals down into the coldest areas where food remained covered by snow due to sub-freezing temperatures. Formerly passable fences became insurmountable barriers when the pronghorn could no longer pass under the bottom wire. Fence corners turned into death traps as coyotes and roving dogs from nearby Winslow attacked the weakened animals (R. A. Jantzen, personal communication). Junipers were again "high-lined" by starving pronghorn and efforts to feed pronghorn by air-dropping bales of alfalfa hay and food concentrates had little effect (R. A. Jantzen, personal communication). The only rations that appeared to do any good was a truckload of roadside clippings hauled in by agency personnel to a few hand-caught captives (Ken Clay, personal communication). Counts along fence-lines the following January tallied more than 300 carcasses, and Region I Supervisor Bob White (1969) estimated that 85 percent of the Anderson Mesa herd had perished.

The remaining pronghorn population in Unit 5 was so low that the hunting season was closed for 2 years. Pronghorn numbers had soon rebounded from similar storms in 1937, 1949, and 1952, but not this time. Even though fawn recruitment in 1969 and 1970 exceeded 80 fawns per 100 females, the animals failed to recoup their pre-storm numbers. Then, in 1971, a major drought occurred, and fawn recruitment dropped (see Table 1). Although sheep were being phased out, cattle numbers remained relatively static, and the elk population was increasing rapidly (Table 3). Meanwhile, other potential problems for pronghorn on Anderson Mesa included burgeoning recreational use of accessible habitats, the construction of a 345 kV power-line through some of the best foraging and fawning sites, the continued invasion of grasslands by junipers and pines, additional fencing of pastures, and conversion of forage-rich cienegas to stock tanks and fishing lakes (Figure 3). Intentionally or not, Forest Service management strategies favored grass over nutritious perennial forbs.

In 1977, the Grapevine Allotment was opened to livestock grazing; this was the year of the last significant fire (80 ha, 200 acres) that would affect Anderson Mesa's pronghorn until the Lizard and Mormon fires of 2003. But the greatest concern of many biologists was that coyote control had practically ceased due to changing management priorities and to a Presidential Executive Order banning the use of Compound-1080 in 1972. By 1976, pronghorn fawn recruitment had plummeted to 20 fawns per 100 females, prompting the AGFD to recommend an aggressive government-trapping program. Neff and Woolsey (1979:32) opined that "the mountain grasslands on Anderson Mesa provide antelope with excellent spring forage, but poor escape cover

for young fawns. It appears that nothing can be done about this problem that is not already being done by USFS range conservationists and their livestock permittes."

But even though scores of coyotes were trapped on Anderson Mesa between 1977 and 1981, fawn recruitment rates remained below 40:100. Pronghorn numbers on Anderson Mesa continued to decline until 1981, when only 21 animals were harvested. More drastic action was clearly called for. Accordingly, a state law was passed that allowed the use of aircraft to shoot animals for management purposes, and a new program to aerial-gun coyotes was initiated. This action, although expensive and controversial, offered two biological advantages: such an effort could be limited to coyotes denning in and near the fawning areas at a time when human access to the mesa was often difficult or impossible, and the killing would be limited to just prior to the fawning season (Neff and Woolsey 1980).

The winters of 1979, 1980, and 1981 were all wetter than average, and after 3 years of aerial gunning, pronghorn fawn survival rates on Anderson Mesa rose from 37 to 57 fawns per 100 females. Once again, the annual harvest in Unit 5 surpassed 100 male pronghorns (Table 1). Later, Neff and other biologists (1985) conducting the research concluded that aerial gunning was necessary to successfully manage the mesa's pronghorn population.

But aerial gunning was no panacea. Such control measures were economically infeasible on an annual basis, and the procedure was unpopular with much of the public. Moreover, aerial gunning in 1987, 1988, and 1989 failed to result in the desired increase in pronghorn fawn recruitment. By the 1990s, recruitment rates were so low as to preclude minimum herd maintenance levels of 20–25 fawns per 100 females. That Anderson Mesa was having another pronghorn crisis was becoming clear to both wildlife managers and sportsmen.

It was also becoming apparent that this crisis was the result of more than coyote predation on neonates. Neff's study had been conducted during the mid-1980s when

Table 3. Estimated pronghorn, elk, mule deer, bison, cattle, and sheep populations and biomass for Game Management Unit 5, 1950–2002. Elk population estimates based on AGFD model from 1988 through 2002; estimates < 1988 = 17% of annual harvest (PR = pronghorn).

Year	No. of PR	PR Biomass in 100 kg	Est. Elk per AGFD Model	Est. Elk	Elk Biomass in 100 kg	Est. Deer	Deer Biomass in 100 kg	Bison	Bison Biomass in 100 kg	Cattle	Cattle Biomass in 100 kg	Sheep	Sheep Biomass 100 kg	Total Biomass in 100 kg
2002	380	169	8184	7,240	17,231	804	482	88	393	5,289	25,810	0	0	44,085
2001	719	320	9192	10,890	25,918	1,356	814	100	447	7,496	36,580	0	0	64,079
2000	400	178	10828	9,340	22,229	1,440	864	90	402	7,729	37,718	0	0	61,391
1999	780	347	11464	10,905	25,954	2,400	1,440	92	411	7,668	37,420	0	0	65,572
1998	809	360	10970	8,825	21,004	1,968	1,181	103	460	7,608	37,127	0	0	60,132
1997	796	354	10274	6,435	15,315	2,796	1,678	92	411	7,150	34,892	0	0	52,650
1996	601	267	10493	5,975	14,221	3,144	1,886	71	317	7,094	34,619	0	0	51,310
1995	909	405	11685	12,730	30,297	4,752	2,851	85	380	7,428	36,249	0	0	70,182
1994	973	433	12804	15,215	36,212	4,224	2,534	113	505	7,367	35,951	0	0	75,635
1993	1199	534	12647	9,970	23,729	4,776	2,866	116	519	7,306	35,653	0	0	63,300
1992	1017	453	11696	9,950	23,681	3,852	2,311	119	532	7,247	35,365	0	0	62,342
1991	1377	613	11268	9,480	22,562	4,716	2,830	114	510	7,389	36,058	0	0	62,573
1990	1699	756	10218	7,485	17,814	4,212	2,527	111	496	7,893	38,518	0	0	60,112
1989	1637	728	10141	7,820	18,612	5,136	3,082	113	505	5,995	29,256	0	0	52,182
1988	1659	738	9565	6,235	14,839	5,724	3,434	116	519	6,087	29,705	0	0	49,235
1987	1354	603	5437	4,665	11,103	5,484	3,290	113	505	6,186	30,188	0	0	45,688
1986	1736	773	5938	5,095	12,126	6,912	4,147	109	487	6,291	30,700	0	0	48,233
1985	1986	884	6439	5,525	13,150	6,012	3,607	107	478	6,392	31,193	0	0	49,312
1984	1581	704	5402	4,635	11,031	5,604	3,362	107	478	6,493	31,686	0	0	47,261
1983	1566	697	4213	3,615	8,604	6,636	3,982	93	416	6,595	32,184	0	0	45,881
1982	1233	549	5029	4,315	10,270	6,228	3,737	89	398	6,697	32,681	0	0	47,634
1981	963	429	4044	3,470	8,259	4,344	2,606	91	407	6,798	33,174	0	0	44,875
1980	880	392	4499	3,860	9,187	2,604	1,562	103	460	6,900	33,672	0	0	45,273
1979	720	320	3840	3,295	7,842	2,976	1,786	97	434	7,332	35,780	0	0	46,162
1978	634	282	3770	3,235	7,699	3,432	2,059	96	429	6,988	34,101	0	0	44,571
1977	674	300	3246	2,785	6,628	3,984	2,390	120	536	7,379	36,010	0	0	45,865
1976	510	227	3347	2,875	6,843	4,428	2,657	139	621	7,770	37,918	0	0	48,265

Table 3. (continued)

Year	No. of PR	PR Biomass in 100 kg	Est. Elk per AGFD Model	Est. Elk	Elk Biomass in 100 kg	Est. Deer	Deer Biomass in 100 kg	Bison	Bison Biomass in 100 kg	Cattle	Cattle Biomass in 100 kg	Sheep	Sheep Biomass 100 kg	Total Biomass in 100 kg
1975	583	259	2028	1,740	4,141	4,740	2,844	167	746	9,067	44,247	0	0	52,238
1974	619	275	2069	1,775	4,225	2,868	1,721	208	930	9,000	39,780	0	0	46,931
1973	370	165	1882	1,615	3,844	3,552	2,131	214	957	9,262	40,938	0	0	48,034
1972	556	247	2115	1,815	4,320	3,372	2,023	275	1,229	9,167	40,518	0	0	48,338
1971	743	331	2815	2,415	5,748	3,024	1,814	215	961	8,685	38,388	0	0	47,241
1970	591	263	1678	1,440	3,427	4,452	2,671	239	1,068	8,044	35,554	0	0	42,984
1969	469	209	1789	1,535	3,653	3,624	2,174	227	1,015	9,547	42,198	0	0	49,249
1968	219	97	1952	1,675	3,987	3,456	2,074	215	961	9,262	40,938	0	0	48,057
1967	1494	665	2220	1,905	4,534	2,928	1,757	220	983	9,000	39,780	0	0	47,719
1966	960	427	2045	1,755	4,177	2,268	1,361	223	997	8,312	36,739	0	0	43,701
1965	830	369	1399	1,200	2,856	2,964	1,778	192	858	8,550	37,791	0	0	43,653
1964	701	312	1253	1,075	2,559	2,628	1,577	202	903	7,785	34,410	0	0	39,760
1963	973	433	1061	910	2,166	4,872	2,923	212	948	7,785	34,410	1490	1,430	42,310
1962	876	390	1131	970	2,309	7,596	4,558	253	1,131	8,977	39,678	1600	1,536	49,601
1961	1240	552	612	525	1,250	9,384	5,630	218	974	8,325	36,797	1750	1,680	46,883
1960	1621	721	1078	925	2,202	10,428	6,257	208	930	9,082	40,142	1547	1,485	51,737
1959	1929	858	420	360	857	6,204	3,722	206	921	9,220	40,752	1593	1,529	48,640
1958	1269	565	268	230	547	5,148	3,089	212	948	9,298	41,097	1640	1,574	47,820
1957	1376	612	227	195	464	4,862	2,917	205	916	9,372	41,424	1687	1,620	47,954
1956	1730	770	152	130	309	4,564	2,738	192	858	9,449	41,765	1773	1,702	48,143
1955	1984	883	903	775	1,845	4,270	2,562	172	769	9,525	42,101	2740	2,630	50,789
1954	2269	1,010	501	430	1,023	3,976	2,386	152	679	9,595	42,410	2786	2,675	50,183
1953	2094	932	466	400	952	3,682	2,209	132	590	9,675	42,764	2833	2,720	50,166
1952	2280	1,015	408	350	833	3,388	2,033	112	501	8,546	37,773	3770	3,619	45,774
1951	1897	844	350	300	714	3,094	1,856	92	411	8,761	38,724	2926	2,809	45,358
1950	2589	1,152	291	250	595	2,800	1,680	358	1,600	9,902	43,767	3450	3,312	52,106

pronghorn fawn recruitment rates had been improving generally throughout Arizona. Winter rains had been above average, and although fawn recruitment rates on Anderson Mesa during the years of aerial gunning were greater than in some adjacent units, the differences were neither unique nor particularly robust. Moreover, the treated study units were never reversed, and Neff et al. (1985) did not monitor the health of pronghorn fawns or evaluate the availability of alternate prey species.

In 1999, the Arizona Antelope Foundation commissioned Harley Shaw to conduct an evaluation of Arizona's aerial gunning efforts as it related to pronghorn fawn recruitment. Shaw (2000) reviewed the whole history of coyote control in Arizona and challenged some long-held assumptions. Shaw and his colleagues (O'Gara and Shaw 2004; Yoakum et al. 2004) noted that the state's mean annual pronghorn fawn recruitment rate had actually declined through the 1950s and 1960s when Compound-1080 and other predator control measures were most in use. Furthermore, this general decline continued from a drought in the mid-1950s through the drought of 1971 and into the droughts of the late 1990s. As for Anderson Mesa, Shaw (2000) showed that fawn recruitment during the control years was not statistically significantly greater than in adjacent untreated units, and declined to below pre-gunning levels once the study was completed. Although gunning had resulted in increased fawn recruitment in 18 of 22 instances, such gains were minor and short lived. Any benefits disappeared 2 years after gunning ceased. Moreover, recruitment rate improvements only averaged about 6 fawns per 100 females in any given year of gunning. Shaw concluded that to be effective, aerial gunning must be continuous or nearly so, and that predator control, to be effective, must be widely applied and conducted no less frequently than every other year.

That the 1990s contained several drought years was also an imperfect explanation of the problem. Although there is a significant correlation between winter precipitation and fawn recruitment in arid areas (Brown et al. 2002), Anderson Mesa is a relatively wet area by pronghorn habitat standards. Only after the driest winters, such as 1996 and 2002, could poor fawn recruitment in Unit 5 be attributed solely to a lack of moisture. Furthermore, fawn-to-female recruitment rates were low in 1998 and 2001 despite relatively wet springs and/or aerial gunning (Table 1).

There were also other problems with attempting to reduce coyote numbers. Fawn recruitment appeared density dependent in that doe-to-fawn ratios generally declined with 50 years of increasing pronghorn numbers despite high and low rainfall years, high and low fur prices, 1080 applications, and aerial gunning programs. Although fawn mortality studies have shown that predators kill anywhere from 9 to 94 percent of fawns less than 3 weeks of age (O'Gara and Shaw 2004), and that most of this predation was by coyotes, these figures do not mean that fawns not killed by predators survive. Other factors that need to be considered, besides weather, predator population size, and alternate prey availability, were the nutritional condition of the doe during pregnancy and the health of the fawn at birth (Yoakum et al. 2004). Birth synchrony may also be important as healthy pronghorn populations tend to have more fawns born at the same time with a lesser percentage taken by coyotes. All in all, fawn recruitment appeared to depend more on sufficient winter precipitation and the availability of nutritious forage than on predator control.

Upon several visits to Anderson Mesa, we hypothesized that the area was experiencing a lack of nutritious forage more extreme than during previous drought periods. We further reasoned that the mesa's pronghorn had become increasingly dependent on annual forbs, and that the drought-stressed animals were suffering from poor-quality perennial forage—a conclusion also reached by Yoakum (2003), who had been contracted by the Arizona Wildlife Federation to study the pronghorn situation on Anderson Mesa. We also noted, however, that although cattle numbers had remained relatively static for 60 years, elk survey and harvest numbers increased nearly every year from 1986 to 1994, when more than 3000 elk were harvested in Unit 5 (Table 1). When stockmen successfully pressured the Arizona Game and Fish Commission into authorizing a special "depredation" elk hunt in 1994, the Arizona Wildlife Federation objected and the hunt was stopped by a suit filed by an animal-rights organization.

We suspected that interspecific competition on Anderson Mesa, despite the drought, was greater in the 1990s than at any time since World War I, and that the pronghorn's poor reproductive rate was based on a lack of nutritious forage due to unprecedented forage consumption by elk and livestock—a situation analogous to other areas having large ungulate populations and declining pronghorn numbers (Bennett 2002; Boccadori and Garrott 2002. That these increased levels of competition were taking place during a time of drought had exacerbated the problem to crisis levels.

Using big game survey and hunt information provided by the AGFD, coupled with livestock numbers provided by the

USFS and Diablo Trust, we calculated ungulate population levels and biomass figures in Unit 5 for each year since 1950. Although survey procedures varied over time, sufficient samples were available to model both population and biomass trends (Brown et al. 2004; Table 3). In addition to these estimates, we also considered an elk population model devised by AGFD that estimated elk numbers in Unit 5 after 1988 (Brown et al. 2004).

Our analysis of the Unit 5 database (Table 3) clearly indicates a negative correlation of pronghorn recruitment with total ungulate biomass (excluding pronghorn) during the previous autumn ($r^2 = 0.29$, $P < 0.001$). The majority of this negative relationship was attributable to a 90-fold decrease in pronghorn recruitment concurrent with a 117-fold increase in elk biomass ($r^2 = 0.48$, $P < 0.001$; Figure 5).

Midsummer drought, as measured by the regional Palmer Drought Severity Index for July, showed a mild but significantly increasing relationship with both pronghorn fawn recruitment ($r^2 = 0.31$, $P = 0.10$) and the number of pronghorn observed on that year's surveys ($r^2 = 0.47$, $P < 0.02$; Brown et

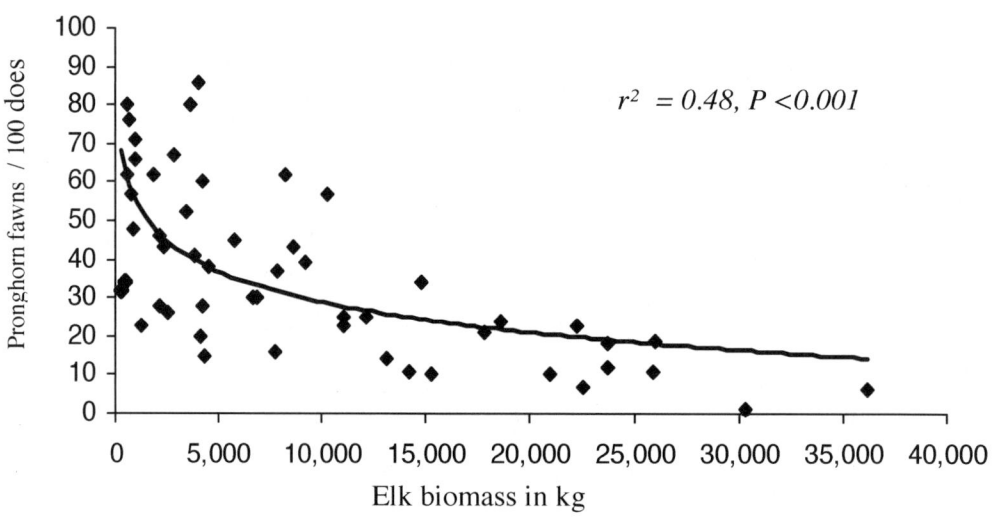

Figure 5. Estimated elk biomass compared with pronghorn fawn recruitment, 1950–2003.

al. 2004; Brown et al. 2006). This relationship could indicate that pronghorn recruitment and population size are increasingly dependent upon annual forage conditions due to the decline in perennial forage brought on by an increase in ungulate use after 1980 (Brown et al. 2004; Brown et al. 2006b).

Over the 53 years covered by the data, elk, mule deer, and pronghorn exhibited different responses to changes in environmental conditions such as drought and grazing intensity. Elk recruitment rates, however, almost always exceeded pronghorn recruitment rates—almost exclusively so during the last few decades (Figure 6). The only years in which the pronghorn recruitment rate greatly exceeded elk recruitment were 1950–56, 1959, 1969, and 1970. The declining recruitment rates and biomass

of pronghorn after 1950 would indicate that fawn recruitment was an important contributor to population size and trends, as postulated by Neff and Woolsey (1979).

Although aerial gunning of coyotes appeared to result in improved pronghorn recruitment in 1982 and 1983, little improvement was noted after treatments in 1988 and 2002. Comparison between areas treated for coyotes and similar-sized control areas on Anderson Mesa in 2002 and 2003 showed no significant ($P > 0.05$) differences in 2002, and both positive and negative differences in 2003 so that the relationship between these two variables was ambiguous (Richard Miller, AGFD, personal communication). We interpret these data as supporting the hypothesis that nutritious forage availability, especially during late winter and early

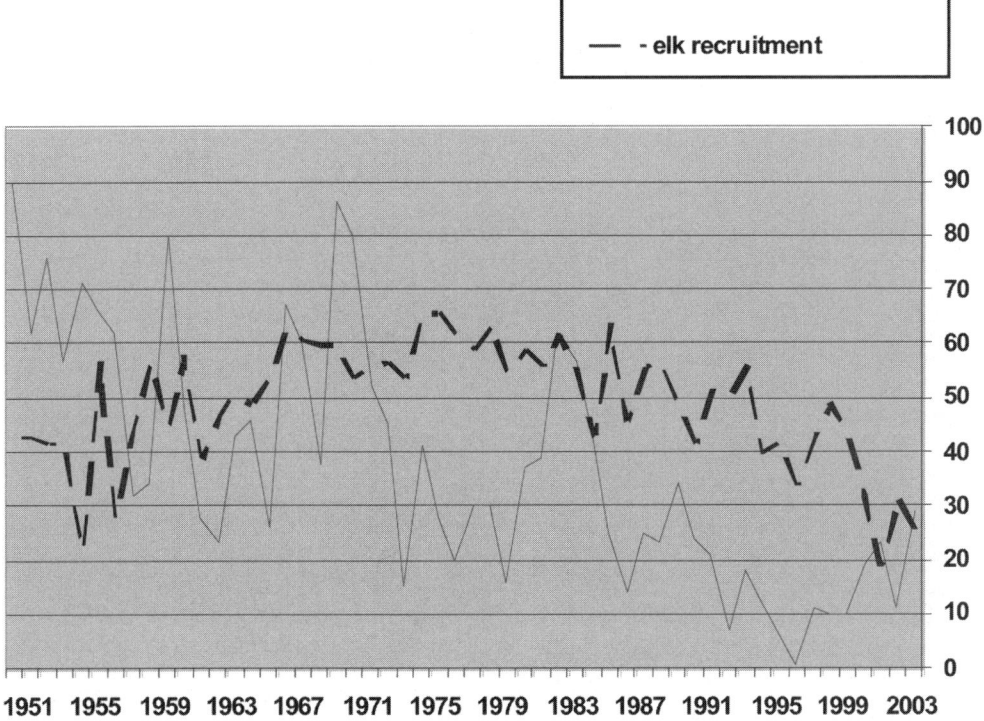

Figure 6. Relative biomass of ungulates in Game Management Unit 5, 1950-2002

spring, limits pronghorn recruitment and population size in Unit 5. Pronghorn recruitment in Unit 5 is significantly reduced when elk biomass is high, and although livestock and bison constitute the largest collective biomass (Figure 7), it is increased elk numbers that most closely correlate with the decline in pronghorn recruitment. This low recruitment, coupled with the impacts of drought, appears to be responsible for the decline of the Anderson Mesa pronghorn population during the 1990s.

CHANGE IN RANGE CONDITIONS

Photographs and descriptions of range conditions on Anderson Mesa indicate a progressive deterioration over time (USFS 1931; Knipe 1940; Peterson 1956; Wallmo 1951; Miller 2001). Changes have included a decline in the diversity of cool-season grasses, an increase in warm-season grasses such as blue grama, a decrease in succulent forbs, an increase in noxious forbs, and an increase in pinyons, junipers, and ponderosa pine. This paucity in variety and quality of nutritious forage caused Yoakum (2003) and

Miller and Drake (2005) to unfavorably compare the nutritional qualities of Anderson Mesa with analogous areas such as Garland Prairie west of Flagstaff despite summer sheep grazing persisting longer on the latter area. We believe that this lack of nutritious vegetation, exacerbated by drought and continued heavy grazing by livestock and elk, has been the primary cause for the decline to crisis levels of pronghorn fawn recruitment and population numbers.

Elk and other ungulates remove all, or nearly all of the annual growth in perennial browse plants, and plant predation is depleting the actual browse (Figure 8). This elk-caused damage to browse prior to the spring arrival of pronghorn on Anderson Mesa may also be compounded by the physiological and spatial advantages of elk over pronghorn on winter and intermediate ranges (Keegan and Wakeling 2003; Wisdom et al. 2004).

Pronghorn populations on Anderson Mesa have declined several times before and rebounded. The species is highly adaptable and, given proper assistance, there is no

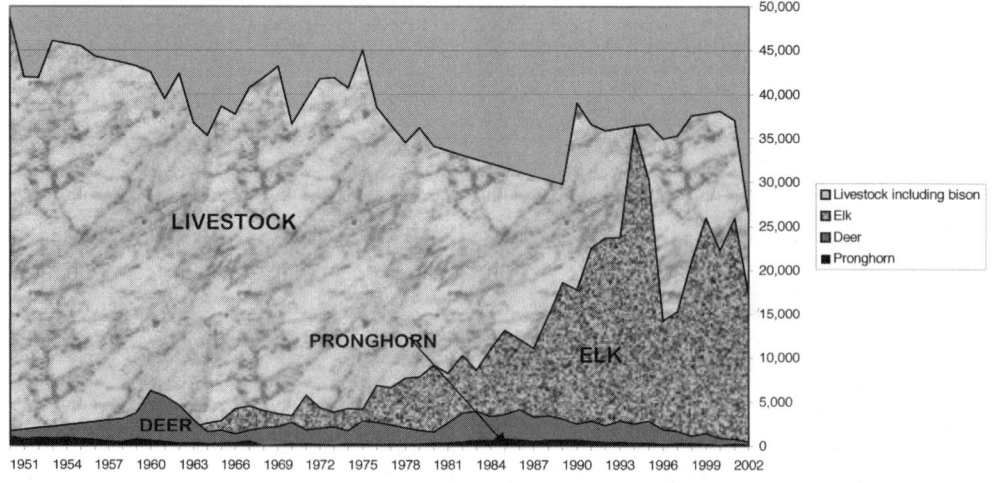

Figure 7. Recruitment rates of pronghorn compared with elk, 1950–2004

reason to think that pronghorn recruitment rates cannot again improve. But whether pronghorn on Anderson Mesa can again attain their former numbers is problematic, as these animals are now subsisting on a declining forage base due to excessive elk and livestock use. Sportsmen, ranchers, and the general public will have to press management agencies to reduce ungulate pressures and improve forage quality if mean annual pronghorn recruitment rates are to again exceed maintenance levels.

ACKNOWLEDGMENTS

Highly important to the preparation of this historical analysis were Don Neff's unpublished Arizona Game and Fish Department reports and personal conversations with Tom Britt, Ken Clay, Don Farmer, John Goodwin, Mike Hannemann, Rick Miller, John Phelps, Harley Shaw, Paul Webb, Steve Gallizioli, and Jim Yoakum. Henry Provencio drafted Figure 1 and provided important livestock information, and Amber Munig of the AGFD supplied the all-important survey and hunt data. The graphs were prepared by Elaine Brown and Manuela Suarez-Gonzalez.

a 2002

Figure 8 (a, b, and c; see next page). Annual appearance of a stand of wolfberry, 2002–2004.

b 2003

c 2003

LITERATURE CITED

Arrington, O. N. 1945. Raymond Ranch Antelope Refuge Utilization and Development Survey. Special Report, 15 April 1945.

Arrington, O. N. 1947. Predator Control as a Management Factor in Antelope Reproduction. Project 22-R, 20 August 1947. Arizona Game and Fish Commission, Phoenix.

Arrington, O. N., and A. E. Edwards. 1951. Predator control as a factor in antelope management. Transactions North American Wildlife Resource Conference 16: 179–193.

Arizona Game and Fish Department and Coconino National Forest. 2004. Anderson Mesa Pronghorn Plan—Implementation Plan Accomplishments and Planned Activities. 31 January 2004. Arizona Game and Fish Department, Region II, Flagstaff.

Barnes, W. C. 1913. Western grazing grounds and forest ranges. The Breeders Gazette, Chicago.

Barnes, W. C. 1988. Arizona Place Names. University of Arizona Press, Tucson.

Bennett, M. 2002. Population and growth rates of feral horses on Sheldon National Wildlife Refuge. Typed report. U.S. Fish and Wildlife Service, Sheldon–Hart Mountain Refuge Complex, Lakeview, Oregon.

Boccadori, S. J., and R. A. Garrott. 2002. Effects of winter range on a pronghorn population in Yellowstone National Park. In Proceedings Biennial Pronghorn Workshop, Kearny, Nebraska vol. 20, p. 114.

Brockway, D. G., R. G. Gatewood, R. B. Paris. 2002. Restoring fire as an ecological process in shortgrass prairie ecosystems: Initial effects of prescribed burning during the dormant and growing seasons. Journal Environmental Management 65: 135–152.

Brown, D. E. (editor). 1983. The wolf in the Southwest: The making of an endangered species. University of Arizona Press, Tucson.

Brown, D. E. (editor). 1994. Biotic communities: Southwestern United States and northwestern Mexico. University of Utah Press, Salt Lake City.

Brown, D. E. (editor). 2007. Wildlife in Arizona Territory, 1865–1912. Arizona Game and Fish Department, Phoenix.

Brown, D. E., W. F. Fagan, R. M. Lee, H. G. Shaw, and B. Turner. 2002. Winter precipitation and pronghorn fawn survival in the Southwest. In Proceedings Biennial Pronghorn Workshop, Kearny, Nebraska vol. 20, pp. 115–122.

Brown, D. E., W. F. Fagan, J. Louie, and H. Provencio. 2004. Elk as a factor affecting pronghorn productivity and population levels on Anderson Mesa, Arizona. Proceedings Pronghorn Workshop 21: 38–53.

Brown, D. E., and C. D. Mitchell. 2005. A comparison of pronghorn horn size in relation to environmental factors. Proceedings of the Symposium, Managing Wildlife in the Southwest: New Challenges for the 21st Century, Alpine , Texas, pp. 49–54. Southwest Section of The Wildlife Society.

Brown, D. E., M. Gonzalez-Suarez, and J. Handka. 2006a. Factors affecting variation in pronghorn horn growth. In Proceedings Pronghorn Symposium 2006, edited by K. A. Cearley and S. Nelle, pp. 77–84. Texas Cooperative Extension, College Station.

Brown, D. E., D. Warnecke, and T. McKinney. 2006b. Effects of midsummer drought on mortality of doe pronghorn (*Antilocapra americana*). Southwestern Naturalist 51: 220–225.

Byers, C. R., and G. A. Bettas (editors). 1999. Records of North American Big Game. 11th ed. Boone and Crockett Club, Missoula, Montana.

Cooper, C. T. 1960. Changes in vegetation, structure and growth of southwestern pine forests since settlement. Ecological Monographs 30 (2): 129–164.

Davis, G. P., Jr. 2001. Man and wildlife in Arizona: The American exploration period, 1824–1865. Arizona Game and Fish Department, Phoenix.

Dodson, M. M. 1991. Russell Culbreath: A legend born in our time. American Trapper (Jan– Feb): 45–52 and (Mar–Apr): 42–48.

Edwards, A. C. 1947. Antelope airplane survey. Project 26-R-1, Job 1. Arizona Game and Fish Commission, Phoenix.

Edwards, A. C. 1950. Antelope investigations. Completion Report, May 1, 1950. Project 26-R-3, Job 1. Arizona Game and Fish Commission, Phoenix.

Erling, H. G. 1956a. Report on a study of reproduction in antelope. Special Report, January 1956. Project W-53-R-5, WP2, J1.

Erling, H. G. 1956b. Report on a study of reproduction in antelope. Project W-53R-6,WP2, J1 . Special Report, April 1956.

Frost, M. H., Jr. 1945a. 1945 antelope survey by plane. Project 9-R, Special Report, July 1945. Arizona Game and Fish Commission, Phoenix.

Frost, M. H., Jr. 1945b. Vitamin deficiency as a possible factor in the decrease of antelope on Anderson Mesa. Project 9-R, Arizona Game and Fish Commission, Phoenix.

Gay, S. M. 1984. Winter range forage availability and utilization of range forage by pronghorn (*Antilocapra americana*) near Anderson Mesa. Master's thesis, Northern Arizona University, Flagstaff.

Goldman, E. A. 1931. General remarks. Report supplementing the report of the committee investigating big game range on certain national forests in Arizona, April 7–26, 1931, inclusive. USDA Forest Service, Apache-Sitgreaves National Forest, Springerville, Arizona.

Hamblin, J. 1881. A narrative of his personal experience, as a frontiersman, missionary to the Indians and explorer. LDS Juvenile Instructor Office, Salt Lake City, Faith Promoting Series 5.

Hitchcock, J. R. W. 1886. In the San Francisco Forests. Outing Magazine (January): 398–405.

Keegan, T. W., and B. F. Wakeling. 2003. Elk and deer competition. In Mule Deer Conservation: Issues and Management Strategies, edited by J. C. deVos, Jr., M. R. Conover, and N. E. Headrick, pp. 139–150. Berryman Institute Press, Utah State University, Logan.

Kartchner, K. C. 1931. Arizona's pronghorn. Arizona Wildlife 3 (Sept): 11–13.

Knipe, T. 1940. Antelope survey on Anderson Mesa, April 1 to August 1, 1940. Arizona 9-R: Antelope. Arizona Game and Fish Commission, Phoenix.

Knipe, T. 1941. The Anderson Mesa antelope hunt. Project 9-R, Special Report, Arizona Game and Fish Commission, Phoenix.

Knipe, T. 1942a. Anderson Mesa antelope count, January 20–22, Arizona 9-R, Antelope. Arizona Game and Fish Commission, Phoenix.

Knipe, T. 1942b. Report on the summer survey of the antelope herds in the Plateau region of northern Arizona. Project 9-R, Special Report, April 15–July 31, 1942 Arizona Game and Fish Commission, Phoenix.

Knipe, T. 1942c. Report to the Arizona Game and Fish Commission on antelope in northern Arizona. Project 9-R, Special Report, Antelope Survey, July 7, 1942. Arizona Game and Fish Commission, Phoenix.

Knipe, T. 1942d. Statistical report on the 1942 antelope hunt in northern Arizona. Project 9-R, Special Report: Antelope. Arizona Game and Fish Commission, Phoenix.

Knipe, T. 1944a. The status of the antelope herds of northern Arizona. P-R Project 9R. Arizona Game and Fish Commission, Phoenix.

Knipe, T. 1944b. Statistical report on the 1943 antelope hunt. Project 9-R, Special Report, Antelope, March 1944. Arizona Game and Fish Commission, Phoenix.

Lauver, M. E. 1938. A history of the use and management of the forested lands of Arizona, 1862–1936. Master's thesis, University of Arizona, Tucson.

Leiberg, J. B., T. F. Rixon, and A. Dodwell. 1904. Forest conditions in the San Francisco Forest Reserve, Arizona. USDI Geological Survey Professional Paper 22. Washington, D. C.

Lewis, N. L. 2000. Arizona wildlife trophies. Arizona Wildlife Federation, Mesa.

Loftfield, J. V. C. 1924. Quantitative studies of the vegetation of the grazing ranges of northern Arizona. Doctoral dissertation, University of Arizona, Tucson.

Mcguire, E. 1941a. Report on the results of antelope investigations in the Anderson Mesa area, November 22–January 18, 1941. Project 9-R: Antelope. Arizona Game and Fish Commission, Phoenix.

Mcguire, E. 1941b. Short summary of statewide antelope work to date. Arizona-9R, July 15, 1941. Arizona Game and Fish Commission, Phoenix.

Mcguire, E. 1941c. Report on the results of antelope investigations in the Anderson Mesa area, November 22—January 13, 1941. Project 9-R: Antelope. Arizona Game and Fish Commission, Phoenix.

Merrick, D. G. 1932. The effect of grazing on the density of forage grasses of the Coconino Plateau in northern Arizona. Unpublished report. Rocky Mountain Forest and Range Experiment Station, Flagstaff, Arizona.

Miller, R. 2001. Vegetation change in forest grasses on the Coconino National Forest since 1880. Review draft. Arizona Game and Fish Department, Region II, Flagstaff, Arizona.

Miller, W. H., and M. Drake. 2005. Nutritional concerns of pronghorn antelope on Anderson Mesa and Garland Prairie, Arizona. Final report to Arizona Game and Fish Department, Phoenix.

Munig, A. 2004. Pronghorn Translocation History. Game Management Branch, Arizona Game and Fish Department, Phoenix.

Neff, D. J. 1974a. Notes on the land use history of Anderson Mesa and the Canyon Diablo Plains in the Forest Service era, 1906 to 1940. Part 1. The homesteaders. Arizona Game and Fish Department, Phoenix. Unpublished manuscript on file.

Neff, D. J. Notes on the land use history of Anderson Mesa in the Forest Service era, 1906 to 1940. Part II. The livestock industry. Arizona Game and Fish Department, Phoenix. Unpublished on file.

Neff, D. J. 1974c. The life of a homesteader. Arizona Game and Fish Department, Phoenix. Unpublished manuscript on file.

Neff, D. J., R. H. Smith, and N. G. Woolsey. 1985. Pronghorn antelope mortality study. Arizona Game and Fish Department Project W-78-R. Final Report 1-22. Phoenix.

Neff, D. J., and N. G. Woolsey. 1979. Effect of predation by coyotes on antelope fawn survival on Anderson Mesa. Arizona Game and Fish Department Project W-78-R. Special Report 8: 1–36.

Neff, D. J., and N. G. Woolsey. 1980. Coyote predation on neonatal fawns on Anderson Mesa, Arizona. Proceedings Pronghorn Antelope Workshop vol. 9, pp. 80–93. Rio Rico, Arizona.

Nelson, E. W. 1925. Status of the pronghorned antelope: 1922–1924. USDA Bulletin 1346. Washington DC.

Nichol, A. A. 1931. Informal criticism. Report on a survey of the game refuges in the Tusayan, Coconino, Sitgreaves, Apache and Crook National Forests, April 9 to May 1, 1931. USDA Forest Service, Apache-Sigreaves National Forest, Springerville, Arizona.

Nichol, A. A. March 3, 1941. Memorandum to the Director, Division of Federal-Aid.

Niehuis, C. C. 1943. Trapping of antelope. Arizona Wildlife and Sportsman (Aug): 6.

O'Connor, J. 1934. Arizona's antelope problem. Outdoor Life (May): 32–33.

O'Connor, J. 1939a. Antelopes to Order. Outdoor Life (April): 40–41, 121, 124.

O'Connor, J. 1939b. Game in the Desert (Swallow of the Plains). Derrydale Press, New York.

O'Connor, J. 1940. Conversation on Conservation. Arizona Wildlife and Sportsman 11 (Feb): 2, 4, 5, 8.

O'Connor, J. 1943. The incredible antelope. Outdoor Life (Sept): 22–23, 61.

O'Connor, J. 1944. On hunters and game. Arizona Wildlife and Sportsman (Dec): 11.

O'Connor, J. 1945. Antelope Aplenty … pick your buck. Outdoor Life (May): 28, 29,107, 108.

O'Connor, J. 1961. The pronghorn. In The Big Game Animals of North America, pp. 67–76. E. F. Dutton, New York.

O'Gara, B. W., and H. G. Shaw. 2004. Predation. In Pronghorn Ecology and Management, edited by B. W. O'Gara and J. G. Yoakum, pp. 337–377. University Press of Colorado and the Wildlife Management Institute, Boulder.

Palmer, W. C. 1965. Meteorological drought. Research Paper No. 45. U.S. Weather Bureau, National Oceanic and Atmospheric Administration, Library and Information Services, Washington DC.

Peterson, W. 1956. Sixty miles an hour on the hoof. Arizona Highways 32 (June): 30–36.

Plummer, F. G. 1904. Forest conditions in the Black Mesa Forest Reserve. U.S. Geological Survey, Washington DC.

Powell, L. E. 1952a. Antelope hunt information. Completion Report, September, 1952. Project W-53-R-2, J-1. Arizona Game and Fish Commission, Phoenix.

Powell, L. E. 1952b. Annual northern Arizona antelope survey. Completion Report, August, 1952. Project 53-R-2, Work Plan 3, Job 1. Arizona Game and Fish Commission, Phoenix.

Powell, L. E. 1952c. Annual northern Arizona antelope surveys. Completion Report, December 1952. Project W-53-R-3, Work Plan 3, Job 1. Arizona Game and Fish Commission, Phoenix.

Powell, L. E. 1953. 1952 antelope hunt information. Completion Report, February, 1953. Project W-53R-3, Work Plan 2, Job 1.Arizona Game and Fish Commission, Phoenix.

Powell, L. E. 1954a. Northern Arizona antelope survey. Completion Report, April, 1954. Project W- 53R-4, WP3, J1. Arizona Game and Fish Department, Phoenix.

Powell, L. E. 1954b. Antelope hunt information. Completion Report, February 1954. Project W-53R-4, WP2, J1. Arizona Game and Fish Department, Phoenix.

Powell, L. E. 1955a. Antelope hunt information, Completion Report, April, 1955. Project W-53R-5, WP2, J1. Arizona Game and Fish Department, Phoenix.

Powell, L. E. 1955b. Northern Arizona antelope survey. Completion Report, July, 1955. Project W-53-R-5, WP3, J1. Arizona Game and Fish Department, Phoenix.

Powell, L. E. 1956. Antelope hunt information. Completion Report, February 1956. Project W-53-R-6, WP2, J1. Arizona Game and Fish Department, Phoenix.

Powell, L. E. 1957a. Antelope hunt information. Completion Report, February 1957. Project W-53R-7, WP2, J1. Arizona Game and Fish Department, Phoenix.

Powell, L. E. 1957b. Arizona antelope survey. Completion Report, April 1957. Project W-53-7, WP3, J1. Arizona Game and Fish Department, Phoenix.

Powell, L. E., and P. M. Webb. 1956. Northern Arizona antelope survey. Completion Report, May 1956. Project W-53-R-6, WP3, J1. Arizona Game and Fish Department, Phoenix.

Reid, J. J., and D. E. Doyel. 1986. Emil W. Haury's Prehistory of the American Southwest. University of Arizona Press, Tucson.

Shaw, H. G. 2000. Assessment of Arizona pronghorn research needs as related to aerial gunning of coyotes. Report to Arizona Antelope Foundation.

Talbert, M. W., and R. R. Hill. 1922. Progress report on the range study plots in the Coconino National Forest, comprising a description of the project and a digest. Unpublished report, Rocky Mountain Forest and Range Experiment Station, Flagstaff, Arizona.

U.S. Forest Service. 1931. Report of the committee investigating big game ranges on certain national forests in Arizona. USDA Forest Service, Apache-Sitgreaves National Forest, Springerville, Arizona.

U.S. Forest Service Region 3. 1932. Report on wildlife within National Forests of Arizona and New Mexico, 1931. USDA Forest Service, Washington DC.

Wallmo, O. C. 1951. Antelope range preference study. Completion Report, July 24, 1951. Project 46-R-2, J-5, Arizona Game and Fish Commission, Phoenix.

Webb, P. M. 2005. The Anderson Mesa antelope drift trap. The Pronghorn (Arizona Antelope Foundation newsletter) 12 (3): 12–13.

Welsh, G. W. 1958a. Arizona antelope survey. Completion Report, March 1958. Project W-53R-8, WP3, J1. Arizona Game and Fish Department, Phoenix.

Welsh, G. W. 1958b. Antelope hunt information. Completion Report, April, 1958. Arizona Game and Fish Department, Phoenix.

Welsh, G. W. 1958c. Arizona antelope survey. Project W-53R9, WP3, J1. Completion Report, June 16–July 19, 1958. Arizona Game and Fish Department, Phoenix.

Welsh, G. W. 1959. Antelope hunt information. Completion Report. Project W-53R-9, WP2, J1. Arizona Game and Fish Department, Phoenix.

Whipple, A. M. 1941. A Pathfinder in the Southwest, edited by G. Foreman. University of Oklahoma Press, Norman.

White, R. W. 1969. Antelope winter kill, Arizona style. Proceedings Western Conference of Game and Fish Commissioners vol. 49, pp. 251–254.

Wilkins, A. S. 1944. 1944 antelope survey by plane. Project 9-R, Special Report, Antelope. Arizona Game and Fish Commission, Phoenix.

Wilkins, A. S., and P. Welles. 1944. Antelope survey—1944. Project 9-R, Special Report. May 22 through June 1, 1944. Arizona Game and Fish Commission, Phoenix.

Wisdom, M. J., N. J. Cimon, B. K. Johnson, E. O. Garton, and J. W. Thomas. 2004. Spatial partitioning of mule deer and elk in relation to traffic. Transactions of the North American Wildlife and Natural Resource Conference 69: 509–530.

Wright, J. T. 1946. Coyote vs. antelope, Anderson Mesa, Coconino County, Arizona. Project 9-R. Arizona Game and Fish Commission, Phoenix.

Yoakum, J. D. 2004. Foraging ecology, diet studies and nutrient values. In Pronghorn Ecology and Management, edited by B. W. O'Gara and J. D. Yoakum, pp. 447–502. University Press of Colorado and Wildlife Management Institute, Boulder.

Yoakum, J. D., R. Barrett, H. G. Shaw, and T. J. Pojar. 2004. Pronghorn neonates, predators, and predator control. Proceedings Pronghorn Workshop vol. 21, pp. 73–95.

EFFECTS OF FENCED TRANSPORTATION CORRIDORS ON PRONGHORN MOVEMENTS AT PETRIFIED FOREST NATIONAL PARK, ARIZONA

Jan V. Hart, Charles van Riper III, David J. Mattson, and Terence R. Arundel

Pronghorn (*Antilocapra americana*) are a species of concern over most of the open landscapes of western North America that they inhabit (O'Gara and Yoakum 2004). Preferred habitats for pronghorn are grasslands and shrubland-steppes, and their current range includes a large portion of the western United States, a much smaller area of adjacent southern Canada, and some isolated parts of northern Mexico. Although once present in far greater numbers than occur today, pronghorn have rebounded from near extirpation since the beginning of the twentieth century (O'Gara and Yoakum 2004). Several factors contributed to the precipitous decline in pronghorn populations, but beginning with the arrival of Europeans the chief driver was overhunting. Protection for pronghorn, in the form of hunting bans of limited effectiveness, began as early as 1883, but their numbers continued to decline until about 1920 (O'Gara and Yoakum 2004). During the following decade, numerous pronghorn refuges were established, though most were closed to hunting but not livestock grazing. In some areas large-scale predator control, mostly of coyotes, was also undertaken, and this too may have helped pronghorn recovery (O'Gara and Yoakum 2004).

Even before over-hunting began to result in pronghorn declines, other factors had emerged that would largely govern the fate of this animal throughout the remainder of the twentieth century and on to the present. Beginning in the mid-1800s, domestic livestock were being introduced into much of the semi-arid rangeland of the West. With the coming of the railroads during the 1880s, large cattle herds were soon encroaching upon and displacing pronghorn from their native ranges. The cattle were followed by sheep, which totaled as many as 40 million during the 1890s in the western United States (Wagner 1978), and these competed directly with pronghorn for forage. Although competition for resources with livestock in the late 1800s likely had adverse effects on pronghorn, perhaps the greatest detriment to their survival was the fencing of pasture and rangeland designed to restrict livestock movement (Brown 1994).

Fences, especially the woven fences that were used to restrict sheep, form impassable barriers to pronghorn, and greatly restrict seasonal movements. Pronghorn evolved on the open plains where speed was important for predator avoidance, but jumping ability was of little value. Consequently, although some animals have demonstrated the ability to jump effectively over fences, most pronghorn seldom do. Instead, pronghorn may travel many "miles" along a fence line searching for a gap within the fence or a space between the fence and ground that is large enough to crawl under. Where fence lines are impervious, pronghorn can become trapped, suffering mortality from lack of suitable forage or the effects of weather (Spillett et al. 1967). In the extreme, the barrier imposed by fences, coupled with particularly harsh winter conditions, can result in the loss of entire populations (e.g., Martinka 1967).

Today, fences remain throughout pronghorn range, and human development, highways, and railways have further fragmented their habitat. Although direct mortality of pronghorn from collisions on highways and railways is rare, it can under certain circumstances result in significant losses (Ward et al. 1976; Mitchell 1980). The presence of fences along transportation corridors, often to exclude livestock, creates a particularly difficult combination of barriers for pronghorn to safely negotiate.

Pronghorn movements are generally governed by forage and water requirements, as influenced by seasonal conditions (Allen et al. 1984; O'Gara and Yoakum 1992; Ockenfels et al. 1997; Bright and van Riper 2000). Yoakum (1978) found that weather, latitude, altitude, and rangeland condition were key factors influencing the timing and length of seasonal movements. In one extreme, where winter conditions were particularly severe, seasonal pronghorn movements were documented in excess of 200 miles (320 km; Riddle 1990). Conversely, in a Colorado study Firchow (1986) documented a greatest mean distance traveled of 4.8 miles (7.7 km) between winter and spring ranges by pronghorn does, which traveled further than bucks.

Where conditions require that pronghorn move long distances seasonally to satisfy nutritional requirements, fenced roads and railways serve to isolate these populations and can substantially exacerbate existing constraints on carrying capacity (O'Gara and Yoakum 1992; van Riper and Ockenfels 1998; Ockenfels et al. 2000). During their 3-year study of pronghorn at Wupatki National Monument in northern Arizona, Bright and van Riper (2000) found that a fenced two-lane highway was a significant barrier to pronghorn movements. There were no documented crossings of the fenced highway that transected the monument, but there were numerous crossings of the monument's primary paved, but unfenced road. Even where habitat requirements and seasonal conditions do not necessitate long-distance movements, fragmentation by fences, roads, and railways can disrupt traditional movement patterns and cause populations to become isolated.

Fences can be readily modified to facilitate pronghorn crossings while maintaining the intended function of livestock exclusion. On multi-strand wire fences, this is most easily accomplished by raising the height of the bottom strand, which should be of smooth wire, to at least 16 inches above the ground. Raising the bottom strand to about 20 inches (0.5 m) at regular intervals and, if barbed, sheathing it with split PVC (polyvinyl chloride) pipe (a.k.a. wildlife bars) can further enhance pronghorn fence crossings (B. Cordasco, personal communication).

Karsky (1988) has studied the effects of fence modifications to facilitate pronghorn movement, but no study has examined the effects of fence modifications along high-volume transportation corridors. Modifying fences adjacent to those corridors, such as interstate highways, might have deleterious effects when pronghorn are able to access the highway corridor and be exposed to traffic. However, in 2000 an opportunity to modify fences along a high-volume railway corridor presented itself, which created a unique opportunity to study this phenomenon in cooperation with several public agencies and the railway's owners. This report provides details of that study and information about pronghorn movements that may prove useful for future pronghorn management.

STUDY AREA

Our study took place within the environs of Petrified Forest National Park (PEFO) in east-central Arizona (Figure 1). Since completion of our study, the park's size has increased substantially from approximately 94,000 acres (37,400 ha) to more than 218,000 acres (87,400 ha). In addition to preserving significant geological, paleontological, and cultural resources, PEFO protects a significant remnant of shortgrass prairie habitat, and pronghorn are frequently seen in many parts of the park. Before the recent expansion, PEFO was roughly hourglass shaped,

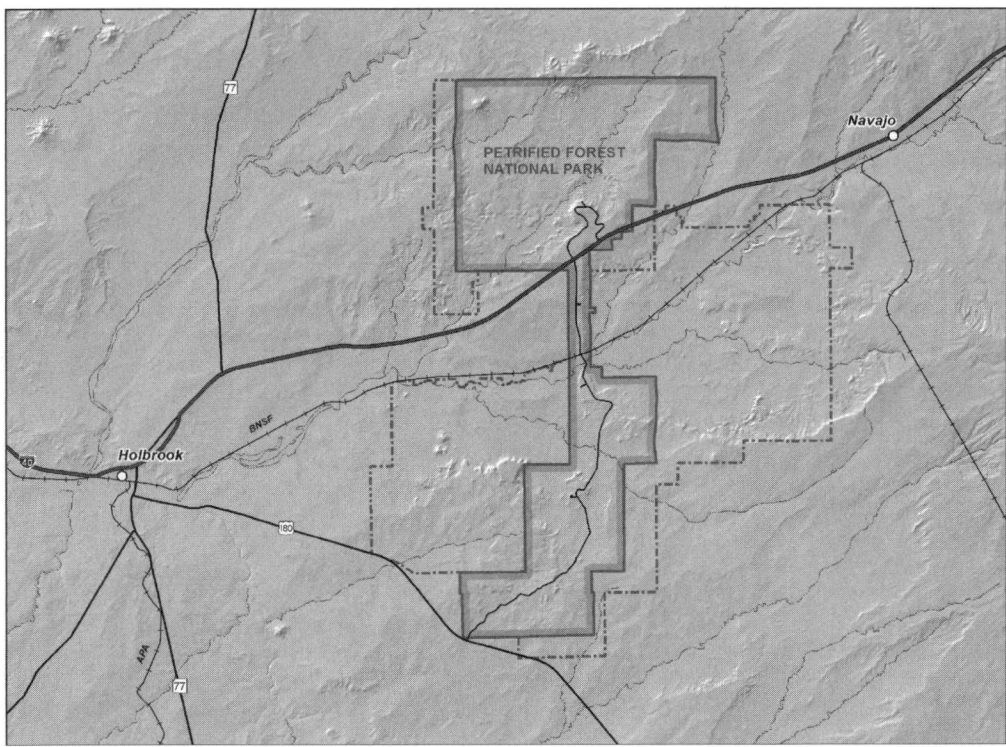

Figure 1. Map of the study area, showing Petrified Forest National Park, Interstate 40, and the BNSF railroad, with the present (2006) boundaries represented as a dashed line and those that existed during this study as the solid line. The area enclosed by these transportation corridors, extending from Holbrook on the west to the Navajo exit of I-40 on the east, constituted the range available to the isolated pronghorn population in this study.

with large northern and southern tracts joined by a narrow neck approximately 1.6 km wide and 7 km long. The main park road runs through this neck, which is traversed at its northern end by Interstate 40 (I-40) and at its southern end by the Burlington Northern and Santa Fe Railway (BNSF; Figure 2).

From a terrestrial wildlife perspective, these two double-fenced, high-volume transportation corridors effectively divide PEFO into three unequal parts, with the smallest section being the narrow neck between I-40 and the railroad. Fences restrict pronghorn travel between the sections, with the only open corridor being the main two-lane paved park road. Daily traffic volume crossing PEFO on I-40 averages almost 18,000 vehicles (Arizona Department of Transpor-

tation 2006), and approximately 100 high-speed freight and passenger trains per day pass through PEFO on the railroad. From Holbrook, some 32 km west of PEFO, to the Navajo exit on I-40 about 16 km to the east, the interstate and the railway are less than 8 km apart, with the land between them constituting the range available for the pronghorn population in this study (Figure 3).

PURPOSE

In an earlier study at PEFO, Ockenfels et al. (1997) had radio-collared 17 pronghorn (13 females, 4 males) within the area of PEFO bounded by the transportation corridors, and also in the open range south of the railway. After 1736 relocations they found no evidence that animals north or south of

Figure 2. Aerial photograph of the "neck" section at Petrified Forest National Park, showing the fence sections along the BNSF right-of-way (between the arrows) that were modified in Treatments 1 and 2. The park boundary is the vertical line at left and the main park road is at right.

the railway had crossed that right-of-way. Pronghorn north of the BNSF railway had home range shapes that were highly elongated, reflecting the constricted area available to them between the two transportation corridors, whereas pronghorn south of the BNSF exhibited home ranges of a more typical shape (O'Gara and Yoakum 1992). The pronghorn north of the BNSF were thus effectively trapped and isolated from their free-ranging counterparts to the south.

Our study was initiated to determine if modification of the fences within PEFO along the BNSF right-of-way would enable pronghorn confined north of the railway to cross the railroad tracks and move to the south. Our study design had three successive components: (1) repeat the methodology of Ockenfels et al. (1997) to verify that

pronghorn were still isolated; (2) modify sections of the fences along the railroad inside PEFO by raising the bottom wires and installing wildlife bars, then evaluating whether crossings occurred; and (3) remove the fences inside PEFO by removing the wire, and then monitoring the effects. To accomplish the last component (fence removal) it was necessary to secure the cooperation of not only the National Park Service at PEFO, but also the BNSF Railroad. It is not common for a railway to compromise on control of their right-of-way, or for the NPS to allow manipulation of their park boundaries. We are therefore extremely fortunate to have obtained the necessary permission, and are indebted to our other cooperator, the Arizona Game and Fish Department, for their efforts in building

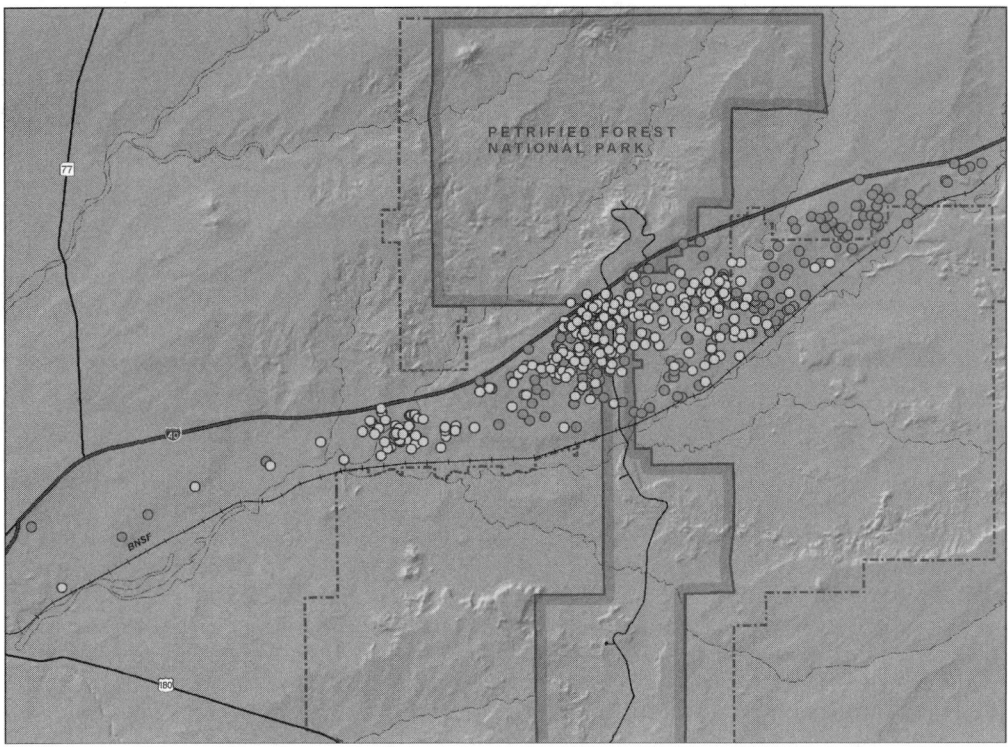

Figure 3. Map of the study area, showing 638 relocations of radio-tagged pronghorn obtained between December 2000 and September 2002 (light gray = male and dark gray = female).

consensus among all parties, facilitating the capture effort, and providing biweekly aerial telemetry of collared pronghorn.

METHODS
Capture and Relocation

Personnel from the Arizona Game and Fish Department (AGFD) captured pronghorn near PEFO on 13 December 2000, using a net gun fired from a helicopter. Each pronghorn was fitted with a VHF radio-collar and individually numbered ear tags and then released at the capture location. They were subsequently relocated using radio telemetry approximately weekly, alternating each week between aerial and ground-based methods. Ground-based relocations were plotted on 7.5' USGS topographic maps, from which Universal Transverse Mercator coordinates (UTMs) were derived to the

nearest 100 m. Aerial relocations were obtained by AGFD and recorded using Global Positioning System (GPS) fixes taken at the point of greatest VHF beacon signal strength as the aircraft passed over the target animal. These GPS fixes were collected in UTMs recorded to the nearest 100 m. All location data were collected in reference to North American Datum 1983 (NAD 83). Although radio-telemetry was used to relocate pronghorn, during both air and ground relocations tagged-animal identifications were confirmed visually and group sizes were noted. During ground-based locations, group composition by sex and age class was also recorded. These data were entered into a Microsoft Access database and then imported into ArcGIS 9.1 and ArcView 3.3 Geographic Information Systems (GIS) to conduct spatial analyses.

GIS Data Layers

We created a new spatial database from the pronghorn relocation data. Other data layers used for GIS analyses included U.S. Geological Survey digital elevation models (DEM), Arizona landcover (land use/ownership) from the USGS Southwest Regional GAP Analysis Project (GAP), and 30 m resolution vegetation type coverages, also from GAP. We further processed DEMs to develop a terrain roughness index (TRI) raster (Riley et al. 1999). We then used these base data layers to derive new spatial data and to analyze the spatial relationships between pronghorn and features of their habitat.

Treatments

The initial phase of location monitoring (pre-treatment) lasted for 143 days, beginning with the capture on 13 December 2000 and ending in early May 2001. During this time pronghorn were relocated weekly to determine whether they would cross the unmodified existing fences and if those fences still served as a barrier to movement as Ockenfels et al. (1997) had previously determined. Additionally, the ground adjacent to the unmodified fence lines was examined every 2 weeks for pronghorn tracks within or to the south of the right-of-way that might indicate crossing of the fences and possibly the railway.

Treatment 1

A section of right-of-way fence 1 km long on both sides of the BNSF railway within PEFO was modified in early May 2001 using methods similar to those described by Autenrieth et al. (2006). On every fourth span of standard four-strand barbed-wire fence, the bottom strand was detached, raised from a nominal 10–12 inches (0.25–3.0 m) to 16–20 inches (0.4–0.5 m) above ground, and then reattached. On each raised section a 4-foot (1.3 m) length of split 1.5-inch (3.8 cm) PVC water pipe was used to sheath the barbed bottom strand. The result was a larger gap under the fence with a smooth and larger diameter bottom "strand." The fence sections that were modified extended across the central portion or "neck" of PEFO, from the main park road bridge over the BNSF on the east to the park boundary on the west (see Figure 2). Weekly monitoring continued as during the pre-treatment period. Treatment 1 continued for 213 days into early December of 2001.

Treatment 2

In early December 2001, all the fence wires were dropped from both 1 km sections of right-of-way fence previously modified in Treatment 1. The wire was then "rolled" or gathered up on both sides of the modified sections, leaving a wire-free fence line, but with the majority of the posts still standing (at PEFO manager's request). Although these fences were constructed primarily with steel t-posts, some posts (mostly on the north side of the railway) were of scrap railroad ties that were rotten below ground. About 20 of these wooden posts collapsed when the wires were removed. Weekly monitoring continued as previously, and Treatment 2 lasted for 295 days, until September 2002.

Home Range, Movement, and Habitat Selection Analyses

Determining whether actual crossings of fenced barriers occurred under the influence of fence treatments was our primary goal. However, we also used relocation data to conduct further analyses that examined changes in home range, movement patterns, and habitat selection with respect to the transportation corridors and to the treatments. These analyses considered movements and expected outcomes with respect to I-40 and the railroad, to test for potential responses by pronghorn to both barriers. Our goal through these analyses was to determine whether more subtle aspects of pronghorn activity were correlated with these human features. We examined changes in home range size and the length of home range intersection with the railroad and I-40, by treatment period. We tested whether pronghorn crossings of the transportation corridors, as determined by relocation data, were similar to what would be expected if movements were random. We developed linear models to examine whether the distance between successive pronghorn locations was related to nearness to transportation corridors, as well as to the potential effects of the treatments. We conducted an analysis of vegetation-type selection by pronghorn based upon GAP vegetation types. And finally, we constructed multi-variable models for habitat selection that included a variety of physical factors, vegetation, and land use (ownership). We undertook these last two analyses to provide additional insight into behaviors and habitat relations of a population of pronghorn isolated under unique circumstances. This analysis established a baseline against which future changes may be evaluated.

Changes in Home Range

We estimated home range configurations and 95 percent contour areas using fixed kernels (Rodgers et al. 2005) and the Animal Movement 2.0 extension in ArcView (Hooge and Eichenlaub 2000). We conducted paired t-tests (Zar 1984) of home range areas among all combinations of pre-treatment, Treatment 1, and Treatment 2, testing the null hypothesis that differences between home range size and length of intersections between paired time periods did not differ from zero.

Crossings of BNSF and I-40

We used Fisher's Exact Test (Zar 1984) to test whether "observed" crossings of the railroad and I-40 by pronghorn were less than what one would expect if pronghorn movements were random relative to these linear human features. We determined 95th percentiles of distances between successive locations for each animal using Hawth's Analysis Tools extension in ArcGIS (Beyer 2004). We assumed that any pronghorn located within this distance of the railroad and interstate would have had the "opportunity" to cross these linear features prior to the next location.

Any pronghorn location within this cut-point distance of either the railroad or the

interstate was used as the initiating point of a vector of a length and compass direction that were randomly determined, but with the constraint that length had to be less than the 95th percentile of movements observed for the animal of interest. Numbers of randomly generated vectors that crossed and did not cross either the railroad or the interstate were counted, along with numbers of vectors connecting successive pronghorn locations that crossed or did not cross these features, considering only pronghorn locations within cut-point distances of the railroad and interstate. A simple four-cell matrix arraying type of vector (random vs. pronghorn) against outcome (crossed vs. didn't cross) was the basis for Fisher's Exact Test. We used number of crossings under the random model as the null hypothesis.

Distances Between Locations

We used Hawth's Analysis Tools extension in ArcGIS (Beyer 2004) to determine the distances between pronghorn locations and several other parameters of potential interest. We used GLM (general linear models; Weisberg 1985) to specify models explaining variation in distance between successive locations of pronghorns, specifically to isolate the effects of nearness to either the interstate or the railroad, as well as potential effects of the treatments. We considered distance to either the interstate or the railroad (DIST), distance to the railroad alone (DISTrr), an interaction of these nearness measures with treatment period (TRT), an interaction of sex and period of territoriality (TERR; March–September vs. the remainder of the year), an interaction of sex with fawning period (FAWN; April–May vs. the remainder of the year), and distance to home range centroid (CENT).

Vegetation Type and Land Ownership

Even though vegetation type and ownership were not included in multi-variable models explaining habitat selection by pronghorn during this study, we undertook a separate analysis focused on these variables because of potential implications for management. We considered five types that constituted some degree of consolidation of GAP types mapped for the study area (see below). We used logistic regression to test whether observed use deviated from use expected "at random" (see below) and to calculate statistics pertaining to goodness-of-fit and predictive efficiency for the model based solely on vegetation type. We tested for interactions of vegetation use with fawning season (April–May), territorial behavior by males (March–September), and sex.

Habitat Selection Models

We used logistic regression and maximum likelihood methods to specify multi-variable models explaining habitat use by pronghorn during the entire study and during treatment periods. We used logistic regression to discriminate between observed habitat use by pronghorn and use expected at random. The null hypothesis of random use was specified according to the construct of discrete choice. In practice, this meant determining the 95th percentile of observed distances between successive locations for each animal, and generating a random point for each pronghorn location constrained to fall within a buffer of a radius equal to the 95th percentile distance, centered on the animal location. We excluded random points from our analysis that fell on the far side of the railroad or interstate because they represented areas effectively unavailable to collared pronghorn.

This resulted in random points paired to each pronghorn location, constituting locations where animals could plausibly have been located if they were orienting towards modeled landscape features randomly rather than selectively. We used the Akaike Information Criterion, corrected for sample size effects, to select "best" models, estimated model parameters by maximum likelihood, and assessed model fit and performance by R^2L, area under the ROC (receiver operating characteristic) curve, and the Hosmer-Lemeshow (2000) goodness-of-fit test. In most cases, numbers of random points equaled numbers of pronghorn locations, meaning that back-transformed logits (i.e., p-values) greater than 0.5 indi-

cated some degree of selection for a locale based on modeled habitat effects.

Because the study area was bounded on opposite sides, north and south, by Interstate 40 and the railroad, respectively, distances of locations from the railroad and the interstate were intrinsically negatively correlated. We dealt with this potential collinearity of candidate explanatory effects in two ways. For the general model of habitat selection, we considered distance to either the interstate or the railroad, whichever was nearer, without differentiating between the two feature types. For investigating the potential effect of nearness to the railroad, differentiated by treatment period, we only considered locations that were nearer to the railroad than they were to the interstate. Distance from the centroid of pronghorn ranges was also theoretically correlated with distance to both the railroad and the interstate, because pronghorn were confined to the space between these two features. However, correlations were, in fact, low ($r < 0.2$).

Candidate explanatory effects included VEG (vegetation type based on mapped GAP types; see above); TRI (terrain roughness index); DIST (distance to either the railroad or the interstate, whichever was nearer); DISTrr (distance to the railroad); DISTint (distance to the interstate); CENT (distance to the centroid of pronghorn range); ELEV (elevation); and OWN (ownership, differentiating between tribal, private, BLM, and NPS).

RESULTS
Capture and Relocation

Nine pronghorn (6 females and 3 males) were captured and radio-collared. During the first aerial relocation effort 1 week following capture, Arizona Game and Fish Department personnel detected a mortality signal from one of the females. An AGFD game warden located the pronghorn remains and estimated that the animal died soon after its release, but he could not determine cause of death because the carcass had been scavenged. The eight remaining pronghorn (5 females and 3 males) were relocated a total of 638 times during this study (see

Figure 3). One other female died before the end of the study, early in Treatment 2, after having been relocated 48 times. USGS personnel located the remains of this pronghorn, which was also transmitting a mortality signal, during a ground-based relocation session. Again, cause of death could not be determined due to extensive scavenging of the carcass. The remaining seven pronghorn (4 females and 3 males) survived the study and were relocated a total of 84–85 times each.

Treatments

During the pre-treatment phase of the study each pronghorn was relocated ca. 21 times (166 total relocations). During this phase, no pronghorn was visually confirmed to be south of the railroad tracks, from which we inferred that no radio-marked pronghorn had crossed the tracks. On one occasion, during inspection of the fence lines for evidence of pronghorn crossings, we observed several sets of pronghorn hoof prints within the railroad right-of-way—that is, within the double fences along the north side of the tracks. It was evident that these pronghorn had gone under the north fence and walked along the fence for 40–50 m inside the right-of-way. Judging from their tracks it appeared that the pronghorn were attracted to pooled water in the ruts of the service vehicle road that parallels the railroad. The tracks departing the service road crossed back under the fence. There was no evidence that any of the pronghorn had gone beyond the service road towards the tracks, so we determined that no crossing of the tracks had occurred.

Treatment 1

During Treatment 1, following fence modification, each pronghorn was relocated 24 times (192 total locations). Again, no locations were recorded south of the railroad tracks, from which we inferred that no radio-marked pronghorn had crossed the tracks. Careful examination bi-weekly of the modified fences failed to reveal evidence suggesting that pronghorn had investigated the fence modifications, but the presence un-

derneath the wildlife bars of other mammal tracks (e.g., coyote, fox, and bobcat) showed that mesocarnivores were taking advantage of improved access to the railroad right-of-way.

Treatment 2

During this phase, following fence removal, each pronghorn was relocated 24 times, except for female 140, which died near 2 January 2002 (see above) and had only been relocated three times since Treatment 2 began. Data for this animal were deleted from analyses with respect to Treatment 2. During this phase, there were 277 total locations. No locations were recorded conclusively south of the railroad tracks, despite the fence wire being down for nearly 10 months. There was also no other evidence that radio-marked pronghorn had crossed the line of remaining posts into the right-of-way.

Summary of Treatment Effects

During the pre-treatment period of nearly 5 months we determined that pronghorn still appeared to be confined to the area north of the railway and south of I-40. This finding was consistent with the determination by Ockenfels et al. (1997) that these high volume transportation corridors were nearly complete barriers to pronghorn movement. Modifying the fences for 6 months to be more "pronghorn friendly" did not result in any conclusive movements by pronghorn across the fence line and railroad, nor did removing the fence wire entirely along 1 km sections of BNSF right-of-way fence on both sides of the railroad tracks. On just one occasion during the 41 ground-based relocation sessions, which included thorough scrutiny of fence lines, did we find evidence that pronghorn had briefly crossed one fence, but not the railroad tracks.

There were 12 individual pronghorn locations that, when plotted, were shown to be a short distance (≤ 100 m) north of I-40 (10 locations) or south of the railroad (2 locations), which could be interpreted as crossings. However, despite good visibility, observers did not visually confirm any of these putative crossings during the associated relocation efforts. The consensus among observers was that these locations were due to recording errors (north of the highway, by observers in the aircraft) and plotting errors (south of the railroad, by an observer on the ground).

Home Range, Movement, and Habitat Selection Analyses

Changes in Home Range

Home ranges were smaller during Treatment 1 compared to the pre-treatment period (Table 1). The length of home range intersection with both the railroad and the interstate tracked this difference in home range size, with intersections of both the railroad and the interstate also tending to be less during Treatment 1 compared to the pre-treatment period (Figure 4).

Crossings of BNSF and I-40

We retained 12 hypothetical "crossings" of the railroad and I-40 for comparison of observed crossings by pronghorn to "crossings" expected if pronghorn were orienting randomly with respect to these linear features. This choice (as opposed to using only confirmed crossings) introduced a conservative bias in our comparison of observed crossings versus number of crossings expected by chance. Even so, the discrepancy between observed "crossings" and those expected at random was striking (Figure 5). Results of Fisher's Exact Tests suggest that this discrepancy was virtually impossible by chance alone, supporting the conclusion that the interstate and the railroad both formed nearly complete barriers to these pronghorn (Table 2).

Distances Between Locations

Neither the model that included all radio-relocations nor the model considering only locations nearer to the railroad than to the interstate explained much variation in distances between successive locations (Table 3). The only consistently substantial effect in these models of distance between successive locations was that of distance from range

centroid (Figure 6). As distance from centroid increased, so did distance between successive locations. No other effect was consistent between these models, including those of treatments, nearness to transportation corridors, sex, and seasonality.

Vegetation Type and Land Ownership

Vegetation type did not provide a basis for predicting pronghorn habitat selection, but pronghorn use of NPS lands at PEFO was greater than that expected by chance. There were no strong effects of sex, fawning season, or seasonal behavior by males on the use of GAP vegetation types or land ownership, so the estimated models included only single variables (Figure 7, Table 4).

Habitat Selection Models

Some results of the multi-variable models of habitat selection were anticipated, but other results were not. Although in isolation vegetation type was significantly related to pronghorn habitat selection, there was no strong effect of VEG on habitat selection in context of any multi-variable model that we considered. Similarly, land use (ownership) was not a good predictor of habitat selection within the multi-variable framework. Consequently, land ownership was also not retained in any model despite the fact that pronghorn use varied from that expected at random when response to this variable was analyzed in isolation.

Candidate variables retained in the models (Table 5) included TRI (terrain roughness index); DIST, expressed as 1/DIST (distance to either the railroad or the interstate, whichever was nearer); DISTrr (distance to the railroad), expressed as 1/DISTrr; DISTint (distance to the interstate), expressed as 1/DISTint; CENT (distance to the centroid of pronghorn range), expressed as ln(CENT + 1); and ELEV (elevation).

As might be expected, pronghorn tended to under-use areas in rougher terrain (Figure 8, top), but the positive relation of pronghorn habitat use to elevation was not anticipated (Figure 8, bottom). Less surprising was the avoidance by pronghorn of areas near the railroad and the interstate, and of

Table 1. Home range size and length of intersection with the railroad and with Interstate 40, by treatment period. Statistical results are for paired t-tests among all combinations of pre-treatment, Treatment 1, and Treatment 2, testing the null hypothesis that differences in home range size and length of intersections between paired time periods did not differ from 0. Figures in parentheses are for n = 8; comparisons with home range size for Treatment 2 deleted animal 140, reducing n to 7 for these tests.

	Mean (ha)	Lower 95% CI	Upper 95% CI
Home Range Size			
Pre-treatment	6589 (6766)	4176 (4703)	9003 (8828)
Treatment 1	3826 (3632)	1717 (1808)	5936 (5456)
Treatment 2	6658	1672	11645
Intersection with Railroad			
Pre-treatment	8672	3293	14050
Treatment 1	3302	−106	6711
Treatment 2	5006	309	9702
Intersection with Interstate 40			
Pre-treatment	14217	8293	20142
Treatment 1	7440	2479	12402
Treatment 2	14577	−208	29362

	df	t-value	p-value
Home Range Size			
Pre-treatment vs. Treatment 1	7	4.67	0.0023
Pre-treatment vs. Treatment 2	6	−0.03	0.978
Treatment 1 vs. Treatment 2	6	−1.16	0.289
Intersection with Railroad			
Pre-treatment vs. Treatment 1	7	2.12	0.072
Pre-treatment vs. Treatment 2	6	1.19	0.272
Treatment 1 vs. Treatment 2	6	−0.70	0.506
Intersection with Interstate 40			
Pre-treatment vs. Treatment 1	7	2.84	0.025
Pre-treatment vs. Treatment 2	6	−0.06	0.956
Treatment 1 vs. Treatment 2	6	−1.12	0.299

areas nearer the periphery of their home ranges (farther from centroid; Figure 9). The parameter for 1/DIST was of greatest relevance to examining treatment effects because this measure specified the extent to which pronghorn were avoiding both the interstate and the railroad. Avoidance of both I-40 and the railroad was greatest during the pre-treatment period (Table 6).

As for the results above, the results below pertain to parameter estimates for the effect of nearness to a linear feature, estimated in context of multi-variable models, but in this case isolating the effect of the railroad (DISTrr). To achieve this, these models used only locations nearer to the railroad than to the interstate. Pronghorn expressed less avoidance of the railroad during Treatment 1 and Treatment 2 compared to the pre-treatment period (Table 7 and Figure 9).

In the results given below, the variable DIST was again evaluated within the multi-variable model, but in this case we isolated the effect of I-40 (DISTint) by using only locations nearer to the interstate than to the railroad. Because fences along the interstate

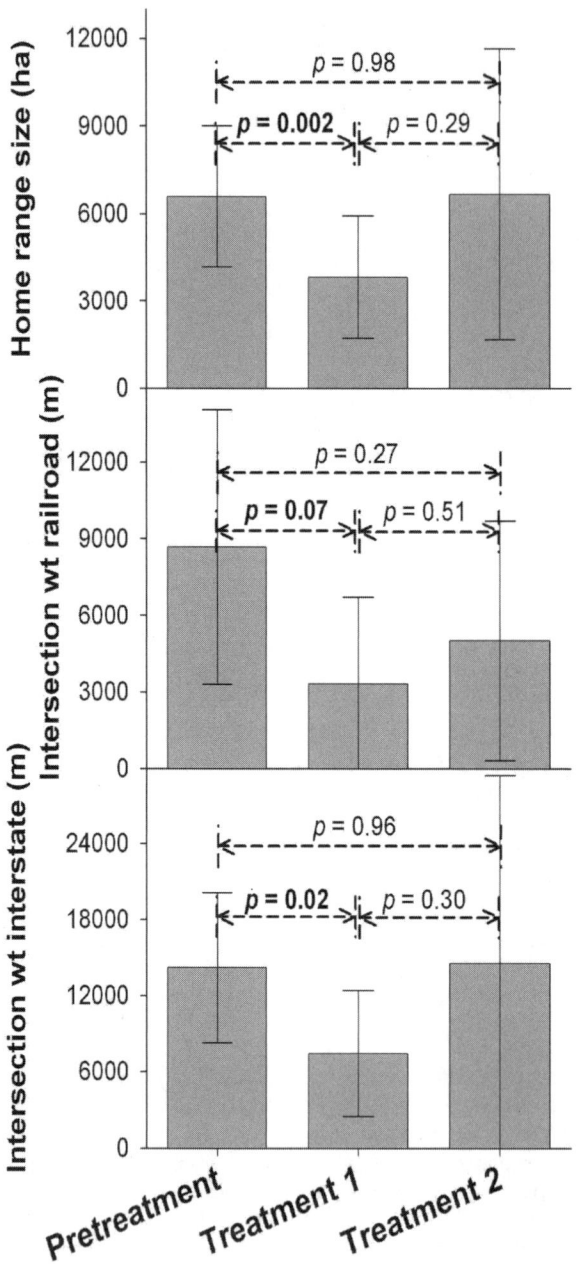

Figure 4. Pronghorn home range size (95% fixed kernel estimates) and length of home range intersection with Interstate 40 and the BNSF railroad by treatment period. Between-treatment *p*-values are shown for each paired t-test. Error bars show 95% CIs.

Table 2. Observed vs. expected crossings of the railroad and Interstate 40, with percent crossings and related 95% CIs as a percentage of total opportunities in parentheses. *P*-values are for results of Fisher's Exact Test regarding the probability that observed patterns were obtained by chance alone; 95th percentiles of distances between successive locations are also given for each animal. These distances defined the threshold at which we considered an animal to have had the "opportunity" to cross.

	Pronghorn	Random
Railroad Crossings:		
Crossed	4 (1.9% [95% CI = 0.9 – 5.1])	105 (50.2% [95% CI = 43.1 – 57.4])
Didn't cross	205	104

Probability of observing this outcome by chance alone: $p = 5.26 \times 10^{-34}$

Interstate Crossings:		
Crossed	12 (2.8% [95% CI = 1.8 – 5.0])	167 (39.7% [95% CI = 36.5 – 43.0])
Didn't cross	409	254

Probability of observing this outcome by chance alone: $p = 2.68 \times 10^{-44}$

95th percentiles of successive movements or each animal:

Animal ID	95th Percentile (m)
140 F	12041
141 M	5907
142 F	8699
144 M	4148
145 F	5827
146 M	5716
148 F	4199
149 F	7122

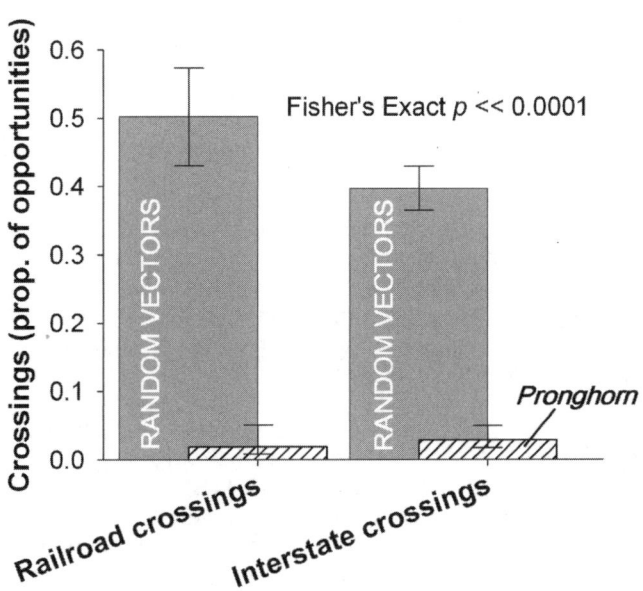

Figure 5. Comparison of "crossings" by pronghorn of BNSF railway and Interstate 40 with what would be expected if pronghorn were orienting to these linear features in a random manner ("random vectors").

Table 3. Models based on GLM analysis of factors influencing movement distances. The General Model applies to all pronghorn locations, over all treatment periods. The Railroad Model applies to all locations nearer to the railroad than to the interstate; distance to the railroad and treatment were not retained in the Railroad model, as either direct or interaction effects. Only the effect of distance to centroid was significantly related to pronghorn movement in these analyses.

Parameter	Point estimate	Lower 95% CI	Upper 95% CI	F-value
General Model:				
Intercept	−3029.9	−5991.4	−66.5	
SEX*FAWN				4.31
females, not fawning	609.8	−180.3	1400.0	
females, fawning	1398.3	483.5	2313.1	
males, not fawning	675.1	−132.6	1482.8	
male, fawning	0.0			
ln(CENT + 1)	612.2	362.9	861.4	23.26
TRT*ln(DIST + 1)				6.34
pre-treatment	82.81	−179.7	345.3	
treatment 1	−59.94	−321.0	201.1	
treatment 2	−64.58	−325.6	196.4	
$R^2 = 0.098$				
F-test, df = 7/614, F = 9.56, $p < 0.0001$				
Railroad Model :				
Intercept	−3533.6	−7247.2	180.0	
SEX*TERR				2.71
females, other	1095.8	−25.8	2217.3	
females, territorial	205.1	−787.5	1197.8	
males, other	1224.2	72.2	2376.3	
male, territorial	0.0			
ln(CENT + 1)	709.0	238.6	1179.5	8.83
$R^2 = 0.088$				
F-test, df = 4/201, F = 4.84, p = 0.0010				

were not "treated," response of pronghorn to the interstate provides some measure of control for responses to the railroad among treatment periods. As with the railroad, pronghorn expressed more avoidance of the interstate during the pre-treatment period (Table 8).

DISCUSSION

Capture and Relocation

We judged our sample size and sampling interval (weekly) to be adequate for determining course-grained responses of pronghorn to fence modifications, but only marginally sufficient for robust inferences about habitat selection. However, the primary objective of the study was simply to see if pronghorn would respond to fence modifications, which would provide useful information for management and, had it been successful, could have contributed to ending the isolation of this pronghorn population.

Treatments

Some aspects of the treatments would be worth improving for future efforts of this kind. First, it would be desirable to modify a longer extent of fence barrier along the railroad right-of-way, which was not practical during our study given the short length of right-of-way within PEFO's boundaries at the time, but it would be feasible now given the substantial enlargement of the "neck" section within current boundaries.

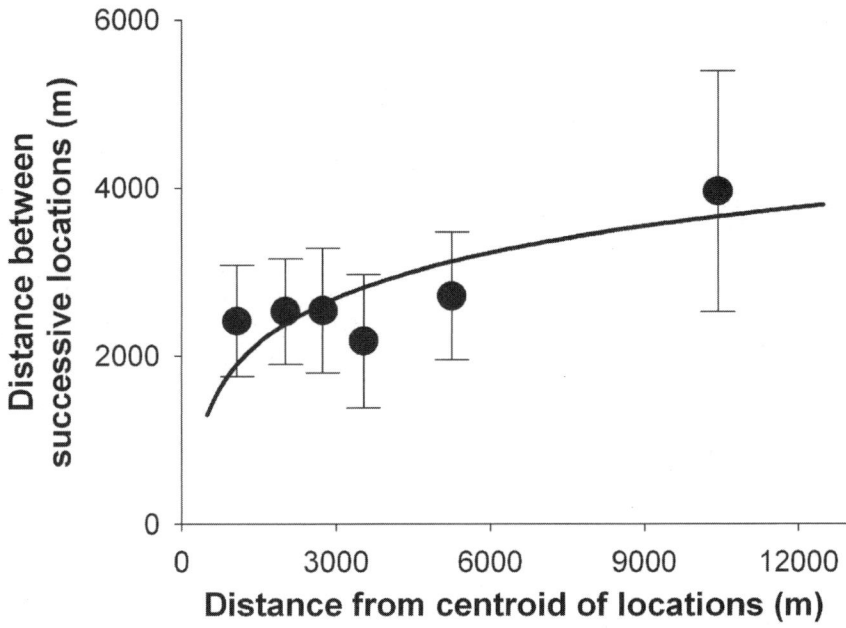

Figure 6. Distance between successive pronghorn movements in relation to distance from centroid of pronghorn locations. Filled circles and error bars show means and associated 95% CIs for septiles of the data. The solid line shows the modeled relation.

Second, fences would ideally remain modified, as in Treatment 1, for a considerably longer time than was implemented during this study. The BNSF railroad likely represents an imposing barrier to pronghorn, being both difficult to negotiate and potentially very dangerous. The prospect of being trapped inside the right-of-way with mile-long trains rumbling through approximately every 15 minutes is conceivably a powerful deterrent to exploration. Pronghorn will likely cross this sort of barrier only if allowed substantial amounts of time and perhaps additional inducements.

Inducements or attractants might take several forms. At a rudimentary level, flagging or some other device might be used to catch the pronghorns' attention. We speculate that over the years pronghorn have learned to avoid areas near the fence and railroad, with no incentive for exploratory

investigations. Strategically placed supplemental high-quality foods might also be used to attract pronghorn to areas near and beyond modified fence sections, if deemed appropriate. If the opportunity to drop the fence arises again, as in Treatment 2, it should come down entirely and stay down for as long as possible. Leaving the posts up mitigated PEFO's concern for surface disturbance impacts, but also likely made it difficult for pronghorn to notice from a distance that the fence had changed. We have also considered that if the fences were down, then pronghorn could be guided across by erecting temporary "drift fences" to block their normal movement, directing them instead towards the railroad.

Other effects of the railroad that potentially deterred pronghorn movements across the right-of-way, and that could not be controlled for in this study, include the raised

Table 4. Results for the logistic regression model discriminating observed pronghorn use of vegetation types and land ownership from use expected "at random". We used vegetation types for this analysis that were consolidations of GAP types mapped for the study area, described below in terms of the name used for this analysis and corresponding Southwest GAP Analysis Project identifiers.
Definition of vegetation types

Name for this Analysis	GAP Numeric ID	GAP Name
Grassland	76	Semi-desert grassland
Shrub steppe	67	Semi-desert shrub steppe
Shrubland	53	Blackbrush–Mormon tea shrubland
	58	Mixed salt desert shrub
	82	Greasewood flat
	108	Sand shrubland
Juniper	36	Pinyon-juniper woodland
	64	Juniper savannah
Rock	9	Bedrock canyon and tableland
	10	Shale badland

Model Results:

Vegetation type	Pronghorn locations (proportion of total vs. random points)
Grassland	0.429 [95% CI = 0.429 – 0.521]
Shrub steppe	0.522 [95% CI = 0.497 – 0.546]
Shrubland	0.485 [95% CI = 0.447 – 0.523]
Juniper	0.667 [95% CI = 0.059 – 0.980]
Rock	0.765 [95% CI = 0.434 – 0.921]

$R^2L = 0.008$; Score test, df = 4, $\chi^2 = 7.6$, $p = 0.106$

Area under the ROC (receiver operating characteristic) curve = 0.53

Hosmer-Lemeshow goodness-of-fit test, df = 2, $\chi^2 = 0.0$, $p = 1.00$

Ownership	Pronghorn locations (prop. of total vs. random points)
BLM	0.430 [95% CI = 0.363 – 0.500]
Tribal	0.390 [95% CI = 0.271 – 0.527]
NPS	0.586 [95% CI = 0.513 – 0.655]
Private/state	0.506 [95% CI = 0.505 – 0.506]

$R^2L = 0.014$; Score test, df = 3, $\chi^2 = 13.4$, $p = 0.004$

Area under the ROC (receiver operating characteristic) curve = 0.55

Hosmer-Lemeshow goodness-of-fit test, df = 2, $\chi^2 = 0.0$, $p = 1.00$

berm and gravel substrate of the rail bed. The tracks in the area where the fences were modified are raised on a gravel berm that often limits or obscures line-of-sight visibility from one side of the tracks to the other. When designing the treatments prior to the study we considered that good visibility was important, and planned to remove excess brush as necessary to create travel corridors with good visibility. When implementing the treatments we decided that vegetation was not limiting to visibility, but noted that pronghorn attempting to cross the tracks along most of the treatment area would have to climb up the berm nearly to the tracks before they would be able to see the fence line on the other side. Limited visibility across the tracks could act as a deterrent to pronghorn crossings independent of the potential barriers posed by the fences. Similarly, the coarse gravel substrate that composes the rail bed and the complete lack of vegetation across the right-of-way may also act as deterrents to exploration.

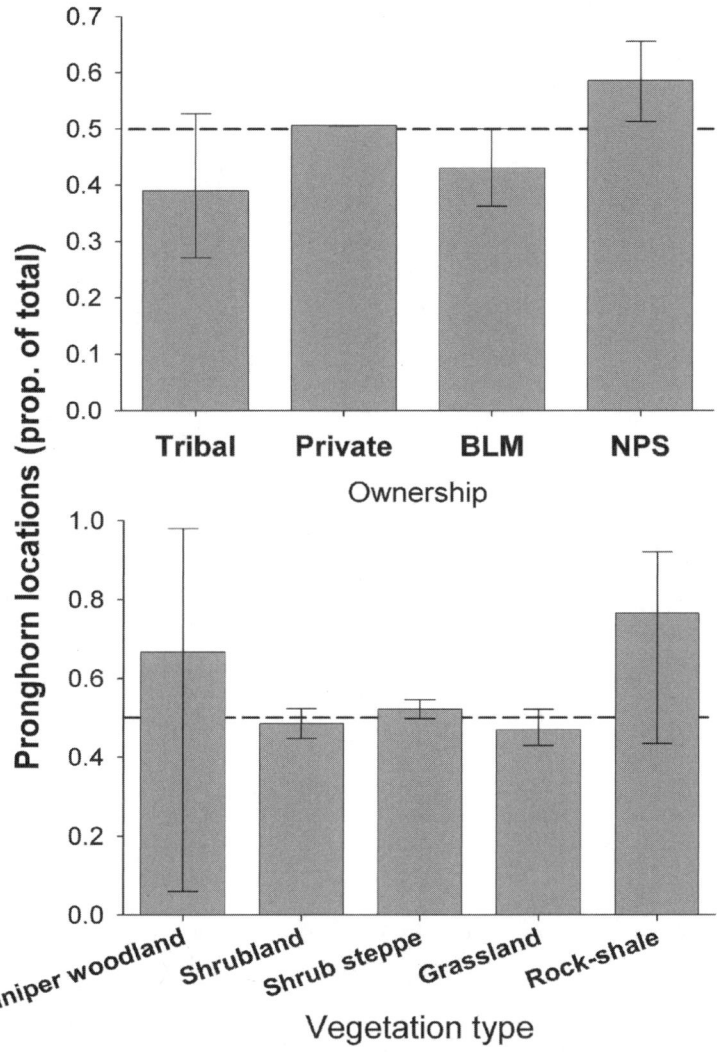

Figure 7. Results of logistic regression analysis of pronghorn locations with respect to land ownership (above) and GAP vegetation type (below). Deviation of observed use by pronghorn from use expected at random is indicated by values above (more use) or below (less use) 0.5. With respect to land ownership, in this interpretation, pronghorn used NPS lands (especially) and private/state lands more than expected if use was random. Use by vegetation type was not different from what would be expected at random (also see Table 4).

Table 5. Results for the multi-variable logistic regression model of pronghorn habitat selection, utilizing all pronghorn relocations, over all treatment periods.

Parameter	Point estimate	Lower 95% CI	Upper 95% CI
Intercept	−2.2508	−5.407	0.892
TRI	−0.0541	−0.0943	−0.0156
1/DIST	−66.9	−120.3	−24.5
ln(CENT + 1)	−0.509	−0.667	−0.355
ELEV2	0.000002354	0.00000136	0.00000336

$R^2L = 0.093$

Score test, df = 4, $\chi^2 = 86.0$, $p < 0.0001$

Area under the ROC curve = 0.65

Hosmer-Lemeshow goodness-of-fit test, df = 8, $\chi^2 = 13.72$, $p = 0.031$

Table 6. Parameter estimates for 1/DIST by treatment period, and statistics for performance of the related comprehensive multi-variable model within which these parameters were derived. 1/DIST expresses the effect of nearness to both transportation corridors. Due to inverse transformation, larger negative parameter values suggest greater avoidance, and a positive upper confidence interval suggests "avoidance" not appreciably different from zero.

Treatment period	1/DIST Point estimate	Lower 95% CI	Upper 95% CI
Pre-treatment	−416.3	−748.4	−156.2
Treatment 1	−31.1	−134.2	44.5
Treatment 2	−44.1	−109.0	8.8

Pre-treatment model

  $R^2L = 0.137$

  Score test, df = 4, $\chi^2 = 23.7$, $p < 0.0001$

  Area under the ROC curve = 0.67

  Hosmer-Lemeshow goodness-of-fit test, df = 8, $\chi^2 = 17.0$, $p = 0.031$

Treatment 1 model

  $R^2L = 0.141$

  Score test, df = 4, $\chi^2 = 41.4$, $p < 0.0001$

  Area under the ROC curve = 0.69

  Hosmer-Lemeshow goodness-of-fit test, df = 8, $\chi^2 = 10.4$, $p = 0.237$

Treatment 2 model

  $R^2L = 0.075$

  Score test, df = 4, $\chi^2 = 30.7$, $p < 0.0001$

  Area under the ROC curve = 0.62

  Hosmer-Lemeshow goodness-of-fit test, df = 8, $\chi^2 = 10.6$, $p = 0.224$

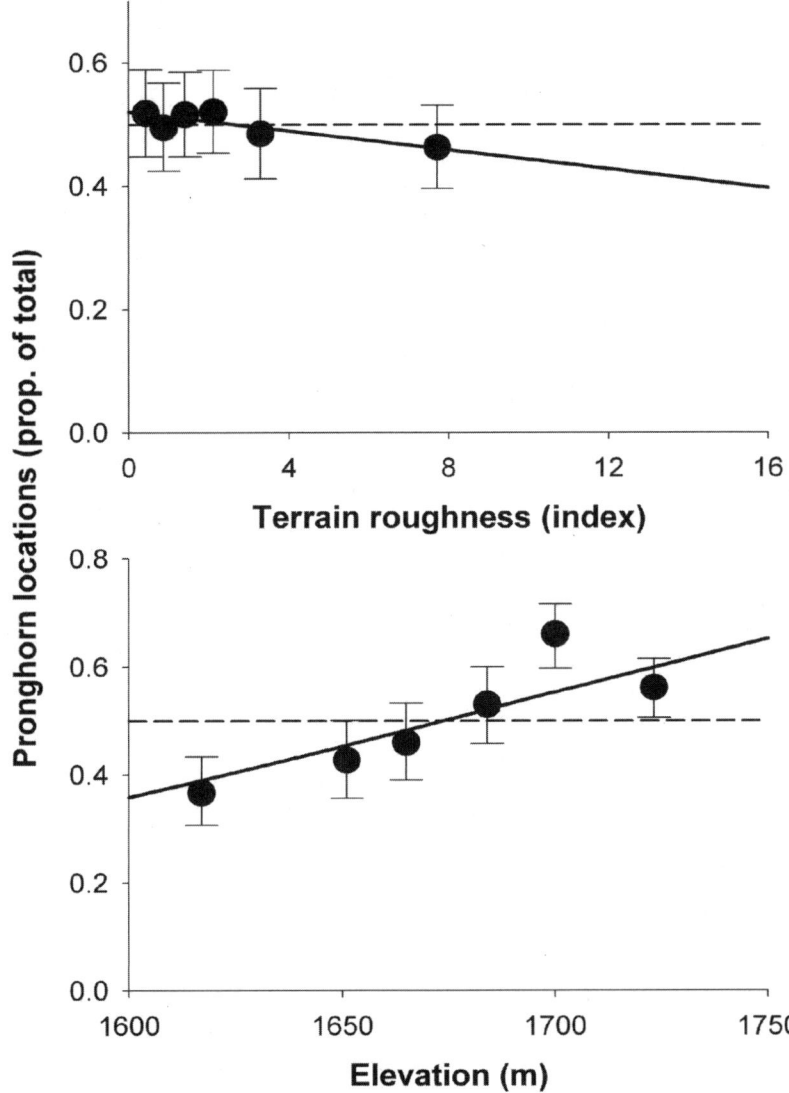

Figure 8. Relationship between pronghorn relocations, as a proportion of all random and observed locations, and terrain roughness (TRI; top) and elevation (bottom). Filled circles and associated error bars denote means and 95% CIs for septiles of the data. Solid lines denote modeled relations and dashed lines denote $p = 0.5$. Values above $p = 0.5$ indicate positive selection whereas values below indicate negative selection.

Table 7. Parameter estimates for 1/DISTrr, by treatment period, and statistics for performance of the related comprehensive multi-variable models. 1/DISTrr expresses the effect of nearness to the railway. Due to inverse transformation, larger negative parameter values suggest greater avoidance, and a positive upper confidence interval suggests "avoidance" not appreciably different from zero.

Treatment period	1/DISTrr Point estimate	Lower 95% CI	Upper 95% CI
Pre-treatment	–375.9	–784.6	–104.6
Treatment 1	–112.4	–495.4	53.9
Treatment 2	–25.8	–161.1	74.1

Pre-treatment model
  $R^2L = 0.110$
  Score test, df = 3, $\chi^2 = 9.59$, $p = 0.0224$
  Area under the ROC curve = 0.60
  Hosmer-Lemeshow goodness-of-fit test, df = 8, $\chi^2 = 5.72$, $p = 0.679$

Treatment 1 model
  $R^2L = 0.251$
  Score test, df = 3, $\chi^2 = 27.3$, $p < 0.0001$
  Area under the ROC curve = 0.75
  Hosmer-Lemeshow goodness-of-fit test, df = 6, $\chi^2 = 25.2$, $p = 0.001$

Treatment 2 model
  $R^2L = 0.052$
  Score test, df = 3, $\chi^2 = 8.5$, $p = 0.037$
  Area under the ROC curve = 0.60
  Hosmer-Lemeshow goodness-of-fit test, df = 8, $\chi^2 = 25.1$, $p = 0.002$

Table 8. Parameter estimates for DISTint, by treatment period, and statistics for performance of the related comprehensive multi-variable models. DISTint expresses the effect of nearness to the interstate. Due to inverse transformation, larger negative parameter values suggest greater avoidance, and a positive upper confidence interval suggests "avoidance" not appreciably different from zero.

Treatment period	1/DISTint Point estimate	Lower 95% CI	Upper 95% CI
Pre-treatment	–757.7	–1405.7	–197.3
Treatment 1	26.0	–99.8	157.4
Treatment 2	–63.2	–151.1	1.9

Pre-treatment model
  $R^2L = 0.316$
  Score test, df = 9, $\chi^2 = 32.8$, $p < 0.0001$
  Area under the ROC curve = 0.76
  Hosmer-Lemeshow goodness-of-fit test, df = 8, $\chi^2 = 18.5$, $p = 0.018$

Treatment 1 model
  $R^2L = 0.124$
  Score test, df = 10, $\chi^2 = 21.3$, $p = 0.019$
  Area under the ROC curve = 0.68
  Hosmer-Lemeshow goodness-of-fit test, df = 8, $\chi^2 = 19.4$, $p = 0.013$

Treatment 2 model
  $R^2L = 0.180$
  Score test, df = 10, $\chi^2 = 42.8$, $p < 0.0001$
  Area under the ROC curve = 0.71
  Hosmer-Lemeshow goodness-of-fit test, df = 8, $\chi^2 = 9.3$, $p = 0.317$

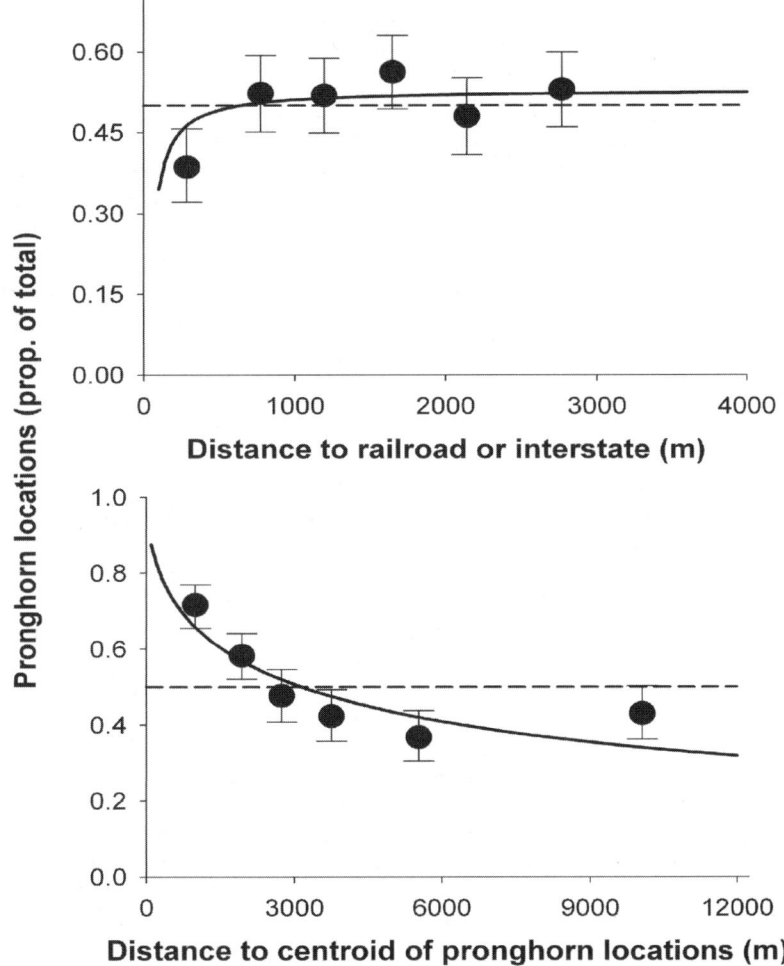

Figure 9. Relations between pronghorn relocations, as a proportion of total random and observed locations, and nearness to either Interstate 40 or the BNSF railroad (top) and to the centroid of pronghorn locations (bottom). Filled circles and associated error bars denote means and 95% CIs for septiles of the data. Solid lines denote modeled relations and dashed lines denote $p = 0.5$. Values above $p = 0.5$ indicate positive selection whereas values below indicate negative selection.

Given the frequency of the train traffic and its inherent disturbance, as well as the other potential physical and psychological deterrents associated with the right-of-way, it may ultimately be necessary to use over- or underpasses to enable pronghorn to negotiate these barriers. However, at this time we believe that efforts to modify the right-of-way to enhance the potential for pronghorn crossings, such as those employed in this study, may still have merit if the scope of the effort can be expanded both spatially and temporally. Given the high costs associated with creating structures to span or tunnel beneath the railroad, we encourage further investigation of the potential to enhance direct crossings of the right-of-way before more complicated and costly measures are pursued.

Home Range, Movement, and Habitat Selection Analyses

Our analyses of home range size, distances between locations, crossings, and habitat selection provided additional insight into the effects of Interstate 40 and the BNSF railroad, potential non-effects of our treatments, and the basic ecology of pronghorn isolated under comparatively unique conditions. These results are not only of near-term relevance to management; they also constitute a baseline against which future changes in behavior and population status can be reckoned and understood. The unfolding fate of our study population will potentially be of considerable interest to those who are more broadly concerned with factors that determine the survival of small isolated populations.

Changes in Home Range

Home ranges were larger during the pre-treatment period than during Treatment 1. The amount of home range intersection with both the railroad and the interstate tracked this difference in home range size, with intersections of both the railroad and the interstate also tending to be less during Treatment 1 compared to the pre-treatment period. Intersection with the interstate in some measure provided a control for the

potential effect of the treatments on location of ranges with respect to the railroad. If pronghorn were responding positively to the treatment(s), we would have expected overlap to increase between home ranges and the railroad, and not between home ranges and the interstate, but this was not the case. In short, variation in amount of overlap between home ranges and both the interstate and the railroad was very likely an artifact of variation in home range sizes, for reasons probably related more to other conditions than to any treatment of the fence along the railroad. Even so, pronghorn probably had greater "opportunity" to cross the railroad during Treatment 2, compared to Treatment 1, because of larger home ranges during the later period and related greater range overlap with the railroad.

Crossings of BNSF and I-40

Because of the long intervals between pronghorn locations, an unknown number of crossings could have gone undetected, but only if crossings were reciprocal (over and then back again prior to the next telemetry location). This potential error constitutes a bias of unknown magnitude leading to underestimation of Exact Test p-values. This scenario (crossing back and forth between detections) seems very unlikely, and any associated potential bias does not cast doubt on the conclusions supported by the tests. Pronghorn, if they ever crossed at all, were clearly crossing both the railroad and interstate far less than would be expected if they were not avoiding or otherwise being repelled by these human features. This result is consistent with the conclusion by Ockenfels et al. (1997, 2000) that both Interstate 40 and the BNSF railroad, together with paralleling fences, are virtually impenetrable barriers for pronghorn.

Distances Between Locations

The only certain effect in the models of distance between successive locations was that of distance from range centroid. As distance from centroid increased, so did distance between successive locations. This was consistent with the model of habitat selection, which suggested that likelihood of use diminished with greater distance from range centroid, in concert suggesting that pronghorn not only used the periphery of their ranges less intensively, but also tended to be more mobile when there. Otherwise, an effect of fawning season showed up in one model but not the other, and an effect of seasonal territoriality similarly varied between models. An effect of nearness to railroad, alone or interacting with treatment period, was not evident for the model that focused only on locations nearer to the railroad than to the interstate. However, in the general model, an interaction effect was evident, with distances between locations being greater when farther from either the interstate or railroad during the pre-treatment period, but not during both treatment periods. These models do not provide a basis for reaching unambiguous conclusions about how movements varied with nearness to the railroad, alone, or interacting with treatment period. Furthermore, there is no basis here for concluding that movements did vary near the railroad as a function of treatment period.

Selection of Vegetation Types

The lack of pattern in vegetation use was unexpected. Error in pronghorn locations and map accuracy potentially bias this analysis, and these errors and related biases are of unknown magnitude and nature. No observed use of a GAP type deviated from that expected at random. Pronghorn used especially NPS lands greater than expected by chance. The ability of both vegetation type and ownership to predict pronghorn habitat use was quite low (area under the ROC curve was < 0.6). We had anticipated seeing a pattern of habitat use by pronghorn consistent with avoidance of woodlands (Ockenfels et al. 1994; Pyle and Yoakum 1994; Bright and van Riper 2000), preference for grasslands (Yoakum 1972; Kindschy et al. 1982; Bright and van Riper 2000), and perhaps also preference for shrub-steppe as fawning cover (Clemente et al. 1995; Bright and van Riper 2000). If vegetation use does not explain the apparent preference for NPS

lands within PEFO, another possible reason is competition with cattle (McNay and O'Gara 1982), which are seasonally present throughout much of the study area outside of PEFO.

Habitat Selection Models

All of the multi-variable habitat selection models were adequate in terms of not likely being the results of chance alone (all Score p-values were < 0.0001), of having good predictive efficiency (area under the ROC curve > 0.7 for all), and of providing good fit to the underlying data (all results for the Hosmer-Lemeshow (2000) test were > 0.05). The models therefore potentially provide a statistically sound basis for explaining and predicting habitat selection by pronghorn in the study area, as well as a good basis for controlling for effects extraneous to nearness to the railroad.

As might be expected, pronghorn tended to under-use areas in rougher terrain, as well as areas nearer the periphery of their home ranges (farther from the centroid). Pronghorn are generally known as denizens of flat open terrain (O'Gara and Yoakum 2004), and the under-use of home range margins is a pervasive biological phenomenon. Other patterns were more difficult to interpret. Vegetation type and ownership, as such, were not retained in any model despite the fact that pronghorn use varied from that expected at random when response to ownership was analyzed in isolation. The positive relation of pronghorn habitat use to elevation and terrain roughness may have been implicitly related to ownership and masked the effect of this variable. Avoidance of areas less than 700 m away from the railroad and the interstate is noteworthy. We did not include the explicit effects of fences, for lack of comprehensive information on these features, which is a major deficiency in the analysis given the known effect of fences on pronghorn movements and habitat use elsewhere (Spillett et al. 1967; Zobell 1968; Howard et al. 1990; Bright and van Riper 2000).

Of relevance to the study design, pronghorn more strongly avoided the railroad during the pre-treatment period (when estimated parameters were largest for the effect of nearness to the railroad), in contrast to Treatments 1 and 2, when parameter estimates did not differ appreciably from zero (i.e., no effect of nearness to the railroad ≈ no avoidance of the railroad). However, this same pattern held for differences in responses of pronghorn to the interstate among treatment periods, suggesting that observed changes in response to the railroad were not attributable to treatment effects.

CONCLUSION

The isolated pronghorn herd at PEFO appears likely to remain so for the foreseeable future. Under existing conditions, Interstate 40 and the BNSF railroad constitute impenetrable barriers. The consistent avoidance of these features by pronghorn moreover diminishes the odds that exploratory behavior will result in chance crossings. We were fortunate to have the opportunity to partner with a diverse group of cooperators who sought to better understand and, if possible, to improve conditions for pronghorn at PEFO. If in the future it again becomes possible to modify fences along the railroad, we encourage a more extensive and sustained program to more effectively test the efficacy of treatments.

LITERATURE CITED

Arizona Department of Transportation (ADOT). 2006. Average annual daily traffic reports. Available at http://tpd.azdot.gov/datateam/aadt.php.

Allen, A. W., J. G. Cook, and M. J. Armbruster. 1984. Habitat Suitability Index Models: Pronghorn. FWS/OBS-82/10.65. U.S. Fish and Wildlife Service, Fort Collins, Colorad.

Autenrieth, R. E., D. E. Brown, J. Cancino, R. M. Lee, R. A. Ockenfels, B. W. O'Gara, T. M. Pojar, and J. D. Yoakum, editors. 2006. Pronghorn Management Guides, 4th ed. Pronghorn Workshop and North Dakota Game and Fish Department, Bismarck.

Beyer, H. L. 2004. Hawth's Analysis Tools for Arc GIS. Available at http://www.spatialecology.com/htools.

Bright, J. L., and C. van Riper III. 2000. Pronghorn home ranges, habitat selection, and distribution around water sources in northern Arizona. USGS Technical Report FRESC/COPL/2000/18, U.S. Geological Survey, Flagstaff, Arizona.

Brown, D. E. 1994. History of pronghorn antelope in Arizona. Pronghorn 2(2): 1, 5, 7.

Clemente, F. R., J. Valdez, L. Holechek, P. J. Zwank, and M. Cardenas. 1995. Pronghorn home range relative to permanent water sources in southern New Mexico. Southwest Naturalist 40 (1): 38–41.

Firchow, K. M. 1986. Ecology of pronghorn on the Piñon Canyon Maneuver Site, Colorado. Master's thesis, Virginia Polytechnical Institute and St. University, Blacksburg.

Hooge, P. N., and B. Eichenlaub. 2000. Animal movement extension to Arcview, version 2.0. Alaska Science Center, Biological Science Office, U.S. Geological Survey, Anchorage.

Hosmer, D. W., and S. Lemeshow. 2000. Applied logistic regression, 2nd ed. Wiley and Sons, New York.

Howard, W. V., K. Green-Hammond, M. Cardenas, and S. L. Beasom. 1990. Habitat requirements for pronghorn on rangelands impacted by livestock and net wire in eastcentral New Mexico Bulletin 750, Agricultural Experimental Station , New Mexico State University, Las Cruces.

Karsky, R. 1988. Fences. U.S. Forest Service Tech. and Dev. Ctr., Missoula, Montana.

Kindschy, R. R., C. Sundstrom, and J. D. Yoakum. 1982. Wildlife habitats in managed rangelands—The Great Basin of southeastern Oregon: Pronghorns. General Technical Report PNW-145. Pacific Northwest Forest and Range Experimental Station, U.S. Forest Service, Portland, Oregon.

Martinka, C. J. 1967. Mortality of northern Montana pronghorns in a severe winter. Journal of Wildlife Management 31(1): 159–164.

McNay, M. E., and B. W. O'Gara. 1982. Cattle-pronghorn interactions during the fawning season in northwestern Nevada. In Wildlife-Livestock Relationships Symposium, edited by J. M. Peek and G. D. Dlake, pp. 593–606. University of Idaho Forest and Wildlife Experimental Station, Moscow, Idaho.

Mitchell, G. J. 1980. The pronghorn antelope in Alberta. Alberta Department of Lands and Forests, Fish and Wildlife Division and University of Regina, Saskatchewan.

Ockenfels, R. A., A. Alexander, C. L. D. Ticer, and W. K. Carrel. 1994. Home ranges, movement patterns, and habitat selection of pronghorn in central Arizona. Res. Branch. Technical Report 13. P-R Project W-78-R. Arizona Game and Fish, Phoenix.

Ockenfels, R. A., W. K. Carrel, and C. van Riper III. 1997. Home ranges and movements of pronghorn in northern Arizona. In Biennial Conference on Research on the Colorado Plateau, volume 3, pp. 45–61.

Ockenfels, R. A., W. K. Carrel, J. S. deVos Jr., and C. L. D. Ticer. 2000. Highway and railroad effects on pronghorn movements in Arizona and Mexico. Pronghorn Antelope 1996 Workshop Proceedings, volume 17, p. 104.

O'Gara, B. W., and J. D. Yoakum, editors. 1992. Pronghorn management guidelines. In Pronghorn Antelope Workshop 15.

O'Gara, B. W., and J. D. Yoakum. 2004. Pronghorn: Ecology and Management. University Press of Colorado, Boulder.

Pyle, W. H., and J. D. Yoakum. 1994. Status of pronghorn management at Hart Mountain Antelope Refuge, Oregon. In Pronghorn Antelope Workshop Proceedings, volume 16, pp. 22–34.

Riddle, P. 1990. Wyoming antelope status report—1990. Pronghorn Antelope Workshop Proceedings, volume 14, p. 24.

Riley, S. J., S. D. DeGloria, and R. Elliot. 1999. A terrain roughness index that quantifies topographic heterogeneity. Intermountain Journal of Sciences 5: 23–27.

Rodgers, A. R., A. P. Carr, L. Smith, and J. G. Kie. 2005. HRT: Home range tools for ArcGIS. Ontario Ministry of Natural Resources, Centre for Northern Forest Ecosystem Research, Thunder Bay, Ontario, Canada.

Spillett, J. J., J. B. Low, and D. Sill. 1967. Livestock fences—How they influence pronghorn antelope movements. Bulletin 470, Utah Agricultural Experiment Station, Logan.

van Riper, C. III, and R. A. Ockenfels. 1998. The influence of transportation corridors on the movement of pronghorn antelope over a fragmented landscape in northern Arizona. In Proceedings of the Second International Conference on Transportation and Wildlife Ecology, Ft. Meyers, Florida, edited by D. Zeigler, pp. 241–248.

Wagner, F. H. 1978. Livestock grazing and the livestock industry. In Wildlife and America, edited by H. P. Brokaw, pp. 121–145. Council Environmental Quality, U.S. Government Printing Office, Washington, D.C.

Ward, A. L., J. J. Cupal, G. A. Goodwin, and H. D. Morris. 1976. Effects of highway construction and use on big game populations. FHWA-RD-76-174. U.S. Federal Highway Admininstration, Washington, D.C.

Weisberg, S. 1985. Applied Linear Regression, 2nd ed. John Wiley, New York.

Yoakum, J. D. 1972. Antelope-vegetative relationships. Antelope States Workshop Proceedings, volume 5, pp. 171–177.

Yoakum, J. D. 1978. Pronghorn. In Big Game of North America: Ecology and Management, edited by J. L. Schmidt and D. L. Gilbert, pp. 103–121. Stackpole Books, Harrisburg, Pennsylvania.

Zar, J. H. 1984. Biostatistical Analysis, 2nd ed. Prentice Hall, Englewood Cliffs, New Jersey.

Zobell, R. 1968. Field studies of antelope movements on fenced ranges. Transactions of the North American Wildlife and Natural Resource Conference, volume 33, pp. 211–216.

RELATIONSHIP BETWEEN MEASURED HABITAT QUALITY AND FIRST AND SECOND YEAR SURVIVAL FOR TRANSPLANTED BIGHORN SHEEP IN ARIZONA

Brian F. Wakeling and Erin Riddering

Translocation programs have been largely responsible for the reestablishment of bighorn sheep (*Ovis canadensis*) populations throughout Arizona (Cunningham et al. 1989). Since the inception of the translocation program, Arizona's bighorn sheep population has increased from an estimated 1500 animals in 1960 to about 6000 animals in 2003 (Wakeling 2003). Bighorn sheep translocations have occurred throughout Arizona, primarily following evaluations of potential habitat using systematic evaluation criteria that allow managers to rank and prioritize habitats based on measured suitability. Many appraisals have been conducted to determine how well a variety of systematic habitat evaluations function in predicting bighorn sheep habitat use and many provide functional expectations for predicting suitable habitat (e.g., Cunningham 1989; Wakeling and Miller 1990).

Recently, mountain lion (*Puma concolor*) predation on transplanted bighorn sheep has been suggested to be the cause for declining bighorn sheep survival in Arizona (Kamler et al. 2002). Kamler et al. (2002) concluded that increased predation by mountain lions on bighorn sheep in more recent translocated populations was directly influenced by increasing mountain lion abundance and distribution within Arizona. An alternate hypothesis that was not explored by Kamler et al. (2002) was that more recent bighorn sheep transplants occurred in habitats with lower habitat suitability scores, which may affect predation rate by mountain lions or overall bighorn sheep survival because these habitats may be less suitable for bighorn sheep occupation.

METHODS

We used the numerical rankings of habitat described by Cunningham (1989), using the technique known as the Cunningham-Brown method. This technique derives a numerical score for factors, including historic occurrence, land status, topography, vegetative cover, exotic ungulates, native ungulates, human disturbance, water availability, habitat discreteness, and range expansion. Habitats could score from 0 (unsuitable) to 80 (excellent). Cunningham (1989) speculated that habitats scoring below 50 were probably unsuitable for bighorn sheep occupation. Most potential mountain ranges that historically contained or potentially could sustain bighorn sheep were evaluated in Arizona during the late 1990s, and rankings were developed by a team of trained observers (Lee et al. 2000).

We examined annual survival and cause-specific mortality rates (Heisey and Fuller 1985) for 2 years after 54 translocations of bighorn sheep that occurred between 1960 and 2002. We employed the same base data set used by Kamler et al. (2002). Each of these translocations released 5–25 radio-marked bighorn sheep. Post-monitoring fixed-wing telemetry flights occurred about once every 1–3 months for at least 2 years following release. Cause of death was investigated and a portion attributed to mountain lion predation based on characteristics of the carcass. Although we recognized that cause of death was at times difficult to determine using this approach, we used the same data set used by Kamler et al. (2002) to investigate whether measured habitat quality was an important overriding factor in bighorn sheep survival. We also compared overall survival rates to contrast relationships with cause-specific mortality. We then used linear correlation to test for an association between measured habitat quality and annual survival or mortality caused by mountain lions during the first 2 years following release (Zar 1996).

RESULTS

Neither survival nor mortality caused by mountain lions proved to have linear relationships with measured habitat quality for either the first or second year following release ($P > 0.5$; Figures 1–4). The correlation equations explained extremely little of the variation in the data sets ($r^2 < 0.04$ in each). There was no evidence to support the theory that habitat quality within the release sites selected to date had influenced predation on bighorn sheep by mountain lions.

DISCUSSION

Because bighorn sheep were being released in habitats based on a priority ranking system, it seemed logical to assume that later releases in lower quality habitats would have lower survival and higher cause-specific mortality. Our analysis failed to support this assumption. A possible reason for the lack of a relationship is that habitat quality must fall below a critical threshold before survival is directly affected, but to date bighorn sheep releases have occurred primarily within the range of suitable habitats above this threshold. For this to be true, Cunningham's (1989) original speculation that habitat must score more than 50 to be suitable must be incorrect, and bighorn sheep must be capable of sustaining themselves in habitats that score as low as 40. An alternate explanation is that the habitat quality scores are not good predictors of habitat suitability. This latter rationale seems unlikely because several studies have tested the suitability ranking method and found that these techniques are fairly reliable in selecting suitable habitats (e.g., Wakeling and Miller 1990). Most of the habitats that received bighorn sheep releases scored over 40 points using the Cunningham-Brown score and the highest quality habitat in this evaluation received a score of 55. Based on our

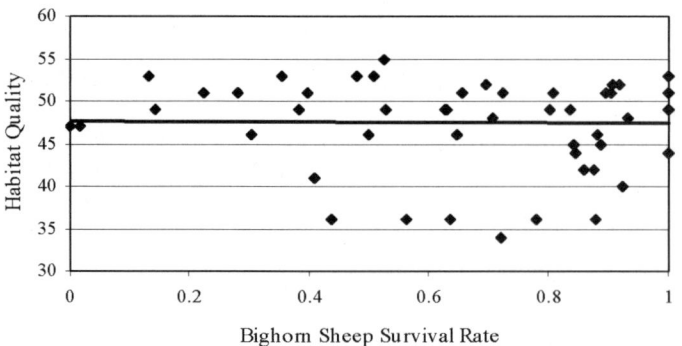

Figure 1. Habitat quality scores (possible 0–80) and bighorn sheep survival rates (possible 0–1.0) for the first year following release in Arizona.

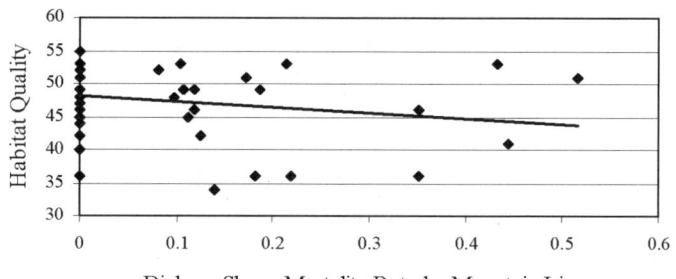

Figure 2. Habitat quality scores (possible 0–80) and bighorn sheep mortality rates (possible 0–1.0) attributed to mountain lion predation for the first year following release in Arizona.

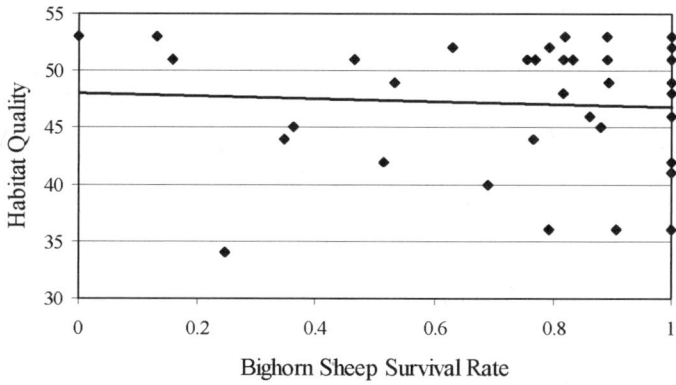

Figure 3. Habitat quality scores (possible 0–80) and bighorn sheep survival rates (possible 0–1.0) for the second year following release in Arizona.

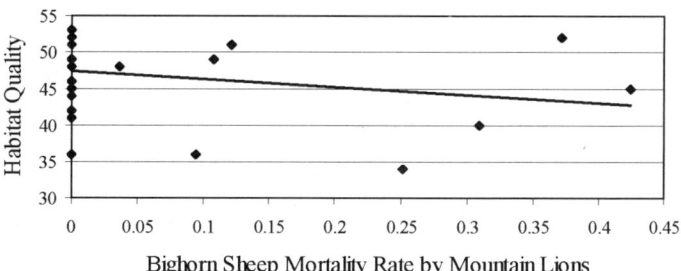

Figure 4. Habitat quality scores (possible 0–80) and bighorn sheep mortality rates (possible 0–1.0) attributed to mountain lion predation for the second year following release in Arizona.

analysis, habitats scoring more than 34 were not correlated with survival or mortality.

We initiated this analysis following the conclusion by Kamler et al. (2002) that increases in mountain lion predation were causing declines in bighorn sheep numbers, based on measured survival rates for translocated bighorn sheep populations. However, Kamler et al. (2002) inadvertently reported inaccurate population estimates for bighorn sheep in Arizona, based on incomplete data sets. We also questioned the effectiveness of using translocated animals as a surrogate population to investigate the survival of resident bighorn sheep. Despite our speculation, our analysis on survival and habitat quality did not support the hypothesis that translocations into lower ranking habitats influenced mountain lion predation on or survival of bighorn sheep.

LITERATURE CITED

Cunningham, S. 1989. Evaluation of bighorn sheep habitat. In The Desert Bighorn Sheep in Arizona, edited by R. M. Lee, pp. 135–160. Arizona Game and Fish Department, Phoenix.

Cunningham, S., N. Dodd, and R. Olding. 1989. Arizona's bighorn sheep reintroduction program. In The Desert Bighorn Sheep in Arizona, edited by R. M. Lee, pp. 203–239. Arizona Game and Fish Department, Phoenix.

Heisey, D. M., and T. K. Fuller. 1985. Evaluation of survival and cause-specific mortality rates using telemetry data. Journal of Wildlife Management 49:668–674.

Kamler, J. F., R. M. Lee, J. C. deVos Jr., W. B. Ballard, and H. A. Whitlaw. 2002. Survival and cougar predation of translocated bighorn sheep in Arizona. Journal of Wildlife Management 66:1267–1272.

Lee, R. M., A. A. Munig, S. B. Zalaznik, D. J. Godec, B. D. Broscheid, S. E. Wagner, W. P Burger, and S. P. Barber. 2000. Evaluation of Bighorn Sheep Habitat in Arizona. Arizona Game and Fish Department, Phoenix, USA.

Wakeling, B. F. 2003. Status of bighorn sheep in Arizona, 2002–2003. Desert Bighorn Council Transactions 47:18–19.

Wakeling, B. F., and W. H. Miller. 1990. A modified habitat suitability index for desert bighorn sheep. In Managing Wildlife in the Southwest, edited by P. R. Krausman and N. S. Smith, pp. 58–66. Arizona Chapter of the Wildlife Society, Phoenix.

Zar, J. H. 1996. Biostatistical Analysis. 3rd edition. Prentice-Hall, Upper Saddle River, New Jersey.

MONITORING THE RIPARIAN BREEDING BIRD COMMUNITY ALONG THE COLORADO RIVER IN THE GREATER GRAND CANYON REGION: STATISTICAL POWER AND IMPLICATIONS FOR LONG-TERM MONITORING

John R. Spence

Avian communities along the Colorado River have changed substantially since the completion of Glen Canyon Dam in 1963 (Carothers and Brown 1991). Pre-dam vegetation along the river consisted of a thin riparian strip controlled primarily by spring flooding. Following completion of the dam, the largest tracts of riparian vegetation (in Glen Canyon) were destroyed, while an extensive "new high water zone" (NHWZ) community developed downstream of the dam through the Grand Canyon. In addition, extensive stands of riparian habitat have become established on silt terraces where the Colorado River drains into Lake Mead. These recent habitat modifications have caused changes in the avian community (Brown et al. 1987; Carothers and Brown 1991).

Because of concerns over loss of sediment, declines in native fish species, and erosion of archeological sites, the Grand Canyon Protection Act was passed in 1992. This Act specified that an environmental impact analysis (EIS) was to be conducted on the effects of Glen Canyon Dam operations on resources along the Colorado River in Glen Canyon National Recreation Area and Grand Canyon National Park. In 1995, the EIS was completed, and the Record of Decision (ROD) on the revised operations of Glen Canyon Dam was signed. Among other things, the ROD established a long-term science program to monitor the effects of dam operations on biological, cultural,

and physical resources along the Colorado River from the dam to the head of Lake Mead, a distance of 410 km. This monitoring program was established to provide the necessary data to adaptively manage dam operations in order to minimize impacts to selected resources (National Research Council 1999).

Various monitoring programs had been established as part of the environmental impact studies since 1982, under management of the Glen Canyon Environmental Studies Program (GCES; National Research Council 1987, 1996). These monitoring and baseline data, including studies on the avifauna, were used to develop management alternatives. In 1996 the Grand Canyon Monitoring and Research Center was established to oversee scientific monitoring of the resources as laid out in the ROD and to develop long-term monitoring and research strategic plans. Birds—principally riparian breeding birds, bald eagle (*Haliaeetus leucocephalus*), and the endangered southwestern willow flycatcher (*Empidonax traillii extimus*)—have been an integral part of past and ongoing monitoring studies along the river corridor.

Birds are an important and conspicuous component of the lacustrine and riparian ecosystems along the Colorado River and on Lake Powell and Lake Mead (Rosenberg et al. 1991; Carothers and Brown 1991; Spence and Bobowski 2003; Holmes et al. 2005). Birds are considered good indicators of change in ecosystems, as they can respond quickly. Such changes could be in response to climatic variation, invasion of the ecosystem by a new exotic species, recreational-based disturbances, changes in the prey-base, shifts in management practices, or some combination of these factors. Due to the strong tendency of passerine birds to exhibit pronounced habitat selection (Hilden 1965; Cody 1985), they can be a useful group of organisms for monitoring effects on habitat in a dynamic system such as the Colorado River (Perrins et al. 1991). Two of the major forcing variables controlling the riparian system are quantity and timing of dam releases, so it is likely that most breeding birds (other than those two or three species that nest right at the water's edge) are responding to changes in vegetation rather than to fluctuating flows. By monitoring avian populations, changes in other components of the riparian ecosystem may be detected, and management practices can be developed to address potential problems.

The principal goal of this study was to determine whether a long-term monitoring program with adequate statistical power could be developed to detect trends in the riparian breeding bird community along the Colorado River. Power analysis is a necessary and important tool in the establishment of any monitoring program (Steidl et al. 1997; Gibbs 1998). It is particularly critical in the case of endangered species monitoring, as the failure to detect a decline may have disastrous consequences (Taylor and Gerrodette 1993).

Most natural wildlife populations vary in abundance from year to year; this variation can result from numerous complex and interacting factors. In more temporarily variable species, it is often difficult to detect subtle long-term trends because of the "noise" (natural variability) in the species' populations. A power analysis provides a measure of how well a monitoring program can detect a trend through such "noise" in the data. Without an estimate of the power in a monitoring program, resource managers and scientists cannot always know if change in a population or species of interest is sta-

tistically significant. Furthermore, without adequate power, they may not be able to detect a significant change in a rare species that may be of management importance. This study used the approach of "prospective" power analysis (cf. Steidl et al. 1997), in which preliminary baseline data on population numbers and variability are gathered over a period of time and then used to design an effective long-term monitoring program, examining factors like sample size considerations, sampling protocols, and duration of data collection.

I used 5 years of breeding bird survey data from the Colorado River between Glen Canyon Dam and upper Lake Mead to develop a baseline data set on the relative abundance of riparian species, to develop a standardized methodology to monitor birds in riparian vegetation, and to examine aspects of statistical power in the data. Data from selected species are used to illustrate the relationships between relative abundance measures, abundance variability over time, and statistical power.

METHODS

Study Area Description

The Colorado River flows 410 km from Glen Canyon Dam to Separation Canyon on upper Lake Mead. The elevation of the river at the dam is 955 m and where it reaches upper Lake Mead the elevation is ca. 365 m, for a total drop of 590 m. This river corridor, which along with the surrounding uplands of the region can be considered the greater Grand Canyon region (cf. LaRue et al. 2001), is managed by three units of the National Park Service (Glen Canyon National Recreation Area, Grand Canyon National Park, and Lake Mead National Recreation Area) as well as by the Navajo Nation and the Hualapai Tribe.

The geology and climate of the region have been well described elsewhere (Beus and Morales 1990; Spence 2004). The climate is arid–warm temperate, with hot summers and cool winters. Precipitation is bimodal, with a late winter peak and a late summer–early fall peak. The most important geological factors relevant to the study of bird communities along the river corridor are the types of bedrock geology present and the presence of major side canyons (Turner and Karpiscak 1980; Stevens et al. 1996). The principal canyons along the Colorado River include Glen Canyon, Marble Canyon, and the Grand Canyon. Reaches where the bedrock consists of Precambrian schist and granite are relatively narrow and tend not to support significant riparian vegetation except at the mouths of tributaries. Reaches where sandstones and shales predominate tend to be wider, and riparian vegetation is often well established along river margins. Where major tributaries enter the river, additional sediment loads occur, and in particular return channel–eddy complexes form that trap finer sediments such as sand. Vegetation on these complexes includes many of the largest and better developed riparian patches in the study area.

The vegetation of the Colorado River corridor is complex and extremely dynamic, changing in response to climate, flooding, and the invasion of new exotic species. Spring flooding originally controlled the abundance and distribution of riparian vegetation, with a distinct "trim-line" at about the 100,000–125,000 cfs level. Above this line occurred an extensive "old high water zone" (OHWZ) community that consisted of a variety of species, of which the most important were apache plume (*Fallugia paradoxa*), net-leaf hackberry (*Celtis reticulata*), mesquite (*Prosopis glandulosa*), and catclaw (*Acacia greggii*). Below this line was sparse vegetation consisting of coyote willow (*Salix exigua*), tamarisk (*Tamarix chinensis*), and rushes, grasses, and annuals. This lower zone was flooded and scoured during most years. Spring flooding ceased after completion of the dam. The area below ca. 60,000 cfs filled in rapidly with riparian species, with tamarisk being the most abundant. This new vegetation community, termed the new high water zone (NHWZ), greatly increased in abundance beginning in 1963 (Pucherelli 1986). Both the NHWZ and OHWZ are variable in composition along the Colorado River, with a major change in the latter

occurring at river kilometer 64, where mesquite and catclaw first appear. Another feature of this vegetation is that it is very discontinuous, with patches of varying sizes from a few square meters up to 10 hectares or more. These patches tend to be isolated from one another by stretches of cliffs and rocky shorelines where little riparian vegetation occurs.

Bird Surveys

A breeding bird trip was launched from Lee's Ferry downstream on or near the first day of April, May, and June each year between 1996 and 2000. Trips averaged 16 days. Each trip included two primary bird surveyors, as well as boat operators and botanists. Patches in the Glen Canyon reach were sampled after the end of each trip.

The breeding bird program selected a subset of riparian vegetation patches in 1996. The choice of which patches to survey was based primarily on patch size and logistics considerations. Given a 16-day trip and work that needed to be completed by 0830–0900, the patches selected necessarily occurred in groups at and downstream of each night's camp. Based on the results of Sogge et al. (1998), larger patches, most greater than one hectare in size, were subjectively selected over smaller ones in order to maximize the number of species and individuals detected.

Felley and Sogge (1997) have compared a variety of techniques that could be used to inventory and monitor riparian breeding birds on the Colorado River corridor. The two principal methods, area surveys and point counts, were shown to give roughly similar results. However, point counts were recommended because they provide stricter control of survey effort and reduce observer variability (cf. Fellow and Sogge 1997). Because previous avian monitoring data had been collected using fixed-radius point counts, that technique was continued during this study. During the course of the project, new distance estimation methods came into more widespread use for avian community monitoring; these methods were utilized in the 2000 field season and are now being advocated because they control for differences in detectability rates among species (Rosenstock et al. 2002).

One to 10 point count stations within each patch were positioned at least 250 m apart, with the number of stations proportional to patch size. At each station a single surveyor recorded all birds heard or seen for 5 minutes. Birds that left the circle as the surveyor approached were counted, and birds were recorded as either within or outside a fixed 50 m radius circle around the center, depending on where they were first detected. Unbounded point count data include all birds detected at a station, both inside and outside the 50 m radius. Aerial species like swallows and swifts were also recorded. Other data recorded included temperature, cloud cover and wind speed, disturbances, bird singing or not, and bird sex and behavior when possible. Only the unbounded point count data are analyzed here.

Determining Breeding Status of Riparian Birds

The decision rules suggested by Sogge et al. (2005) were adopted to determine the breeding vs. non-breeding status of non-aerialist species in the riparian zone of the Colorado River. They provided species-specific rules to determine whether individuals being detected at point count stations or by other count methods during monthly surveys were likely to be either local breeders or migrants. Relative abundance data were adjusted by these rules to the extent possible to reflect only local breeding individuals. Bird nomenclature follows the American Ornithologists Union (7th edition, 1998).

Data Analysis

Included in this study are data from point counts conducted between 1996 and 2000 on 27 trips in Glen Canyon, Marble Canyon, and the Grand Canyon above Diamond Creek. Surveys were not conducted above Lees Ferry in 2000; data are thus reported only from the 1996 through 1999 surveys from the Glen Canyon stretch, whereas 1996 through 2000 data are reported for the rest of the study area.

The patch was used as the sampling unit to calculate mean relative abundance and to determine statistical power. Individual point counts could not be used as the sampling units because of the spatial autocorrelation problems that could occur when two or more point counts are located in the same local patch of vegetation (S. Urquhart, personal communication 2000). To compare patch differences statistically, point count data were summed within a patch for those with more than one point count station, and mean detection rates were calculated for each patch survey. Then the mean number of individuals for each species detected per patch was calculated based on three trips per year. Values were computed for all patches that were sampled on all three trips for at least 4 years between 1996 and 2000 (minimum 12 counts per point per patch). This resulted in a sample of 46 patches. Data were then compared across years for each species to determine if there were significant differences in relative abundance between years using a repeated-measures analysis of variance model.

Two important assumptions had to be made for these preliminary analyses. First, it was necessary to assume that between-patch variability in trend over years was not significant. Second, and perhaps more important, the computed mean relative bird abundance per patch was assumed to be representative of between-patch differences resulting from potentially uneven sampling within patches. Both of these assumptions are examined here, along with other aspects of the sampling design and methodology.

Power analysis was performed on the data from point counts for the 16 most commonly detected terrestrial riparian species in order to determine if adequate power exists for the purposes of long-term monitoring. The program MONITOR was used for the analyses (Gibbs 1995). Power is defined as

$$Power = 1 - \beta$$

with β the probability of making a Type II error (accepting a false null hypothesis). Power is the ability of a statistical test to correctly reject a false null hypothesis.

Adequate power levels are generally 80 percent or higher. A power of 80 percent indicates that, on average, a change that is actually occurring will be detected 80 percent of the time. The inverse is that a change that is actually occurring will not be detected 20 percent of the time. The Type I error (α or rejection of a true null hypothesis) was set at 0.05 for the power tests. The analysis used a Monte Carlo simulation to generate simulated sets of count data, which were then compared with the actual inputs through a route-regression approach. Replications were set at 1000. Trend projections were set at 5, 10, 15, 20, and 25 percent (change in bird detections per year) for a future specified timeframe (e.g. 5 or 10 yrs). A two-tailed test was used, testing the null hypothesis that the trend does not differ from zero. Power analyses were conducted on a patch basis, based on overall means across years for each patch.

There is a trade-off between power and α such that stringent levels of the latter reduce power. An α of 0.10 could be selected rather than the current level of 0.05. The selection of levels of α and power in a study depends on numerous factors. In many instances, particularly in monitoring changes in wildlife populations, a Type I error may be less costly to management than a Type II error. This follows from the concept that it is less costly to reject the null hypothesis of no change ("crying wolf") than to accept it if it is false. A manager runs the danger of not detecting change if α is set too high and power is thus too low. Many researchers advocate the relationship of $\alpha = \beta = 0.10$, and a power of 0.90.

RESULTS

Between 1996 and 2000, 32 riparian species that either bred or potentially could have bred were detected in the study area (Spence 2004). Breeding records (primarily nests) exist for 24 of these species; the remaining 8 were considered possible breeders—that is, suitable habitat occurred and the species probably breeds but there were no records, or apparently suitable habitat was available but there were no breeding records.

From 1996 to 2000, a total of 1700 point counts were completed in 76 patches of riparian vegetation between Glen Canyon Dam and Lake Mead. Table 1 shows the 16 most common species detected during the 5-year period. Lucy's warbler (*Vermivora lucae*) was the most commonly detected species in all years (see Figure 1), followed by house finch (*Carpodacus mexicanus*), Bewick's wren (*Thryomanes bewickii*), and Bell's vireo (*Vireo bellii*). Lucy's warbler, house finch, and black-chinned hummingbird (*Archilochus alexandri*) were the most widespread, occurring in the largest number of riparian vegetation patches. Four species—ash-throated flycatcher (*Myiarchus cinerascens*), black-chinned hummingbird, mourning dove (*Zenaida macroura*), and yellow-breasted chat (*Icteria virens*)—showed significant differences in relative abundance between years using a repeated measures ANOVA.

Power analyses for the 16 species that occurred at 10 or more patches per year are shown in Table 2. Trends in species detection rates per year with 10 percent effect sizes (10% change in detections per year), α of 0.05, and power of 0.80 or greater could only be determined for eight species: Lucy's warbler, house finch, Bell's vireo, Bewick's wren, black-chinned hummingbird, ash-throated flycatcher, yellow-breasted chat, and blue-gray gnatcatcher. At a larger 20 percent change per year, adequate power also existed to detect change in common yellowthroat (*Geothlypis trichas*), yellow warbler (*Dendroica petechia*), and possibly song sparrow (*Melospiza melodia*). For the remaining five species—mourning dove, blue grosbeak (*Passerina caerulea*), Bullock's oriole (*Icterus bullockii*), lesser goldfinch (*Carduelis tristis*), and brown-headed cowbird (*Molothus ater*)—power was below 0.80, even at a less stringent α of 0.10. Years of monitoring needed to reach a power of 0.80 for each species varied from 5 for Lucy's warbler to 30 for brown-headed cowbird. At a shorter period of 10 years, the number of patches occupied in order to reach sufficient power varies from 15 for house finch to 60 for brown-headed cowbird. However, only the eight most common species occupied

Table 1. Mean total detections per year, mean patch detection per year, and mean number of patches occupied by at least one individual per year are listed for 16 species of riparian breeding birds based on data generated from point counts sampled between 1996 and 2000 in 46 patches of riparian vegetation between Glen Canyon Dam and Diamond Creek. For three species found only below Lee's Ferry (Bell's vireo, song sparrow, and lesser goldfinch), data from 36 patches in the Grand Canyon were used.

Species	Mean Total Detections (1 SD)	Mean Patch Detections (1 SD)	Mean No. Patches Occupied Per Year
Lucy's warbler (*Vermivora lucae*)	802 (215)	2.29 (1.35)	42
House finch (*Carpodacus mexicanus*)	317 (27)	1.06 (0.93)	40
Bewick's wren (*Thryomanes bewickii*)	276 (43)	0.73 (0.59)	30
Bell's vireo (*Vireo bellii*)	235 (85)	0.79 (0.82)	23
Yellow-breasted chat (*Icteria virens*)	165 (64)	0.35 (0.45)	31
Ash-throated flycatcher (*Myiarchus cinerascens*)	132 (26)	0.30 (0.41)	33
Blue-gray gnatcatcher (*Polioptila caerulea*)	122 (59)	0.26 (0.37)	26
Black-chinned hummingbird (*Archilochus alexandri*)	120 (60)	0.41 (0.64)	37
Song sparrow (*Melospiza meolida*)	117 (82)	0.22 (0.30)	14
Yellow warbler (*Dendroica petechia*)[1]	94 (10)	0.27 (0.40)	24
Common yellowthroat (*Geothlypis trichas*)	82 (20)	0.18 (0.28)	19
Mourning dove (*Zenaida macroura*)	65 (41)	0.12 (0.23)	14
Brown-headed cowbird (*Molothus ater*)	50 (14)	0.10 (0.18)	12
Lesser goldfinch (*Carduelis tristis*)	48 (12)	0.16 (0.27)	14
Blue grosbeak (*Passerina caerulea*)	48 (24)	0.12 (0.20)	19
Bullock's oriole (*Icterus bullockii*)	29 (14)	0.07 (0.12)	10

[1]Based on May–June data only.

enough patches for power to be sufficient to detect change.

The power curves for three species selected to show different patterns are shown in Figures 2 through 4. Lucy's warbler was sufficiently common that good power existed, even at the small rate of decline of 5 percent or less in relative abundance over only 7 years of monitoring. For common yellowthroat, the power was only adequate to detect major declines of –20 to –25 percent per year over a 10-year period of monitoring. Declines could not be detected for mourning dove at any level of power, and it would take 17 years to reach a power of 0.80 and for a 10 percent decline per year. The number of years under the sampling program to achieve enough power to detect change varied between 5 and 30 (see Table 2). Increasing the length of the monitoring program, the number of patches

sampled, or the number of surveys per year, or changing to a one-tailed test, would increase power and allow the use of lower effect sizes.

To be confident of detecting changes in rarer species, the 1996 through 2000 monitoring program would have to be modified either by increasing the number of patches or surveys per year or by changing the parameters of the statistical power tests. For several species, however, it is unlikely that any modifications could be made that would achieve sufficient power. Some species, such as brown-headed cowbird or Bullock's oriole, were simply too rare and patchily distributed along the Colorado River between Glen Canyon Dam and Lake Mead to be effectively monitored with point count surveys. This was also true for the other 14 species of riparian breeders listed but not analyzed in Table 1.

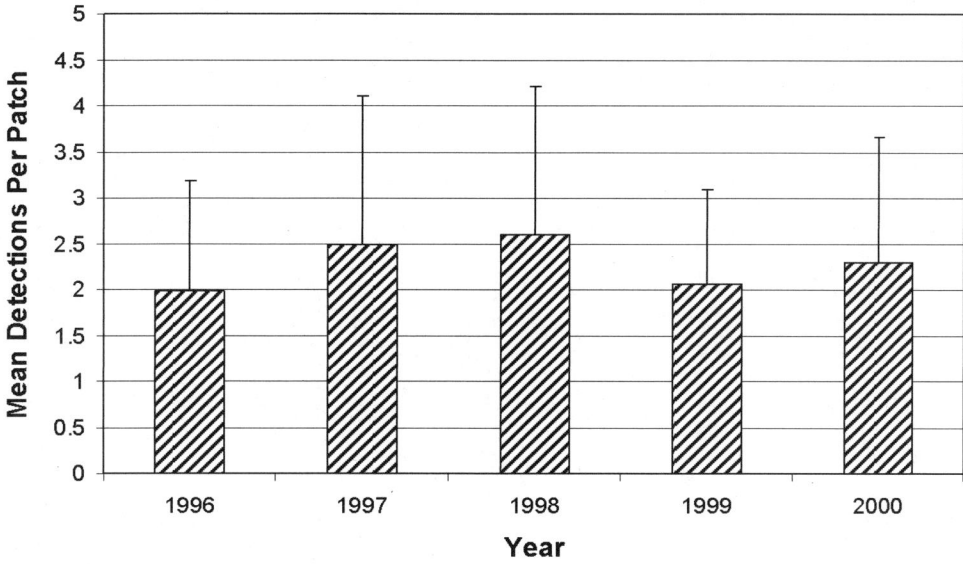

Figure 1. Mean detection rate (± 1 SD) per year per patch for Lucy's warbler along the Colorado River between Glen Canyon Dam and Diamond Creek. The data are from point counts sampled between 1996 and 2000 in 46 patches of riparian vegetation. Difference between years was not significant based on a repeated-measures ANOVA (F = 0.46, P = 0.768).

DISCUSSION

Study Design

Several methodological and statistical problems were revealed by the analyses of this study, with important implications for the establishment of a future long-term monitoring program for riparian birds along the Colorado River. The principal concerns are based on three major considerations: (1) the appropriate sampling approach to be used, including the proper field methods and parameters used to compute bird abundance, (2) the study-wide sampling design, and (3) problems of sampling associated with the narrow, linear, and discontinuous nature of the riparian vegetation.

At the time this monitoring study was initiated, distance estimation had not yet been widely advocated as a method to sample bird populations. Although the use of fixed-radius point counts has been widespread, theoretical considerations suggest that it is not an appropriate method for tracking bird abundance over time. These problems, including differences in species detectabilities, have been reviewed by Rosenstock et al. (2002) and Ellingson and Lukacs (2003). The adoption of distance estimation, however, remains problematic in dense multi-layered vegetation, where it may be extremely difficult to estimate distance to vocalizing birds.

Distance estimation was tested in the study area in 2000 (summarized in Spence 2004). Because the sampling design violated several assumptions of the method, the results obtained were at best suggestive of the efficacy of distance estimation as a method to monitor breeding bird abundance in the riparian vegetation along the Colorado River. Only four bird species were abundant enough for distance estimation to be feasible: Lucy's warbler, Bell's vireo, Bewick's wren, and house finch.

The use of mean relative abundance values for bird species, based on point count stations in patches of vegetation, is problem-

Table 2. Results of power analyses for 16 species of riparian breeding birds based on data generated from point counts sampled between 1996 and 2000 in 46 patches of riparian vegetation between Glen Canyon Dam and Diamond Creek. For three species found only below Lee's Ferry (Bell's vireo, song sparrow, and lesser goldfinch), data from 36 patches in the Grand Canyon were used. Species followed by an asterisk showed significant differences between years in relative abundance using a repeated-measures ANOVA.

Species	10%[1]	20%[2]	$\alpha = 0.10$[3]	Years to Sample[4]	Patches Occupied[5]	Patches Needed[6]
Lucy's warbler	1.000	1.000	1.000	5	45	20
House finch	0.996	1.000	0.998	7	46	15
Bewick's wren	0.999	1.000	1.000	6	44	20
Bell's vireo	0.996	1.000	1.000	7	29	25
Black-chinned hummingbird*	0.957	0.998	0.987	8	45	35
Ash-throated flycatcher*	0.947	0.998	0.969	9	46	30
Yellow-breasted chat*	0.904	0.995	0.956	9	41	40
Blue-gray gnatcatcher	0.900	0.982	0.931	10	39	40
Mourning dove*	0.388	0.604	0.503	17	31	55
Blue grosbeak	0.458	0.725	0.581	15	35	55
Common yellowthroat	0.573	0.810	0.724	14	36	50
Yellow warbler	0.695	0.907	0.809	12	35	50
Bullock's oriole	0.251	0.373	0.357	28	21	55
Lesser goldfinch	0.502	0.725	0.517	18	26	45
Song sparrow	0.550	0.789	0.853	14	22	40
Brown-headed cowbird	0.231	0.394	0.365	30	22	60

[1]Power to detect a 10% change over 10 years for a two-tailed test, $\alpha = 0.05$, three surveys per year.
[2]Power to detect a 20% change over 10 years for a two-tailed test, $\alpha = 0.05$, three surveys per year.
[3]Power to detect a 10% change over 10 years for a two-tailed test, $\alpha = 0.10$, three surveys per year.
[4]Years needed to monitor 46 patches in order to reach a power of ≥ 0.80 to detect a 10% change for a two-tailed test, $\alpha = 0.05$, three surveys per year.
[5]Number of patches where each species was detected at least once between 1996 and 2000.
[6]Number of patches to monitor in order to reach a power of ≥ 0.80 to detect a 10% change over 10 years for a two-tailed test, $\alpha = 0.05$, three surveys per year.

atic for several reasons. Sampling a patch by placing one or more point counts, with the number proportional to patch size, may produce a biased estimate of relative abundance, one that does not reflect the potential true abundance. In addition, both structural (spatial) and temporal variability within and between patches can affect estimation of means and variances, in particular by potentially over-inflating coefficients of variance (B. Powell, personal communication 2006). Differences between patches in trends over time comprise a real concern, especially in light of the significant differences in vegetation structure and climate that occur along the river corridor between Glen Canyon Dam and Lake Mead (Spence 2004).

Because of problems of high spatial and temporal variability and rarity, an alternative to point counts would be to use some form of occupancy monitoring. Recent theoretical developments in occupancy theory have provided a solid basis for its use in monitoring (MacKenzie et al. 2003; Field et al. 2005; Royle and Link 2006). Determining presence/absence is much easier than attempting to determine number of individuals. The nature of the riparian vegetation in the study area, consisting as it does of numerous different sized and generally isolated patches, is probably conducive to an occupancy-based approach.

Because of the initial subjective selection of larger riparian vegetation patches, infer-

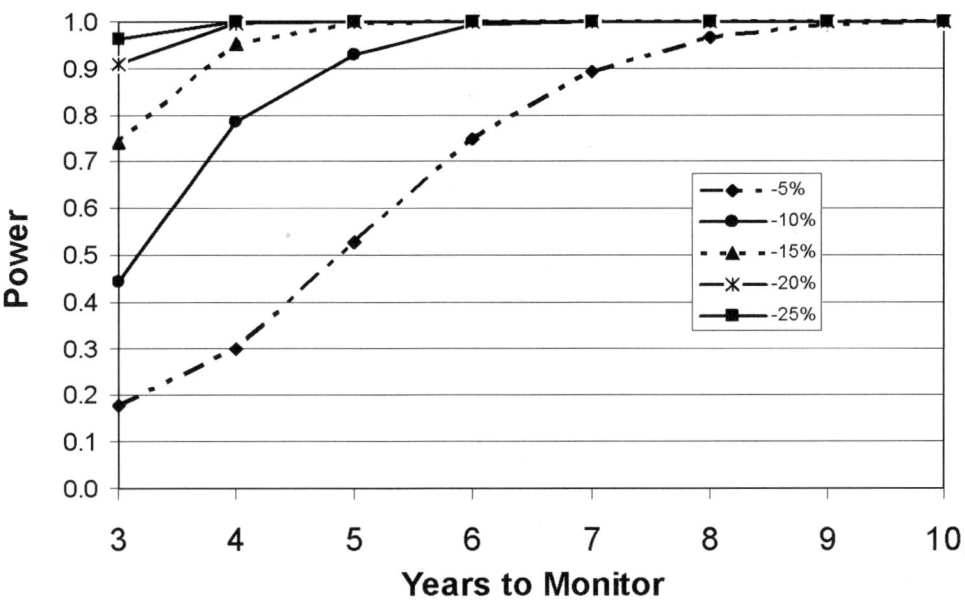

Figure 2. Power curves for declines of –5 to –25% per year over 10 years of monitoring, at α = 0.05 and a two-tailed test for Lucy's warbler dove along the Colorado River between Glen Canyon Dam and Diamond Creek. The data are from point counts sampled between 1996 and 2000 in 46 patches of riparian vegetation.

ences cannot be made about the entire river corridor in the study area for riparian breeding bird species. In addition, non-random selection of study patches violates assumptions of parametric statistical tests, and can bias estimates of means and variances in bird abundance data. In order to make inferences to the entire study area, a sampling design where all riparian vegetation is pooled and sites are randomly selected is necessary, perhaps by developing a grid-based stratification of the corridor and combining this with a random selection process using a panel design.

Finally, the nature of the study area, with its 410 km long river corridor, with riparian vegetation typically less than 100 m wide, and with an elevation drop of nearly 600 m, creates numerous logistical and sampling problems. Phenological differences in both the vegetation and bird communities will be pronounced along this elevational gradient, with birds already breeding at lower elevations and not yet present at higher elevations. The narrow and patchy nature of the vegetation, interspersed with upland vegetation, violates the assumptions in many monitoring techniques (such as distance estimation) that require fairly uniform vegetation composition and structure.

Statistical Power

This study provides 5 years of baseline information on relative abundance variability in breeding bird species in the study area. Five years of data may be adequate to capture much of the year-to-year variability, although this is an assumption that may be complicated by longer-term directional trends related to either successional patterns in the riparian vegetation along the river corridor or changes in migration patterns

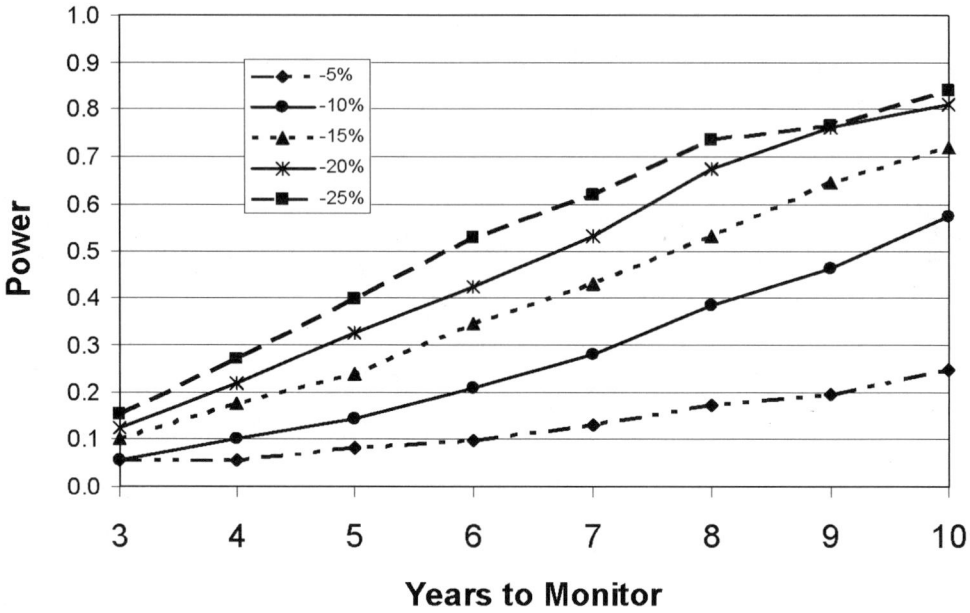

Figure 3. Power curves for declines of –5 to –25% per year over 10 years of monitoring, at α = 0.05 and a two-tailed test for common yellowthroat along the Colorado River between Glen Canyon Dam and Diamond Creek. The data are from point counts sampled between 1996 and 2000 in 46 patches of riparian vegetation.

and critical migratory and wintering habitats. The data are probably adequate to develop upper and lower limits in detection rates and relative abundance for many breeding species, and thus can be used to develop thresholds for long-term monitoring goals and for statistical power considerations.

The power analyses revealed that the 1996–2000 monitoring program had adequate power to detect change in relative abundance in some but not all riparian bird species. Using the parameters set in the analyses, there was adequate power to detect change in Lucy's warbler, house finch, Bell's vireo, Bewick's wren, black-chinned hummingbird, ash-throated flycatcher, yellow-breasted chat, and blue-gray gnatcatcher. These species represented only 25 percent of the total breeding community in the study area, but more than 80 percent of

the total number of individuals detected per year. However, there was insufficient power in the program to monitor many other bird species, including several relatively widespread but uncommon ones like common yellowthroat, blue grosbeak, mourning dove, and yellow warbler. Changing aspects of the monitoring program and statistical methods would improve power for a few additional species, but many species were simply too rare to be monitored without a large increase in the number of patches and point counts sampled. Clearly, other ways of monitoring these species would be required if they are considered to be of management importance by the federal and tribal agencies involved in managing the river corridor.

One aspect of the current study that merits additional emphasis is that the modeled effect sizes are very large. A decline of 10–20 percent per year is a major decline for a

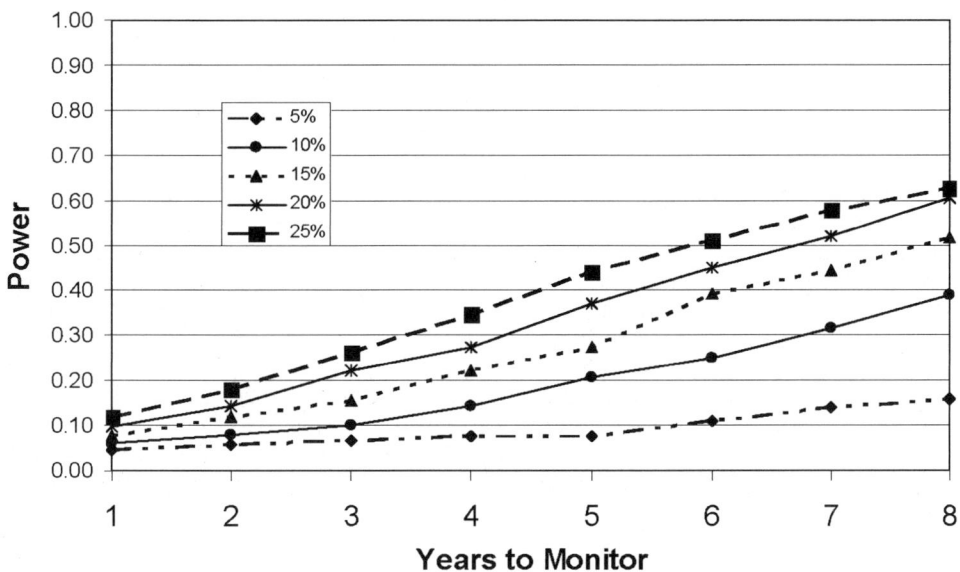

Figure 4. Power curves for declines of –5 to –25% per year over 10 years of monitoring, at α = 0.05 and a two-tailed test for mourning dove along the Colorado River between Glen Canyon Dam and Diamond Creek. The data are from point counts sampled between 1996 and 2000 in 46 patches of riparian vegetation.

species. Effect sizes of 1–5 percent are more typical targets for a long-term monitoring program, but these cannot be achieved with this data set over the 10-year examined time interval. A decline of 10 percent per year in a population of 1000 individuals means that at 10 years the population has dropped to 349, and at 20 years to 122 individuals. These would be catastrophic declines for most vertebrate species including birds. A bird population that is declining by 10–20 percent each year is likely to be in serious trouble in a very short time because of the magnitude of the change. In these cases, monitoring may be too late for management decisions that could potentially reverse such a decline. However, extending the power analyses out to 20–30 years provides sufficient power for the eight most common species in the study area at much lower effect changes, on the order of 1–5 percent change per year (Spence 2004).

There are at least four additional ways in which power could be increased to "capture" more of the bird species in the study area: (1) increasing the sampling effort in the current program, (2) examining power to detect change in combined "guilds" of species, (3) changing the bird abundance parameter, or (4) changing the specified parameters in the power analyses. Of these four, increasing the sampling effort will cause the greatest increases in logistical costs but is most likely to lead to the best improved power. Using guilds of species, based on such factors as diet, foraging strategies, or nest site selection, could provide adequate power for trend detection, but would provide only coarser-grained data on population trends and would potentially miss important changes in particular species. However, there may be some instances in which a guild-based approach may be useful. Rather than using the mean number of detections per year, the highest count could be used, as was done by Sogge et al. (2005).

Finally, rather than using a null hypothesis of no change in a population over time, one could use a one-tailed test to test the hypothesis of no declines over time. This would follow from the concept that declines are of greater management concern than increases for native species, excluding possibly the brown-headed cowbird. Beta and alpha values could also be manipulated to change power, although doing so depends strongly on the contrasting costs and benefits of Type I and Type II errors and their consequences in bird monitoring within the study area.

Two species of neotropical migrants, Lucy's warbler and Bell's vireo, are common in the study area and can be relatively easily monitored with sufficient power using distance estimation. These two species are on the Partner's in Flight (PIF) list for Arizona, as species of concern with relatively small geographic ranges. Most other riparian species in the study area are common and not on regional PIF lists. In addition to these two migrant species, two other common permanent residents, Bewick's wren and house finch, can also be monitored with sufficient power. Thus a monitoring program could be established that focuses on these four species—two neotropical migrants and two residents.

Management Implications

This study has implications for the management of selected wildlife populations in the Grand Canyon region, especially those such as birds and amphibians that show considerable year to year variation in numbers. Many past monitoring efforts have collected data on populations that have shown trends over time. However, in the absence of power tests the significance of such trends is often unknown. In many cases past programs have not had sufficient power to detect trends at all when analyzed. This has been a serious problem in particular with amphibian monitoring (cf. Hayes and Steidl 1997).

The 5-year monitoring program was intensive and costly, with annual budgets of more than $100,000 when logistics support was included. Although increasing the number of surveys per year would increase statistical power, this would make any such monitoring program prohibitively expensive under most circumstances. The number of patches sampled on a trip could also be increased over the typical 45–60 patches that were visited per trip in the past, which could also increase power significantly.

Analyzing the power in a monitoring data set highlights one critical aspect of monitoring—the time component. In order to detect moderate changes in bird species abundance, up to 20 or more years of monitoring on an annual basis may be required. In the current study, the power to detect change over shorter timeframes exists only for a few species, such as Lucy's warbler and Bewick's wren, and only for relatively large effect sizes. Hence, long-term commitment of resources would be needed to detect trends in the riparian bird community along the Colorado River (cf. Dunning and Kilgo 2000). Such long-term commitments of time and resources are still rare in bird monitoring programs and ecology, although the National Research Council has long advocated just this for the Adaptive Management Program (National Research Council 1987, 1996, 1999).

Bird populations, especially those that migrate during the non-breeding season, are affected by numerous environmental factors away from their breeding areas. Thus long-term monitoring of birds in the Grand Canyon region must take into account other aspects of the species' life cycles, wintering and migrational movements, and associated habitat. Conceptual modeling is one way to examine these various interacting factors (cf. Karr 1991). Breeding bird abundance within the study area is affected by numerous other variables outside the river corridor. Principal among these are winter and migration habitat changes and winter habitat climate variability, which can strongly influence bird survivorship. Dam operations affect birds primarily through effects on breeding habitat and recreational use. Under normal ROD operations, these impacts are likely to be fairly minor compared with climate and habitat changes outside the Colorado River corridor. The major impacts of dam operations are the planned or unplanned floods, including those in 1983 and 1996. These floods can scour out some or much of the riparian vegetation in the study area. Past flooding, particularly that in 1983, may explain many of the differences found in the breeding bird communities between the mid-1980s and the present study, as the extent of riparian vegetation was much reduced after the 1983 event (Spence 2004; Holmes et al. 2005).

For managers, who typically work with relatively limited budgets, it is critical that all aspects of a monitoring program are examined before committing scarce resources. Conceptual modeling and prospective statistical power analyses are among the most important tools in the development of long-term monitoring programs, and their use should be advocated as often as possible. Certainly, power should be addressed prior to launching into a potentially costly and long-term monitoring effort (Taylor and Gerrodette 1993).

ACKNOWLEDGMENTS

The first 2 years of the study (1996–1997) were conducted with a grant from Glen Canyon Environmental Studies. In 1998, funding was obtained from Grand Canyon Monitoring and Research Center to continue the earlier studies and to expand their scope and integrate them with habitat characterization and Glen Canyon NRA projects. These two programs owe much to the earlier USGS project that conducted preliminary monitoring feasibility studies on the river corridor from 1993 to 1995 (Sogge et al. 2005). Thanks are due to many people who helped on this project, including Nikolle Brown, Lara Dickson, Brian Dierker, Jennifer Holmes, Chuck LaRue, and numerous boatmen and volunteers. Reviews and criticism by two anonymous reviewers, Brian Powell, Mark Sogge, and Charles van Riper III are gratefully acknowledged.

LITERATURE CITED

American Ornithologists Union. 1998. Checklist of North American Birds. 7th edition. American Ornithologists Union, Washington, D.C.

Beus, S. S., and M. Morales. 1990. Grand Canyon Geology. Oxford University Press and Museum of Northern Arizona Press, Flagstaff.

Brown, B. T., S. W. Carothers, and R. R. Johnson. 1987. Grand Canyon Birds. University of Arizona Press, Tucson.

Carothers, S. W., and B. T. Brown. 1991. The Colorado River Through Grand Canyon. University of Arizona Press, Tucson.

Cody, M. L. 1985. An introduction to habitat selection in birds. In Habitat Selection in Birds, edited by M. L. Cody, pp. 3–56. Academic Press, San Diego, California.

Dunning, J. B. Jr., and J. C. Kilgo (editors). 2000. Avian Research at the Savannah River Site: A Model for Integrating Basic Research and Long-Term Management. Studies in Avian Biology No. 21. Cooper Ornithological Society.

Ellingson, A. R., and P. M. Lukacs. 2003. Improving methods for regional landbird monitoring: A reply to Hutto and Young. Wildlife Society Bulletin 31: 896–902.

Felley, D. L., and M. K. Sogge. 1997. Comparison of techniques for monitoring riparian birds in Grand Canyon National Park. In Proceedings of the Third Biennial Conference of Research on the Colorado Plateau, edited by C. van Riper III and E. Deshler, pp. 73–83. USDI National Park Service Transactions and Proceedings Series NPS/NRNAU/NRTP-97/12.

Field, S. A., A. J. Tyre, and H. P. Possingham. 2005. Optimizing allocation of monitoring effort under economic and observational constraints. Journal of Wildlife Management 69: 473–482.

Gibbs, J. P. 1995. MONITOR Version 6.2: Software for estimating the power of population monitoring programs to detect trends in plant and animal abundance. Department of Biology, Yale University, New Haven, Connecticut.

Gibbs, J. P. 1998. Integrating Monitoring Objectives with Sound Sampling Design. A Pilot Review of Selected Monitoring Programs at Shenandoah National Park. State University of New York, Syracuse. Final report to the National Park Service.

Hayes, J. P., and R. J. Steidl. 1997. Statistical power analysis and amphibian population trends. Conservation Biology 11: 273–275.

Hilden, O. 1965. Habitat selection in birds: A review. Ann. Zool. Fenn. 2: 53–75.

Holmes, J., J. R. Spence, and M. K. Sogge. 2005. Birds of the Colorado River in Grand Canyon: A synthesis of status, trends and dam operation effects. In The State of the Colorado River Ecosystem, edited by S. P. Gloss, J. E. Lovich, and T. E. Melis, pp. 123–138. U.S. Geological Survey Circular 128.

Karr, R. 1991. Biological integrity: A long neglected aspect of water resources management. Ecological Applications 1: 66–84.

LaRue, C. T., L. L. Dickson, N. L. Brown, J. R. Spence, and L. E. Stevens. 2001. Recent bird records from the Grand Canyon region, 1974–2000. Western Birds 32: 101–118.

MacKenzie, D. I., J. D. Nichols, J. E. Hines, M. G. Knutson, and A. B. Franklin. 2003. Estimating site occupancy, colonization, and local extinction when a species is detected imperfectly. Ecology 84: 2200–2207.

National Research Council. 1987. River and Dam Management: A Review of the Bureau of Reclamation's Glen Canyon Environmental Studies. National Academy Press, Washington, D.C.

National Research Council. 1996. River Resource Management in the Grand Canyon. National Academy Press, Washington, D.C.

National Research Council. 1999. Downstream. Adaptive Management of Glen Canyon Dam and the Colorado River Ecosystem. National Academy Press, Washington, D.C.

Perrins, C. M., J. D. Lebreton, and G. J. M. Hirons. 1991. Bird Population Studies; Relevance to Conservation and Management. Oxford University Press, London.

Pucherelli, M. J. 1986. Evaluation of Riparian Vegetation Trends in the Grand Canyon Using Multitemporal Remote Sensing Techniques. Final report to USDI, Bureau of Reclamation. Glen Canyon Environmental Studies GCES/18/87. NTIS PB88-183470.

Rosenberg, K. V., R. D. Ohmart, W. C. Hunter, and B. W. Anderson. 1991. Birds of the Lower Colorado River Valley. University of Arizona Press, Tucson.

Rosenstock, S. S., D. R. Anderson, K. M. Giesen, T. Leukering, and M. F. Carter. 2002. Landbird counting techniques: Current practices and alternatives. Auk 119: 46–53.

Royle, J. A., and W. A. Link. 2006. Generalized site occupancy models allowing for false positive and false negative errors. Ecology 87: 835–841.

Sogge, M. K., D. Felley, and M. Wotawa. 1998. Riparian Bird Community Ecology in the Grand Canyon. Final report to the Bureau of Reclamation, Grand Canyon Monitoring and Research Center. U.S. Geological Survey, Colorado Plateau Field Station, Flagstaff.

Sogge, M. K., D. Felley, and M. Wotawa. 2005. A quantitative model of avian community and habitat relationships along the Colorado River in the Grand Canyon. In The Colorado Plateau II: Biophysical, Socioeconomic, and Cultural Resources, edited by C. van Riper III and D. J. Mattson, pp. 163–192. University of Arizona Press, Tucson.

Spence, J. R. 2004. The Riparian and Aquatic Bird Communities Along the Colorado River from Glen Canyon Dam to Lake Mead, 1996–2000. Final report to the USGS Grand Canyon Monitoring and Research Center, Flagstaff. Resource Management Division, Glen Canyon NRA.

Spence, J. R., and B. R. Bobowski. 2003. 1994–1997 water bird surveys of Lake Powell, a large oligotrophic reservoir on the Colorado River, Utah and Arizona. Western Birds 34: 133–148.

Steidl, R. J., J. P. Hayes, and E. Schauber. 1997. Statistical power analysis in wildlife research. Journal of Wildlife Management 61: 270–279.

Stevens, L. E., J. C. Schmidt, T. J. Ayers, and B. T. Brown. 1996. Flow regulation, geomorphology, and Colorado River marsh development in the Grand Canyon, Arizona. Ecological Appl. 5: 1025–1039.

Taylor, B. L., and T. Gerrodette. 1993. The uses of statistical power in conservation biology: The vaquita and northern spotted owl. Conservation Biology 7: 489–500.

Turner, R. M., and M. M. Karpiscak. 1980. Recent Vegetation Changes Along the Colorado River Between Glen Canyon Dam and Lake Mead, Arizona. USDI, U.S. Geological Survey, Professional Paper 1132.

AVIAN COMMUNITY RESPONSES TO FOREST THINNING AND PRESCRIBED SURFACE FIRE, ALONE AND IN COMBINATION

Sarah Hurteau, Brett G. Dickson, Thomas D. Sisk, and William M. Block

The U.S. Forest Service and other federal agencies treated more than 2.8 million ha of forested land between 2001 and 2003 under the Healthy Forest Restoration Act (USDA Forest Service 2004). Because many more areas across our landscape are slated for fuel reduction treatments, understanding how wildlife communities will respond is essential to conserving biodiversity, yet little is understood about the ecological consequences of such treatments. Prior to Euro-American settlement in the late 1800s, ponderosa pine–dominated forests of the Southwest were characterized by open stands with a dense herbaceous understory (Stone et al. 1999). Low-intensity surface fires, with return intervals of 2–12 years on average, maintained this stand structure (Covington and Moore 1994). However, fire suppression, livestock grazing, and selective logging of old fire-resistant trees have significantly altered forest structure (Harrington and Sackett 1988; Covington et al. 1997; Stone et al. 1999; Youngblood et al. 2004). Today's forests typically are characterized by dense, closed-canopy forests (Cooper 1960; Covington and Moore 1994), leaving them susceptible to stand-replacing fires, pathogen and pest outbreaks, and non-native species invasions (Fulé et al. 2002; Bock and Block 2005). In the Southwest, fuel reduction treatments employ mechanical thinning and prescribed surface fire to mitigate excessive fuel accumulations and, in some cases, to re-create forest structural characteristics that existed prior to the regular interruption of natural disturbance regimes (Meyer et al. 2001). However, the broader ecological effects of these techniques are not well studied.

Much of the research on the effects of fuel reduction and restoration treatments has been focused on one dominant tree species (ponderosa pine, *Pinus ponderosa*) and the response of understory vegetation after thinning and/or prescribed fire (Harrington and Sackett 1988; Covington et al. 1997; Stone et al. 1999; Franklin et al. 2002; Fulé et al. 2002). Research conducted to determine the effects of these treatments on wildlife has consisted largely of observational studies, generally addressing a single treatment alternative (e.g., King and DeGraaf 2000; Saab et al. 2004) or focusing on a single species (e.g., Germaine and Germaine 2002), and not the response of the avian community as a whole. Although literature reviews have been conducted to identify how avian communities respond to mechanical thinning, wildfire, and post-fire salvage logging (e.g., Kotliar et al. 2002; Saab and Powell 2005), no study has experimentally tested the effects of multiple fuel reduction treatment types on avian communities.

Our research was conducted on the Fire and Fire Surrogates (FFS) program Southwest Plateau sites (Edminster et al. 2000). The national FFS program is an integrated network of 13 long-term research sites designed to quantify the effects of prescribed fire and mechanical thinning on a set of ecosystem health response variables, including wildlife (S. Zack and W. F. Laudenslayer, unpublished report). Although the FFS program was not developed to focus on wildlife, its experimental design permits studies that can improve our understanding of how various treatment types affect sensitive and ecologically important species. However, there are several inherent limitations in the FFS experimental design, especially when focusing on such mobile taxa as birds (Dickson et al. 2004). For example, the treatment units are small in size, limiting the number of avian detections gathered and our ability to estimate detection probabilities for multiple species.

Our general objective for this study was to evaluate the effects of fuel reduction treatments on the organization of the avian community, but we also explored a finer-scaled response at the level of individual species. We evaluated the change in density for a suite of "focal" species, representing a range of foraging modes, to provide examples of how various guilds may respond to future fuel reduction treatments.

To examine short-term effects of different forest fuel reduction treatments on the avian community, we sought to address the following research questions: (1) Does the rank abundance or organization of bird species differ among treatment types? (2) Do bird densities change in response to fuel treatments? Our main objectives were to evaluate how the avian community responds to various experimental fuel reduction treatments and to provide baseline information that could be used in the development of forest management guidelines for wildlife.

METHODS
Study Area

Our three study sites were located on the Kaibab (K.A. Hill) and Coconino National Forests (Powerline and Rudd's Tank, hereafter KA, PL, and RT, respectively), west of Flagstaff, Arizona. Overstory composition at each site was dominated by ponderosa pine and occasionally included Gambel oak (*Quercus gambelii*), one-seed juniper (*Juniperus monosperma*), and alligator-bark juniper (*J. deppeana*). Small-diameter (< 25 cm) ponderosas were common at each site, and larger yellow-bark trees occurred in small clumps that typically resulted from previous timber harvests (Dickson et al. 2004). Com-

mon understory and grassland vegetation included Arizona fescue (*Festuca arizonica*) and blue grama (*Bouteloua gracilis*). Mean site elevation was approximately 2200 m and the topography was mostly flat. For the period 1971–2000, average annual precipitation was 58 cm and average annual temperature was 7.9° C (Staudenmaier et al. 2005).

Experimental Design

We used a modified before-after/control-impact experimental design (Green 1979; Stewart-Oaten et al. 1986; Stewart-Oaten et al. 1992) to assess the response of bird community attributes to fuel reduction treatments. Each study site was established in a block design and represented a single replicate ($n = 3$). We divided each replicate into four units, including three experimental fuel reduction treatments (thin-only, prescribed burn-only, and thin followed by prescribed burn) and a single control unit (Figure 1). This design was used, in part, to efficiently explore treatment alternatives that are highly representative of the project-scale (e.g. < 1000 ha) treatments currently being implemented in many of the Southwest's ponderosa pine forests. Although our control units could not be characterized as having been dominated by a natural disturbance regime, they do capture the forest structural conditions that are typical for ponderosa pine–dominated forests in the region. Thinning treatments were conducted during the fall of 2002, and prescribed surface fire treatments were completed in the fall of 2003. Each treatment unit consisted of a 10 ha core sampling area surrounded by a 30 m buffer. Units ranged in total size from 16 ha (RT) to 30 ha (PL). We established a 36-point array of sample points within each treatment unit, typically as a 6 x 6 grid, with 50 m between points (but see Figure 1). Regardless of array shape, we maintained a minimum of 50 m spacing between grid points and a buffer zone of ≥ 50 m between points and the edge of the treatment unit.

Avian Community Sampling and Analysis

We sampled the avian community during the summer breeding season (late May

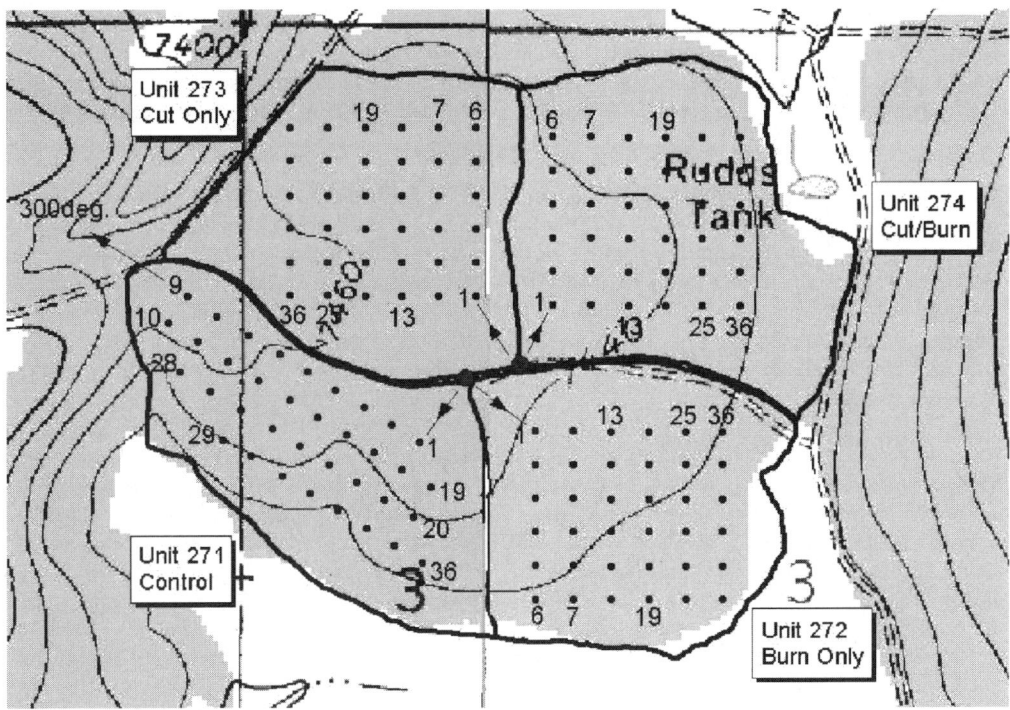

Figure 1. The Power Line (PL) FFS study site in northern Arizona. Map illustrates our experimental design, including the 36 point sampling array in each treatment unit, and the close proximity of treatment units to one another.

through early July) of each survey year. We conducted pre-treatment surveys between 2000 and 2002 and post-treatment surveys between 2003 and 2005. Additional details on the national FFS standardized wildlife sampling protocols, and specific modifications made for the Southwest Plateau sites, are described by Dickson et al. (2004).

To assess the numerical response of avian species to treatments, we used a sampling design that permitted density estimation by distance methods (Buckland et al. 2001). We conducted 50 m fixed-radius counts of all species detected by sight or sound during a 5-minute period at all points on a 36-point treatment unit array. We estimated the distance from the observer to each bird detected and mapped its location. Over the course of the breeding season, we counted birds at each point on a single occasion,

usually before 10 a.m., and visited 9–12 points per day. This sampling scheme resulted in 3–4 visits to each treatment unit per season. We spaced out our visits over the breeding season to account for temporal heterogeneity in the detectabilities of birds. We randomly selected points in a way that minimized sampling radius overlap or double-counting of the same individual on a given day. Because distance-based models (see below) are robust in the detection of the same individual(s) at > 1 point, or during different sampling periods, we were able to relax any assumption of complete spatial independence among sampling points (Buckland et al. 2001).

For the most abundant species, we generated a rank abundance table to examine differences in the avian community between treatments, and to monitor change in community organization over time. A rank was then assigned to each species on the list based on the number of detections within each treatment. If there was a tie in the number of detections, each tied species was given the average rank. Because of the small size of the FFS treatment blocks, we had too few detections for most species and were only able to estimate densities for the five most common species. We assessed the significance of rank abundance and community composition among treatments using Spearman's rank correlation procedure (rs = test statistic; $\alpha = 0.05$).

Focal Species Analysis

We reviewed the published literature to generate hypotheses about the directional (positive, negative, or neutral) response to treatments expected for five common species belonging to different foraging guilds. This suite of focal species included western bluebird (*Sialia mexicana*; insectivore generalist), mountain chickadee (*Poecile gambeli*; foliage insectivore), pygmy nuthatch (*Sitta pygmaea*; bark insectivore), dark-eyed junco (*Junco hyemalis*; ground insectivore), and yellow-rumped warbler (*Dendroica coronata*; foliage insectivore). Western bluebirds are well-known fire specialists (Bock and Block 2005; Saab and Powell 2005), so we hypothesized that their densities would increase in burned areas. As an insectivorous species, western bluebirds can benefit from increases in insect populations following fire (Guinan et al. 2000). Because mountain chickadees are more abundant in unburned areas, and respond negatively to fire (Kotliar et al. 2002; Bock and Block 2005; Saab and Powell 2005), we expected a negative response by this species to all treatment types. For the remaining three species (dark-eyed junco, pygmy nuthatch, and yellow-rumped warbler) we expected a neutral response to the fuel reduction treatments because previous research has shown a mixed response by these species (Kotliar et al. 2002; Bock and Block 2005; Saab and Powell 2005).

To quantify the response of each of these species, we used two different ad hoc approaches. First, we visually evaluated the estimated standard errors associated with post-treatment density estimates for control vs. treatment type or treatment vs. treatment for each species, and considered nonoverlapping standard errors to be indicative of a response to treatment. Second, we developed an index value to describe the magnitude of change observed in density estimates, within and among treatment types. Given the subjectivity of both approaches, our results are more qualitatively suggestive than quantitatively definitive.

For the five focal species, we used the program DISTANCE (version 5.0; Thomas et al. 2005) to estimate density (individuals per hectare) by treatment type, or "stratum." Because the thin & burn treatments were implemented in different years, we calculated bird densities separately for three periods: P1, a pre-treatment period (2000–2002), P2, a post-thin period (2003), and P3, a post-thin and post-burn period (2004–2005). We used multiple covariate distance sampling analysis (Buckland et al. 2004; Marques and Buckland 2004) and considered year, study site, and treatment type as factors in estimating a global detection probability for each species. Since we required species- and stratum-specific estimates, this approach was more robust to small sample size than conventional distances sampling analyses (Marques and Buckland 2004). Because we used a global detection function to estimate species densities by treatment type, our stratified estimates are not independent, have associated variance estimates that are biased toward small, and violate the assumptions of traditional statistical tests (Buckland et al. 2001). We used Akaike's Information Criterion (AIC) to assess model fit and to select the best model of density from a candidate set that included all possible factor combinations (Burnham and Anderson 2002). We averaged density estimates when AIC difference values for competing models were ≤ 2, and we computed unconditional standard errors (Burnham and Anderson 2002).

To quantify the magnitude of the directional treatment response by our focal species, we used the "raw" differences in density (Δ), which we calculated by subtracting the control density from the treatment density, for both pre- (Δ-pre) and post-treatment (Δ-post) density estimates output by program DISTANCE. We then calculated a standardized difference (Δ^*) as Δ-post treatment minus Δ-pre treatment to evaluate the cumulative response to treatments by species. We considered a large response to be one with $\Delta^* > 0.40$, a small response to be one with $\Delta^* = 0.10–0.40$, and a neutral response to be one with $\Delta^* < 0.10$. Although these designations are arbitrary, we have included them to standardize our language used to describe the magnitude of the responses observed in the focal set of species analyzed. For both the raw and cumulative difference calculations, positive and negative differences indicated a positive and negative response to treatment, respectively.

RESULTS
Community Response
We recorded 5839 detections among 62 total species during the 2000–2005 breeding seasons. Our species list for analysis was generated for the 25 most frequently detected species during the 6-year study period. Table 1 illustrates how some species increased in rank abundance while others decreased. These results suggest that there is no difference in the avian community structure among treatments. Pygmy nuthatch, for example, dropped in rank from first to fourth in P3, and mountain chickadee dropped from fourth to nineteenth. Western bluebird remained relatively constant though time, whereas after treatments the violet-green swallow moved from the bottom of the list to within the top 10. The Spearman rank correlation test, comparing the species present in the control and the thin-only treatment in P2, showed that avian ranked abundances were correlated among treatments, and were significant for all four comparisons of interest (for post-thin P2, control vs. thin-only, $rs = 0.754$, $P < 0.0001$; for post-thin and post-burn P3, control vs. thin-only, $rs = 0.720$, $P < 0.0001$; for control vs. burn-only, $rs = 0.789$, $P < 0.0001$; and for control vs. thin & burn, $rs = 0.728$, $P <$

0.0001). Similar analyses were not conducted for the pre-treatment period because previous research has shown that the structure was not significantly different among treatment units within each study area (Dickson et al. 2004).

Focal Species Response
Response to treatment types during the post-treatment period varied considerably among species. Only those with ≥ 340 detections were considered for the focal species analysis. Our estimated standard errors associated with post-treatment density indicated a positive treatment response by western bluebird to the thin-only and thin & burn treatments (Figure 2). The density of dark-eyed junco increased on the burn-only treatment, indicating a positive treatment response. Pygmy nuthatch also showed a positive response to the burn-only treatment in the post-treatment period, and yellow-rumped warbler exhibited a negative response to the thin-only treatment. Mountain chickadee was the only species for which the standard error estimates did not suggest a treatment response.

For all species except yellow-rumped warbler, the direction of the change in density on the control unit was identical to the direction on the treatment units. On the control units, we observed a small increase in western bluebird ($\Delta^* = 0.17$) and dark-eyed junco ($\Delta^* = 0.18$) densities, and a large ($\Delta^* = 0.41$) increase in yellow-rumped warbler density (Table 2). In contrast, we observed a small decrease in mountain chickadee ($\Delta^* = -0.19$) and pygmy nuthatch ($\Delta^* = -0.26$) densities. For the thin-only treatment, we observed a small positive increase in western bluebird ($\Delta^* = 0.16$) and dark-eyed junco ($\Delta^* = 0.24$) densities, but a small decrease in mountain chickadee ($\Delta^* = -0.39$) and yellow-rumped warbler ($\Delta^* = -0.24$) densities. We observed a very small ($\Delta^* = -0.09$) decrease in pygmy nuthatch density, which we considered a neutral response. In the burn-only treatment, patterns were similar to those described above. We observed a small increase in western bluebird ($\Delta^* = 0.16$) and yellow-

Table 1. The 25 most frequently detected species by treatment type during two sampling periods on the three FFS study sites in northern Arizona, 2000–2005. Species ranked according to total number of detections for each species within each treatment type. For species with tied abundance values, we determined rank by computing the average of these values. All community comparisons were significantly correlated among treatments within a given period (see Results section).

Species	P2 (Post-Thin Only)		P3 (Post-Thin and Post-Burn)			
	Control	Thin Only	Control	Thin Only	Burn Only	Thin & Burn
Pygmy nuthatch (*Sitta pygmaea*)	1	1	4	4	4	4
Yellow-rumped warbler (*Dendroica coronata*)	2	3	2	3	1	2
Western bluebird (*Sialia mexicana*)	3	4	3	1	2.5	1
Mountain chickadee (*Poecile gambeli*)	4	14.5	6	19	13	18.5
Dark-eyed junco (*Junco hyemalis*)	5	2	1	2	2.5	3
Chipping sparrow (*Spizella passerina*)	6	40.5	23.5	15	11	12.5
American robin (*Turdus migratorius*)	9.5	5	8.5	10	7	5
Brown creeper (*Certhia americana*)	9.5	14.5	14	13	8	23
Grace's warbler (*Dendroica graciae*)	9.5	10.5	5	15	5	14
Plumbeous vireo (*Vireo plumbeus*)	9.5	6	11	5	6	8
Steller's jay (*Cyanocitta stelleri*)	9.5	10.5	20.5	18	20.5	12.5
White-breasted nuthatch (*Sitta carolinensis*)	9.5	10.5	8.5	17	18	10
Western wood-pewee (*Contopus sordidulus*)	13	7	18	8	13	6.5
Brown-headed cowbird (*Molothrus ater*)	16	8	8.5	9	13	6.5
Gray flycatcher (*Empidonax wrightii*)	16	18.5	19	26.5	25.5	45.5
Hairy woodpecker (*Picoides villosus*)	16	18.5	20.5	15	16.5	17
Pine siskin (*Carduelis pinus*)	16	10.5	15	6.5	16.5	16
Western tanager (*Piranga ludoviciana*)	16	13	13	11.5	10	15
Common raven (*Corvus corax*)	40	40.5	47.5	24	20.5	45.5
Hermit thrush (*Catharus guttatus*)	40	40.5	23.5	46.5	22.5	45.5
Mourning dove (*Zenaida macroura*)	40	16	27	26.5	46.5	22
Northern flicker (*Colaptes auratus*)	40	40.5	16.5	20.5	25.5	18.5
Red crossbill (*Loxia curvirostra*)	40	40.5	16.5	20.5	22.5	20
Violet-green swallow (*Tachycineta thalassina*)	40	40.5	8.5	6.5	9	9

Table 2. Pre- and post-treatment density (birds per hectare) estimates, raw (Δ) and standardized (Δ*) estimated differences, and directional response by treatment type for five species detected on the three FFS study areas, 2000–2005. Overall (observed) and expected (based on Kotliar et al. 2002; Bock and Block 2005) responses are indicated as positive (+), negative (–), or neutral (0). Since measures of variance for our density estimates are biased (see text), they are not reported here.

	Control [a]	Δ	Thin Only [b]	Δ	Burn Only [c]	Δ	Thin & Burn [c]	Δ	Expected	Observed
Western bluebird										
Pre-treatment	0.37		0.45	0.08	0.37	0.00	0.44	0.07		
Post-treatment	0.46/0.61		0.70	0.24	0.77	0.16	0.95	0.34		
Δ*		0.17		0.16		0.16		0.28	+	+
Mountain chickadee										
Pre-treatment	0.34		0.71	0.37	0.53	0.19	0.50	0.16		
Post-treatment	0.15		0.13	-0.02	N/A	N/A	N/A	N/A		
Δ*		-0.19		-0.39		N/A		N/A	–	–
Pygmy nuthatch										
Pre-treatment	0.69		0.90	0.21	0.98	0.29	0.83	0.14		
Post-treatment	0.47/0.39		0.59	0.12	0.62	0.23	0.50	0.11		
Δ*		-0.26		-0.09		-0.06		-0.03	0	–
Dark-eyed junco										
Pre-treatment	0.54		0.42	-0.12	0.26	-0.28	0.30	-0.24		
Post-treatment	0.64/0.80		0.76	0.12	2.65	1.85	0.80	0.00		
Δ*		0.18		0.24		2.13		0.24	0	+
Yellow-rumped warbler										
Pre-treatment	0.53		0.41	-0.12	0.56	0.03	0.67	0.14		
Post-treatment	0.85/1.02		0.49	-0.36	1.25	0.23	0.83	-0.19		
Δ*		0.41		-0.24		0.20		-0.33	0	+

N/A signifies treatment units for which no estimates are available due to insufficient detections to produce a reliable density estimate.

[a] Treatments were staggered; we report two estimates for post-treatment control density: P2P3, a pooled estimate over P2, and P3/P3. We were unable to produce a reliable estimate for P3, so mountain chickadee estimates are from the P2P3 pooled estimate. We computed Δ* as the average of P2P3 and P3 post-treatment control density – pre-treatment control density; e.g. western bluebird: [(0.46 + 0.61)/2] – 0.37.

[b] Thin-only Δ computed as P2P3 pre-treatment density – P2P3 pre-treatment control density (e.g. western bluebird, 0.45 – 0.37).

[c] Burn-only and thin & burn Δ computed as P3 pre-treatment density – P3 pre-treatment control density (e.g. western bluebird, 0.37 – 0.37). We calculated Δ* as the difference of the Δs for each treatment (i.e., post-treatment Δ – pre-treatment Δ).

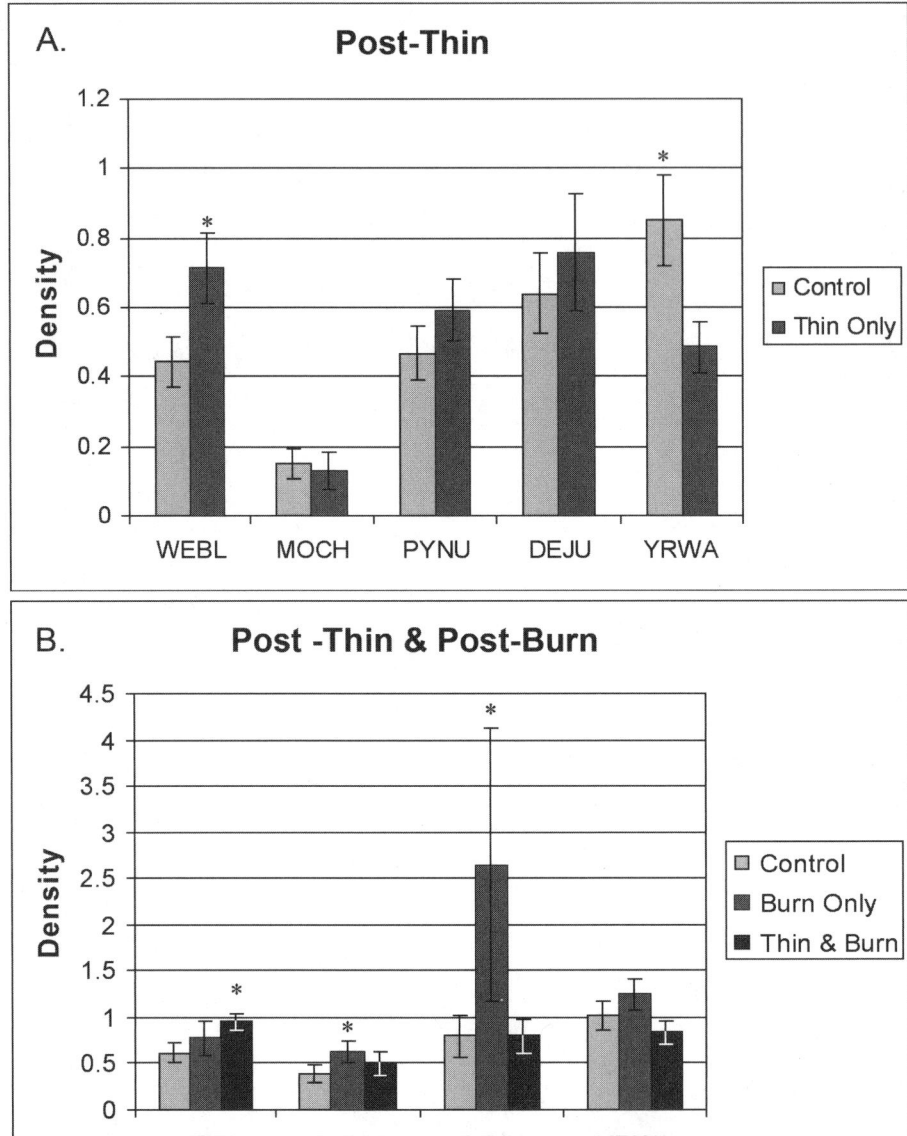

Figure 2. Post-treatment density estimates (individuals/ha) and unconditional standard errors for five species detected on the three FFS study sites in northern Arizona, 2003–2005. * indicates possible treatment effect on densities. Because the thin & burn treatments were implemented in different years, we calculated densities separately for each period. (a) Post-thin densities for P2 (2003–2005) were estimated using data for five focal species in the control and thin-only treatment units: WEBL, western bluebird ($n = 189$); MOCH, mountain chickadee ($n = 35$); PYNU, pygmy nuthatch ($n = 146$); DEJU, dark-eyed junco ($n = 171$); and YRWA, yellow-rumped warbler ($n = 176$). (b) Post-thin and post-burn densities for P3 (2004–2005) were estimated using data for four focal species in the control, burn-only, and thin & burn treatment units: WEBL, $n = 217$; PYNU, $n = 136$; DEJU, $n = 204$; YRWA, $n = 233$.

rumped warbler ($\Delta^* = 0.20$) densities and a large ($\Delta^* = 2.13$) increase in the density of dark-eyed junco (Figure 2). We observed a small ($\Delta^* = -0.06$) but neutral response by pygmy nuthatch. Mountain chickadee detections were so few ($n = 13$) in the burn-only treatment type that we were unable to generate a reliable density estimate. In the thin & burn treatment type, we observed a small increase in density for western blue-bird ($\Delta^* = 0.28$) and dark-eyed junco ($\Delta^* = 0.24$), a small ($\Delta^* = -0.33$) negative response in yellow-rumped warbler (Table 2) density, and a neutral decrease ($\Delta^* = -0.03$) in density in pygmy nuthatch. Again, we had too few ($n = 8$) mountain chickadee detections to produce a reliable density estimate for this treatment type.

DISCUSSION
Community Response

Our results indicate that the short-term (3-year) impacts of forest fuel reduction treatments, including thinning and prescribed fire, had minimal influence on the avian community attributes we measured. With the exception of violet-green swallow and mountain chickadee, all other species ranks (i.e. species composition) remained relatively constant, within and among treatment types, through time (Table 1). This trend suggests that the basic community organization was probably unaffected by the fuel reduction treatments.

Focal Species Response

The positive response by western bluebird to treatment is consistent with the results of other studies (Germaine and Germaine 2002; Bock and Block 2005; Saab and Powell 2005). We observed the largest increase in western bluebird density in the thin & burn treatment, indicating that the combination or interaction of thinning and fire likely benefits this species. Previous research suggests that this open-habitat species may benefit from burning (Kotliar et al. 2002) because prescribed fire can increase insect densities (Holmes 1990; Guinan et al. 2000).

Mountain chickadees appear to be especially sensitive to the burn-only and the thin

& burn treatment types. Previous research on this species has shown that many timber management practices can have deleterious effects on long-term population stability (McCallum et al. 1999). Although we were unable to test how these treatments will influence population persistence, our results demonstrate a negative post-treatment numerical response of mountain chickadee to all treatments when compared to the densities present during the pre-treatment period. Indeed, the number of detections decreased to the point that we were unable to produce density estimates for two of the four treatment types. Although there was very little difference in the densities present in the thin-only and control units during the post-treatment period, one might conclude that thinning has little impact on mountain chickadee densities. However, when the densities we estimated in these treatment units during the pre-treatment period are taken into account, it is reasonably clear that the disturbance associated with the implementation of the fuel reduction treatments has negatively impacted this species.

The pygmy nuthatch is one of the most abundant species in many ponderosa pine systems (Kingery and Ghalambor 2001); we thus expected pygmy nuthatch to have a neutral response to the treatments. Previous research has shown that this species tends to be affected by logging, fire, grazing, or development if suitable nesting sites are removed (Kingery and Ghalambor 2001), which was not the case on our small treatment blocks surrounded by extensive untreated forest. Our results do indicate that fuel reduction treatments may have had a negative impact on this species (negative cumulative response, Δ^*; Table 2), but the pattern was largely driven by the stronger negative response on the control units. The larger declines in density recorded on the control units, when compared to the treatment units, suggest that some background environmental change extraneous to the treatments may be influencing the relative abundance of this species at our study sites.

Since dark-eyed junco are ground omnivores, we expected to see an increase in density in both the burn-only and the thin & burn treatments due to the opening up of understory habitat and increased foraging opportunities after burning. Although dark-eyed junco exhibited the largest magnitude of response recorded for any of the species that we analyzed, we observed an increased density only in the burn-only treatment. These results are consistent with previous research conducted in Arizona on larger study units (Short 2003; Dickson 2006). It seems clear that this species is sensitive to burning and is not simply responding to the increased openness of the forest understory. A post-burn increase in seed and insect production may be responsible for the increases we observed, and additional research on this mechanism is warranted.

We speculate that the mixed response to treatments observed in yellow-rumped warbler may be a result of changes in forest structure. The shift from a closed to an open canopy in the thin-only and thin & burn treatments likely influenced the availability of foraging habitats for this foliage-gleaning species (Saab and Powell 2005). The removal of trees by treatments effectively increases the distance between foraging patches and may decrease yellow-rumped warbler densities. Yellow-rumped warbler exhibited a marked increase in the control units over the course of the study, suggesting that the more variable numbers for the treatment types reflect a smaller response to treatment than control for this species.

Design Limitations and Recommendations

Due to the analytical constraints imposed by the small size of the FFS treatment units, we were unable to detect changes in population density for several of the most common species. The 10 ha core sampling areas within treatment blocks were simply too small to allow for sufficient detections of most species. We attempted to overcome the inherent limitations of small treatment units and low replication by maximizing the number of sampling points per treatment ($n = 36$) and adjusting the sampling sequence to minimize the possibility of double-counting birds on a single day. However, with most

species represented by a dozen or fewer individuals per unit, analytical possibilities were severely limited.

The size and adjacency of the treatment units presents concerns that the among-unit movements of individuals could have affected our density estimates. Since individual birds can move readily among the different treatment units within a short period of time, our samples were not spatially independent and individuals may have used multiple post-treatment forest conditions. We used DISTANCE to analyze our data as a way to increase the precision of our density estimates and minimize the influence of sample size and the aforementioned spatial constraints. Furthermore, DISTANCE is robust to repeatedly counting the same individual, and does not cause bias if counts are associated with multiple visits or if an unconditional variance estimator is used (Buckland et al. 2001).

The landscape-level implementation of forest treatments is likely to have "winners," such as western bluebird, and "losers," such as mountain chickadee. Whether these trends are significant with respect to population persistence in a given location cannot be determined by examining changes in population density on small units, such as those used in the FFS study. Background environmental change, such as annual fluctuations in precipitation, has been shown to significantly affect bird density in northern Arizona (Szaro and Balda 1986), and this change may have influenced the patterns we observed. However, our focal species analysis may be capturing the annual variation in the abundance of these species since many of the standard errors overlap for post-treatment density estimates (Figure 2).

The implications of our research for broader forest management scenarios are limited, but not without management and conservation value. Our results from small-scale treatments suggest that a mosaic of treatments may be suitable to a wider range of forest avian species. Manipulative experiments involving treatments of 10 ha or more, even with low replication, are rare, and offer excellent opportunities for finer-scale research that focuses on individual bird movements, foraging behavior, nest success, response to microclimate, and other factors that likely underlie changes in density. We have utilized the FFS study sites in Arizona to explore and link these factors and encourage other investigators to expand their focus on these and other ecological relationships that may be more appropriate to this experimental design. Although we found design constraints to be significant on our low-productivity ponderosa pine sites, this may be less of an issue at other FFS locations. Indeed, it is important to recognize that, in our experience, such design constraints are typical of avian community studies. Furthermore, the FFS program focuses on the responses of ecosystem components and processes that are manifested at scales different from wildlife. The effective study of wildlife taxa requires considerable effort and larger unit sizes, which may be precluded because of funding or logistical constraints. Combining such different response variables into one study presents inherent challenges to any experimental design.

CONCLUSIONS

While recognizing that limitations in our study design prevent us from making broad-scale forest management recommendations, we believe we can offer both land managers and researchers lessons learned from our experience. We provide evidence that specific treatment types can affect different species in different ways. For example, western bluebirds increase in response to thinning and burning treatments, yet mountain chickadees respond negatively to this treatment type. Although we were unable to identify a single "best" treatment type, our results can provide land managers with much-needed project-scale information, while informing future studies that attempt to link the directional response of avian communities to forest management activities. Notably, the results of our study are consistent with emerging, large-scale studies of avian community response to forest treatments in the region (e.g., Short 2003; Dickson 2006). Our research further

demonstrates that wildlife studies should be designed at scales that encompass the space-use and ecological requirements of the focal taxa. In order to understand how fuel reduction treatments affect wildlife, including birds, treatments should be carried out at spatial and temporal scales large enough to effectively estimate their impacts. Moreover, future designed experiments will improve our knowledge of how forest management practices can protect human communities and, at the same time, restore ecological processes.

ACKNOWLEDGMENTS

We thank the Fire and Fire Surrogates Program, the Interagency Joint Fire Sciences Program, and C. Edminster of the USFS Rocky Mountain Research Station, Flagstaff, for funding this research. The helpful comments of M. D. Hurteau, R. Hutto, M. D. Meyer, C. van Riper III, and two anonymous reviewers improved drafts of our manuscript. We also thank the Sisk Lab of Applied Ecology, J. Bailey (Northern Arizona University, School of Forestry), L. Dickson, K. Bratland, and the numerous field technicians who worked on this project.

LITERATURE CITED

Bock, C. E., and W. M. Block. 2005. Fire and birds in the southwestern United States. Studies in Avian Biology 30: 14–32.

Buckland, S. T., D. R. Anderson, K. P. Burnham, J. L. Laake, D. L. Borchers, and L. Thomas. 2001. Introduction to Distance Sampling. Oxford University Press, London.

Buckland, S. T., D. R. Anderson, K. P. Burnham, J. L. Laake, D. L. Borchers, and L. Thomas, editors. 2004. Advanced Distance Sampling. Oxford University Press, London.

Burnham, K. P. and D. R Anderson. 2002. Model Selection and Multimodel Inference: A Practical Information-Theoretic Approach. Springer-Verlag, New York.

Cooper, C. F. 1960. Changes in vegetation, structure, and growth of southwestern ponderosa pine forests since white settlement. Ecological Monographs 30: 129–164.

Covington, W. W., and M. M. Moore. 1994. Southwestern ponderosa pine forest structure: Changes since Euro-American settlement. Journal of Forestry 92: 39–47.

Covington, W. W., P. Z. Fulé, M. M. Moore, S. C. Hart, T. E. Kolb, J. N. Mast, S. S. Sackett and M. R. Wagner. 1997. Restoring ecosystem health in ponderosa pine forests of the southwest. Journal of Forestry 95: 23–29.

Dickson, B. G. 2006. Multi-scale response of avian communities to prescribed fire: Implications for fuel management and restoration treatments in southwestern ponderosa pine forests. Dissertation, Colorado State University, Fort Collins.

Dickson, B. G., W. M. Block, and T. D. Sisk. 2004. Conceptual framework for studying the effects of fuels treatments on avian communities in ponderosa pine forests of northern Arizona. In The Colorado Plateau: Cultural, Biological and Physical Research, edited by C. van Riper III and K. Cole, pp. 193–200. University of Arizona Press, Tucson.

Edminster, C. B., C. P. Weatherspoon, and D. G. Neary. 2000. The fire and fire surrogates study: Providing guidelines for fire in future watershed management decisions. In Land Stewardship in the 21st Century: The Contributions of Watershed Management, Tucson, March 13–16, 2000, pp. 312–315. Proceedings RMRS-P-13, USDA Forest Service, Fort Collins, Colorado.

Franklin, J. F., T. A. Spies, R. Van Pelt, A. B. Carey, D. A. Thornburgh, D. R. Berg, D. B. Lindenmayer, M. E. Harmon, W. S. Keeton, D. C. Shaw, K. Bible, and J. Chen. 2002. Disturbances and structural development of natural forest ecosystems with silvicultural implications, using Douglas-fir forests as an example. Forest Ecology and Management 155: 399–423.

Fulé, P. Z., W. W. Covington, H. B. Smith, J. D. Springer, T. A. Heinlein, K. D. Huisinga, and M. M. Moore. 2002. Comparing ecological restoration alternatives: Grand Canyon, Arizona. Forest Ecology and Management 170: 19–41.

Germaine, H. L., and S. S. Germaine. 2002. Forest restoration treatment effects on the nest success of western bluebirds (Sialia mexicana). Restoration Ecology 10: 362–367.

Green, R. H. 1979. Sampling Design and Statistical Methods for Environmental Biologists. Wiley, New York.

Guinan, J. A., P. A.Gowaty, and E. K. Eltzroth. 2000. Western Bluebird (Sialia mexicana). In The Birds of North America, No. 510, edited by A. Poole and F. Gill. The Birds of North America, Inc., Philidelphia.

Harrington, M. G., and S. S. Sackett. 1988. Conference: Effects of Fire in Management of Southwestern Natural Resources. Tucson, Arizona, November 14–17.

Holmes, R. T. 1990. Ecology and evolutionary impacts of bird predation on forest insects: An overview. Studies in Avian Biology 13: 6–13.

King, D. I., and R. M. DeGraaf. 2000. Bird species diversity and nesting success in mature, clearcut and shelterwood forest in northern New Hampshire, USA. Forest Ecology and Management 129: 227–235.

Kingery, H. E., and C. K. Ghalambor. 2001. Pygmy Nuthatch (Sitta pygmaea). In The Birds of North America, No. 567, edited by A. Poole and F. Gill. The Birds of North America, Inc., Philidelphia.

Kotliar, N. B., S. J. Hejl, R. L. Hutto, V. A. Saab, C. P. Melcher, and M. E. McFadzen. 2002. Effects of fire and post-fire salvage logging on avian communities in conifer-dominated forests of the western United States. Studies in Avian Biology 25: 49–64.

Marques, F. C. C., and S. T. Buckland. 2004. Covariate models for the detection function. In Advanced Distance Sampling, edited by S. T. Buckland , D. R. Anderson, K. P. Burnham, J. L. Laake, D. L. Borchers, and L. Thomas. Oxford University Press, London.

McCallum, D. A., R. Grundel, and D. L. Dahlsten. 1999. Mountain Chickadee (Poecile gambeli). In The Birds of North America, No. 453, edited by A. Poole and F. Gill. The Birds of North America, Inc., Philidelphia.

Meyer, C. L., T. D. Sisk, and W. W. Covington. 2001. Microclimatic changes induced by ecological restoration of ponderosa pine forests in northern Arizona. Restoration Ecology 9: 443–452.

Saab, V. A., J. Dudley, and W. L. Thompson. 2004. Factors influencing occupancy of nest cavities in recently burned forests. The Condor 106: 20–36.

Saab, V. A., and H. D. W. Powell. 2005. Fire and avian ecology in North America: Process influencing pattern. Studies in Avian Biology 30: 1–13.

Short, K. 2003. Effects of autumn prescribed fire on understory birds (Junco species) in southwestern ponderosa pine forests. Dissertation, University of Montana, Missoula.

Staudenmaier, M. Jr., R. Preston, and P. Sorenson. 2005. NOAA Technical Memorandum NWS WR-273, Climate of Flagstaff, Arizona. National Weather Service Office, Flagstaff, Arizona. http://www.wrh.noaa.gov/wrh/techMemos/273.pdf Accessed 03/07/2006.

Stewart-Oaten, A., W. W. Murdoch, and K. R. Parker. 1986. Environmental impact assessment: "Pseudoreplication" in time? Ecology 67: 929–940.

Stewart-Oaten, A, J. R. Bence, and C. W. Osenberg. 1992. Assessing effects of unreplicated perturbations: No simple solutions. Ecology 73: 1396–1404.

Stone, J. E., T. E. Kolb, and W. W. Covington. 1999. Effects of restoration thinning on presettlement Pinus Ponderosa in northern Arizona. Restoration Ecology 7: 172–182.

Szaro, R. C., and R. P. Balda. 1986. Relationships among weather, habitat structure, and ponderosa pine forest birds. Journal of Wildlife Management 50: 253–260.

Thomas, L., J. L. Laake, S. Strindberg, F. F. C. Marques, S. T. Buckland, D. L. Borchers, D. R. Anderson, K. P. Burnham, S. L. Hedley, J. H. Pollard, J. R. B. Bishop, and T. A. Marques. 2005. Distance V5.0. Release 5. Research Unit for Wildlife Population Assessment, University of St. Andrews, UK.

USDA Forest Service. 2004. http://www.fs.fed.us/projects/hfi/. Accessed 12/07/2004.

Youngblood, A., T. Max, and K. Coe. 2004. Stand structure in eastside old-growth ponderosa pine forests of Oregon and northern California. Forest Ecology and Management 199: 191–217.

OVERVIEW OF HERPETOFAUNA INVENTORIES IN SOUTHERN COLORADO PLATEAU NATIONAL PARKS

Trevor B. Persons, Erika M. Nowak, and David G. Mikesic

In 2000, the National Park Service (NPS) initiated a nationwide program to inventory vertebrates and vascular plants within the national parks. As part of this inventory effort, 265 National Park units (parks, monuments, recreation areas, historic sites) were identified as having significant natural resources, and these were divided into 32 groups or "networks" based on geographical proximity and similar habitat types. The many NPS areas on the Colorado Plateau of Utah, northern Arizona, northwestern New Mexico, and western Colorado were divided into northern and southern Colorado Plateau networks. Here we summarize the results of amphibian and reptile (herpetofauna) inventories at the 19 parks in the Southern Colorado Plateau Inventory & Monitoring Network (SCPN; Figure 1) and synthesize distribution and habitat information for all amphibian and reptile species across that network. The primary goal of these complete species inventories (sensu Scott 1994) was to document 90 percent of the species present at each park. To evaluate our progress toward that goal we estimated our level of inventory completeness for each park at the end of the project. We also discuss considerations for future inventory work.

STUDY AREA

Our study area covered much of the southern Colorado Plateau, as well as the adjacent Rio Grande Valley area of north-central New Mexico. We surveyed 15 parks in the SCPN: Aztec Ruins National Monument (AZRU), Bandelier National Monument (BAND), Canyon de Chelly National Monument (CACH), El Malpais National Monument (ELMA), Chaco Culture National Historic Park (CHCU), El Morro National Monument (ELMO), Glen Canyon National Recreation Area (GLCA), Navajo National Monument (NAVA), Hubbell Trading Post National Historic Site (HUTR), Petroglyph National Monument (PETR), Salinas Pueblo Missions National Monument (SAPU), Sunset Crater Volcano National Monument (SUCR), Walnut Canyon National Monument (WACA), Wupatki National Monument (WUPA), and Yucca House National Monument (YUHO). In addition, we reviewed existing data for all parks in the SCPN, including those that did not receive funding for new fieldwork, i.e., Grand Canyon National Park (GRCA), Petrified Forest National Park (PEFO), Mesa Verde National Park (MEVE), and Rainbow Bridge National Monument (RABR).

Elevations range from 348 m along the Colorado River in western GRCA to 3081 m atop Cerro Grande at BAND. Except for the three parks encompassing parts of the Colorado River (GLCA, GRCA, and RABR), elevations are generally between 1500 and 2500 m at most parks. Common habitats throughout the study area include Great Basin desertscrub, grassland, juniper savanna, pinyon-juniper woodland, and ponderosa pine forest. Mixed conifer forest occurs in parts of BAND and GRCA, and at the other extreme, Sonoran and Mojave Desert vegetation occurs along the Colorado River in parts of the GRCA. Most parks contain some

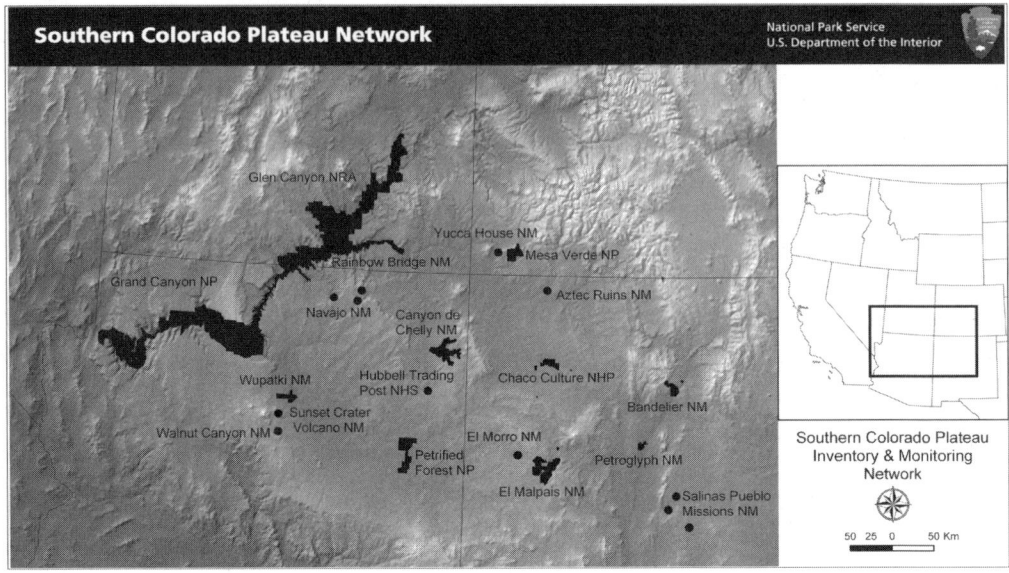

Figure 1. Locations of National Park Service units in the Southern Colorado Plateau Inventory & Monitoring Network.

aquatic habitats in the form of rivers, ponds, seeps and springs, intermittent dry washes with *tinajas*, or temporary rain pools. The habitats and other features of each park unit have been described by Mikesic (2004a, 2004b, 2004c), Thomas et al. (2004), and Persons and Nowak (2006).

METHODS

Because of the dispersed locations of the parks surveyed, the limited time and funding available for fieldwork at any given park, and the objective of providing species presence information, we only used visual survey methods during these inventories. Although trapping methods (such as pitfall traps; Campbell and Christman 1982) often capture rare or secretive species that are difficult to detect using visual surveys, these results are often obtained only after substantial field effort. We used standard survey methods, including time-area constrained searches, generalized visual encounter surveys (both day and night), and nighttime road driving (e.g., Crump and Scott 1994; Rosen and Lowe 1994; Shaffer and Juterbock

1994). We collected voucher specimens at many parks, and also documented species with photographs. We surveyed existing literature and museum specimen databases in an attempt to document species from each park. Detailed information on field methods, voucher specimens, bibliographic references, and museum specimen records is presented by Mikesic (2004a, 2004b, 2004c) and Persons and Nowak (2006). For this overview we have updated the park species lists based on additional surveys at GLCA (Charles Drost, unpublished data) and an opportunistic specimen collection at PEFO in 2006.

To estimate inventory completeness, we developed a master list of species potentially occurring at each park based on literature review, extensive personal knowledge of the distribution and habitats of Southwestern amphibians and reptiles, specimen data from selected museum collections, and results of our fieldwork. Probability of species occurrence was ranked as low (0–33%), medium (34–67%), or high (68–100%). In Table 1 these rankings are coded as 1, 2, and 3, respectively. For quantitative analysis,

Table 1. Amphibian and reptile species known or expected to occur at each park in the Southern Colorado Plateau Inventory & Monitoring Network. Scientific names follow Stebbins (2003). Documented species are coded according to their abundance status in NPSpecies: A = abundant, C = common, U = uncommon, and R = rare. Because we lack abundance data for RABR, documented species are simply shown with an X. Probability ranking of undocumented species: 1 = low, 2 = medium, and 3 = high. Weighted total is equivalent to the total number of species expected to occur, and estimated inventory completeness is simply the number documented divided by the weighted total.

	AZRU	BAND	CACH	CHCU	ELMA	ELMO	GLCA	GRCA	HUTR	MEVE	NAVA	PEFO	PETR	RABR	SAPU	SUCR	WACA	WUPA	YUHO
Ambystomatidae																			
Ambystoma tigrinum	3	C	U	C	U	C	U	U	2	U	3	U	2	1	C	1	R	R	U
Plethodontidae																			
Plethodon neomexicanus	–	U	–	–	–	–	–	–	–	–	–	–	–	–	–	–	–	–	–
Pelobatidae																			
Scaphiopus couchii	–	2	–	–	–	–	–	–	–	–	–	–	3	–	–	–	–	–	–
Spea bombifrons	3	1	U	U	C	C	2	2	C	1	U	U	3	1	U	2	C	C	–
Spea intermontana	–	–	–	–	–	C	C	C	–	–	U	C	–	3	–	–	–	–	–
Spea multiplicata	3	3	C	C	C	C	3	3	C	3	C	A	U	1	U	1	A	A	3
Bufonidae																			
Bufo alvarius	–	–	–	–	–	–	1	1	–	–	–	–	–	–	–	–	–	–	–
Bufo cognatus	1	1	1	1	–	2	2	1	1	–	1	A	2	–	–	–	C	C	–
Bufo punctatus	R	R	C	2	2	A	A	1	1	U	U	U	1	X	U	2	C	C	1
Bufo woodhousii	C	C	C	3	R	A	A	C	C	U	C	U	3	X	U	U	U	U	C
Hylidae																			
Hyla arenicolor	R	C	C	–	U	1	C	A	–	–	R	–	–	1	–	U	U	–	–
Hyla eximia	–	–	–	–	R	2	–	–	–	–	–	–	–	–	–	1	1	–	–
Pseudacris triseriata	C	U	1	–	U	2	–	–	2	2	–	–	–	–	3	3	1	–	C
Ranidae																			
Rana catesbeiana	1	C	–	–	–	–	R	1	–	–	–	–	–	–	–	–	–	–	–
Rana onca	–	–	–	–	–	–	1	1	–	–	–	–	–	–	–	–	–	–	–
Rana pipiens	1	1	1	–	–	U	1	1	–	1	R	–	–	X	–	–	–	–	1
Rana yavapaiensis	–	–	–	–	–	–	2	R	–	–	–	–	–	–	–	–	–	–	–
Emydidae																			
Chrysemys picta	R	1	–	–	–	1	1	–	–	–	–	–	–	–	–	–	–	–	–
Terrapene ornata	–	–	–	–	–	–	–	–	–	–	–	–	R	–	1	–	–	–	–
Testudinidae																			
Gopherus agassizii	–	–	–	–	–	–	R	–	–	–	–	–	–	–	–	–	–	–	–
Trionychidae																			
Trionyx spiniferus	–	3	–	–	–	–	–	–	–	–	–	–	–	–	–	–	–	–	–
Eublepharidae																			
Coleonyx variegatus	–	–	–	–	R	R	U	U	–	–	–	–	–	1	–	–	–	–	–
Iguanidae																			
Dipsosaurus dorsalis	–	–	–	–	–	–	1	1	–	–	–	–	–	–	–	–	–	–	–
Sauromalus obesus	–	–	–	–	–	C	C	C	–	–	–	–	–	2	–	–	–	–	–

Table 1 (continued)

	AZRU	BAND	CACH	CHCU	ELMA	ELMO	GLCA	GRCA	HUTR	MEVE	NAVA	PEFO	PETR	RABR	SAPU	SUCR	WACA	WUPA	YUHO
Crotaphytidae																			
Crotaphytus bicinctores	—	—	—	—	—	—	C	C	—	—	—	—	—	X	—	—	—	—	—
Crotaphytus collaris	C	U	C	C	U	1	U	C	U	C	3	C	U	1	C	—	1	C	U
Gambelia wislizenii	1	1	2	2	—	—	3	3	1	1	1	3	C	3	1	—	1	C	—
Phrynosomatidae																			
Callisaurus draconoides	1	1	—	—	—	—	1	U	—	—	2	—	—	—	—	—	—	—	—
Holbrookia maculata	1	1	R	C	C	R	1	3	U	R	2	C	U	2	R	—	2	A	U
Phrynosoma hernandesi	3	U	C	C	C	U	3	C	3	C	U	C	U	2	3	U	C	A	U
Phrynosoma modestum	—	3	—	—	—	—	—	—	—	—	—	—	—	2	C	—	—	—	—
Phrynosoma platyrhinos	—	—	—	—	—	—	U	U	U	—	U	—	—	2	—	—	—	—	—
Sceloporus graciosus	R	—	A	A	1	2	U	A	A	A	A	A	—	X	—	—	2	2	A
Sceloporus magister	C	A	A	A	A	A	A	C	C	A	A	A	A	X	A	3	C	C	C
Sceloporus undulatus	—	—	A	A	A	A	A	A	A	A	A	A	A	—	A	—	A	A	A
Urosaurus graciosus	1	—	—	—	—	—	1	1	1	—	—	—	1	2	C	—	A	—	1
Urosaurus ornatus	1	U	C	U	A	U	A	A	1	C	C	C	—	X	A	3	A	A	U
Uta stansburiana	2	—	A	A	—	—	A	A	1	U	U	—	A	—	—	—	—	—	—
Xantusiidae																			
Xantusia vigilis	—	—	—	—	—	—	R	R	—	—	—	—	1	1	—	—	—	—	—
Scincidae																			
Eumeces multivirgatus	—	C	U	U	C	C	R	R	R	R	3	—	1	—	C	3	R	R	3
Eumeces obsoletus	—	C	—	—	U	2	2	—	—	—	—	—	—	—	—	—	2	—	—
Eumeces skiltonianus	—	—	—	—	—	—	C	C	—	—	—	—	C	—	—	—	—	—	—
Teiidae																			
Cnemidophorus exsanguis	C	C	C	2	2	—	—	—	—	—	—	—	1	—	A	—	R	—	—
Cnemidophorus inornatus	1	1	1	1	—	2	U	U	—	—	—	R	U	—	A	—	C	C	A
C. neomexicanus	3	3	—	—	—	—	—	—	—	—	—	R	A	—	2	—	—	—	—
Cnemidophorus tesselatus	R	—	—	—	—	—	—	—	—	—	—	—	—	—	—	—	—	—	—
Cnemidophorus tigris	C	U	2	2	A	A	A	A	A	A	U	A	—	X	1	U	A	A	A
Cnemidophorus velox	A	A	A	A	A	A	U	U	C	C	A	A	—	—	1	U	C	A	A
Anguidae																			
Elgaria kingii	—	—	—	—	—	—	—	—	—	—	—	—	—	—	—	—	2	—	—
Helodermatidae																			
Heloderma suspectum	—	—	—	—	—	—	R	—	—	—	—	—	—	—	—	—	—	—	—
Leptotyphlopidae																			
Leptotyphlops dulcis	3	—	3	—	—	—	—	—	—	—	—	U	—	—	1	—	—	—	—
Leptotyphlops humilis	—	—	—	—	—	—	R	R	—	—	—	—	—	—	—	—	1	—	—
Colubridae																			
Arizona elegans	3	3	—	—	—	R	3	3	3	3	3	C	U	2	U	U	1	C	1
Coluber constrictor	1	1	R	—	—	2	—	—	1	1	1	1	1	—	U	—	1	—	1
Diadophis punctatus	1	R	1	1	3	R	R	1	1	1	1	—	R	—	3	—	R	—	1

Table 1 (continued)

	AZRU	BAND	CACH	CHCU	ELMA	ELMO	GLCA	GRCA	HUTR	MEVE	NAVA	PEFO	PETR	RABR	SAPU	SUCR	WACA	WUPA	YUHO
Colubridae (continued)																			
Elaphe guttata	–	3	–	–	–	–	–	–	–	–	–	–	–	–	–	–	–	–	–
Gyalopion canum	–	1	–	1	–	–	–	–	–	–	–	–	3	–	–	–	–	–	–
Heterodon nasicus	2	1	1	1	–	–	–	–	–	–	–	–	U	–	3	–	–	3	3
Hypsiglena torquata	3	3	3	C	U	U	U	U	3	U	3	C	U	X	U	2	3	U	3
Lampropeltis getula	1	3	1	1	–	U	U	U	–	–	1	U	3	2	–	–	–	U	3
Lampropeltis pyromelana	–	1	–	–	R	R	U	U	1	R	–	–	–	–	–	3	C	–	–
Lampropeltis triangulum	2	1	2	2	U	–	1	1	1	–	1	U	1	–	3	–	1	R	2
Masticophis flagellum	–	3	–	–	–	U	U	U	–	C	–	C	C	–	–	–	–	–	–
Masticophis taeniatus	U	U	U	U	U	C	C	C	3	C	U	C	C	3	C	3	3	C	C
Opheodrys vernalis	C	3	–	–	–	–	–	–	–	R	–	–	–	–	–	–	–	–	C
Pituophis catenifer	C	C	U	C	C	C	C	C	C	C	U	C	C	3	C	U	C	C	C
Rhinocheilus lecontei	–	2	–	–	–	–	R	R	C	–	–	–	U	1	C	–	–	1	1
Salvadora grahamiae	C	C	–	3	3	1	U	–	–	–	–	–	–	3	–	–	–	–	–
Salvadora hexalepis	–	–	–	–	–	R	R	U	–	–	–	1	–	3	–	3	3	U	–
Sonora semiannulata	–	–	–	–	–	3	3	R	–	–	–	1	–	1	–	–	–	R	–
Tantilla hobartsmithi	–	–	–	–	–	–	–	R	–	–	–	–	U	1	–	–	–	1	–
Tantilla nigriceps	–	–	–	–	–	–	3	–	–	–	–	–	U	1	3	–	–	1	–
Thamnophis cyrtopsis	2	R	1	R	3	3	3	1	1	1	1	R	1	1	C	1	1	1	1
Thamnophis elegans	C	C	C	U	C	C	R	C	C	1	3	2	1	2	C	1	C	2	A
Thamnophis marcianus	–	–	–	–	–	–	–	–	–	–	–	1	1	2	–	–	–	–	–
Thamnophis sirtalis	–	–	–	–	–	–	R	R	–	–	–	1	1	1	–	1	–	1	–
Trimorphodon biscutatus	–	–	–	–	–	–	–	–	–	–	–	–	–	–	U	–	–	–	–
Tropidoclonion lineatum	–	–	–	–	–	–	–	–	–	–	–	–	–	–	U	–	–	–	–
Viperidae																			
Crotalus atrox	–	C	–	–	C	C	–	3	–	–	–	–	C	–	C	–	–	–	–
Crotalus mitchellii	–	–	–	–	–	–	–	C	–	–	–	–	–	–	–	–	–	–	–
Crotalus molossus	–	–	–	–	3	1	C	U	–	–	.	–	–	–	2	–	3	–	–
Crotalus scutulatus	–	–	–	–	1	–	U	3	–	–	–	–	–	–	–	–	–	–	–
Crotalus viridis	U	3	U	C	C	U	C	C	U	C	U	C	U	3	C	3	U	C	C
Sistrurus catenatus	–	–	–	–	–	–	–	–	–	–	–	–	R	–	2	–	–	–	–
Total documented	9	23	20	17	20	17	31	43	11	19	16	24	25	9	22	5	14	27	14
Weighted total	21.0	35.2	23.4	20.9	26.2	21.0	36.8	50.3	16.0	21.2	22.5	25.8	31.7	19.4	30.2	9.3	21.2	28.5	18.2
Estimated inventory completeness	57%	65%	85%	81%	76%	81%	84%	85%	69%	90%	71%	93%	79%	46%	73%	54%	66%	95%	77%

Aztec Ruins National Monument (AZRU), Bandelier National Monument (BAND), Canyon de Chelly National Monument (CACH), Chaco Culture National Historic Park (CHCU), El Malpais National Monument (ELMA), El Morro National Monument (ELMO), Glen Canyon National Recreation Area (GLCA), Grand Canyon National Park (GRCA), Hubbell Trading Post National Historic Site (HUTR), Mesa Verde National Park (MEVE), Navajo National Monument (NAVA), Petrified Forest National Park (PEFO), Petroglyph National Monument (PETR), Rainbow Bridge National Monument (RABR), Salinas Pueblo Missions National Monument (SAPU), Sunset Crater Volcano National Monument (SUCR), Walnut Canyon National Monument (WACA), Wupatki National Monument (WUPA), Yucca House National Monument (YUHO).

these rankings were converted to the mid-point of their percentage range: 0.17, 0.50, and 0.83. These values were used as weighting factors for species not yet documented. For example, two species with rankings of medium probability of occurrence would combine to equal one full expected species (0.50 x 2 = 1.00 species), whereas six species of low probability of occurrence would be required to equal one full expected species (0.17 x 6 = 1.02 species). Species found by us during the inventory, or known from previously collected specimens or reliable observations, were weighted as 1.

Such weighting of categorical probability data is generally not recommended for statistical applications; however, we think it is justifiable because we are not using the resulting inventory completeness estimates for statistical probability or hypothesis testing. Instead, we are generating locally specific estimates of percent inventory completeness as mandated by the NPS Inventory and Monitoring Program, in a manner that integrates a range of information including inventory results, pre-existing information, and professional knowledge. These considerations should be kept in mind when interpreting the inventory completeness estimates.

Scientific and common names follow Stebbins (2003). Recent studies have proposed taxonomic changes for some species found in SCPN parks, and interested readers should consult Crother (2000), Crother et al. (2003), and Frost et al. (2006) for summaries of most of these proposals. Because many of these proposed changes remain controversial we have chosen to adhere to the more familiar taxonomy of Stebbins (2003), which is a standard reference for herpetologists and non-herpetologists alike.

RESULTS AND DISCUSSION
Overview of Inventory Results

We found 50 amphibian and reptile species in the 15 NPS units we surveyed in 2001–2003, plus one additional species at GLCA in 2006. We documented 8 more species from these parks through literature and museum specimen review and communications with NPS staff. We also confirmed 13 additional species from the four parks we did not survey (most from GRCA), for a total of 72 species documented in one or more SCPN parks. The park-by-park status of each species is shown in Table 1, including the probability rankings for undocumented species. Table 1 also includes the relative abundance of each species within each park, using abundance designations as in the Park Service's NPSpecies database (Persons and Nowak 2006). Habitat associations of each species at each of the 15 parks we surveyed, as well as documentation details (e.g., specimen records, literature citations) for all species at all 19 parks, are presented in Mikesic (2004a, 2004b, 2004c) and Persons and Nowak (2006).

The eastern fence lizard (*Sceloporus undulatus*) is the only species documented from all SCPN parks. Other species that probably occur at most or all parks include tiger salamander (*Ambystoma tigrinum*), Mexican spadefoot (*Spea multiplicata*), greater short-horned lizard (*Phrynosoma hernandesi*), plateau striped whiptail (*Cnemidophorus velox*), night snake (*Hypsiglena torquata*), striped whipsnake (*Masticophis taeniatus*), gopher snake (*Pituophis catenifer*), western terrestrial garter snake (*Thamnophis elegans*), and western rattlesnake (*Crotalus viridis*). Only one species—the Jemez Mountains salamander (*Plethodon neomexicanus*)—is endemic to the SCPN; it occurs only in the Jemez Mountains of New Mexico, including portions of BAND.

Because the SCPN includes parks in and near the Rio Grande Valley of New Mexico (i.e., BAND, PETR, SAPU), the network-wide list includes many species not characteristic of the Colorado Plateau. Eleven such documented species occur only in these three parks, although a few of them are considered hypothetical at ELMA, which lies near the southeastern edge of the Colorado Plateau. In addition, the Great Plains skink (*Eumeces obsoletus*) and western diamond-backed rattlesnake (*Crotalus atrox*) are known from one or more Rio Grande Valley area parks, and both have been documented from the eastern portion of ELMA. The New Mexico whiptail (*Cnemidophorus neomexicanus*) is considered a Rio Grande Valley species, as the population at PEFO is likely introduced (Persons and Wright 1999). Similarly, eight documented species characteristic of the Sonoran and Mojave Deserts occur only in desert habitats associated with the Colorado River corridor in western GRCA. The western diamond-backed rattlesnake and black-tailed rattlesnake (*Crotalus molossus*) have each only been documented from one of these areas, but both species probably occur in western GRCA and at parks in central New Mexico. A list of the species documented only from these two regions is presented in Table 2.

Most (10 of 12) of the undocumented species suspected to occur in one or more parks are also marginal to the region, their distributions approaching the SCPN area along the Mogollon Rim in northern Arizona, in the middle Rio Grande Valley, or in desert habitats near the western Grand Canyon. Only the racer (*Coluber constrictor*) and corn snake (*Elaphe guttata*) occur within the interior of the Colorado Plateau, and both are uncommon, with spotty distributions in the region (Degenhardt et al. 1996; Hammerson 1999; Stebbins 2003).

Among the highlights of these inventories was the discovery of mountain treefrogs (*Hyla eximia*) at ELMA, which constituted a substantial range extension in New Mexico and substantiated an unverified record from the Zuni Mountains region from the 1870s (Monatesti et al. 2005). Another exciting discovery was the occurrence of the western banded gecko (*Coleonyx variegatus*) at WUPA, a new record for the Little Colorado River basin and only the second record from the Colorado Plateau away from the Colorado River (Persons and Nowak 2004). At NAVA we discovered northern leopard frogs (*Rana pipiens*), a regionally declining species virtually eliminated from the Navajo Nation. We also documented new county records for red-spotted toad (*Bufo punctatus*) and roundtail horned lizard (*Phrynosoma*

modestum) from SAPU and night snake from ELMO (Persons and Nowak 2005a, 2005b, 2005c). At WACA we discovered an isolated population of little striped whiptail (*Cnemidophorus inornatus*), a species with a limited distribution on the southern Colorado Plateau (Persons 2005; Brennan and Holycross 2006). During additional fieldwork at GLCA in 2006 we discovered a population of desert night lizard (*Xantusia vigilis*), a new park record and one of only a handful of known populations in Utah (Charles Drost, unpublished data). Finally, black-necked garter snake (*Thamnophis cyrtopsis*) was recorded opportunistically at PEFO in 2006. This is one of only a few low-elevation records from northern Arizona (Persons and Rosen 2001) and a new record for a well-inventoried park (Drost et al. 2001).

Although our ability to document many species was probably compromised by the persistent drought that severely impacted the southern Colorado Plateau during the inventory period (Webb et al. 2004), we nonetheless obtained baseline data on amphibian and reptile species occurrence at all 15 SCPN parks we surveyed. For parks with little or no previous herpetofauna inventory work (i.e., AZRU, HUTR, SAPU, SUCR, WACA, YUHO), we now have data on occurrence of common species. We also documented new species at most of the other parks, all of which have had some prior herpetofauna inventory work.

Estimates of Inventory Completeness

We calculated a mean estimated inventory completeness of 75 percent (range 46–95%) for all species and parks combined (Table 3); inventory completeness was 70 percent for amphibians (range 0–100%), 85 percent for lizards (range 60–100%), and 65 percent for snakes (range 15–93%). Turtles are only known from four SCPN parks (AZRU, GLCA, GRCA, and PETR), and may occur at two others (BAND and SAPU). For these six parks, estimated inventory completeness for turtles was 50 percent (range 0–100%). Estimated inventory completeness values for each taxonomic group at each park are pre-

Table 2. Amphibian and reptile species only documented from SCPN parks and habitats not typical of the Colorado Plateau. Species marked with an asterisk have also been documented from eastern ELMA, on the southeastern edge of the Colorado Plateau. The New Mexico whiptail also occurs at PEFO, but that population is probably introduced.

Rio Grande Valley Region (BAND, PETR, SAPU)	Desert Habitats in Western GRCA
Western box turtle (*Terrapene ornata*)	Lowland leopard frog (*Rana yavapaiensis*)
Great Plains skink (*Eumeces obsoletus*)*	Desert tortoise (*Gopherus agassizii*)
Roundtail horned lizard (*Phrynosoma modestum*)	Zebra-tailed lizard (*Callisaurus draconoides*)
Chihuahuan spotted whiptail (*Cnemidophorus exsanguis*)	Gila monster (*Heloderma suspectum*)
New Mexico whiptail (*Cnemidophorus neomexicanus*)	Western blind snake (*Leptotyphlops humilis*)
Checkered whiptail (*Cnemidophorus tesselatus*)	Western lyre snake (*Trimorphodon biscutatus*)
Texas blind snake (*Leptotyphlops dulcis*)	Speckled rattlesnake (*Crotalus mitchellii*)
Western hog-nosed snake (*Heterodon nasicus*)	Black-tailed rattlesnake (*Crotalus molossus*)
Mountain patch-nosed snake (*Salvadora grahamii*)	
Plains black-headed snake (*Tantilla nigriceps*)	
Lined snake (*Tropidoclonion lineatum*)	
Western diamond-backed rattlesnake (*Crotalus atrox*)*	
Massasauga (*Sistrurus catenatus*)	

sented in Table 3. The relatively high success rate for lizards is likely because most lizard species are diurnal and conspicuous, and our efforts (as well as those of most previous workers) were biased toward daytime searches, which easily detect such species. Many amphibians on the southern Colorado Plateau breed during the summer monsoon season, and are often active on only a few nights per year, making them difficult to locate. In addition, given that these inventories took place during a severe regional drought (Webb et al. 2004), amphibian populations and/or activity may have been reduced during the project period. As with amphibians, many snake species are primarily nocturnal, and many are extremely secretive in their habits, so a low completeness for snakes is not surprising.

For both amphibians and snakes, however, the single most important factor limiting inventory completeness was the lack of extensive networks of roads in most of the SCPN parks surveyed. Based on data from our own studies in the region, nighttime road surveys are by far the most effective method for detecting both amphibians and snakes. For example, at PEFO (Drost et al. 2001) and WUPA (Persons 2001; Persons and Nowak 2006) the combination of general daytime foot surveys for lizards and extensive nighttime road surveys for amphibians and snakes resulted in an overall estimated inventory completeness of more than 90 percent at both parks.

AMPHIBIANS AND REPTILES IN SCPN PARKS

This annotated list includes 72 species documented from the 19 SCPN parks, as well as 12 undocumented species that may occur in one or more parks. Arrangement is the same as in Table 1; that is, within families species are listed alphabetically by scientific name. In summarizing distribution and habitat information we relied heavily on the excellent regional publications by Degenhardt et al. (1996), Hammerson (1999), Stebbins (2003), and Brennan and Holycross (2006).

Table 3. Estimated inventory completeness (in percent) of different taxonomic groups of amphibians and reptiles at each park in the Southern Colorado Plateau Inventory & Monitoring Network.

Park	Amphibians	Turtles	Lizards	Snakes	Overall
AZRU	40	100	68	52	57
BAND	82	0	83	47	65
CACH	92	–	97	61	85
CHCU	67	–	87	84	81
ELMA	86	–	92	56	76
ELMO	71	–	84	86	81
GLCA	78	0	89	81	84
GRCA	71	100	90	86	84
HUTR	82	–	79	50	69
MEVE	64	–	100	93	90
NAVA	86	–	75	49	71
PEFO	100	–	92	90	93
PETR	20	100	97	83	79
RABR	67	–	60	15	46
SAPU	86	0	83	62	73
SUCR	0	–	83	24	54
WACA	69	–	78	55	66
WUPA	100	–	96	90	95
YUHO	72	–	88	67	77
Mean	70	50	85	65	75

Aztec Ruins National Monument (AZRU), Bandelier National Monument (BAND), Canyon de Chelly National Monument (CACH), Chaco Culture National Historic Park (CHCU), El Malpaís National Monument (ELMA), El Morro National Monument (ELMO), Glen Canyon National Recreation Area (GLCA), Grand Canyon National Park (GRCA), Hubbell Trading Post National Historic Site (HUTR), Mesa Verde National Park (MEVE), Navajo National Monument (NAVA), Petrified Forest National Park (PEFO), Petroglyph National Monument (PETR), Rainbow Bridge National Monument (RABR), Salinas Pueblo Missions National Monument (SAPU), Sunset Crater Volcano National Monument (SUCR), Walnut Canyon National Monument (WACA), Wupatki National Monument (WUPA), Yucca House National Monument (YUHO).

Family Ambystomatidae

Tiger Salamander (*Ambystoma tigrinum*). Documented from most SCPN parks, this widespread species occurs in all habitats from Great Basin desertscrub through mixed conifer forest. Tiger salamanders breed in a variety of permanent or temporary water sources, such as springs, cattle tanks, and temporary monsoon rain pools.

Family Plethodontidae

Jemez Mountains Salamander (*Plethedon neomexicanus*). This small, secretive salamander is endemic to the Jemez Mountains in north-central New Mexico, including the northwest corner of BAND. It occurs in high-elevation mixed-conifer forest and spends most of the year in underground rock fissures, generally coming to the surface only during the summer monsoon season, when it can be found under rocks and rotting logs (Ramotnik 1988). We did not observe this species, but previous research has documented it from a number of sites within BAND. The recent discovery (Cummer et al. 2005) of a Jemez Mountains salamander infected with Chytrid fungus is of concern, as this pathogen has been implicated in amphibian declines and die-offs worldwide.

Family Pelobatidae

Couch's Spadefoot (*Scaphiopus couchii*). This is the common spadefoot of the Sonoran and Chihuahuan deserts, and its distribution extends up the Rio Grande Valley where it may occur at PETR and BAND. It also occurs as disjunct populations within grassland habitats in the Little Colorado River basin in eastern Arizona, including at PEFO (Drost et al. 2001). Like other spadefoots, it breeds in temporary pools during the summer monsoon season.

Plains Spadefoot (*Spea bombifrons*). The distribution of this Great Plains species extends through New Mexico onto the southeastern Colorado Plateau, and it has been documented at many SCPN parks. The plains spadefoot occupies a variety of habitats from desertscrub through pinyon-juniper woodland, but is most common in grasslands where it breeds in temporary pools during the summer monsoon season. Throughout its range in the SCPN this species occurs together with the similar Mexican Spadefoot, which is usually more abundant.

Great Basin Spadefoot (*Spea intermontana*). This relative of the plains and Mexican spadefoot occurs throughout the Great Basin and much of the northern Colorado Plateau. The Great Basin spadefoot has only been documented west of the Colorado River, and in the SCPN it has been found only at GLCA and GRCA, where it occurs from desertscrub to ponderosa pine forest. Unlike other spadefoots in the SCPN, which generally breed during the summer monsoon season, Great Basin spadefoots at GLCA (and probably also at GRCA) often breed in spring before intermittent streams dry up in mid-summer. Elucidating the distribution of all three *Spea* spadefoot toads at GRCA and GLCA would greatly add to our understanding of the regional distribution and biogeography of these species.

Mexican Spadefoot (*Spea multiplicata*). This is the common spadefoot over most of the southern Colorado Plateau and adjacent areas in Arizona and New Mexico. It has been documented from 11 SCPN parks and probably occurs at most others. The Mexican spadefoot ranges from Great Basin desertscrub to ponderosa pine forest, where it breeds in temporary pools during the summer monsoon season. At most parks, this species occurs together with the similar plains spadefoot, which it usually outnumbers. The Mexican and plains spadefoot are known to interbreed (e.g., Simovich 1994), and we encountered suspected hybrids at some parks.

Family Bufonidae

Sonoran Desert Toad (*Bufo alvarius*). This large toad is restricted to the Sonoran Desert region, and formerly occurred along much of the lower Colorado River below Lake Mead. Stebbins (2003) mapped and mentioned a sight record from the Colorado River at Seventy-five Mile Creek in GRCA, but we suspect this report may be in error and that this toad does not occur in SCPN parks.

Great Plains Toad (*Bufo cognatus*). Like the plains spadefoot, this is a Great Plains species whose distribution extends into the Southwestern desert and Colorado Plateau regions. It has a fragmented distribution on the southern Colorado Plateau, and in the SCPN has only been documented at PEFO and WUPA, where it occurs in grasslands. Like most other anurans at these parks it breeds in temporary pools during the summer monsoon period.

Red-spotted Toad (*Bufo punctatus*). The red-spotted toad, which occurs throughout the region, has been found at more than half the SCPN parks. It favors rocky habitats, especially rocky canyons and washes, from desertscrub to pinyon-juniper woodland. Its absence from many parks is likely due to the lack of suitable rocky habitats and associated temporary pools. Our collection of a red-spotted toad at SAPU constituted a new Torrence County record (see Persons and Nowak 2005a).

Woodhouse's Toad (*Bufo woodhousii*). This widespread species probably occurs at almost every SCPN park, although it appears to be absent from SUCR and WACA. It occurs in a variety of habitats, but is most common in the sandy floodplains of large streams and rivers, including the Colorado River (GLCA, GRCA), Little Colorado River (WUPA), and Rio Grande (BAND).

Family Hylidae

Canyon Treefrog (*Hyla arenicolor*). The canyon treefrog occurs throughout the region but it has a spotty distribution and has only been documented from seven SCPN parks. The species occupies a variety of habitats, but its distribution is limited by its preference for permanent or near-permanent pools in rocky canyons. Within the SCPN it is most abundant in the many side canyons of the Colorado River in GRCA. We discovered an isolated population of canyon treefrogs in Cherry Canyon at WACA.

Mountain Treefrog (*Hyla eximia*). We collected mountain treefrogs from a temporary pool in ponderosa pine forest near the Continental Divide at ELMA (Monatesti et al. 2005). These specimens were the first documented record for the region, extending the known distribution about 143 km north-northeast of the nearest locality in the Gila River basin in Catron County, New Mexico. Our specimens lend credence to an unverified nineteenth-century record from near Nutria (northwest of ELMA), and suggest that mountain treefrogs may be more widely distributed in the Zuni Mountains region of New Mexico; they could possibly occur at ELMO. They also occur in the Flagstaff, Arizona area, and could possibly occur at WACA. Mountain treefrogs breed in temporary pools during the summer monsoon season.

Western Chorus Frog (*Pseudacris triseriata*). This tiny species breeds in spring in temporary or semi-permanent pools (cattle tanks, flooded fields, roadside ditches) with emergent vegetation. The western chorus frog, which is widespread over much of North America, is absent from much of the Colorado Plateau. It is however quite abundant throughout the many ponds across the flat, open high elevations of the Chuska Mountains on the Navajo Nation. They occur in a variety of habitats from grassland through mixed conifer forest, but have only been found in a handful of SCPN parks.

Family Ranidae

Bullfrog (*Rana catesbeiana*). This species, native to the eastern United States, is widely regarded as a predatory scourge to native aquatic herpetofauna in the Southwest (e.g., Rosen and Schwalbe 2002). During our initial surveys we documented this species only at BAND, where it is common along the Rio Grande. However, during supplemental fieldwork in 2006 we heard bullfrogs calling from extensive cattail marshes near Hite Marina on Lake Powell at GLCA (Charles Drost, unpublished data).

Relict Leopard Frog (*Rana onca*). This imperiled species is found in desert streams and springs in the Lake Mead–Virgin River region just west of GRCA (Bradford et al. 2005). Lowland leopard frogs (*R. yavapaiensis*) discovered recently in Surprise Canyon in western GRCA were initially thought, based on geographic proximity, to be this species (Charles Drost, unpublished data). So although not yet documented within the SCPN, isolated populations of relict leopard frogs may yet be discovered in extreme western GRCA.

Northern Leopard Frog (*Rana pipiens*). This widespread species has declined throughout the Southwest (Rorabaugh 2005), and has apparently been extirpated from CACH, GRCA, and MEVE. Northern leopard frogs were collected in the 1930s from RABR, before the construction of Glen Canyon Dam and the creation of Lake Powell, and it seems doubtful the species still occurs there. They persist at a number of sites along and near the Colorado River at GLCA, but may be declining there as well (Charles Drost, unpublished data). During our surveys we discovered an apparently isolated population of this species near Inscription House Ruin at NAVA. Northern leopard frogs usually breed in marshes, ponds, and semi-permanent canyon pools.

Lowland Leopard Frog (*Rana yavapaiensis*). This species occurs in deserts, grasslands, and open woodlands below the Mogollon

Rim, and its distribution extends northwestward toward the Lake Mead area. Like other leopard frogs in the Southwest, lowland leopard frogs have undergone population declines in recent decades (Sredl 2005). The distributions of this species and the similar relict leopard frog in the western Grand Canyon region are not fully understood. Leopard frogs discovered recently in Surprise Canyon north of the Colorado River in western GRCA were initially thought, based on geographic proximity, to be relict leopard frogs, but genetic analysis identified them as lowland leopard frogs (Charles Drost, unpublished data).

Family Emydidae

Painted Turtle (*Chrysemys picta*). The highly aquatic painted turtle has been collected at AZRU and GLCA, and may possibly occur along the Rio Grande at BAND. The species is known from the San Juan River basin near AZRU, and although the single specimen was collected more than half a century ago, suitable habitat still exists along the Animas River at AZRU. At GLCA, however, the few records predate the creation of Lake Powell, and the species may be extirpated there.

Western Box Turtle (*Terrapene ornata*). The distribution of this terrestrial species does not include the Colorado Plateau, but its distribution in New Mexico extends up the Rio Grande Valley as far as Albuquerque, where it occurs in sandy grassland and shrubland habitats at PETR. Suitable sandy habitat also exists at the Quarai unit of SAPU, but western box turtles have not been reported from there. A specimen found near a developed area at PEFO likely represents an abandoned pet.

Family Testudinidae

Desert Tortoise (*Gopherus agassizii*). This large, terrestrial species occurs in the Sonoran and Mojave Deserts within its range in the United States. Desert tortoise scat was collected from near the Colorado River in extreme western GRCA in the early 1970s (Tom Van Devender, personal communication), but its current status within the park is unknown.

Family Trionychidae

Spiny Softshell (*Trionyx spiniferus*). This highly aquatic turtle usually inhabits rivers with sandy or muddy bottoms. Spiny softshells are distributed throughout much of the Rio Grande in New Mexico, and probably occur in the river along the southern boundary of BAND.

Family Eublepharidae

Western Banded Gecko (*Coleonyx variegatus*). Until recently this nocturnal lizard, which is largely restricted to the Sonoran and Mojave Desert regions, was known on the Colorado Plateau only from along the Colorado River in GRCA. Brennan et al. (2002) reported a western banded gecko from Antelope Point in GLCA, and Persons and Nowak (2004) discovered the species at WUPA. Both of these reports suggest that this species may be more widespread in low elevations in the Colorado River basin. All of these new records were of individuals found under sandstone slabs in Great Basin desertscrub habitats.

Family Iguanidae

Desert Iguana (*Dipsosaurus dorsalis*). This large, thermophilic lizard is restricted to the Sonoran and Mojave Deserts, and is usually found in open habitats dominated by creosote bush (*Larrea tridentata*). It has not been documented in the SCPN, but could possibly occur in desert habitats near the Colorado River in extreme western GRCA.

Common Chuckwalla (*Sauromalus obesus*). This large, herbivorous lizard is found throughout the Sonoran and Mojave Deserts, and its distribution follows the Colorado River through GRCA and GLCA as far north as Hite, Utah. Common chuckwallas inhabit canyons, lava rockpiles, outcrops, and other rocky habitats. Lizards from GLCA were formerly considered a separate subspecies, the Glen Canyon chuckwalla (*S. o. multiforamininatus*).

Family Crotaphytidae

Great Basin Collared Lizard (*Crotaphytus bicinctores*). This collared lizard is distributed throughout the Mojave and Great Basin desert regions, and on the Colorado Plateau it primarily occurs north of the Colorado River. Like other collared lizards, this species lives in rocky areas in a variety of habitats. It is common at GLCA and GRCA, and has been documented previously from RABR.

Eastern Collared Lizard (*Crotaphytus collaris*). This colorful lizard is found in rocky habitats throughout the southern Colorado Plateau, ranging from Great Basin desertscrub to pinyon-juniper woodland. It only occurs south of the Colorado River, and is largely replaced by the Great Basin collared lizard at GLCA and GRCA. Eastern collared lizards have been documented from most SCPN parks.

Long-nosed Leopard Lizard (*Gambelia wislizenii*). This relative of the collared lizards is distributed throughout the region, but is usually found in open, shrubby habitats at lower elevations. It has been documented only from GLCA, PETR, and WUPA, but probably occurs at a few other parks, especially PEFO (Drost et al. 2001).

Family Phrynosomatidae

Zebra-tailed Lizard (*Callisaurus draconoides*). This desert species does not occur on the Colorado Plateau, but extends its distribution along the Colorado River corridor into western GRCA. Zebra-tailed lizards prefer open habitats such as sandy dry washes and unpaved desert roads.

Lesser Earless Lizard (*Holbrookia maculata*). This small, terrestrial species occurs in open grasslands and woodlands throughout the southern Colorado Plateau. Lesser earless lizards prefer sparse, open habitats, and are often abundant in prairie dog towns, as at PEFO (Drost et al. 2001). They have been documented from 11 SCPN parks, and may occur at others.

Greater Short-horned Lizard (*Phrynosoma hernandesi*). This is the common horned lizard over most of the Colorado Plateau, occurring from grasslands to mixed conifer forest. This species appears to be absent from Great Basin desertscrub, even in areas (such as WUPA) where the desert horned lizard does not occur. It has been documented from most SCPN parks, and is one of the few species that may occur in every park in the network. Like other horned lizards, the greater short-horned lizard is an ant-eating specialist.

Roundtail Horned Lizard (*Phrynosoma modestum*). This small horned lizard species is considered a "rock mimic." It is primarily distributed in the Chihuahuan Desert, and only occurs in the Rio Grande Valley area of the SCPN, where it has been documented from PETR and SAPU. Our collection of a specimen at SAPU constituted a new Torrence County record (Persons and Nowak 2005b). The species may also occur in lower elevations of BAND.

Desert Horned Lizard (*Phrynosoma platyrhinos*). This is the common horned lizard in most of the Sonoran, Mojave, and Great Basin Deserts, and it extends its distribution along the Colorado River into GRCA and GLCA. It is so far the only species of horned lizard documented from GLCA, and it occurs throughout lower elevation Great Basin desertscub habitats there.

Sagebrush Lizard (*Sceloporus graciosus*). Sagebrush lizards are widely distributed in the west, and reach their southernmost distributional limits at the southern edge of the Colorado Plateau. They have been documented from most SCPN parks on the Colorado Plateau, but are absent from those associated with the Rio Grande Valley (BAND, PETR, and SAPU). They occupy grasslands, shrublands, and woodlands with sandy soils and scattered shrubs (such as sagebrush). They are much more terrestrial than other spiny lizards (*Sceloporus*) in the region.

Desert Spiny Lizard (*Sceloporus magister*). This large spiny lizard is found in all four North American deserts and throughout lower elevations across much of the southern Colorado Plateau. It occurs in rocky habitats in Great Basin desertscrub at GLCA, GRCA, RABR, and WUPA.

Eastern Fence Lizard (*Sceloporus undulatus*). This is the most commonly observed amphibian or reptile in the SCPN, and it is the only species documented from all 19 parks. Eastern fence lizards occur in all habitats from Great Basin desertscrub to mixed conifer forest. They are usually found on rocks, but also climb on trees, fence posts, and buildings. Proposed taxonomic changes (Leaché and Reeder 2002) treat eastern fence lizards from most of the Colorado Plateau as the plateau lizard (*S. tristichus*) and those in the Rio Grande Valley region as the southwestern fence lizard (*S. cowlesi*). The distributions of these proposed species are largely unknown in northwestern New Mexico (Leaché and Reeder 2002), and the two species (defined using DNA analysis) are almost impossible to distinguish in the field (Brennan and Holycross 2006).

Long-tailed Brush Lizard (*Urosaurus graciosus*). This relative of the ornate tree lizard is found in the Sonoran and Mojave Deserts, and could possibly occur along the Colorado River corridor in extreme western GRCA. Long-tailed brush lizards live on large desert shrubs and trees such as creosote bush and desert willow (*Chilopsis linearis*).

Ornate Tree Lizard (*Urosaurus ornatus*). This species occurs widely in rocky habitats (e.g., canyons, lava flows) throughout the Southwest; it has been documented from 13 SCPN parks. The ornate tree lizard ranges from desertscrub to ponderosa pine forest, and across much of the region it is usually found climbing on rocks, not trees.

Side-blotched Lizard (*Uta stansburiana*). This widespread species is distributed across the Colorado Plateau, ranging from the lowest deserts to pinyon-juniper woodland. Although sometimes found on the ground amongst shrubs, in our area it is usually found on rocks. The side-blotched lizard is apparently absent from many higher elevation SCPN parks (e.g., BAND, ELMA, ELMO, SUCR).

Family Xantusiidae

Desert Night Lizard (*Xantusia vigilis*). Previously, this tiny, secretive lizard was known in the SCPN only from Ribbon Falls at GRCA. In 2006 we discovered a population of desert night lizards along Last Chance Creek at GLCA, one of only a handful of populations known from Utah. It is unlikely this species occurs elsewhere in the SCPN.

Family Scincidae

Many-lined Skink (*Eumeces multivirgatus*). This small skink is distributed across the southern Colorado Plateau region, and has been documented from seven SCPN parks. Habitats in the SCPN include pinyon-juniper woodlands and ponderosa pine forests, where it is often found in leaf litter or under logs.

Great Plains Skink (*Eumeces obsoletus*). This large skink primarily occurs to the south and east of the southern Colorado Plateau. Within the SCPN it is found in shrublands, grasslands, and woodlands in the Rio Grande Valley parks (BAND, PETR, SAPU) and at ELMA. Great Plains skinks occur in ponderosa pine forest along the upper edge of the Mogollon Rim south of Flagstaff, Arizona, and may occur nearby at WACA.

Western Skink (*Eumeces skiltonianus*). The distribution of this species extends south from the Great Basin onto the Colorado Plateau of southwestern Utah and the Arizona Strip. Western skinks are common in ponderosa pine forests on the GRCA's North Rim.

Family Teiidae

Chihuahuan Spotted Whiptail (*Cnemidophorus exsanguis*). This all-female whiptail is distributed throughout the Chihuahuan Desert region. It has only been documented in the SCPN at BAND and SAPU, although it may also occur at ELMA or PETR. Chihuahuan spotted whiptails in the SCPN are usually found in broken country (e.g., canyons, dry washes, rocky hillsides) in pinyon-juniper woodlands and ponderosa pine forests. Like other all-female whiptails, this species reproduces by parthenogenesis, whereby a female lays unfertilized eggs that hatch into more females, all genetically identical to their mother.

Little Striped Whiptail (*Cnemidophorus inornatus*). This stunningly handsome species is widespread over much of New Mexico, including the Rio Grande Valley, but it has just a spotty distribution on the southern Colorado Plateau. In most areas, the typical small forms of this species are found in sandy grasslands, including at PEFO, PETR, and WUPA. At SAPU it occurs in open pinyon-juniper woodlands with markedly sandy soil. A larger form has a relictual distribution in chaparral, pinyon-juniper woodland, and ponderosa pine forest in central and northern Arizona, including on the south rim at GRCA. We discovered an isolated population of this larger form at WACA during our inventories. Lizards from the northern Arizona portion of the range are sometimes treated separately as the Pai striped whiptail (*C. pai*), but the composition and validity of this taxon is unresolved.

New Mexico Whiptail (*Cnemidophorus neomexicanus*). This all-female species is native to the Rio Grande Valley, where it is abundant at PETR and probably also occurs along the Rio Grande at BAND. Its distribution extends eastward from the Rio Grande Valley to near the Abó unit of SAPU, and it may eventually be found there. Persons and Wright (1999) discovered a small, probably introduced population along the Puerco River at PEFO. Throughout its range the New Mexico whiptail occupies sandy, disturbed habitats (such as flood plains and sandy washes), and it is even common in weedy fields and vacant lots in and around cities (e.g., Albuquerque).

Checkered Whiptail (*Cnemidophorus tesselatus*). This all-female whiptail has a spotty distribution throughout its range, which includes the Rio Grande Valley region of New Mexico. The checkered whiptail has been documented from open, rocky slopes near the Rio Grande at BAND. It probably does not occur elsewhere in the SCPN.

Western (Tiger) Whiptail (*Cnemidophorus tigris*). This widespread species is found in all four North American deserts and across the Colorado Plateau, where it has been documented from seven SCPN parks. In the SCPN the western whiptail occupies open habitats from Great Basin desertscrub to pinyon-juniper woodland. It is the only whiptail found throughout most of GLCA and GRCA.

Plateau Striped Whiptail (*Cnemidophorus velox*). This all-female species is distributed on the Colorado Plateau and adjacent areas of New Mexico, and has been documented from 16 SCPN parks. It is often abundant in grassland, pinyon-juniper woodland, and ponderosa pine forest habitats. The plateau striped whiptail is actually a complex of several mostly undescribed species (Wright 1993). Throughout the SCPN we found lizards referable to *C. velox* proper as well as to multiple undescribed species.

Family Anguidae
Madrean Alligator Lizard (*Elgaria kingii*). This species is distributed along and below the Mogollon Rim in Arizona and New Mexico. We have found it in rocky ponderosa pine forest habitat south of Flagstaff, Arizona, and it could occur in similar habitat nearby at WACA.

Family Helodermatidae
Gila Monster (*Heloderma suspectum*). This species, the only venomous lizard in the United States, is found in the Sonoran Desert and parts of the Mojave Desert. The gila monster penetrates the Grand Canyon in desertscrub along the Colorado River, where it has been found as far upstream as the vicinity of Diamond Creek.

Family Leptotyphlopidae
Texas Blind Snake (*Leptotyphlops dulcis*). This small, burrowing species only occurs in the SCPN region in the Rio Grande Valley, and it has only been documented at PETR, although it almost certainly occurs at BAND as well. Found in a variety of habitats throughout its range, the individual we collected from PETR came from sandy shrubland in Rinconada Canyon.

Western Blind Snake (*Leptotyphlops humilis*). This species occurs in desert areas below the Mogollon Rim, and extends its distribution into GRCA along the Colorado River, where it has been collected near Phantom Ranch and in the Little Colorado River gorge. An old specimen collected at WACA may have been mislabeled. Although an isolated population occurs in the Verde Valley in central Arizona, it has only been found below about 1524 m (5000 ft) in elevation, much lower than WACA.

Family Colubridae

Glossy Snake (*Arizona elegans*). This species occurs in a variety of habitats from desertscrub to pinyon-juniper woodland, but seems to be most common in grasslands such as those at PEFO and WUPA. Although only documented from six SCPN parks, the largely nocturnal glossy snake likely occurs at many others. It is most easily found by night driving.

Racer (*Coluber constrictor*). Although this relative of the whipsnakes (*Masticophis*) is widespread, it has very limited distribution on the southern Colorado Plateau, and has not been found in any SCPN park. It occurs in the Rio Grande Valley and in the San Juan River basin of New Mexico; apparently suitable habitat is present at a number of parks in those regions (i.e., AZRU, BAND, MEVE, PETR, SAPU, YUHO). Racers are usually found in more mesic habitats than whipsnakes, including grasslands, meadows, and riparian areas.

Ring-necked Snake (*Diadophis punctatus*). This small, secretive snake is widespread and common over much of North America, but has a spotty distribution in the arid Southwest. Permanent subsurface moisture is probably required, and isolated populations occur in mesic microhabitats (e.g., forests, canyons, seeps and springs) across the southern Colorado Plateau region. Ring-necked snakes have been documented from BAND, ELMO, GRCA, PETR, and WACA. The individual we found at PETR, a new park record, was abroad in grassland during the summer monsoon season.

Corn Snake (*Elaphe guttata*). This species of rat snake (*Elaphe*) is distributed in mesic habitats throughout much of New Mexico away from the Colorado Plateau, and a disjunct population occurs on the northern Colorado Plateau in east-central Utah and west-central Colorado. It has not been documented from any SCPN park, but almost certainly occurs in well-watered canyons at BAND.

Chihuahuan Hook-nosed Snake (*Gyalopion canum*). This secretive, burrowing species of the Chihuahuan Desert has not been documented in any SCPN park. It ranges northward along the Rio Grande Valley to near BAND, where it may possibly occur. It likely occurs in the sandy, shrub-dominated habitats below the escarpment at PETR.

Western Hog-nosed Snake (*Heterodon nasicus*). This species is widespread throughout the Great Plains and Chihuahuan Desert regions, where it favors sandy grassland habitats. It has only been documented from PETR, but we also received reliable reports from the Gran Quivira unit of SAPU, where it undoubtedly occurs. A disjunct population in the San Juan River basin in northwestern New Mexico suggests that it could occur at AZRU or other parks on the southern Colorado Plateau.

Night Snake (*Hypsiglena torquata*). This small, strictly nocturnal species has been documented from 10 SCPN parks, and probably occurs at all of them. Night snakes are found from Great Basin desertscrub to ponderosa pine forest. Our collection of a night snake at ELMO constituted a new Cibola County record (Persons and Nowak 2005c).

Common Kingsnake (*Lampropeltis getula*). Two very different-looking subspecies of this snake occur in the SCPN region. The dark colored, light-speckled desert kingsnake (*L. g. splendida*) of New Mexico and Texas ranges up the Rio Grande Valley and probably occurs at BAND and PETR. The boldly patterned black and white crossbanded California kingsnake (*L. g. californiae*) extends eastward onto the southern Colorado Plateau, and has been documented in Great Basin desertscrub and grassland at GLCA, GRCA, PEFO, and WUPA.

Sonoran Mountain Kingsnake (*Lampropeltis pyromelana*). This species is primarily distributed to the north and south of the Colorado Plateau in the mountains of central Utah and the Mogollon Rim–Madrean Archipelago regions, respectively. Sonoran mountain kingsnakes occur in ponderosa pine and mixed conifer forest on the North Rim at GRCA and in pinyon-juniper woodland and ponderosa pine forest at WACA. They probably also occur in ponderosa pine forest at SUCR.

Milk Snake (*Lampropeltis triangulum*). This widespread species has a very spotty distribution on and near the Colorado Plateau, but has nonetheless been documented from four SCPN parks (ELMO, MEVE, PEFO, WUPA), and may occur at others. In Arizona, including at PEFO and WUPA, milk snakes are generally known from sandy grasslands. However, at ELMO and MEVE they are found in higher elevation pinyon-juniper woodlands.

Coachwhip (*Masticophis flagellum*). This large, fast diurnal snake is generally restricted to arid, lower elevation habitats in the region. Its distribution more or less circumscribes the southern boundary of the Colorado Plateau, and it occurs in the Rio Grande Valley. This species is common in shrublands and grasslands at PETR, and is probably also present in the lowest elevations at BAND. It also follows desertscrub into western GRCA along the Colorado River, occurring upstream to the vicinity of Diamond Creek. It should be noted that Stebbins (2003) incorrectly depicts the range of the coachwhip as extending across much of southern Utah, which it does not.

Striped Whipsnake (*Masticophis taeniatus*). This relative of the coachwhip is distributed throughout the Colorado Plateau region. It has been documented at 15 SCPN parks, and almost certainly occurs at all of them. Striped whipsnakes range from Great Basin desertscrub to ponderosa pine forest.

Smooth Green Snake (*Opheodrys vernalis*). This small, well-camouflaged species has a spotty distribution in the Southwest, and in the SCPN has only been recorded from MEVE. It is known from the Jemez Mountains in New Mexico and probably occurs in the upper elevations of BAND. Smooth green snakes prefer grassy areas with permanent moisture, such as wet meadows, open patches in forests, and along streams.

Gopher Snake (*Pituophis catenifer*). This widespread, generalist species is found in every habitat in the region from the lowest deserts to mixed conifer forest. Gopher snakes have been documented from every SCPN park except RABR, where they undoubtedly occur as well.

Long-nosed Snake (*Rhinocheilus lecontei*). This species is primarily distributed in desert regions surrounding the southern Colorado Plateau, but it has been found at a handful of lower elevation sites across northern Arizona and southern Utah, including in western GRCA and southern GLCA. Long-nosed snakes also occupy the Rio Grande Valley in New Mexico, and we documented them from PETR during our surveys.

Mountain Patch-nosed Snake (*Salvadora grahamii*). This species is distributed south and east of the Colorado Plateau, and in the SCPN has only been documented from BAND. It almost certainly occurs at ELMA and SAPU as well. In our area, mountain patch-nosed snakes inhabit broken country (canyons, rocky hillsides) in pinyon-juniper woodland, usually at higher elevations than the similar western patch-nosed snake.

Western Patch-nosed Snake (*Salvadora hexalepis*). This fast, diurnal species occurs primarily in desert areas in the Southwest, and extends its distribution along the Colorado River corridor into southern Utah, where it has been documented from GRCA and southern GLCA. It also follows the Little Colorado River valley upstream to WUPA, where it occurs in Great Basin desertscrub and grassland habitats. A specimen collected recently from pinyon-juniper woodland near WACA suggests that the species probably occurs at that park as well.

Western Ground Snake (*Sonora semiannulata*). This small, secretive species is distributed primarily in deserts and grasslands away from the Colorado Plateau, but it follows the Colorado River corridor across the plateau into southern Utah. Until recently, western ground snakes were known in the SCPN only from GRCA. In his surveys at Grand Staircase–Escalante National Monument, Oliver (2003) found a western ground snake within GLCA (near Last Chance Creek), and Persons (1999) reported two specimens collected in 1951 at WUPA.

Southwestern Black-headed Snake (*Tantilla hobartsmithi*). This small, secretive snake has a spotty distribution on the Colorado Plateau, and has only been documented from GRCA. Records from southern Utah near GLCA suggest that it probably occurs there as well, and isolated populations may eventually be discovered elsewhere. Southwestern black-headed snakes occur in a variety of low- and middle-elevation habitats (to around 2000 m), including Great Basin desertscrub, grassland, and pinyon-juniper woodland.

Plains Black-headed Snake (*Tantilla nigriceps*). This relative of the southwestern black-headed snake is distributed throughout much of the Chihuahuan Desert and Great Plains regions, and occurs in the SCPN area only in and near the Rio Grande Valley of New Mexico. It has been documented from Rinconada Canyon at PETR, and probably also occurs at SAPU. Habitats include grasslands, shrublands, and woodlands.

Black-necked Garter Snake (*Thamnophis cyrtopsis*). This aquatic species is primarily distributed south and east of the Colorado Plateau, with apparently disjunct populations in the San Juan River basin and near Moab, Utah. Black-necked garter snakes have only been documented from BAND, ELMA, PEFO, and SAPU. The PEFO record (collection of a road-killed specimen) was obtained opportunistically in 2006. Although usually associated with foothill and mountain streams, the new record from PEFO, as well as other Little Colorado River

basin records reported by Persons and Rosen (2001), suggest that the species may be more widely distributed in lower elevation habitats within the interior of the southern Colorado Plateau.

Western Terrestrial Garter Snake (*Thamnophis elegans*). This garter snake is widespread across the West, including all of the Colorado Plateau. It has been documented from 13 SCPN parks, and likely occurs at others. As the name implies, this species is more terrestrial than other garter snakes, although it is still most commonly found in streams, ponds, or other water bodies. Although uncommon and local at lower elevations, the western terrestrial garter snake occurs in all habitats through mixed-conifer forest. An old specimen from WUPA is problematical, as it has imprecise locality data and the species has not been reported there since.

Checkered Garter Snake (*Thamnophis marcianus*). This species does not occur on the Colorado Plateau, and has not been found in any SCPN park. It primarily inhabits grasslands south and east of the region, but its occurrence in the Rio Grande Valley suggests that it could possibly occur, at least as a transient, at the disjunct unit of PETR located near the Rio Grande.

Common Garter Snake (*Thamnophis sirtalis*). This species, widespread throughout eastern North America, occurs as an isolated population in the Rio Grande Valley of New Mexico. Although not yet documented within the SCPN, like the checkered garter snake, its presence along the river within Albuquerque suggests that it could occur at the disjunct unit of PETR located near the Rio Grande.

Western Lyre Snake (*Trimorphodon biscutatus*). This species is distributed throughout the desert regions surrounding the southern Colorado Plateau, and has been documented from the Colorado River corridor at GRCA. Western lyre snakes occur from desertscrub through pinyon-juniper and pine-oak woodland, and are usually found in extensively rocky habitats. Suitable-looking habitat is present at WACA, and the species, which is known from nearby Oak Creek Canyon to the south, could possibly occur there.

Lined Snake (*Tropidoclonion lineatum*). This small relative of the garter snakes primarily occupies the Great Plains, but also has a patchy distribution in eastern New Mexico. We documented lined snake from the Quarai unit of SAPU, and this is likely the only place where it occurs in the SCPN. Lined snakes prefer grassy areas with adequate moisture, such as the fields surrounding the permanent stream at Quarai.

Family Viperidae

Western Diamond-backed Rattlesnake (*Crotalus atrox*). This large rattlesnake primarily occurs below the Mogollon Rim in Arizona and throughout eastern New Mexico, including the Rio Grande Valley. It is common at BAND and PETR, and also occurs at ELMA and SAPU. Brown (2000a) reported the species from the confluence of the Colorado River and Diamond Creek in Grand Canyon, just outside the park boundary. Western diamond-backed rattlesnakes occupy a variety of habitats from deserts to pinyon-juniper woodlands, and often occur in rocky or brushy areas.

Speckled Rattlesnake (*Crotalus mitchellii*). This species of the Sonoran and Mojave Desert regions occurs in the SCPN only in western GRCA, where it follows desert habitats along the Colorado River as far upstream as Havasu Canyon. Speckled rattlesnakes are usually found in rocky habitats, especially in canyons in the vicinity of seeps and springs.

Black-tailed Rattlesnake (*Crotalus molossus*). This species occurs widely to the south of the Colorado Plateau, and its distribution in northwestern Arizona includes the western portion of GRCA (Lowe et al. 1986; Brown 2000b). It probably also occurs at ELMA, the Abó unit of SAPU, and WACA. Black-tailed rattlesnakes range from desertscrub to pine forests, and are usually found in rocky habitats.

Mojave Rattlesnake (*Crotalus scutulatus*). This species, which is widespread throughout the Sonoran and Mojave Desert regions, has not been documented from any SCPN park. However, Brown (2003) reported a record from near the Colorado River in extreme western Grand Canyon, just outside the park boundary. Mojave rattlesnakes, which inhabit desertscrub and grassland habitats, probably occur within a limited area in western GRCA.

Western Rattlesnake (*Crotalus viridis*). This is the common rattlesnake of the Colorado Plateau. It has been documented from 16 SCPN parks, and probably occurs in all of them. Habitats of western rattlesnakes in the SCPN range from desertscrub to ponderosa pine forest. Recent research has proposed assigning the numerous traditional subspecies to an eastern and western species (Pook et al. 2000; Ashton and de Queiroz 2001) or elevating most western subspecies to full species status (Douglas et al. 2002). Under any of these new arrangements the species occurring at most SCPN parks is the "eastern" prairie rattlesnake (*C. viridis*), comprising the traditional Hopi (*C. v. nuntius*) and prairie (*C. v. viridis*) subspecies.

Under the arrangement of Pook et al. (2000) and Ashton and de Queiroz (2001), which is endorsed by Crother et al. (2003), the remaining "western" subspecies comprise a new species (*C. oreganus*), still called western rattlesnake. The Arizona black rattlesnake (*C. v. cerberus*), which is endemic to the central Arizona and western New Mexico highlands along and south of the Mogollon Rim, probably occurs at WACA (where Hopi rattlesnakes have been documented), and also possibly at SUCR. The Arizona black rattlesnake is treated as a full species (*C. cerberus*) by Douglas et al. (2002) and Brennan and Holycross (2006).

In addition to Hopi rattlesnakes above the GRCA south rim, that park harbors the Grand Canyon rattlesnake (*C. v. abyssus*) within the canyon and the Great Basin rattlesnake (*C. v. lutosus*) above the North Rim. Both of these subspecies are considered full species by Douglas et al. (2002). Finally, another "western" subspecies, the midget-faded rattlesnake (*C. v. concolor; C. concolor* of Douglas et al. 2002), is found throughout GLCA. The Grand Canyon rattlesnake has also been documented from the southern end of GLCA, and Hopi or prairie rattlesnakes probably occur in eastern GLCA along the San Juan River arm.

Massasauga (*Sistrurus catenatus*). This small rattlesnake of the southern Great Plains and upper Midwest reaches the western edge of its distribution in grasslands in New Mexico and southeastern Arizona. It is only known to occur in the SCPN at PETR, where it is found in the sandy grasslands in the western "Volcanoes" portion of the monument. Suitable sandy habitat exists at the Gran Quivira unit of SAPU, and the species should be looked for there.

CONSIDERATIONS FOR FUTURE INVENTORY WORK

The SCPN has not yet reached its stated goal of 90 percent estimated inventory completeness for amphibians and reptiles at many parks. Therefore, the NPS may wish to consider additional efforts toward the goal of establishing a more complete baseline of species occurrence at SCPN parks, perhaps in conjunction with planned monitoring activities. It should be noted that inventory and monitoring are not mutually exclusive endeavors. In fact, many methods designed to monitor more inclusive subsets of the herpetofauna (e.g., pitfall trapping for lizards, road driving surveys for snakes) are the same techniques used for inventory. Thus, a well-rounded monitoring program that focuses on each major subset of the herpetofauna will not only monitor the population status of particular focal species within each group, but will simultaneously monitor the community as a whole, and in the process will document new species and lead toward a more complete baseline of species occurrence. Park-specific considerations for further inventory work can be found in Persons and Nowak (2006).

Aside from turtles, which only occur in a handful of parks (Table 3), snakes are the group least well documented in the SCPN, with an overall inventory completeness of only 65 percent (Table 3). Most snake species are effectively sampled by night driving (e.g. Drost et al. 2001; Rosen and Lowe 1994). Additional night driving surveys at parks with available roads (e.g., CACH, CHCU, ELMA, and WUPA) may yield new snake records. For parks without extensive roads, snakes will most likely be documented opportunistically by researchers and NPS staff, unless targeted trapping methods are employed.

Amphibians, with an overall estimated inventory completeness of 70 percent (Table 3), are also effectively sampled by night driving on the southern Colorado Plateau (Drost et al. 2001). Additional night driving surveys in non-drought years would likely document new species at many parks. In addition, because most amphibians (i.e., anurans) produce advertisement mating calls during the breeding season, targeted nocturnal surveys during appropriate seasons and weather (e.g., after heavy summer monsoon rains) should be able to document most species.

Lizards are the best-documented group in the SCPN, with an overall estimated inventory completeness of 85 percent (Table 3). Most undocumented species would best be searched for using targeted daytime general visual encounter survey methods.

Given that inventory completeness remains low at many parks, one cost-effective method of increasing our knowledge of species presence is through the use of photography. We encourage park staff and visitors to record their amphibian and reptile observations using high-quality, close-up photographs. Digital or hard-copy photographs could be shared with regional herpetofauna experts and archived by NPS Resource Management staff.

ACKNOWLEDGMENTS

Funding for this work was provided by the NPS Inventory and Monitoring Program, and was facilitated by Rod Parnell and Ron Hiebert through the Colorado Plateau Cooperative Ecosystem Studies Unit at Northern Arizona University. Additional funding for report writing was provided by the

USGS Southwest Biological Science Center (SBSC). Our work was greatly aided by Anne Cully, NPS SCPN coordinator. Nicole Tancreto, data manager for the SCPN, cheerfully helped facilitate many aspects of the project. Dori Cawley, Charles Drost, Dave Mattson, Jeanne Pendergast, Marie Saul, and Maureen Stuart of the SBSC were also invaluable. NPS personnel were enthusiastic about our research and provided logistical support, background information, and numerous species sightings. In particular, we thank NPS Resource Management staff at the SCPN parks, including Terry Nichols (AZRU), Steve Fettig (BAND), Scott Travis (CACH), Brad Shattuck (CHCU), Herschel Schultz (ELMA), Sarah Beckwith (ELMO), John Spence (GLCA), Nancy Stone (HUTR), George San Miguel and Marilyn Colyer (MEVE, YUHO), Roger Moder (NAVA), Karen Beppler-Dorn, Pat Thompson, and Marit Wilkerson (PEFO), Mike Medrano and Dara Saville (PETR), Phil Wilson (SAPU), and Steve Mitchelson and Paul Whitefield (SUCR, WACA, and WUPA). Andy Holycross provided information on snake occurrences at many parks. Jodi Norris (SCPN) provided the map in Figure 1. Mark Sogge (SBSC) made many valuable suggestions on earlier drafts of the manuscript. Finally, we thank our invaluable field assistants Shawn Knox, A. J. Monatesti, Renata Platenberg, and Eric Zepnewski.

LITERATURE CITED

Ashton, K. G., and A. de Quieroz. 2001. Molecular systematics of the western rattlesnake, *Crotalus viridis* (Viperidae), with comments on the utility of the D-loop in phylogenetic studies of snakes. Molecular Phylogenetics and Evolution 21(2): 176–189.

Bradford, D. F., R. D. Jennings, and J. R. Jaeger. 2005 *Rana onca*. In Amphibian Declines: The Conservation Status of United States Species, edited by M. Lannoo, pp. 567–568. University of California Press, Berkeley.

Brennan, T. C., M. J. Feldner, and H. F. Koenig. 2002. *Coleonyx variegatus* geographic distribution. Herpetological Review 33(4): 320.

Brennan, T. C., and A. T. Holycross. 2006. A Field Guide to Amphibians and Reptiles in Arizona. Arizona Game and Fish Department, Phoenix.

Brown, N. L. 2000a. *Crotalus atrox* geographic distribution. Herpetological Review 31(1): 54–55.

Brown, N. L. 2000b. *Crotalus molossus* geographic distribution. Herpetological Review 31(1): 55.

Brown, N. L. 2003. *Crotalus scutulatus* geographic distribution. Herpetological Review 34(2): 168.

Campbell, H. W., and S. P. Christman. 1982. Field techniques for herpetofaunal community analysis. In Herpetological Communities, by N. J. Scott, Jr., pp. 193–200. U.S. Fish and Wildlife Service, Wildlife Research Report 13.

Crother, B. I., chair. 2000. Scientific and Standard English Names of Amphibians and Reptiles of North America North of Mexico, with Comments Regarding Confidence in our Understanding. Herpetological Circular No. 29, Society for the Study of Amphibians and Reptiles. St. Louis, Missouri.

Crother, B. I., J. Boundy, J. A. Campbell, K. DeQuieroz, D. Frost, D. M. Green, R. Highton, J. B. Iverson, R. W. McDiarmid, P. A. Meylan, T. W. Reeder, M. E. Seidel, J. W. Sites, Jr., S. G. Tilley, and D. B. Wake. 2003. Scientific and standard English names of amphibians and reptiles of North America north of Mexico: Update. Herpetological Review 34(3): 196–203.

Crump, M. L., and N. J. Scott. 1994. Visual encounter surveys. In Measuring and Monitoring Biodiversity: Standard Methods for Amphibians, by W. Heyer, M. Donnelly, R. McDiarmid, L. Hayek, and M. Foster, pp. 84–92. Smithsonian Institution Press, Washington, D.C.

Cummer, M. R., D. E. Green, and E. M. O'Neill. 2005. Aquatic chytrid pathogen detected in terrestrial plethodontid salamander. Herpetological Review 36(3): 248–249.

Degenhardt, W. G., C. W. Painter, and A. H. Price. 1996. Amphibians and Reptiles of New Mexico. University of New Mexico Press, Albuquerque.

Douglas, M. E., M. R. Douglas, G. Schuett, L. Porras, and A. Holycross. 2002. Phylogeography of the western rattlesnake (*Crotalus viridis*) complex, with emphasis on the Colorado Plateau. In Biology of the Vipers, edited by G. W. Schuett, M. Hoggren, M. E. Douglas, and H. W. Greene, editors, pp. 11–50. Eagle Mountain Publishing, Eagle Mountain, Utah.

Drost, C. A., T. B. Persons, and E. M. Nowak. 2001. Herpetofauna survey of Petrified Forest National Park, Arizona. In Proceedings of the Fifth Biennial Conference of Research on the Colorado Plateau, edited by C. van Riper III, K. A. Thomas, and M. A. Stuart, pp. 83–102. U.S. Geological Survey/FRESC Report Series USGS FRESC/COPL/2001/24.

Frost, D. R., T. Grant, J. Faivovich, R. H. Bain, A. Haas, C. F. B. Haddad, R. O. De Sá, A. Channing, M. Wilkinson, S. C. Donnellan, C. J. Raxworthy, J. A. Campbell, B. L. Blotto, P. Moler, R. C. Drewes, R. A. Nussbaum, J. D. Lynch, D. M. Green, and W. C. Wheeler. 2006. The Amphibian Tree of Life. Bulletin of the American Museum of Natural History 297, 370 pp.

Hammerson, G. A. 1999. Amphibians and Reptiles in Colorado, 2nd ed. University Press of Colorado, Niwot.

Leaché, A. D., and T. W. Reeder. 2002. Molecular systematics of the eastern fence lizard (*Sceloporus undulatus*): A comparison of parsimony, likelihood, and bayesian approaches. Systematic Biology 51(1): 44–68.

Lowe, C. H., C. R. Schwalbe, and T. B. Johnson. 1986. The Venomous Reptiles of Arizona. Arizona Game and Fish Department, Phoenix.

Mikesic, D. 2004a. Inventory of amphibians and reptiles at Canyon de Chelly National Monument. Report to Southern Colorado Plateau Inventory & Monitoring Network. Navajo Natural Heritage Program, Navajo Nation Department of Fish & Wildlife, Window Rock, AZ.

Mikesic, D. 2004b. Inventory of amphibians and reptiles at Hubbell Trading Post National Historic Site. Report to Southern Colorado Plateau Inventory & Monitoring Network. Navajo Natural Heritage Program, Navajo Nation Department of Fish and Wildlife, Window Rock, Arizona.

Mikesic, D. 2004c. Inventory of amphibians and reptiles at Navajo National Monument. Report to Southern Colorado Plateau Inventory & Monitoring Network. Navajo Natural Heritage Program, Navajo Nation Department of Fish and Wildlife, Window Rock, Arizona.

Monatesti, A. J., T. B. Persons, and E. M. Nowak. 2005. *Hyla eximia* geographic distribution. Herpetological Review 36(1): 74–75.

Oliver, G. V. 2003. Amphibians and Reptiles of the Grand Staircase–Escalante National Monument: Distribution, Abundance, and Taxonomy. Utah Division of Wildlife Resources Natural Heritage Program, Salt Lake City, Utah.

Persons, T. 1999. *Sonora semiannulata* geographic distribution. Herpetological Review 30(1): 55.

Persons, T. B. 2001. Distribution, activity, and road mortality of amphibians and reptiles at Wupatki National Monument, Arizona. Report to National Park Service. USGS Colorado Plateau Field Station, Flagstaff, Arizona.

Persons, T. B. 2005. Distribution and habitat association of the little striped whiptail (*Cnemidophorus inornatus*) at Wupatki National Monument, Arizona. USGS Open-File Report 2005-1139. Southwest Biological Science Center, Flagstaff, Arizona.

Persons, T. B., and E. M. Nowak. 2004. *Coleonyx variegatus* geographic distribution. Herpetological Review 35(1): 81.

Persons, T. B., and E. M. Nowak. 2005a. *Bufo punctatus* geographic distribution. Herpetological Review 36(2): 198.

Persons, T. B., and E. M. Nowak. 2005b. *Phrynosoma modestum* geographic distribution. Herpetological Review 36(1): 80.

Persons, T. B., and E. M. Nowak. 2005c. *Hypsiglena torquata* geographic distribution. Herpetological Review 36(1): 82.

Persons, T. B., and E. M. Nowak. 2006. Inventory of amphibians and reptiles in southern Colorado Plateau national parks. USGS Open File Report 2006-1132. Southwest Biological Science Center, Colorado Plateau Research Station, Flagstaff, Arizona. Available at http://sbsc.wr.usgs.gov/files/pdfs/ofr_2006-1132.pdf

Persons, T. B., and P. C. Rosen. 2001. *Thamnophis cyrtopsis* geographic distribution. Herpetological Review 32(2): 125.

Persons, T., and J. W. Wright. 1999. Discovery of *Cnemidophorus neomexicanus* in Arizona. Herpetological Review 30(4): 207–209.

Pook, C. E., W. Wuster, and R. S. Thorpe. 2000. Historical biogeography of the western rattlesnake (Serpentes: Viperidae: *Crotalus viridis*), inferred from mitochondrial DNA sequence information. Molecular Phylogenetics and Evolution 15(2): 269–282.

Ramotnik, C. A. 1988. Habitat requirements and movements of Jemez Mountains salamanders, *Plethodon neomexicanus*. Unpublished Master's thesis, Colorado State University, Fort Collins.

Rorabaugh, J. C. 2005. *Rana pipiens*. In Amphibian Declines: The Conservation Status of United States Species, edited by M. Lannoo, pp. 570–577. University of California Press, Berkeley.

Rosen, P. C., and C. H. Lowe. 1994. Highway mortality of snakes in the Sonoran Desert of southern Arizona. Biological Conservation 68: 143–148.

Rosen, P. C., and C. R. Schwalbe. 2002. Widespread effects of introduced species on reptiles and amphibians in the Sonoran Desert region. In Invasive Exotic Species in the Sonoran Region, edited by B. Tellman, pp. 220–240. University of Arizona Press and Arizona-Sonora Desert Museum, Tucson.

Scott, N. J. 1994. Complete Species Inventories. In Measuring and Monitoring Biodiversity: Standard Methods for Amphibians, edited by W. R. Heyer, M. A. Donnelly, R. W. McDiarmid, L. C. Hayek, and M. S. Foster, pp. 78–84. Smithsonian Institution Press, Washington, D.C.

Shaffer, H. B., and J. E. Juterbock. 1994. Night driving. In Measuring and Monitoring Biological Diversity: Standard Methods for Amphibians, edited by W. R. Heyer, M. A. Donnelly, R. W. McDiarmid, L. C. Hayek, and M. S. Foster, pp. 163–166. Smithsonian Institution Press, Washington, D.C.

Simovich, M. A. 1994. The dynamics of a spadefoot toad (*Spea multiplicata* and *S. bombifrons*) hybridization system. In Herpetology of the North American Deserts, edited by P.R. Brown and J.W. Wright, pp. 167–182. Southwestern Herpetologists Society, Special Publication No. 5, Van Nuys, California.

Sredl, M. J. 2005. *Rana yavapaiensis*. In Amphibian Declines: The Conservation Status of United States Species, edited by M. Lannoo, pp. 596–599. University of California Press, Berkeley.

Stebbins, R. C. 2003. A Field Guide to Western Reptiles and Amphibians, 3rd ed. Houghton Mifflin, Boston.

Thomas, L., C. Lauver, M. Hendrie, N. Tancreto, J. Whittier, M. Atkins, J. Miller, and A. Cully. 2004. Vital signs monitoring plan for the Southern Colorado Plateau Network: Phase II report. National Park Service, Southern Colorado Plateau Network, Flagstaff, Arizona.

Webb, R. H., G. J. McCabe, R. Hereford, and C. Wilkowske. 2004. Climatic fluctuations, drought, and flow in the Colorado River. USGS Fact Sheet 3062-04.

Wright, J. W. 1993. Evolution of the lizards in the genus *Cnemidophorus*. In Biology of Whiptail Lizards (Genus *Cnemidophorus*), edited by J. W. Wright and L. J. Vitt, pp. 27–81. Oklahoma Museum of Natural History, Norman.

PERSISTENCE OF APACHE TROUT FOLLOWING WILDFIRES IN THE WHITE MOUNTAINS OF ARIZONA

Jonathan Long

Wildfires are a natural disturbance that can rejuvenate and rebuild trout habitat by stimulating flood scour and deposition (Gresswell 1999; Bisson et al. 2003). However, several case studies from Arizona and New Mexico have shown severe reductions in fish populations shortly after wildfires (Neary et al. 2005). The consequences of wildfire are a particular concern in the mountainous watersheds that harbor Apache trout (*Oncorhynchus gilae* ssp. *apache*) and Gila trout (*Oncorhynchus gilae* ssp. *gilae*). Populations of these federally protected species are scattered among small stream reaches that are isolated by the presence of non-native trouts, intermittent flows, thermal regimes, natural barriers, and barriers to exclude non-native trout (Brown et al. 2001; Rieman et al. 2003; USFWS 2006). Wildfires have extirpated seven populations of Gila trout since 1989, impeding recovery efforts (Propst et al. 1992; USFWS 2006). Emergency evacuation of populations threatened by fire has become an important element of Gila trout recovery (Anderson 1992; USFWS 2006), and has been recommended as a "fundamental management approach" for conserving isolated populations of rare native fishes (Rinne 2004). However, because evacuation is costly and stressful to fish, managers need to know when it is necessary to prevent population losses.

The extent of the threat to Apache trout from wildfire has not been clear, since large wildfires have not struck its ancestral home on Mount Baldy in the White Mountains of Arizona in several decades (Dieterich 1983; Gomez and Tiller 1990). Place names such as Aspen Butte, Aspen Ridge, and Burnt Mountain suggest that this landscape has experienced stand-replacing wildfires, because extensive stands of aspen (*Populus tremuloides*) often reflect past wildfires in mixed-conifer forests (Swetnam et al. 2001; Margolis 2003). However, a recent study of stand-replacing high-elevation fires (Margolis 2003) did not include the White Mountains, in part because other areas offered clearer historical evidence of such fires (Ellis Margolis, personal communication).

Recent climatic conditions and historical management practices have increased the likelihood of large and severe wildfires in the White Mountains. Widespread insect infestations, abetted by warm winters and dry summers, have killed huge swaths of spruce in bands around Mount Baldy (Lynch 2004). The combination of increased forest fuels and warming temperatures over the past century has increased the likelihood of wildfires across vegetation zones in Arizona mountains (Cocke et al. 2005). Researchers have asserted that past timber harvest and road construction activities have rendered habitats more vulnerable to post-fire impacts; this combination of threats has motivated many land managers and community members to express concern that large wildfires in the White Mountains could jeopardize valued resources such as Apache trout (Abrams 2005). Debates over how to manage forests in the face of such risks need to be guided by analyses of wildfire effects within local landscape contexts (Rieman et al. 2003).

Two streams (Grant Creek and KP Creek) whose high-elevation watersheds have recently experienced wildfires (the Steeple fire of July 2003 and KP fire of May 2004) are especially important because they represent ancestral habitat for native trout (Figure 1). Native trout had been collected from KP Creek in 1904, although the taxonomic identity is uncertain because the original specimens have been lost (Miller 1950). KP Creek was later stocked with a strain of Apache trout from an unconnected stream as part of recovery efforts (Dowling and Childs 1992). Researchers have concluded that the present-day trout populations in both KP Creek and Grant Creek are hybridized with rainbow trout (Figure 2; Rinne 1985; Dowling and Childs 1992; USFWS 2003). The current recovery plan is to stock these streams with the Spruce Creek lineage of Gila trout, which appears to be an intermediate form between Gila and Apache trout (Riddle et al. 1998; USFWS 2003). The impact of wildfire on the two creeks was thus not a major concern from a conservation standpoint, but their response to wildfires could yield insights into how wildfires might affect other streams that have been designated for recovery of Apache trout.

The severity of a burn within a stream's watershed is a primary factor in predicting the impact of fires on aquatic systems. Burton (2001) found that 60–88 percent of the watersheds of streams that experienced high levels of post-fire debris flooding had burned at high severity; only 14–37 percent of the watersheds of streams that did not experience post-fire debris flooding were burned at high severity. Cannon (2001) has reported that the percent of watershed area burned at moderate to high severity is an important predictor of debris flows, and Schaffner and Reed (2005) successfully predicted increases in peak flows by calculating area burned at moderate to high severity. Managers can easily evaluate burn severity in the aftermath of a wildfire, making it an important and accessible tool for predicting fish persistence in different settings.

A variety of factors besides burn characteristics affect the responses of stream ecosystems to wildfires, including human influences and natural landscape attributes

(Schaffner and Reed 2005). Cannon et al. (2003) found that slopes greater than 30 percent and soils with increased permeability and organic matter content were associated with increased probability of fire-induced debris flows. Debris flows are also strongly related to the occurrence of post-fire storm events (Cannon 2001), but it is difficult to predict such events. The availability of lightly burned refugia and absence of barriers are two factors that help fish avoid impacts from fires (Burton 2001; Bisson et al. 2003).

Previous research in the White Mountains indicates that landscape attributes such as topography, geology, and vegetation type need to be considered when evaluating fish habitat (Long and Medina 2005). Within a region, mean basin elevation and slope help to explain variation in post-fire flooding (Schaffner and Reid 2005). Therefore, comparing landscape attributes of the study streams to those designated for Apache trout recovery would help in evaluating whether their post-fire responses would be representative. Extrapolating the responses of two streams to a much larger and more diverse region is speculative, but it could help managers who must act on limited information concerning the risk of wildfires to Apache trout.

METHODS

This study examined the upper reaches of KP and Grant Creeks, perennial tributaries of the Blue River (Figure 1). A five-person team from the Rocky Mountain Research Station (RMRS) sampled fish populations and habitat conditions at seven 50 m long sampling sites in KP and Grant Creeks in June of 2004 (Figure 3), after the KP fire was contained. Fish populations were resampled at six of the sites one year later. The team attempted to relocate sites that had been sampled for fish in September of 1995 by the Arizona Game and Fish Department (AGFD) as part of their General Aquatic Wildlife Surveys (GAWS) program. The team successfully located three of those monitoring sites (labeled 22, 20, and 19 from upstream to downstream) in the upper

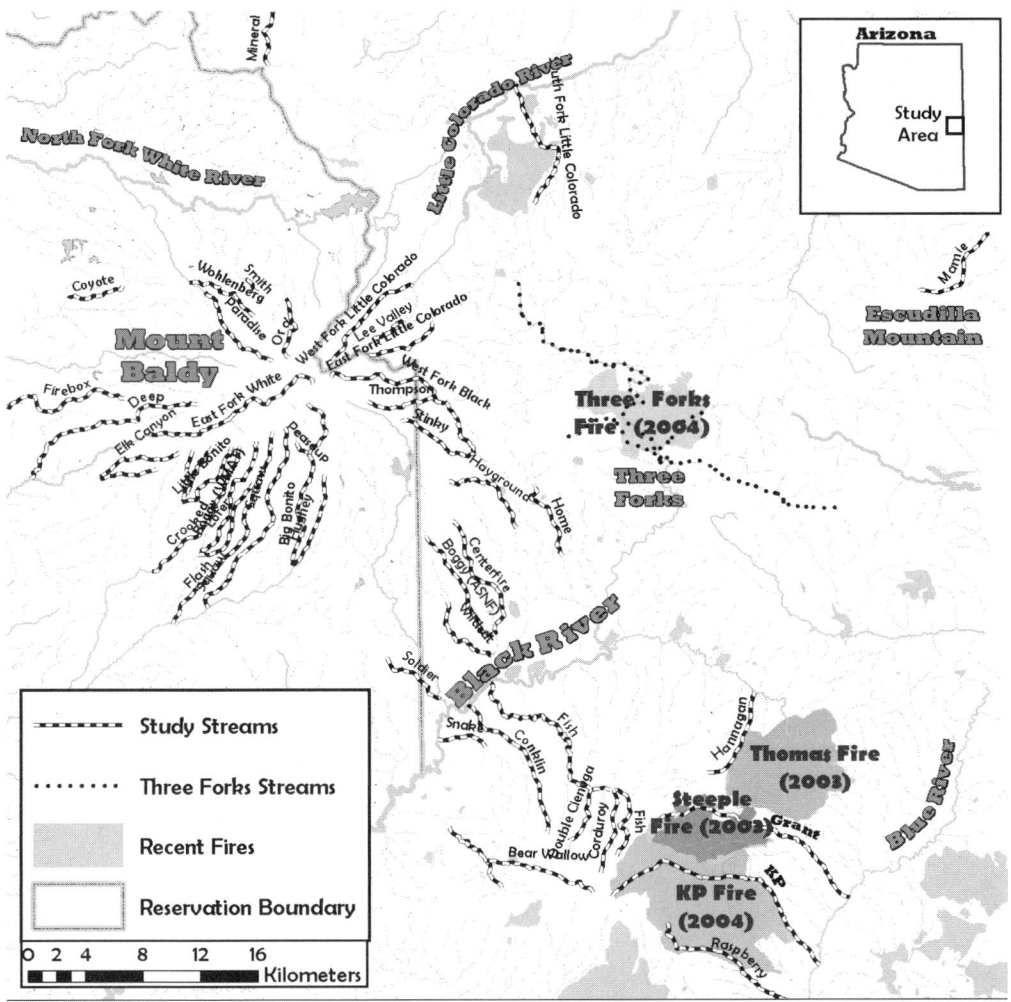

Figure 1. Locations of Apache trout streams and the study streams in the White Mountains of east-central Arizona.

reaches of Grant Creek, although AGFD had sampled fish only at the middle site. In KP Creek, the original site markers could not be located, so the team approximated the three uppermost sites (24, 23, and 22 from upstream to downstream) based on the descriptions of channel gradient and intrastation distances recorded on the 1995 survey forms. Fish populations were sampled using backpack-mounted electroshocking gear. Each reach was blocked off with nets to prevent fish from escaping during sampling and each was sampled three times using the depletion method (Zippin 1958). Lengths of all fishes were recorded and used to calculate mean lengths with 95 percent confidence intervals. The maximum likelihood population estimate and 95 percent confidence interval were calculated at each sampling site using the Microfish 3.0 computer program (Van Deventer and Platts 1989).

Figure 2. Hybridized Apache trout collected in KP Creek, June 2004.

Landscape Attributes of Study Sites

Watersheds of the study sites were delineated using the hydrologic modeling tools in ArcMap version 9 (ESRI, Redlands, CA). Burn severity information (based on differenced normalized burn ratio data calculated from satellite images) provided by the Apache-Sitgreaves National Forest was incorporated into the geographic database, and the data were used to calculate the percent of each watershed burned at moderate to high severity.

Mapping landscape attributes of KP and Grant Creek served to compare them to the 36 streams that the Apache Trout Recovery Team has identified as priorities for conservation of pure strains of Apache trout within its ancestral range (Ruiz and Novy 2000). A coarse-scale map of Arizona (Reynolds 1988) available in digital format from the Arizona Land Resource Information System served to characterize site geology (Figure 4). The two major geologic types are felsic volcanic formations along the upper slopes of Mount Baldy and mafic volcanic formations along the lower slopes and adjacent plateaus. Slopes were calculated using ArcMap to process 10 m digital elevation models of the region (Figure 5). Data for the vegetation layer were provided by the U.S. Geological Survey's GAP analysis program. In the three major mixed-conifer forest types that predominate within the range of Apache trout (Figure 6), the dominant tree species are Engelmann spruce (*Picea engelmannii*), Douglas fir (*Pseudotsuga menziesii*), and ponderosa pine (*Pinus ponderosa*).

Maps and elevations in the draft recovery plan and Arizona Game and Fish Department's GAWS reports indicated the upper and lower limits of habitat in each stream,

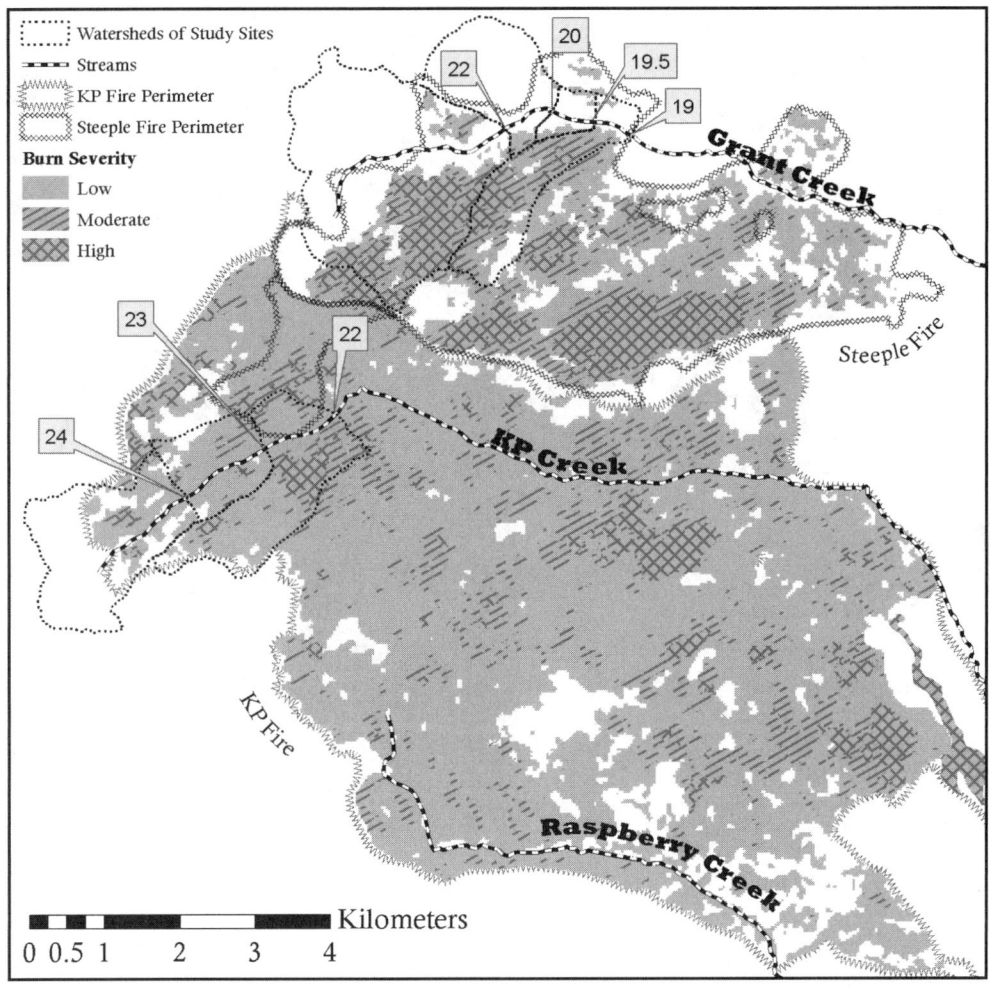

Figure 3. Map of fire impact study sites with overlay of burn severity for the KP and Steeple fires.

and in the absence of upstream limits, the beginning of a solid blue line on topographic maps approximated the start of habitat. The blue-line method provides only a rough approximation of suitable habitat (Svec et al. 2005), but it facilitates comparison of the general landscape characteristics of Apache trout streams. Figure 7 classifies the streams into three landscape types based on the characteristics at the upper end of habitat within a stream. The highest-elevation habitat could serve as a refuge from wildfire im-

pacts if only the lower part of the watershed were burned.

RESULTS
KP Creek

The estimates of trout population at the three reaches sampled in KP Creek in 2005 were higher than or equal to the population counts at their approximate locations when surveyed 10 years earlier (Table 1). The numeric data from 2004 were misplaced and not recovered, but the memories of the

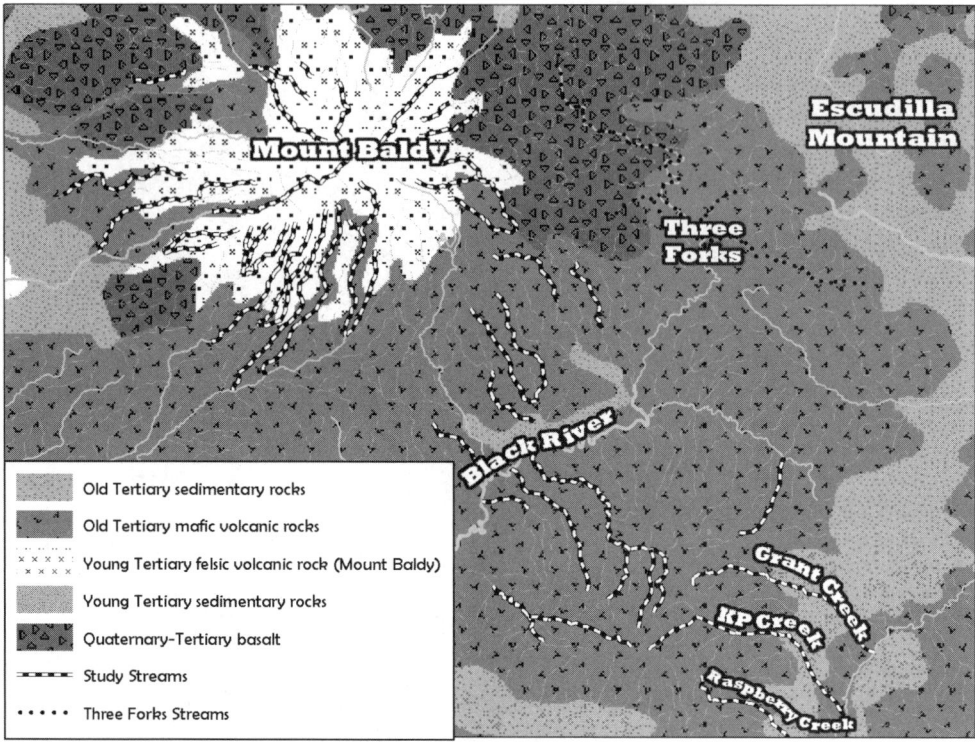

Figure 4. Geology of the White Mountains, revealing that Apache trout streams originate from the felsic slopes of Mount Baldy (light area) or on adjacent mafic plateaus (dark gray and medium gray areas with triangular stippling).

surveyors were that the numbers and sizes of fish were similar to those collected in 2005. Photographs taken during sampling in 2004 indicated that fishes were present in sizes ranging from at least 95 to 195 mm.

The percent of severely and moderately burned watershed above each site increased from 4.7 percent at the upstream site to 8.4 percent at the middle site and to 19.1 percent at the downstream site. The team observed no evidence of recent debris flows or channel incision at the sites in 2004 or 2005, but a very large debris fan was observed in the channel < 100 m downstream from the lowermost site in 2004 (Figure 8). It emanated from a drainage that burned at high severity in the previous year's Steeple fire, which had burned into the KP watershed.

Grant Creek

In Grant Creek, temporal comparisons were limited by the fact that fish had been sampled only at one site (20) prior to the fire. That sample was collected in September of 1995. Resampling was conducted in early June in both 2004 and 2005. The population estimate within reach 20 in 1995 was twice the levels estimated in 2004 and 2005 (Table 1). However, the fish in the 1995 sample were only half the length of those in 2004 and 2005. The apparent abundance of small fish could reflect seasonal variation, since the late summer sampling could have collected young-of-year fish. Sampling at a downstream site (19) yielded higher population estimates and smaller sizes compared

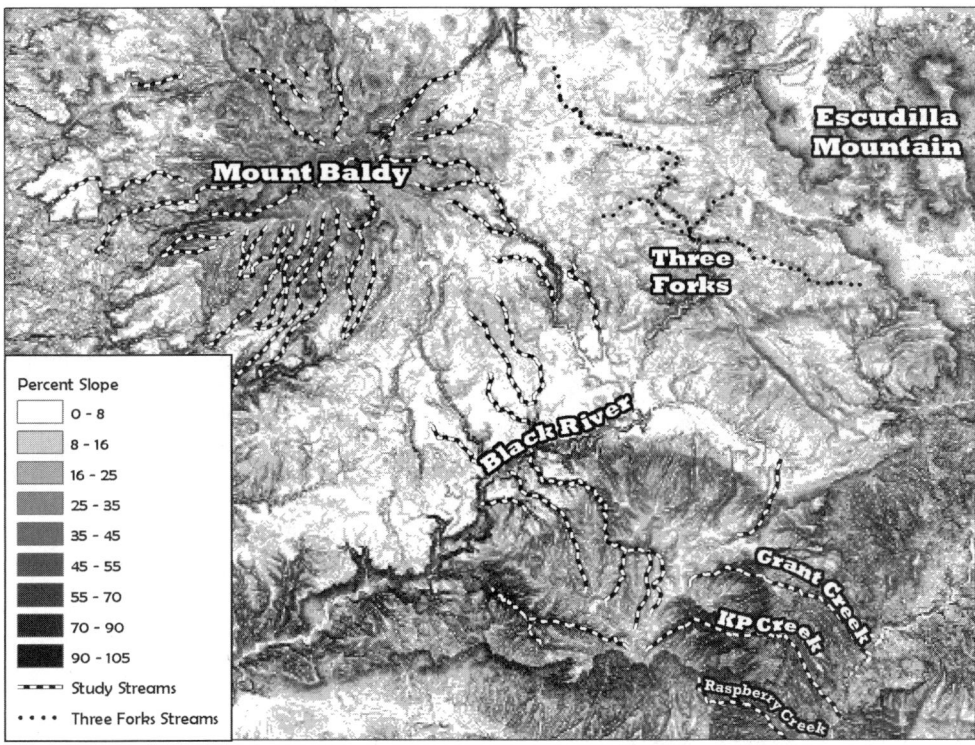

Figure 5. Topography of the White Mountains with darker shading indicating steeper slopes.

to site 20 in both 2004 and 2005 (Table 1). A reach between sites 19 and 20 (19.5) that was sampled only in 2004 had sizes and numbers of trout similar to site 20 (Table 1). Sampling at the uppermost site (22) indicated that no fish were present even though that reach had flowing water in 2004 and 2005.

The percent of watershed burned at moderate to high severity at the sampling sites was higher than at sites in KP Creek (Table 1). The channel along Grant Creek was impacted by debris flows emanating from side canyons in several locations. These flows formed alluvial fans that confined the channel and induced upstream aggradation. Water flowed underneath piles of woody debris in the reach above site 20 (Figure 9). Although the reach itself was not heavily impacted, the pile had the potential to block

fish movements upstream. A debris fan below site 19 also appeared to have formed another potential barrier. Site 22 had the highest burn severity and was the only sampled reach where the channel was incised.

Landscape Comparison

KP and Grant Creeks are similar to most Apache trout streams in that they flow through mixed-conifer forests in mafic canyons. However, their upper watersheds (where the study sites were located) are relatively steeper than most Apache trout watersheds (Figure 5). Habitat in 11 Apache trout streams extends into the high-elevation (over 3000 m) spruce zone on Mount Baldy, and seven others harbor Apache trout only within the lower elevation (below 2500 m) ponderosa pine zone (Figure 7).

Table 1. Trout population estimates, lengths, and burn severity at sites on Grant Creek and KP Creek, White Mountains, Arizona (x = no data, NA = not applicable).

Site	Population Estimate (Total Catch to Upper 95% CI)			Mean Length ± 95% CI (in mm)			% Watershed Burned at Moderate to High Severity
	9/95 (pre-fire)	6/04 (before runoff from fire)	6/05 (1 year after fire)	9/95 (pre-fire)	6/04 (before runoff from fire)	6/05 (1 year after fire)	
Grant 22	x	0	0	NA	NA	NA	29.8%
Grant 20	28 (28-29)	14 (14-15)	15 (15-15)	84 ± 11	174 ± 11	174 ± 28	23.2%
Grant 19.5	x	13 (13-14)	x	x	179 ± 7	x	22.4%
Grant 19	x	20 (18-27)	23 (19-35)	x	158 ± 11	154 ± 21	28.3%
KP 24	4*	x	11 (11-13)	x	x	115 ± 32	4.7%
KP 23	12*	x	12 (12-13)	x	x	129 ± 17	8.4%
KP 22	25*	x	33 (33-35)	x	x	140 ± 12	19.1%

*1995 data for KP creek were reported only as total fish caught, not a maximum likelihood estimate based upon three-pass depletion.

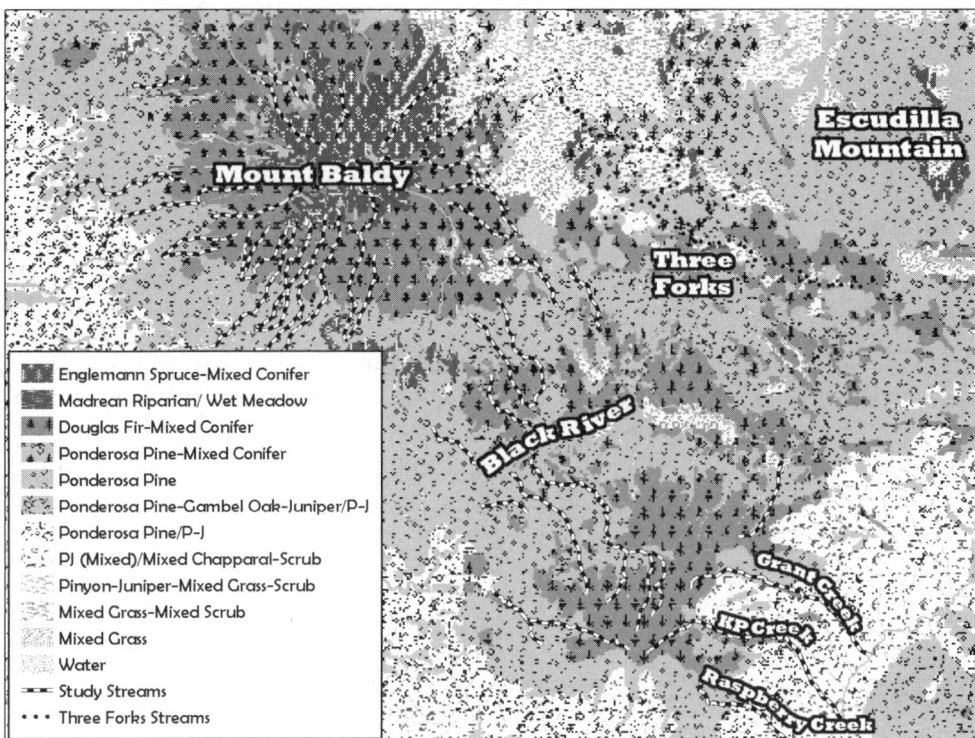

Figure 6. Vegetation types in the White Mountains, with darker shades representing high-elevation communities with naturally less frequent fire regimes.

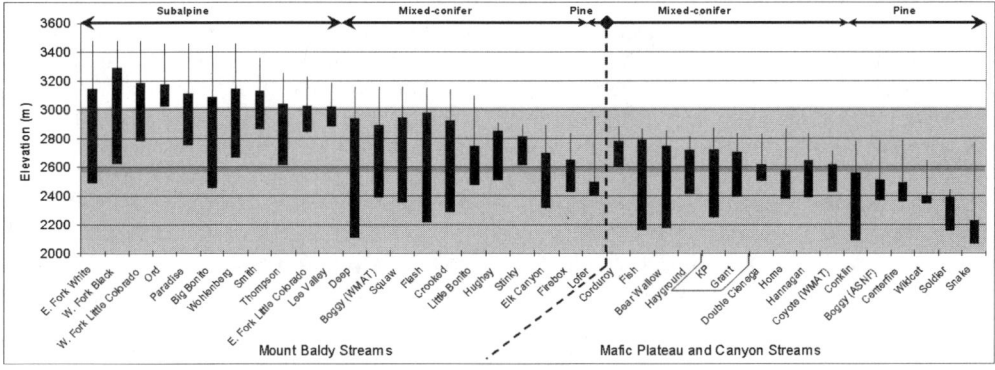

Figure 7. Bar chart representing the approximate elevation ranges of trout habitats in study streams arranged by geologic types and shaded by approximate vegetation zone. The whiskers above each bar denote the elevation at the summit of each watershed.

Figure 8. Large debris flow in KP Creek below site 22, emanating from a small area that was severely burned by the Steeple fire the year previously.

DISCUSSION

The differences in fish populations and lengths in Grant Creek before and after the fires could reflect seasonal variation (i.e., fewer young fish in the June samples) or post-fire impacts that might have inhibited recruitment of young fish. The data from both creeks are insufficient to evaluate whether the fire caused changes in abundance. They do demonstrate that fish persisted in both KP and Grant Creeks for at least the first year of post-fire runoff, which Neary et al. (2005) have suggested is a critical timeframe for extirpations to occur. The Steeple fire appeared to have induced debris flows along both KP and Grant Creeks, resulting in debris piles that could have created barriers to fish movement. Such

flows were not observed in the year after the less severe KP fire. These results are consistent with the hypothesis that the extent of moderate to high burn severity is an important predictor of physical impacts such as debris flows. Since trout did survive in both streams, the results indicate that wildfires that are severe enough to induce debris flows do not necessarily extirpate trout populations.

Table 2 compiles previous research by the Rocky Mountain Research Station and other published reports of wildfire impacts on fish populations within watersheds in eastern Arizona. The percent of each watershed burned at moderate or high severity was calculated using the methods described previously. The KP fire also burned the headwaters of Raspberry Creek, which flows

Table 2. Impacts of fires on fish habitat in east-central Arizona based on various case studies between 1990 and 2004.

Fire	Location (in Arizona)	Streams Examined	Dominant Upland Vegetation	Watershed Area Burned at Moderate–High Severity	Hydro-Geomorphic Impacts	Effects on Fish Populations	References
Dude (1990)	Mogollon Rim	Bonita Creek	Pine	> 75%*	Debris flows and channel incision	Extirpation	Rinne 1996
		Ellison Creek	Pine	> 75%*	Debris flows and channel incision	Extirpation	
		Dude Creek	Pine	> 50%*	Debris flows and channel incision	Extirpation (1 trout found 1 year later)	
White Springs (1996)	Mogollon Rim	White Spring	Pine	~ 75%	Debris flows and channel incision	Extirpation	Long & Endfield 2000
Steeple (2003)	White Mtns	Grant Creek (site 19)	Douglas fir/ pine	28.3%	Debris flows and channel incision	Trout persisted	This study
		Raspberry Creek	Pine	7.0%	—	Trout persisted	
Rodeo-Chediski (2002)	Mogollon Rim	Limestone Canyon	Pine	72%	Debris flows and channel incision	Extirpation	Long & Burnette 2004
Steeple (2003)	White Mtns	Grant Creek (site 19)	Douglas fir/ pine	30%	Debris flows and channel incision	Trout persisted	This study
		Raspberry Creek	Pine	7.0%	—	Trout persisted	USFWS 2006
KP (2004)	White Mtns	KP Creek (site 22)	Douglas fir/ pine	19.1%	Not observed 1 year post-fire	Trout persisted	This study
Three Forks (2004)	White Mtns	N. Fork E. Fork Black River	Douglas fir/pine, mixed grass/shrub	3.5%	Short-term turbidity increase due to ash	Trout persisted	Unpub. data, J. Rinne, C. Carter & J. Long (RMRS)
		Boneyard Creek		4.8%			
		W. Fork E. Fork of Black River		6.4%			

*Estimated based on fire perimeter maps and observations by J. Leonard, an RMRS technician who has studied the Dude fire.

Figure 9. Debris flow in Grant Creek above site 20.

down a steep, south-facing slope through ponderosa pine forest (Figures 6 and 7). In response, federal and state agencies evacuated a portion of the Gila trout population that had been introduced to the stream, and the U.S. Forest Service set a backfire to inhibit the spread of the fire. A maximum of only 7 percent of the watershed burned at moderate to high severity; the U.S. Fish and Wildlife Service reported that trout survived in the creek (USFWS 2006).

The Three Forks fire burned 3200 hectares in the relatively flat headwaters of the East Fork of the Black River (Figures 2 and 6). Only 4–6 percent of each tributary drainage burned at moderate to high severity (Table 2). Historical data from AGFD's GAWS surveys indicate that brown trout and various warmwater species such as desert sucker and speckled dace were present in samples from all periods, although Apache trout were not. The samples were highly variable between sampling periods both before the

fire and afterwards. The post-fire samples in 2004 and 2005 showed that fish persisted and no changes in channel morphology were observed (unpublished data of J. Rinne, C. Carter, and J. Long on file at the RMRS in Flagstaff).

The Steeple, KP, and Three Forks fires had more moderate impacts compared to earlier wildfires on ponderosa pine forests along the Mogollon Rim west of the White Mountains. Several of the earlier fires, including the Dude fire of 1990 (Rinne 1996), the White Springs fire of 1996 (Long and Endfield 2000), and the Rodeo-Chediski fire of 2002 (Long and Burnette 2004; Long et al. 2005), induced fish extirpations and major impacts to stream channels including channel avulsion and rapid incision. In each case, localized extirpations of fish populations occurred within the first couple of years at sites where more than 50 percent of the watershed area burned at moderate to high severity (Table 2).

The Aspen fire of 2003 burned most of the upper Sabino Creek watershed and thus triggered an evacuation of rare Gila chubs from Sabino Creek (Tobin 2004). Post-fire debris flows in the canyon substantially modified aquatic habitats and initially appeared to eliminate all aquatic vertebrates in the canyon (USDA 2004). However, researchers subsequently discovered several adult and numerous young Gila chubs in a large pool (Tobin 2004). Forty-seven percent of the watershed above that reach burned at moderate to high severity.

These case studies of fire impacts are consistent with a hypothesis that the proportion of moderate to high severity burn is a useful predictor of impacts to fishes and their habitats. Specifically, extirpations have been reported for streams where more than half of the watershed burned at moderate to high severity. By consistently collecting pre- and post-fire data, future studies could test such a threshold of burn severity.

Despite its potential utility, an index of burn severity is a simple metric that ignores many factors that are likely to influence post-fire stream impacts. However, landscape classifications can help to control for important factors such as watershed steepness, soil erodibility, and availability of refugia. Managers need to tailor their management strategies to the historic fire regime within particular forest types (Schoennagel et al. 2004). Accordingly, the typology used in Figure 7 may be helpful in planning management of Apache trout habitats.

Spruce Zone

Stand-replacing wildfires in the subalpine zone dominated by spruce are normally infrequent due to high moisture (Veblen et al. 1991; Swetnam and Baisan 1996), but stand-replacing fires have been reported in spruce-fir communities in northern New Mexico during severe droughts in the mid eighteenth century (Margolis 2003). The risk of stand-replacing wildfires in this zone may be increasing due to the combination of drought, warming, and insect outbreaks, which has increased fuel loads in the form of needles and dead trees. A combination of

drought, historical fire suppression, and heavy insect damage fueled the 11,800 ha Nuttall Complex fire in spruce-fir and mixed-conifer forest on Mount Graham in 2004 (Koprowski et al. 2005). The fire, which burned 59 percent of the watershed of Marijilda Creek at moderate to high severity, induced extreme flooding and debris flows (Schaffner and Reed 2005), demonstrating that high-elevation streams in east-central Arizona are vulnerable to high-severity wildfire given the current climate and stand conditions. However, management options may be limited because fuel reduction treatments are unlikely to mitigate fire hazards in this naturally dense forest type (Schoennagel et al. 2004).

The spruce zone on Mount Baldy, which harbors many of the larger populations of Apache trout, was the source of three of the remaining pure lineages of this fish (Ord, East Fork White River, and Smith). Watersheds in this zone have slopes in excess of 30 percent, relatively coarse-textured geologic formations, and soils with high organic-matter content (Long et al. 2003). If the relationships reported by Cannon et al. (2003) apply to this region, then post-wildfire debris flows could be more likely to occur in streams on Mount Baldy than in streams on the adjacent basaltic plateaus. On the other hand, several of the high-elevation streams, including Ord Creek, East Fork of the White River, and Big Bonito Creek, have multiple perennial tributaries that could provide refuge for trout.

Mixed Conifer Zone
(Douglas fir Dominated)

The KP and Steeple fires resulted in relatively large, mixed-severity burns within the Douglas fir zone following several years of drought conditions; mixed-severity fire regimes are typical in this zone (Rollins et al. 2000; Margolis 2003). Dieterich (1983) found that the mean fire return interval in this forest type in the White Mountains is 22 years, and stand-replacing fires appear to be uncommon. However, severe droughts combined with heavy fuel loading and insect outbreaks could trigger more severe fires in this forest type. Two such fires occurred in 1951—the Escudilla fire east of Mount Baldy (Savage and Mast 2005) and the McKnight fire in New Mexico (Rollins et al. 2000). The McKnight fire left a legacy of incised and unstable channels prone to scouring (Medina and Martin 1988), and harsh floods have repeatedly diminished the populations of Gila trout that were introduced into McKnight Creek after the fire (USFWS 2003). That outcome suggests that wildfires in this zone could have more severe consequences than were observed during this study.

Ponderosa Pine Zone

Fire-scar studies have suggested that wildfires in relatively dry ponderosa pine forests along the Mogollon Rim of Arizona and the Gila Mountains of New Mexico have historically been primarily low ground fires with return intervals of less than 10 years (Swetnam and Baisan 1996). More than two-thirds of Apache trout streams extend into the ponderosa pine zone. Meanwhile human alteration of fire regimes in those forest types may have increased the occurrence of high-severity crown wildfires (Brown et al. 2001; Savage and Mast 2005), which caused the fish extirpations reported from the Mogollon Rim. Not all wildfires in ponderosa pine forests in Arizona and New Mexico have caused such severe impacts, though; for example, Earl and Blinn (2004) reported that recent wildfires in ponderosa pine forest in New Mexico generated relatively short-lived impacts to streams. The risks of high-severity wildfire are lower when such stands are kept less dense (Rollins et al. 2000). Nevertheless, as is the case in other forest types, severe fires may also have occurred naturally during warmer, drier periods (Pierce et al. 2004). In the relatively moist ponderosa pine forests found at higher elevations, mixed-severity fire regimes naturally predominate and treatments to reduce the risk of high-severity wildfire are less likely to be effective (Schoennagel et al. 2004).

CONCLUSIONS

KP and Grant Creeks are similar in geology, topography, and vegetation to the majority of streams planned for recovery of Apache trout. These streams extend into mixed coni-fer forests where mixed-severity wildfires such as the KP and Steeple fires are typical. Fish extirpations have been reported from streams in drier, lower-elevation forest types where wildfires have been more severe, but long-term fire history studies suggest that high-severity wildfires do occur in high-elevation forest types during extended dry periods. Trout populations persisted following the mixed-severity wildfires in KP and Grant Creeks, indicating that evacuation of populations may be unnecessary when a watershed is not severely burned. While many factors can influence the likelihood of fish persistence, burn severity can be determined rapidly using satellite imagery. Consequently, until more confirmatory studies are conducted, that metric may help managers evaluate whether to evacuate Apache trout populations that are threatened by wildfires.

ACKNOWLEDGMENTS

The U.S. Forest Service provided funding for this research through the National Fire Plan. W. Wall with the Apache-Sitgreaves National Forest and M. Lopez and K. Meyer with the Arizona Game and Fish Department shared historical data on fish and fires in the region. I give special thanks to C. Carter, J. Leonard, S. Kelly, and J. Rinne for their assistance in collecting the field data for this study. I also thank three anonymous reviewers whose detailed comments substantially improved the manuscript.

LITERATURE CITED

Abrams, J. 2005. Report on a needs assessment for collaborative landscape planning in the White Mountains of Arizona. Ecological Restoration Institute, Northern Arizona University. Available online at http://www.forestera.nau.edu/docs/wm_needs_assessment_final.pdf.

Anderson, B. L. 1992. Fire and Gila trout recovery in wilderness watersheds. In Proceedings of the 36th Annual New Mexico Water Conference: Agencies and Science Working for the Future, pp. 123–128. WRRI Report No. 265. New Mexico Water Resources Research Institute, New Mexico State University, Las Cruces.

Bisson, P. A., B. E. Rieman, C. Luce, P. F. Hessburg, D. C. Lee, J. L. Kershner, G. H. Reeves, and R. E. Gresswell. 2003. Fire and aquatic ecosystems of the western USA: Current knowledge and key questions. Forest Ecology and Management 178: 213–229.

Brown, D. K., A. A. Echelle, D. L. Propst, J. E. Brooks, and W. L. Fisher. 2001. Catastrophic wildfire and number of populations as factors influencing risk of extinction for Gila trout (Oncorhynchus gilae). Western North American Naturalist 61: 139–148.

Burton, T. A. 2001. Effects of uncharacteristically large and intense wildfires on native fish: 14 years of observations on the Boise National Forest. Journal of the Idaho Academy of Science 37: 82–84.

Cannon, S. H. 2001. Debris-flow generation from recently burned watersheds. Environmental and Engineering Geoscience 7: 321–342.

Cannon, S. H., J. H. Gartner, M. G. Rupert, and J. A. Michael. 2003. Emergency assessment of debris-flow hazards from basins burned by the Piru, Simi, and Verdale fires of 2003, Southern California. U.S. Geological Survey. Available online at http://pubs.usgs.gov/of/2003/ofr-03-481/.

Cocke, A. E., P. Z. Fulé, and J. E. Crouse. 2005. Forest change on a steep mountain gradient after extended fire exclusion: San Francisco Peaks, Arizona, USA. Journal of Applied Ecology.

Dieterich, J. H. 1983. Fire history of southwestern mixed conifer: A case study. Forest Ecology and Management 6: 13–31.

Dowling, T. E., and M. R. Childs. 1992. Impact of hybridization on a threatened trout of the Southwestern United States. Conservation Biology 6: 355–364.

Earl, S. R., and D. W. Blinn. 2004. Effects of wildfire ash on water chemistry and biota in southwestern U.S.A. streams. Freshwater Biology 48.

Gomez, A. R., and V. E. Tiller. 1990. Fort Apache Forestry: A History of Timber Management and Forest Protection on the Fort Apache Indian Reservation, 1870–1985. Tiller Research, Albuquerque, New Mexico.

Gresswell, R. E. 1999. Fire and aquatic ecosystems in forested biomes of North America. Transactions of the American Fisheries Society 128: 193–221.

Koprowski, J. L., M. I. Alanen, and A. M. Lynch. 2005. Nowhere to run and nowhere to hide: Response of endemic Mt. Graham red squirrels to catastrophic forest damage. Biological Conservation 126: 491–498.

Long, J. W., and B. M. Burnette. 2004. Effects of wildfires on riparian restoration sites. In Proceedings of the 16th Annual Conference of the Society for Ecological Restoration, Victoria, BC.

Long, J. W., and D. Endfield. 2000. Restoration of White Springs. In Land Stewardship in the 21st Century: The Contributions of Watershed Management, technical coordinators P. F. Folliott, M. B. Baker, C. B. Edminster, B. Carleton, M. C. Dillon, and K. C. Mora, pp. 359–360. USDA Forest Service, Rocky Mountain Research Station P-13. Fort Collins, Colorado.

Long, J. W., and A. L. Medina. 2005. Geologic influences on Apache trout habitat. Journal of the Arizona-Nevada Academy of Science 38(2): 88–101.

Long, J. W., A. Tecle, and B. M. Burnette. 2003. Geologic influences on recovery of riparian wetlands on the White Mountain Apache Reservation. Journal of the Arizona-Nevada Academy of Science 35: 46–60.

Long, J. W., B. M. Burnette, and C. S. Lupe. 2005. Fire and springs: Reestablishing the balance on the White Mountain Apache Reservation. In The Colorado Plateau II: Biophysical, Socioeconomic, and Cultural Research, edited by C. van Riper and D. J. Mattson, pp. 381–396. University of Arizona Press, Tucson.

Lynch, A. M. 2004. Fate and characteristics of Picea damaged by Elatobium abietinum (Walker) (Homkoptera: Aphididae) in the White Mountains of Arizona. Western North American Naturalist 64: 7–17.

Margolis, E. Q. 2003. Stand-replacing fire history and aspen ecology in the Upper Rio Grande Basin. Master's thesis, University of Arizona, Tucson.

Medina, A. L., and S. C. Martin. 1988. Stream channel and vegetation changes in sections of McKnight Creek, New Mexico. Great Basin Naturalist 48: 373–381.

Miller, R. R. 1950. Notes on the cutthroat and rainbow trouts with the description of a new species from the Gila River, New Mexico. Occasional Papers of the Museum of Zoology 529: 1–43. University of Michigan, Ann Arbor.

Neary, D. G., K. C. Ryan, and L. F. DeBano. 2005. Wildland fire in ecosystems: Effects of fire on soils and water. General Technical Report RMRS-GTR-42 (vol. 4). USDA Forest Service, Ogden, Utah.

Pierce, J. L., G. A. Meyer, and A. J. T. Jull. 2004. Fire-induced erosion and millennial scale climate change in northern ponderosa pine forests. Nature 432: 87–90.

Propst, D. L., J. A. Stefferud, and P. R. Turner. 1992. Conservation and status of Gila trout, Oncorhynchus gilae. The Southwestern Naturalist 37: 117–125.

Reynolds, S. J. 1988. Arizona geologic map. Map 26, scale 1:1000000. Arizona Geological Survey, Tucson.

Riddle, B. R., D. L. Propst, and T. L. Yates. 1998. Mitochondrial DNA variation in Gila trout, Oncorhynchus gilae: Implications for management of an endangered species. Copeia 1998: 31–39.

Rieman, B. E., D. C. Lee, D. Burns, R. E. Gresswell, M. K. Young, R. Stowell, J. N. Rinne, and P. Howell. 2003. Status of native fishes in the Western United States and issues for fire and fuels management. Forest Ecology and Management 178: 197–211.

Rinne, J. N. 1985. Variation in Apache trout populations in the White Mountains, Arizona. North American Journal of Fisheries Management 5: 146–158.

Rinne, J. N. 1996. Short-term effects of wildfire on fishes and aquatic macroinvertebrates in the southwestern United States. North American Journal of Fisheries Management 16: 653–658.

Rinne, J. N. 2004. Forests, fish and fire: Relationships and management implications for fishes in the southwestern USA. In Proceedings of the Forest Land-Fish Conference II—Ecosystem Stewardship through Collaboration, edited by G. J. Scrimgeour, G. Eisler, B. McCulloch, U. Silins, and M. Monita, pp. 151–156.

Rollins, M., T. W. Swetnam, and P. Morgan. 2000. Twentieth-century fire patterns in the Selway-Bitterroot Wilderness Area, Idaho/Montana, and the Gila/Aldo Leopold Wilderness Complex, New Mexico. In Wilderness Science in a Time of Change Conference, Volume 5: Wilderness Ecosystems, Threats, and Management, edited by D. N. Cole, S. F. McCool, W. T. Borrie, and J. O'Loughlin, pp. 283–287. RMRS-P-15 (Vol. 5). USDA Forest Service, Ogden, Utah.

Ruiz, L. D., and J. R. Novy. 2000. Recovery status of Apache trout Oncorhynchus apache. In Proceedings of the Wild Trout VII Symposium: Management in the New Millennium, Are We Ready? Edited by D. Schill, S. Moore, P. Byorth, and P. Hamre, pp. 155–160. Montana State University, Bozeman.

Savage, M., and J. N. Mast. 2005. How resilient are southwestern ponderosa pine forests after crown fires? Canadian Journal of Forest Research 35: 967–977.

Schaffner, M., and W. B. Reed. 2005. Effects of wildfire in the mountainous terrain of southeast Arizona: Post-burn hydrologic response of nine watersheds. Technical Attachment 05-01. NOAA National Weather Service, Weather Forecast Office, Tucson, Arizona.

Schoennagel, T., T. T. Veblen, and W. H. Romme. 2004. The interaction of fire, fuels, and climate across Rocky Mountain Forests. BioScience 54: 661–676.

Svec, J. R., R. K. Kolka, and J. W. Stringer. 2005. Defining perennial, intermittent, and ephemeral channels in Eastern Kentucky: Application to forestry best management practices. Forest Ecology and Management 214: 170–182.

Swetnam, T. W., and C. H. Baisan. 1996. Historical fire regime patterns in the Southwestern United States since AD 1700. In Fire Effects in Southwestern Forests: The Second La Mesa Fire Symposium, edited by C. D. Allen, pp. 11–32. USDA, Rocky Mountain Forest and Range Experiment Station, Los Alamos, New Mexico.

Swetnam, T. W., C. H. Baisan, and J. M. Kaib. 2001. Forest fire histories of La Frontera: Fire-scar reconstructions of fire regimes in the United States/Mexico Borderlands. In Vegetation and Flora of La Frontera: Historic Vegetation Change along the United States/Mexico Boundary, edited by G. L. Webster and C. J. Bahre, pp. 95–119. University of New Mexico Press, Albuquerque.

Tobin, M. 2004. "Doomed" chub survive. Arizona Daily Star (15 July).

USDA. 2004. Coronado National Forest Management Indicator Species Population Status and Trends. USDA Forest Service, Southwestern Region, Coronado National Forest. Available at http://www.fs.fed.us/r3/projects/2004-cor-misreport.pdf.

USFWS. 2003. Gila trout recovery plan, 3rd rev. U.S. Fish and Wildlife Service Region 2, Albuquerque, New Mexico.

USFWS. 2006. Endangered and threatened wildlife and plants; Reclassification of the gila trout (Oncorhynchus gilae) from endangered to threatened; special rule for Gila trout in New Mexico and Arizona. Federal Register 71: 40657–40674.

Van Deventer, J. S., and W. S. Platts. 1989. Microcomputer software system for generating population statistics from electrofishing data—Users guide for Microfish 3.0. General Technical Report INT-254. U.S. Forest Service, Boise, Idaho.

Veblen, T. T., K. S. Hadley, and M. S. Reid. 1991. Disturbance and stand development of a Colorado subalpine forest. Journal of Biogeography 18: 707–716.

Zippin, C. 1958. The removal method of population estimation. Journal of Wildlife Management 22: 82–90.

CAVE-DWELLING INVERTEBRATES OF GRAND CANYON NATIONAL PARK

J. Judson Wynne, Charles A. Drost, Neil S. Cobb, and John R. Rihs

Cave ecosystems are among the most fragile ecosystems on Earth (Elliott 2000; Hamilton-Smith and Eberhard 2000) due, in part, to the sensitivity of cave-dwelling organisms to disturbance. Because many troglomorphic taxa (obligate cave-dwelling organisms) are endemic to a single cave or region (Reddell 1994; Culver et al. 2000; Christman et al. 2005), and are generally characterized by low population numbers (Mitchell 1970; Krajick 2001), many populations are considered imperiled (Reddell 1994; Culver et al. 2000). Most studies of cave invertebrates have been simple inventories, with relatively little data collected on species and community ecology.

For this study we have synthesized all known information on cave-dwelling invertebrates in Grand Canyon National Park (GRCA). There is a paucity of knowledge about caves in GRCA, as well as other areas on the southern Colorado Plateau. The available information is limited to a few intensive studies (where invertebrates were collected and identified) and cave trip reports (where invertebrates were documented visually). Here we determine the extent of knowledge concerning cave-dwelling invertebrate fauna, identify the seemingly most common cave-dwelling invertebrates, and present our preliminary understanding of invertebrate diversity and endemism in Grand Canyon caves.

METHODS

We conducted a literature review that includes all published literature, obtained primarily through Northern Arizona University's Cline Library and Internet searches, and unpublished literature and cave trip reports on file at GRCA Museum Collections. We did not include reports supported by little or no documentation or those in which invertebrate observations are not well described (i.e., above the family taxonomic level).

We divided Grand Canyon cave invertebrates into five cavernicole (cave-dwelling organism) groups and one special case category: (1) Troglobites, which are obligatory terrestrial cave-adapted species occurring only in caves or similar subterranean habitats, (2) troglophiles, which are species occurring facultatively within caves and completing their life cycles there, but probably also occurring in surface environments, (3) trogloxenes, which are taxa that live in caves for shelter and potentially favorable microclimate but that return to epigean habitats to forage (refer to Barr 1968), (4) stygobites, which are obligatory aquatic cave-adapted organisms (refer to Culver and White 2005), (5) unknown cavernicoles, which are organisms not categorized due to a lack of information, and (6) special cases, which include organisms brought into the cave by vertebrate species. Additionally, because organisms known to feed in guano deposits are of interest to cave ecologists, we included the subgroup guanophiles. Troglobites, troglophiles, and trogloxenes are the groups generally known to contain guanophiles. Current taxonomy was verified for most invertebrates using the Integrated Taxonomic Information System (http://www.itis.gov) and Triplehorn and Johnson (2005).

RESULTS

We reviewed the results of nine studies (conducted between 1975 and 2001) representing surveys of 15 GRCA caves (Table 1). These studies identified approximately 37 cave-dwelling invertebrates that are known to occur in Grand Canyon caves: 3 troglobites, 6 trogloxenes, 14 troglophiles, 1 stygobite, 10 unknown cavernicoles, and 3 species representing special cases. Voucher specimens were used to identify 30 (~81%) of theses invertebrates. Of the unknown cavernicoles, Peck (1978) has suggested that one of these might be troglomorphic. Two species are tentatively considered guanophiles—one Collembola (*Tomocerus* sp.) and one troglobitic Leiodid beetle (*Ptomaphagus cocytus*).

In four of the nine studies that we reviewed—Welbourn (1978), Peck (1980), Roth (1982), and Drost and Blinn (1997)—the researchers collected specimens, whereas Triplehorn (1975) analyzed and described specimens collected by another researcher. Welbourn (1978) conducted a regional study of Horseshoe Mesa caves, where he identified 12 invertebrates from eight caves. Peck (1980) identified 15 species (including three previously undescribed troglobitic species) from three stream caves (Roaring Springs, Tapeats, and Thunder River Caves). Roth (1982) identified one pholcid spider (*Psilochorus* sp.) from one cave. Drost and Blinn (1997) identified 19 species from Roaring Springs Cave, including one previously undescribed stygobitic amphipod. Based on specimens from three caves within Grand Canyon, Triplehorn (1975) described a new tenebrionid beetle species, *Eleodes leptoselis*. Specimens were collected from 13 of the 15 caves considered for this review.

Taxonomically, the most diverse groups in the known Grand Canyon cave invertebrate fauna are spiders (order Araneae, at least eight species), beetles (order Coleoptera, four species, including one special case), and flies (order Diptera, at least two species; Tables 2 and 3). Other groups of particular interest due to their ecological and evolutionary relationship to cave environments include amphipods (one endemic deep-cave species, *Stygobromus blinni*), harvestmen (order Opiliones), mites (order Acari, one potentially cave-adapted species), springtails (order Collembola, two species, at least one of which is cave-adapted), and diplurans (order Diplura, with at least one potentially cave-adapted species).

Spiders and beetles are the most widespread invertebrates. There were 14 occurrences of spiders in 12 of 15 caves sampled, and 12 occurrences of beetles in 10 of 15 caves (Table 2). Cave crickets (order Orthoptera, family Raphidophoridae) were also widespread, occurring in 10 of 15 caves. The two most widespread species were the cave cricket *Ceuthophilus yavapai* and the tenebrionid beetle *Eleodes leptoselis*, occurring in eight and six caves, respectively.

Specimens that we examined from Grand Canyon caves also yielded five undescribed species, four potentially undescribed species, and two new state records. The undescribed and newly described species consist of one stygobite (*Stygobromus blinni* Wang and Holsinger 2001; Figure 1) and three troglobites (*Telema* sp. nov., *Tomocerus* sp. nov., and *Ptomaphagus cocytus* nov.) from Roaring Springs and Tapeats Caves (Peck 1980; Drost and Blinn 1997), and one troglophile (*Eleodes* [*Caverneleodes*] *leptoselis* sp. nov.) from Cave of the Domes (Triplehorn 1975; Welbourn 1978; Hill et al. 1998), Cave 68-Olie (Triplehorn 1975), Tse An Cho Cave (Triplehorn 1975; Welbourn 1978), and potentially from Land's End Cave, Scorpion Cave, and Tuning Fork Cave (Welbourn 1978). All of these species are believed to be endemic to the Grand Canyon. Four species have not been identified and may represent new and endemic species; these comprise a mite (*Rhagidia* cf. *hilli*), a dipluran (*Haplocampa* sp.), a collembolan (*Entomobrya* cf. *californica*), and a tineid moth (*Tinea* nov. sp.). Two caddisfly species (*Micrasema onisca* and *Lepidostoma apornum*) documented in Roaring Springs Cave establish new records in Arizona (Drost and Blinn 1997). Also, *Telema* sp. nov. is the only known troglobitic spider in Arizona (Peck 1980).

Table 1. Invertebrate inventories by cave and research effort in Grand Canyon National Park.

Study	Drost & Blinn (1997)[1]	Hill et al. (1998)	Hill et al. (2000)	Hill et al. (2001)	Pape (1998)	Peck (1980)[1]	Roth (1982)[1]	Triplehorn (1975)[1,2]	Welbourn (1976)[1]
Babylon Cave	–	X	–	–	–	–	–	–	X
Bat Cave	–	–	X	–	X	–	–	–	–
Cave 68-Olie	–	–	–	–	–	–	–	X	–
Cave of the Domes	–	X	–	–	–	–	–	X	X
Christmas Tree Cave	–	–	–	–	–	–	X	–	–
Crystal Forest Cave	–	X	–	–	–	–	–	–	X
Land's End Cave	–	–	–	–	–	–	–	–	X
Middle Cave	–	–	–	–	–	–	–	–	X
Rampart Cave	–	–	–	–	–	–	–	–	–
Roaring Springs Cave	X	–	–	–	–	X	–	–	–
Scorpion Cave	–	–	–	–	–	–	–	–	X
Tapeats Cave	–	–	–	–	–	X	–	–	–
Thunder River Cave	–	–	–	–	–	X	–	–	–
Tse An Cho Cave	–	–	–	–	–	–	–	X	X
Tse'an Kaetan Cave	–	–	–	X	–	–	–	–	–
Tuning Fork Cave	–	–	–	–	–	–	–	–	X

[1] Studies in which researchers collected voucher specimens.
[2] Fieldwork was not conducted during this effort; rather, a new species was described.

Table 2. Invertebrates recorded from selected caves in Grand Canyon National Park. When invertebrates were not identified to species level, the lowest taxonomic level noted in the original source is listed.

	Babylon Cave	Bat Cave	Cave 68-Olie	Cave of the Domes	Christmas Tree Cave	Crystal Forest Cave	Land's End Cave	Middle Cave
Stygobromus blinni	—	—	—	—	—	—	—	—
Loxosceles sp.	—	X	—	—	—	—	—	—
Telema sp.	—	—	—	—	—	—	—	—
Kibramoa suprenans	—	X	—	—	—	—	—	—
Psilochorus sp.	—	X	—	—	X	—	—	—
Achaearanea canionis	—	—	—	—	—	—	—	—
Lepthyphantes sp.	—	—	—	—	—	—	—	—
Selenops sp.	—	X	—	—	—	X	—	—
Leiobunum cf. *townsendii*	—	—	—	X	—	X	—	—
Neoallochernes stercoreus	—	X	—	—	—	—	—	X
Family Anystidae	—	X	—	—	—	X	—	—
Chiroptonyssus robustipes	—	X	—	—	—	—	—	—
Rhagidia cf. *hilli*	—	—	—	X	—	—	—	—
Entomobrya californica?	—	—	—	—	—	—	—	—
Tomocerus sp.	—	—	—	—	—	—	—	—
Haplocampa sp.	X	—	—	—	—	X	—	—
Ceuthophilus yavapai	X	—	—	X	—	X	X	X
Hesperoperla pacifica	—	—	—	—	—	—	—	—
Psyllipsocus ramburii	X	—	—	X	—	—	—	—
Psyllipsocus sp.	X	X	—	X	—	—	—	—
Bembidion sp.	—	—	—	—	—	—	—	—
Ptomaphagus cocytus	—	—	—	—	—	—	—	—
Eschatomoxys sp.	—	X	—	—	—	—	X	—
Eleodes (*Caverneleodes*) *leptoselis*	—	—	—	X	—	—	X	—
Smicronyx imbricatus	—	X	X	—	—	—	—	—
Sceliphron sp. or *Chalybion* sp.	—	X	—	—	—	—	—	—
Micrasema onisca	—	—	—	—	—	—	—	—
Lepidostoma apornum	—	—	—	—	—	—	—	—
Tinea n. sp.	—	X	—	—	—	X	—	—
Pronoctua typica	—	—	—	—	—	—	—	—
Family Arctiidae?	—	—	—	—	—	—	—	—
Sternopsylla texana	—	X	—	—	—	—	—	—
Family Tipulidae	—	—	—	X	—	—	—	—
Limonia sp.	—	—	—	—	—	—	—	—
Tipula (*Beilardina*) *rupicola*	—	—	—	—	—	—	—	—
Mycetophila sp.	—	—	—	—	—	—	—	—
Sciara sp.	—	—	—	—	—	—	—	—

Table 2 (continued)

	Roaring Springs Cave	Scorpion Cave	Tapeats Cave	Thunder River Cave	Tse An Cho Cave	Tse'an Kaetan Cave	Tuning Fork Cave
Stygobromus blinni	X	–	–	–	–	–	–
Loxosceles sp.	X	–	–	X	–	–	–
Telema sp.	X	–	–	–	–	–	–
Kibramoa suprenans	–	–	–	–	–	–	–
Psilochorus sp.	X	–	–	–	–	–	–
Achaearanea canionis	X	–	X	–	–	–	–
Lepthyphantes sp.	–	–	–	–	–	–	–
Selenops sp.	–	–	–	–	–	–	–
Leiobunum cf. *townsendii*	X	–	–	–	–	–	X
Neoallochernes stercoreus	–	–	–	–	–	–	–
Family Anystidae	–	–	–	–	–	–	X
Chiroptonyssus robustipes	–	–	–	–	–	–	–
Rhagidia cf. *hilli*	X	–	–	–	–	–	–
Entomobrya californica?	–	–	–	–	–	–	X
Tomocerus sp.	X	–	–	–	–	–	–
Haplocampa sp.	X	–	X	–	–	–	–
Ceuthophilus yavapai	X	X	–	–	X	X	X
Hesperoperla pacifica	–	–	–	–	–	–	–
Psyllipsocus ramburii	–	–	X	–	X	–	X
Psyllipsocus sp.	–	–	–	–	–	–	–
Bembidion sp.	X	–	–	–	–	–	–
Ptomaphagus cocytus	X	–	X	–	–	–	–
Eschatomoxys sp.	–	–	–	X	–	–	–
Eleodes (Caverneleodes) leptoselis	–	X	–	–	X	–	X
Smicronyx imbricatus	–	–	–	–	–	–	–
Sceliphron sp. or *Chalybion* sp.	X	–	–	–	–	–	–
Micrasema onisca	X	–	–	–	–	–	–
Lepidostoma apornum	X	–	–	–	–	–	–
Tinea n. sp.	X	–	–	–	–	–	–
Pronoctua typica	–	–	–	–	–	X	–
Family Arctiidae?	–	–	–	–	–	–	–
Sternopsylla texana	X	–	–	–	–	–	–
Family Tipulidae	X	–	–	–	–	–	–
Limonia sp.	X	–	–	–	–	–	–
Tipula (Bellardina) rupicola	X	–	–	–	–	–	–
Mycetophila sp.	–	–	–	–	–	–	–
Sciara sp.	–	–	X	–	–	–	–

Table 3. Annotated list of cave invertebrates in Grand Canyon National Park.

Class Crustacea
    Order Amphipoda
        Family Crangonyctidae
Stygobromus blinni Wang and Holsinger 2001. Stygobite. Roaring Springs Cave (Drost and Blinn 1997). This cave-adapted species was found in the middle and deep zone of the cave. *Stygobromus blinni* is currently known only from Roaring Springs Cave and its associated waters.

Class Arachnida
    Order Araneae
Welbourn (1978) noted unidentified spiders in Babylon Cave, Cave of the Domes, Crystal Forest Cave, Middle Cave, Scorpion Cave, Tse An Cho Cave, and Tuning Fork Cave. Welbourn (1978) collected several specimens along the walls and ceilings, which may represent at least two species. These spiders were collected within the twilight and near the edge of the dark zone (Welbourn 1978). Similarly, Hill et al. (1998) identified a medium-sized, tan, velvety spider near a wet area near the back of Crystal Forest Cave. These unidentified spiders may or may not be among the species listed below.

        Family Loxoscelidae
Loxosceles sp. Undetermined species. Troglophile (Peck 1980). Bat Cave (Pape 1998; Hill et al. 2001), Roaring Springs Cave (Drost and Blinn 1997), and Thunder River Cave (Peck 1980). One individual was collected within the dark zone of Bat Cave (Pape 1998). Hill et al. (2001) also identified a *Loxosceles* sp. during their Bat Cave survey. A *Loxosceles* sp. was also observed within the twilight zone of Roaring Springs Cave (Drost and Blinn 1997). In Thunder River Cave, Peck (1980) identified a *Loxosceles* sp. in webs along the base of the walls near guano deposits.

        Family Telemidae
Telema sp. Undescribed troglobite (Peck 1980). Roaring Springs Cave (Peck 1980). Peck suggested that this is the only known cave-adapted spider in Arizona.

        Family Plectreuridae
Kibramoa suprenans Chamberlin 1920. Troglophile (Pape 1998). Bat Cave (Pape 1998). Pape observed webs of this species along the base of cave walls, in rock piles, and along walkway and machinery from abandoned guano mining operations.

        Family Pholcidae
Psilochorus sp. Troglophile (Pape 1998). Christmas Tree Cave (Roth 1982) and Bat Cave (Pape 1998). Pape suggested that a small population exists within Bat Cave.

        Family Theridiidae
Achaearanea canionis Chamberlin and Gertsch. Troglophile/trogloxene (Peck 1980; unknown cavernicole). Tapeats Cave (Peck 1980) and Roaring Springs Cave (Peck 1980; Drost and Blinn 1997). This species was observed within the twilight zone of Roaring Springs Cave (Drost and Blinn 1997). It is known to occur in caves, as well as more mesic epigean environments in Arizona, California, New Mexico, and Utah, and "other species within this genus are frequently found near cave entrances" (Peck 1980). We assume that Peck (1980) and Drost and Blinn (1997) observed the same species.

        Family Linyphiidae
Lepthyphantes sp. Unknown cavernicole. Roaring Springs Cave (Drost and Blinn 1997). This species was observed within the twilight zone of the cave.

        Family Selenopidae
Selenops sp. Troglophile (Pape 1998). Bat Cave (Pape 1998) and Crystal Forest Cave (Hill et al. 1998). Several molts of this species were observed within the cave twilight zone (Pape 1998). Pape suggested that their occurrence was probably deep enough within the cave to prey upon guanophiles.

        Family Thomisidae
Unidentified genus and species. Unknown cavernicole. Hill et al. (1998) tentatively identified a spider photographed in Crystal Forest Cave as a crab spider, either a trogloxene or troglophile.

    Order Opiliones
        Family Sclerosomatidae
Leiobunum cf. *townsendii* Weed. Troglobite/troglophile (Peck 1980; troglophile). Roaring Springs Cave (Peck 1980; Drost and Blinn 1997), Cave of the Domes, Tuning Fork Cave (Welbourn 1978), and Crystal Forest Cave (Welbourn 1978; Hill et al. 1998). Drost and Blinn (1997) observed *Leiobunum* sp. within the twilight zone of Roaring Springs Cave. This is likely the same species identified by Peck (1980). Welbourn (1978) suggested that harvestmen are not common in large numbers in Horseshoe Mesa caves.

In Crystal Forest Cave, Hill et al. (1998) identified piles of legs, which they incorrectly identified as "Phalangida." Species of *Leiobunum* are common in the U.S. Southwest, and members of this group are frequently encountered in and around caves (Peck 1980). This species is widely known to den in the light to twilight zones of caves in the Southwest (W. Shear, personal communication; Wynne, personal observation). Thus, we consider this a troglophilic species.

Order Pseudoscorpiones
Family Chernetidae
Neoallochernes stercoreus Turk 1949. Unknown cavernicole. Bat Cave (Pape 1998). Pape suggested that this species is often associated with active Mexican free-tailed bat (*Tadarida brasiliensis*) guano deposits.

Order Acari
Family Anystidae
Undetermined genus and species. Trogloxene? (Welbourn 1978; unknown cavernicole). Crystal Forest Cave (Welbourn 1978; Hill et al. 1998), Middle Cave, and Tuning Fork Cave (Welbourn 1978). Anystids were observed under rocks near the edge of the twilight zone (Welbourn 1978). In Crystal Forest Cave, Hill et al. (1998) observed a small gray mite "running" around the floor "in a wet area near a speleothem drip" and suggested that it was the same anystid observed by Welbourn (1978). Predacious mites of this family are known to occur within the twilight zones of other Arizona and New Mexico caves (Welbourn 1978). Welbourn indicated that these mites have been observed in large numbers beneath rocks within Wupatki National Monument earth cracks.

Family Macronyssidae
Chiroptonyssus robustipes Ewing. Special case: ectoparasite. Bat Cave (Pape 1998). This species is an obligate parasite on the Mexican free-tailed bat (*Tadarida brasiliensis*).

Family Rhagidiidae
Rhagidia cf. *hilli* Strandtmann 1971. Troglobite/troglophile (Peck 1980; unknown cavernicole). Roaring Springs Cave (Peck 1980; Drost and Blinn 1997). Peck indicated that this genus occurs in other caves in the western United States and Mexico.

Class Collembola
Order Entomobryomorpha
Family Entomobryidae
Entomobrya cf. *californica* Schott 1891. Troglophile (Welbourn 1978). Cave of the Domes and Tuning Fork Cave (Welbourn 1978). This species was restricted to the moist wet areas of Horseshoe Mesa caves (Welbourn 1978). K. Christiansen suggested that the Collembolan observed in Horseshoe Mesa caves may be different from other *E. californica* and that further study was needed (Welbourn 1978).

Tomocerus sp. Undescribed troglobite (Peck 1980). Guanophile? Roaring Springs Cave (Peck 1980; Drost and Blinn 1997). Peck (1980) indicated that several individuals were observed within "scattered mouse (bat?) droppings on moist sand" within the cave's dark zone. Drost and Blinn (1997) collected this species within the cave's twilight zone. Four species within this genus are known from caves—three from the United States (Kentucky, Illinois, Missouri, and New Mexico) and one from Japan (Peck 1980).

Order Diplura
Family Campodeidae
Haplocampa sp. Troglobite/troglophile (Peck 1980; unknown cavernicole). Tapeats Cave (Peck 1980), and probably Roaring Springs Cave (Drost and Blinn 1997). In Tapeats Cave, three specimens were found on scat on "silt along a wall in the main trunk passage" (Peck 1980). Peck indicated that various genera of this family are commonly observed in caves. Drost and Blinn (1997) observed diplurans within the dark zone of Roaring Springs; two individuals were found dead and in poor condition within a small pool, and were not identifiable beyond family.

Class Insecta
Order Orthoptera
Family Rhaphidophoridae
Ceuthophilus yavapai Hubbell 1936. Trogloxene (Welbourn 1978; Peck 1980). Cave of the Domes (Welbourn 1978; Hill et al. 1998), Babylon Cave, Crystal Forest Cave, Land's End Cave, Middle Cave, Scorpion Cave, Tse An Cho Cave, Tuning Fork Cave (Welbourn 1978), Roaring Springs Cave (Peck 1980; Drost and Blinn 1997), and Tse'an Kaetan Cave (Hill et al. 2000). Cave crickets were observed on the walls and ceilings near the entrances and near the back areas of Horseshoe Mesa caves (Welbourn 1978). Welbourn (1978) suggested that these crickets leave the caves at night to feed, and observed crickets on the walls outside the entrance of Cave of the Domes. Hill et al. (1998) suggested that their cricket observations in Cave of the Domes are *C. yavapai*. Drost and Blinn (1997) documented a *Ceuthophilus* sp. at the entrance of Roaring Springs.

Order Plecoptera
Family Perlidae

Hesperoperla pacifica Banks 1900. Troglophile. Roaring Springs Cave (Drost and Blinn 1997). Numerous aquatic nymphs and some flying adults were observed from near the entrance through the dark zone.

Order Psocoptera
Family Psocidae

Psyllipsocus ramburii Selys-Longchamps 1872. Troglophile (Peck 1980). Tapeats Cave (Peck 1980), Bat Cave (Pape 1998), Babylon Cave, Cave of the Domes, Tse An Cho Cave, and Tuning Fork Cave (Welbourn 1976). Peck (1980) collected one specimen from a "decayed leaf on moist rock floor in dark zone," but indicated that this species was frequently observed within dry detritus. Pape (1998) suggested that this invertebrate feeds on micro-fungi occurring on guano deposits; Pape also suggested that the psocids observed within Bat Cave were likely the same species collected and identified by Peck (1980) at Tapeats Cave. Welbourn (1978) identified psocids (likely *P. ramburii*) in organic material and guano within the dark zones of these Horseshoe Mesa caves.

Order Coleoptera
Family Carabidae

Bembidion sp. Unknown cavernicole. Roaring Springs Cave (Drost and Blinn 1997). Drost and Blinn observed this species within the cave entrance.

Family Leiodidae

Ptomaphagus cocytus Peck 1973. Troglobite (Peck 1980). Guanophile? Roaring Springs Cave (Peck 1980; Drost and Blinn 1997) and Tapeats Cave (Peck 1980). In Roaring Springs Cave, adults and larvae were observed "feeding on a few scattered mouse (or bat?) droppings on moist sand," while in Tapeats Cave they were on "droppings and decaying cottonwood leaves on moist floor" in the cave's dark zone (Peck 1980).

Family Tenebrionidae

Eschatomoxys sp. Troglophile (Peck 1980). Thunder River Cave (Peck 1980) and Bat Cave (Pape 1998). Peck (1980) collected one specimen within a dry passage of Thunder River Cave. Pape (1998) suggested that the tenebrionid beetle observed within Bat Cave is likely the same species collected and identified from Thunder River Cave.

Eleodes (*Caverneleodes*) *leptoselis* Triplehorn 1975. Troglophile? (Welbourn 1978; unknown cavernicole). Cave of the Domes (Triplehorn 1975; Welbourn 1978; Hill et al. 1998), Cave 68-Olie (Triplehorn 1975), Tse An Cho Cave (Triplehorn 1975; Welbourn 1978), and Land's End Cave, Scorpion Cave, and Tuning Fork Cave (Welbourn 1978). Welbourn observed these beetles on the cave floor and walls within the dark zone, and indicated that they were commonly found in Cave of the Domes. Hill et al. (1998) observed three dead tenebrionid beetles within the entrance area and near the cave register at Cave of the Domes, and suggests that this was likely *E. leptoselis*.

Family Curculionidae

Smicronyx imbricatus Casey 1892. Special case. Bat Cave (Pape 1998). Pape suggested that larvae of this species were brought into the cave and deposited in the seeds contained within *Bassariscus astutus* (ringtail cat) scat.

Order Hymenoptera
Family Sphecidae

Sceliphron sp. or *Chalybion* sp. Trogloxene (Pape 1998). Bat Cave (Pape 1998). Pape indicated that "the only use of the cave by this species is for shelter for nest building and its relationship is that of a trogloxene."

Order Trichoptera
Family Brachycentridae

Micrasema onisca Ross 1947. Troglophile. Roaring Springs Cave (Drost and Blinn 1997). Drost and Blinn recorded an adult within the twilight zone of the cave. *M. onisca* had not previously been recorded in Arizona.

Family Lepidostomatidae

Lepidostoma apornum Denning 1949. Troglophile. Roaring Springs Cave (Drost and Blinn 1997). Drost and Blinn collected an adult of this species within the twilight zone of the cave, which was a first record in Arizona.

Order Lepidoptera
Family Tineidae

Tinea nov. sp. Troglophile (Pape 1998). Bat Cave (Pape 1998) and Crystal Forest Cave (Welbourn 1978). At least two species of tineids were identified within Bat Cave, with one representing a new species (Pape 1998). During the larval stage, they feed on fresh guano (Pape 1998). An unidentified tineid was observed on a wall near the entrance of Crystal Forest Cave (Welbourn 1978).

Family Noctuidae

Pronoctua typica Smith 1894. Trogloxene (Peck 1980). Roaring Springs Cave (Peck 1980; Drost and Blinn 1997). Peck (1980) noted adult moths "on the ceiling and walls ... just inside the dark zone." This is a wide-ranging species in the West occurring from British Columbia to California and Arizona (Peck 1980). Peck indicated that other members belonging to the subfamily Noctuinae are known to hibernate or aestivate in caves.

Family Arctiidae?

Undetermined genus and species. Troglophile. Tse'an Kaetan Cave (Hill et al. 2000). Hill et al. documented casings of pupal moths.

Order Siphonaptera
Family Ichnopsyllidae

Sternopsylla texana Fox 1914. Special case: ectoparasite. Bat Cave (Pape 1998). This species is believed to be present while Mexican free-tailed bats (*Tadarida brasiliensis*) are in residence (Pape 1998).

Order Diptera
Family Tipulidae

Undetermined genus and species. Trogloxene (Welbourn 1978). Cave of the Domes (Welbourn 1978). One specimen was collected from the cave entrance room wall.

Limonia sp. Unknown cavernicole. Roaring Springs Cave (Drost and Blinn 1997). This species was identified within the twilight zone.

Tipula (*Bellardina*) *rupicola* Doane 1912. Trogloxene (Peck 1980). Roaring Springs Cave (Peck 1980; Drost and Blinn 1997). This species was noted in the twilight zone of the cave (Peck 1980; Drost and Blinn 1997). Peck (1980) suggested that this species used the cave as a "daytime retreat."

Family Mycetophilidae

Mycetophila sp. Trogloxene (Peck 1980). Roaring Springs Cave (Peck 1980; Drost and Blinn 1997). This species was identified within the twilight zone (Drost and Blinn 1997), and also in the dark zone (Peck 1980). *Mycetophila* occurs widely in western North America, and has been recorded in other caves (Peck 1980).

Family Sciaridae

Sciara sp. Undetermined species. Troglophile (Peck 1980). Tapeats Cave (Peck 1980). Peck identified one adult on rotting leaves in the cave's dark zone. Fungus gnats of this genus are often found in association with wet caves (Peck 1980).

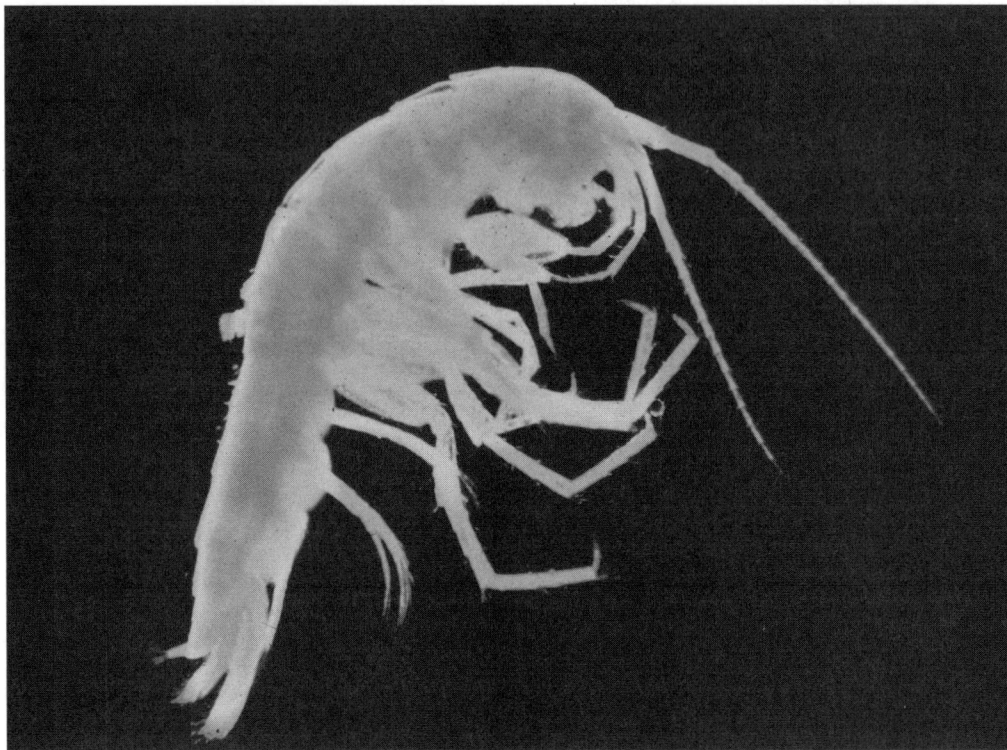

Figure 1. *Stygobromus blinni* Wang and Holsinger 2001. This endemic stygobite was found in the middle and deep zone of Roaring Springs Cave. It is currently known only from Roaring Springs Cave and its associated waters. Photo courtesy D. Blinn.

DISCUSSION

This review underscores the need for additional research on cavernicole invertebrates in the Grand Canyon. Of the 15 caves reported, endemic troglomorphic species have been confirmed from only two caves—Roaring Springs and Tapeats. Additionally, four species may prove to be endemic to Grand Canyon caves, but this awaits further study.

Peck (1978) has suggested that the low numbers of cave-adapted taxa in the U.S. Southwest are due to low nutrient input and the high cryptoaridity associated with Southwest cave systems, but it is evident that the low numbers from Grand Canyon caves also reflect limited sampling efforts.

Although voucher specimens were used to identify 30 of 37 (~81%) invertebrates presented in this review, most research undertaken in Grand Canyon caves has employed visual observation to identify invertebrates. As indicated by Barr and Reddell (1967), cave inventories cannot be considered complete without invertebrate trapping. Grand Canyon species lists for each cave should therefore be considered incomplete. Species identifications made from visual surveys are open to question due to the difficulty in identifying most invertebrate species; they do not provide a strong basis for comparative studies, and are of limited value. We also recognize that taxonomic expertise among the authors likely differs, so it is possible that some species have been previously recorded under multiple names, and some may have been incorrectly identified.

To best further the knowledge of Grand Canyon cave entomology and allow for comparative analyses, we propose the use of a standardized systematic approach. We propose an inventory protocol for sampling invertebrates that employs both a random sampling strategy with specimen collection and a suite of the most effective currently known techniques. Sampling field sessions should last at least 14 days and should take place during at least two seasons (e.g., 7 days during summer and 7 during winter), and representative sampling should be done within the three light zones (light, twilight, and dark). Because baited traps are generally more effective than unbaited traps, and result in higher capture rates (Slaney and Weinstein 1996), we recommend using baited pitfall and leaf litter trapping techniques. Direct intuitive searches should also be conducted because many predacious and winged invertebrates are not likely to be captured using baited trapping techniques. Searches should occur in areas deemed most likely to contain certain invertebrate species. To reduce the risk of overcollecting, only three to five individuals per species should be collected, which is considered adequate for positive identification (Schneider and Culver 2004).

The information presented here represents about 5 percent of the caves known to occur in Grand Canyon National Park (Tom Gilliland, Arizona Cave Survey, personal communication 2006). Thus, by implementing an inventory program with a consistent and rigorous sampling design with judicious specimen collection, using the protocols outlined above, researchers will obtain more reliable estimates of species richness (on a per-cave basis), yielding a stronger basis for conducting comparative analyses across cave systems and likely leading to the identification of more endemic and troglomorphic invertebrates in Grand Canyon caves.

CONCLUSION

Caves are among the most fragile and understudied ecosystems on Earth. Limited research exists on caves in Grand Canyon National Park and the southern Colorado Plateau. We reviewed all available literature, including cave trip reports, comprising just nine studies of 15 caves at Grand Canyon National Park. Approximately 37 cave-dwelling invertebrates are known to occur in Grand Canyon caves (3 troglobites, 6 trogloxenes, 14 troglophiles, 1 stygobite, 10 unknown cavernicoles, and 3 special case species). Currently, only four cave-adapted taxa are known for the Grand Canyon. Because there are published data for only about 5 percent of the known caves in Grand Canyon National Park, more endemic cave-adapted invertebrates are expected to be found in the future.

ACKNOWLEDGMENTS

The staff of the Grand Canyon National Park Museum Collections was most helpful in locating information on cave invertebrates in their holdings. T. Gilleland provided valuable insights into the general level of cave study in the Grand Canyon region. S. B. Peck provided additional information on his stream cave survey. E. Bernard provided insights regarding current Collembola taxonomy. D. Blinn provided the photo of *Stygobromus blinni*. We extend much gratitude to three anonymous reviewers whose comments greatly improved the manuscript.

LITERATURE CITED

Barr, T. C. Jr. 1968. Cave ecology and evolution of troglobites. In Evolutionary Biology, Volume 2, edited by T. Dobzhansky, M. K. Hecht, and W. C. Steere, pp. 35–102. Appleton, Century, Crofts, New York.

Barr, T. C., and J. R. Reddell. 1967. The arthropod cave fauna of the Carlsbad Caverns Region, New Mexico. Southwestern Naturalist 12: 253–274.

Christman, M. C., D. C. Culver, M. K. Madden, and D. White. 2005. Patterns of endemism of the eastern North American cave fauna. Journal of Biogeography 32: 1442–1452.

Culver, D. C., L. L. Master, M. C. Christman, and H. H. Hobbs III. 2000. Obligate cave fauna of the 48 contiguous United States. Conservation Biology 14: 386–401.

Culver, D., and White, W. 2005. Encyclopedia of Caves. Elsevier Academic Press, Burlington, Massachusetts.

Drost, C. A., and D. W. Blinn. 1997. Invertebrate community of Roaring Springs Cave, Grand Canyon National Park, Arizona. Southwestern Naturalist 42: 497–500.

Elliott, W. 2000. Conservation of the North American cave and karst biota. In Subterranean Ecosystems, Ecosystems of the World, Volume 30, edited by H. Wilkens, D. C. Culver, and W. F. Humphreys, pp. 665–669. Elsevier, Amsterdam.

Hamilton-Smith E., and S. Eberhard. 2000. Conservation of cave communities in Australia. In Subterranean Ecosystems, Ecosystems of the World, Volume 30, edited by H. Wilkens, D. C. Culver, and W. F. Humphreys, pp. 647–664. Elsevier, Amsterdam.

Hill, C. A., R. H. Buecher, D. C. Buecher, C. Mosch, and M. Goar. 1998. Trip report—Horseshoe Mesa, March 14–16, 1998, Grand Canyon National Park, Arizona. Unpublished report to Grand Canyon National Park.

Hill, C. A., V. J. Polyak, D. C. Buecher, R. H. Buecher, and P. P. Provencio. 2000. Trip report: Tse'an Kaetan (Cave of Prayer Sticks), Grand Canyon National Park, Arizona. Unpublished report to Grand Canyon National Park.

Hill, C. A., V. J. Polyak, D. C. Buecher, R. H. Buecher, and C. G. Graf. 2001. Trip report: Bat Cave, Grand Canyon National Park, Arizona. Unpublished report to Grand Canyon National Park.

Krajick, K. 2001. Cave biologists unearth buried treasures. Science 283: 2378–2381.

Mitchell, R. W. 1970. Total number and density estimates of some species of cavernicoles inhabiting Fern Cave, Texas. Annales de Spéléologie 25: 73–90.

Pape, R. B. 1998. Bat Cave, Grand Canyon National Park: Baseline Biological Inventory, Final Report, May 1998. Report submitted to Grand Canyon National Park.

Peck, S. B. 1978. New montane *Ptomaphagus* beetles from New Mexico and zoography of southwestern caves (Coleoptera; Leiodidae; Catopinae). Southwestern Naturalist 23: 227–238.

Peck, S. B. 1980. Climatic change and the evolution of cave invertebrates in the Grand Canyon, Arizona. NSS Bulletin 42: 53–60.

Reddell, J. R. 1994. The cave fauna of Texas with special reference to the western Edwards Plateau. In The Caves and Karst of Texas, edited by W. R. Elliott and G. Veni, pp. 31–50. National Speleological Society, Huntsville, Alabama.

Roth, V. D. 1982. Preliminary Survey of Colorado River-side Arachnida, Chilopoda, Diplopoda and Isopoda from mile 0 to mile 225 of the Grand Canyon. Unpublished report submitted to Grand Canyon National Park.

Schneider, K., and D. C. Culver. 2004. Estimating subterranean species richness using intensive sampling and rarefaction curves in a high density cave region in West Virginia. Journal of Cave and Karst Studies 66: 39–45.

Slaney, D. P., and P. Weinstein. 1996. Leaf litter traps for sampling Orthopteroid insects in tropical caves. Journal of Orthoptera Research 5: 51–52.

Triplehorn, C. A. 1975. A new subgenus of *Eleodes* with new cave-dwelling species (Coleoptera: Tenebrionidae). Coleopterists Bulletin 29: 39–43.

Triplehorn, C. A., and N. F. Johnson. 2005. An Introduction to the Study of Insects, 7th ed. Thomson Brooks/Cole, Belmont, California.

Wang, D., and J. R. Holsinger. 2001. Systematics of the subterranean amphipod genus *Stygobromus* (Crangonyctidae) in western North America, with emphasis on the hubbsi group. Amphipacifica 3(2): 39–147.

Welbourn, W. C. 1978. Preliminary report on the cave fauna. In Cave Resources of Horseshoe Mesa, Grand Canyon National Park, pp. 36–41. Unpublished report by the Cave Research Foundation submitted to Grand Canyon National Park.

Addressing Vegetation Issues

VEGETATION OF PETRIFIED FOREST NATIONAL PARK, ARIZONA

Kathryn A. Thomas, M. L. Hansen, and Keith A. Schulz

Petrified Forest National Park in northeast Arizona (Figure 1) is known for its rocks, ruins, and scenic Painted Desert vistas, but the park is equally a premier example of Colorado Plateau grasslands and steppe. The vegetation of the park has been protected from many development disturbances since 1905 and from widespread cattle grazing since 1963. The most recent description of vegetation at Petrified Forest originated as part of the National Park Service's Southern Colorado Plateau Network classification and mapping of vegetation types at the park. We describe the vegetation of the park using the results of this classification work.

A vegetation classification describes assemblages of plants and assumes that there are characteristic and repeated groupings of plant species across the environment. Vegetation classification systems are tools used to describe the patterns observed. In the last decade, there has been a national effort to establish standards and protocols for vegetation studies that will lead to consistency and comparability in describing and classifying those patterns.

In recognizing the need for a federal standard for vegetation classification and reporting of vegetation statistics, the Federal Geographic Data Committee (1997) has adopted the National Vegetation Classification Standard (NVCS) for inventory, mapping, and reporting on vegetation resources. The standard builds on earlier work on vegetation classification done by UNESCO (1973), Driscoll et al. (1984), and Grossman et al. (1998).

The NVCS describes a hierarchical classification system with the upper levels (system, class, subclass, group, subgroup, and formation) describing physiognomic characters of the vegetation and the lower levels (alliance and association) based on floristic characters of the vegetation. The initial standards (FGDC 1997) describe the upper level categories, but not the floristic levels. Identification and description of these finest levels (the alliance and association) has been derived from various mapping projects, such as the Southern Colorado Plateau Network's vegetation mapping efforts and plant community classifications by academic and government researchers.

An alliance is a "physiognomically uniform group of plant associations sharing one or more dominant or diagnostic species, which as a rule are found in the uppermost stratum of the vegetation" (Grossman et al. 1998). An association is "characterized by diagnostic species that occur in all strata (overstory and understory) of the vegetation" (FGDC 1997). Alliances and associations are also identified by the predominant lifeform and the vegetation structure of the assemblage they describe, and are placed hierarchically below the FGDC physiognomic levels.

METHODS

The U.S. Geological Survey, in partnership with the Southern Colorado Plateau Network, NatureServe, and Kansas State University, coordinated the production of a map (spatial database) of vegetation types at

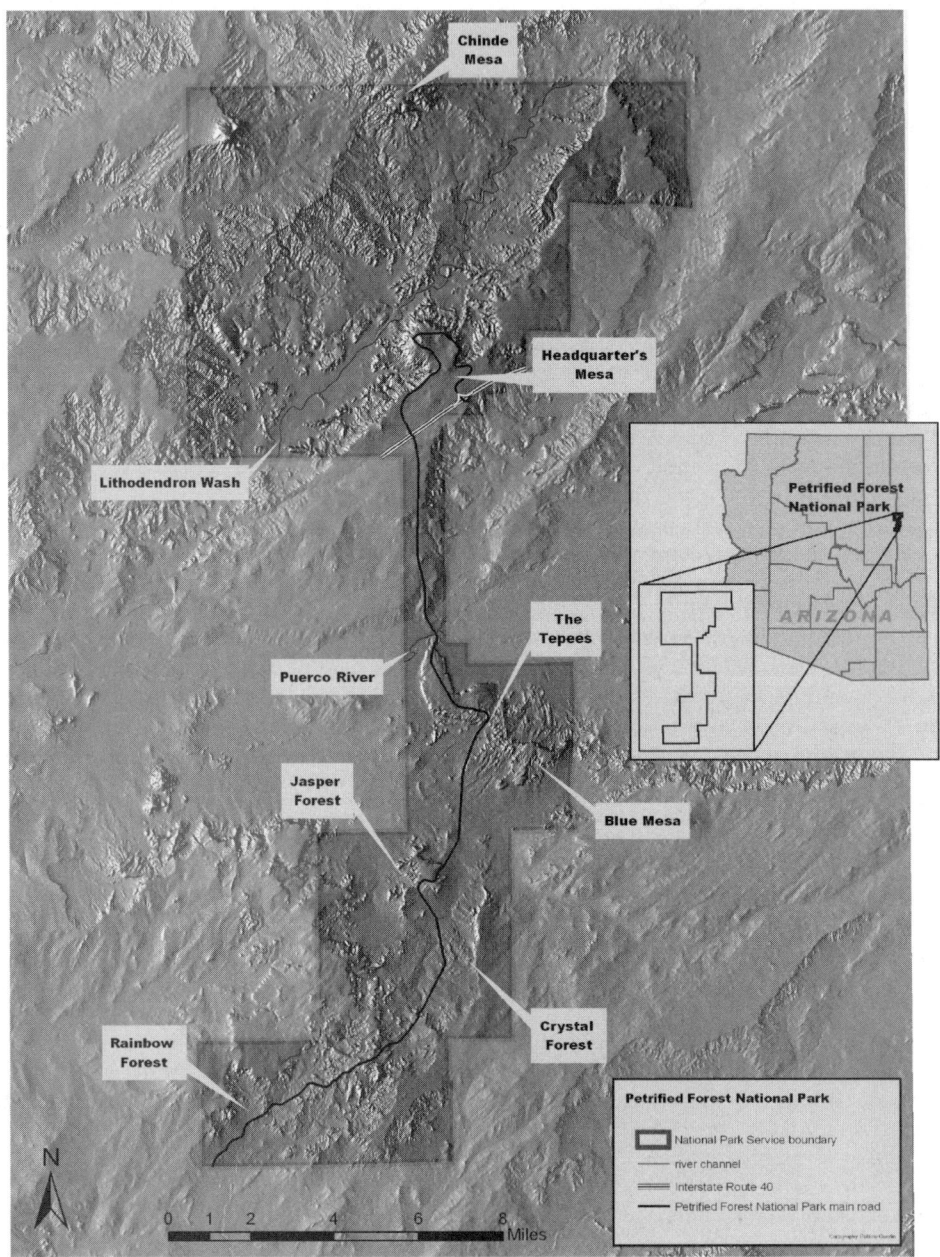

Figure 1. The location of Petrified Forest National Park and common place names within the park.

Petrified Forest National Park. As part of the mapping, vegetation was classified to the finest levels of the National Vegetation Classification—the association and the alliance.

Petrified Forest is a 38,008 hectare park in Apache and Navajo Counties in northeast Arizona. Vegetation has been classified and described both within the park and in a 1 km area surrounding the park. The park supports a variety of landscapes (substrates and topography) that influence the resulting vegetation communities: basaltic uplands, talus slopes found at mesa bases, active and stabilized sand sheets and low dunes, generally flat-topped sandstone-capped mesas and plateaus (with sandy loam to loam soils), erosional badlands consisting of highly erodible bentonite hills and steep mesa side slopes, alkaline swales and flats, washes (including the large Lithodendron Wash), outwash plains characterized by seasonal sheet-flow, and the Puerco River corridor and terrace.

DATA COLLECTION AND ANALYSIS

Our classifications are the result of quantitative assessment of data from 186 relevés collected in 1996 and 1997 for a study that began in 1996 (Thomas et al. 2003a) and data from 149 relevés collected in 2003 for the mapping project. We sampled such that the collection of 335 relevés represents vegetation types throughout the park.

Each relevé was either 400 or 1000 sq m; the larger size was used for vegetation types with 10 percent or less cover. Measurements at each relevé included a listing of perennial plant species, cover by species, strata, substrate and topographic features of the location, the geographic coordinates of the relevé, and elevation, location, and aspect.

We entered all collected data into an Access database and then imported these data into the multivariate statistical analysis program PC-Ord (McCune and Mefford 1999; McCune and Grace 2002). We incorporated the 1996 and 2003 data sets into separate vegetation matrices, relativized the data due to extremely high beta diversity, and transformed the data using an Arcsine square root transformation to approximate common cover class scales. Data were then evaluated using nonmetric multidimensional scaling (NMS) ordination and a cluster analysis and divided into matrices by physiognomic classes for further data analysis.

We then used NMS, cluster analyses, and indicator species analysis (ISA; Dufrêne and Legendre 1997), which is a technique to identify the species or species assemblages that characterize a group of sampling units. The ISA indicator values are a mechanism for choosing a stopping point in cluster analysis. Each stopping point was interpreted as the finest level of ecologically meaningful classification or NVC vegetation association. We conducted an ISA for each matrix and graphically looked at the optimal number of significant indicator species ($p < 0.005$) and the average p value per cluster. We assigned each plot an association name using previously defined association nomenclature (NatureServe 2003) and developed new potential associations if needed.

After classifying association names for each physiognomic class, we compared the results for each physiognomic class and determined if there was overlap between classes. For instance, we identified an *Atriplex obovata* and a *Sporobolus airoides* assemblage as occurring in sparse, shrubland, and steppe communities. We aggregated these types into a single vegetation association, placed the vegetation communities with similar species composition into the modal physiognomic matrix, and reanalyzed these data with the new physiognomic classes using cluster analysis, ISA, and NMS ordinations to determine the most parsimonious vegetation associations based on the NVC.

We performed a multiple-response permutation procedure (MRPP) test, which is a nonparametric procedure used to test the hypothesis that there is no difference between groups and which provides a test statistic and a p value. We used the MRPP to test if the allocated NVC associations were different enough between each other to be statistically significant. Finally, we used the 1996 data to develop vegetation associations using the same statistical techniques mentioned above.

The names assigned to each vegetation type contain much information about the composition and structure of the assemblage. The dominant plants, either having the highest aerial cover or being the most visually dominant, are listed as the primary components of the alliance name, followed by the characteristic lifeform of the assemblage. The association name gives more detail on the co-dominant or characteristically associated species. A slash (/) is used to separate plant species that occur at different strata; for example, the name of the dominant tree is separated from the name of the dominant shrub by the slash. However, if more than one plant in the same stratum is dominant, a dash (–) is used to separate the plant names. Parentheses indicate a plant that is characteristic of the vegetation type but is not necessary for the identity of the assemblage.

RESULTS

The vegetation at Petrified Forest was classified into 28 alliances, 32 associations, and 5 park specials. Some of the classification relevés sampled did not represent existing NVC associations and did not have enough relevé data to support the development of a new vegetation association. We classified these relevés as six different park special vegetation types, each of which may be incorporated into the NVC later, pending additional field data supporting the assemblage as a consistent vegetation type. Six relevés in the 1996 classification effort did not have enough data for classification to the association level and were only classified to the alliance level.

Each of the alliances, associations, and park specials are part of one of five physiog-nomic groups: woodland, shrubland, shrub-herbaceous, herbaceous, and sparse forb (Figure 2). We describe these five groups and their associated vegetation classification results below using the common names. The corresponding scientific names for the vegetation types appear in the accompanying tables (Tables 1–5).

Woodland Vegetation

Two woodland alliances and associations occur at the park (Table 1). The woodland class at Petrified Forest is not common; it occurs on only 72 ha. Woodland tree cover in the park is usually below 25 percent and shrub and grassland understory species are usually present, but with low cover. One-seed juniper/Bigelow's sagebrush woodland has a shrubby aspect and occurs on high-elevation mesas and basaltic uplands, including Chinde and Headquarters Mesas. Fremont cottonwood/rubber rabbitbrush woodland is characteristic of the terraces along the Puerco River.

Shrubland Vegetation

Shrublands are well represented at Petrified Forest National Park, comprising nearly 30 percent of the land cover, and consisting of 12 alliances, 14 associations, and 2 park specials (Table 2). Shrublands have no or few trees and typically have more cover of shrubs than grasses. However, there is a wide range of variability in the expression of shrublands at the park and some vegetation associations classified as shrublands may appear more like a steppe.

The habitats that support intermittent and temporary water flows form distinctive vegetation environments. Salt-tolerant plant species dominate the vegetation associations

Table 1. Petrified Forest National Park has two woodland alliances and associations.

NVC Alliance	NVC Association Name	Common Name
Juniperus monosperma woodland	*Juniperus monosperma/ Artemisia bigelovii* woodland	One-seed juniper/Bigelow's sagebrush woodland
Populus fremontii temporarily flooded woodland	*Populus fremontii/ Ericameria nauseosa* woodland	Cottonwood/rubber rabbitbrush woodland

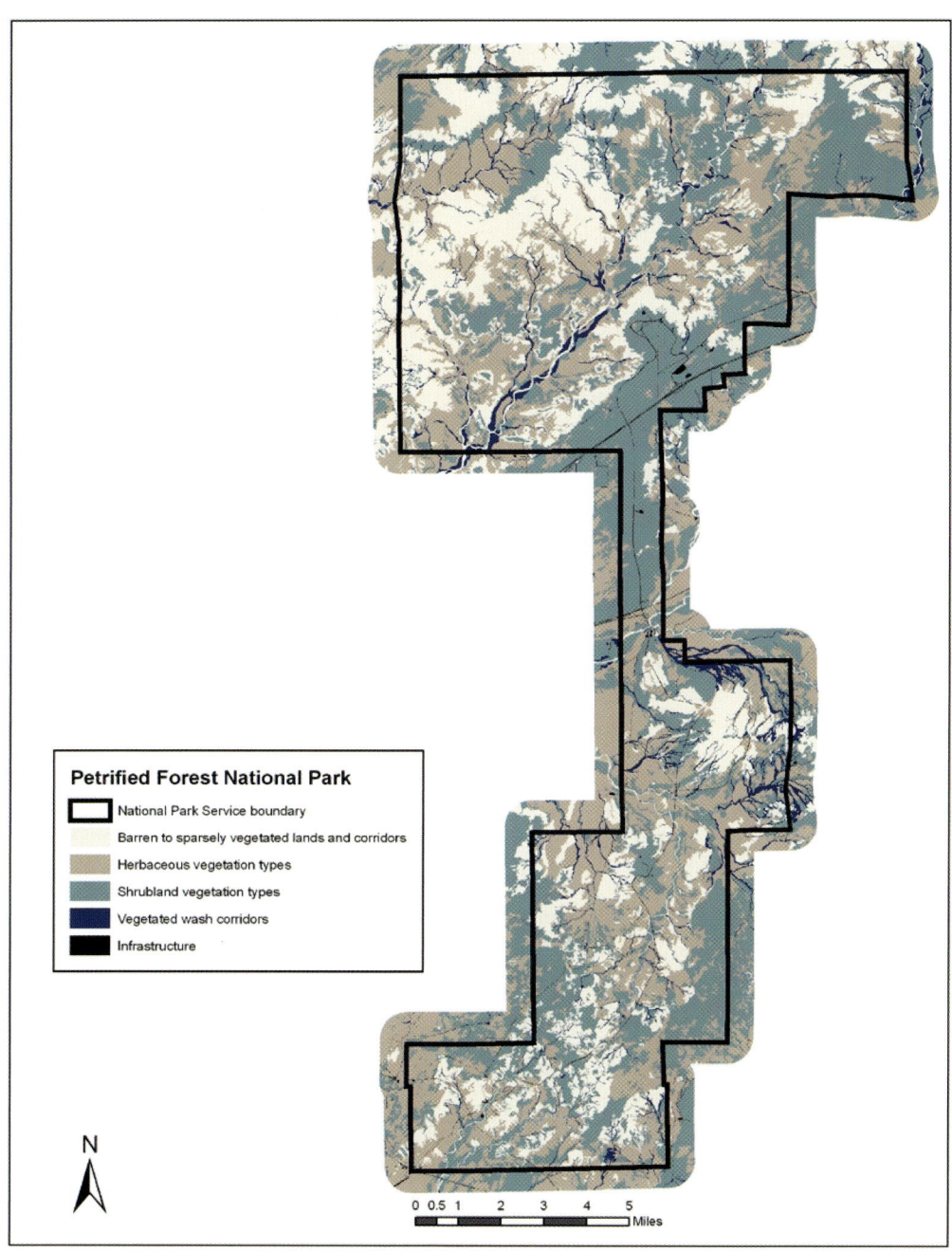

Petrified Forest National Park

- National Park Service boundary
- Barren to sparsely vegetated lands and corridors
- Herbaceous vegetation types
- Shrubland vegetation types
- Vegetated wash corridors
- Infrastructure

N

0 0.5 1   2   3   4   5
Miles

Figure 2. The vegetation at Petrified Forest National Park grouped by physiognomic type.

Table 2. Petrified Forest National Park has 12 shrubland alliances, 14 shrubland associations, and 2 shrubland park specials.

NVC Alliance	NVC Association Name	Common Name
Allenrolfea occidentalis Shrubland	*Allenrolfea occidentalis* Shrubland	Iodine bush shrubland
Artemisia filifolia Shrubland	*Artemisia filifolia/Bouteloua eriopoda* Shrubland	Sand sage / black grama shrubland
	Artemisia filifolia Colorado Plateau Shrubland	Sand sage Colorado Plateau shrubland
Atriplex canescens Shrubland	*Atriplex canescens/Pleuraphis jamesii* Shrubland	Four-wing saltbush / galleta shrubland
Ephedra torreyana Sparsely Vegetated	*Ephedra torreyana–Artemisia bigelovii* Sparse Vegetation	Bigelow's sagebrush– Torrey's jointfir shrubland
Ericameria nauseosa Shrubland	*Ericameria nauseosa* Desert Wash Shrubland	Rubber rabbitbrush desert wash shrubland
Gutierrezia sarothrae Dwarf-Shrubland	*Gutierrezia sarothrae (Opuntia* spp.)/ *Pleuraphis jamesii* Dwarf-Shrubland	Snakeweed–(prickly pear)/ galleta dwarf-shrubland
Park Special	*Iva acerosa/Sporobolus airoides* Shrubland	Copperweed / alkali sacaton shrubland
Park Special	*Salvia pachyphylla* dwarf-Shrubland	Blue sage dwarf-shrubland
Purshia (stansburiana, mexicana) Shrubland	*Purshia stansburiana–Eriogonum corymbosum* Shrubland	Cliffrose–crispleaf Buckwheat shrubland
Rhus trilobata Shrubland	*Rhus trilobata–Ephedra (viridis, torreyana)* Talus Shrubland	Mormon tea–three-leafed sumac talus shrubland
Salix (exigua, interior) Temporarily Flooded Shrubland	*Salix exigua*/Barren Shrubland	Coyote willow shrubland
Sarcobatus vermiculatus Intermittently Flooded Shrubland	*Sarcobatus vermiculatus/Atriplex obovata* Shrubland	Greasewood / New Mexico saltbush shrubland
	Sarcobatus vermiculatus/ Suaeda moquinii Shrubland	Greasewood / shrubby seepweed shrubland
Suaeda moquinii shrubland	*Suaeda moquinii* Shrubland	Shrubby seepweed shrubland
Tamarix spp. Semi-natural Temporarily Flooded Shrubland	*Tamarix* spp. Temporarily Flooded Shrubland	Tamarisk shrubland

Table 3. Petrified Forest National Park has four shrub-herbaceous alliances and associations.

NVC Alliance	NVC Association Name	Common Name
Calamovilfa gigantea Shrub-Herbaceous Alliance	*Calamovilfa gigantea* Desert Wash Shrub-Herbaceous Vegetation	Giant sandreed desert wash shrub-herbaceous vegetation
Ericameria nauseosa Shrub-Short Herbaceous Alliance	*Ericameria nauseosa/Bouteloua gracilis* Shrub-Herbaceous Vegetation	Rubber rabbitbrush/blue grama shrub-herbaceous vegetation
Krascheninnikovia lanata Dwarf-Shrubland Herbaceous Alliance	*Krascheninnikovia lanata/Bouteloua gracilis* Dwarf-Shrubland Herbaceous Vegetation	Winter-fat/blue grama dwarf-shrubland herbaceous
Sporobolus airoides–(Pleuraphis jamesii) Shrub-Herbaceous Alliance	*Atriplex obovata/Sporobolus airoides–(Pleuraphis jamesii)* Shrub-Herbaceous Vegetation	New Mexico saltbush/alkali sacaton–(galleta) shrub-herbaceous vegetation

Table 4. Petrified Forest National Park has four herbaceous alliances and six herbaceous associations.

NVC Alliance	NVC Association Name	Common Name
Bouteloua eriopoda Herbaceous Alliance	*Bouteloua eriopoda–Pleuraphis jamesii* Herbaceous Vegetation	Black grama–galleta herbaceous vegetation
Bouteloua gracilis Herbaceous Alliance	*Bouteloua gracilis–Pleuraphis jamesii* Herbaceous Vegetation	Blue grama–galleta herbaceous vegetation
	Bouteloua gracilis Herbaceous Vegetation	Blue grama herbaceous vegetation
Pleuraphis jamesii Herbaceous Alliance	*Pleuraphis jamesii–Sporobolus airoides* Herbaceous Vegetation	Galleta–alkali sacaton herbaceous vegetation
Sporobolus airoides Sod Herbaceous Alliance	*Sporobolus airoides–Bouteloua gracilis* Herbaceous Vegetation	Alkali sacaton–blue grama herbaceous vegetation
	Sporobolus airoides Southern Plains Herbaceous Vegetation	Alkali sacaton herbaceous vegetation

Table 5. Petrified Forest National Park has two sparsely vegetated alliances and three sparse associations.

NVC Alliance	NVC Association Name	Common Name
Painted Desert Sparsely Vegetated Alliance	*Atriplex obovata* Badland Sparse Vegetation	New Mexico saltbush badland sparse vegetation
Painted Desert Sparsely Vegetated Alliance	*Eriogonum leptophyllum* Sparse Vegetation	Slenderleaf buckwheat sparse dwarf-shrubland vegetation
Zuckia brandegeei Sparsely Vegetated Alliance	*Zuckia brandegeei* Sparse Vegetation	Arizona siltbush sparse dwarf-shrubland vegetation
Park Special	*Salsola tragus* Sand Dune Vegetation	Russian thistle sand dune vegetation

found in swales and flats: iodine bush, greasewood/New Mexico saltbush, greasewood/shrubby seepweed, and shrubby seepweed shrublands. Tamarisk shrubland occurs in these habitats, in washes, and in the Puerco River corridor and its terraces. Coyote willow and rubber rabbitbrush desert wash shrublands occur along the Puerco River terrace. Rubber rabbitbrush desert wash and copperweed/alkali sacaton shrublands occur along washes, including the largest, Lithodendron Wash.

Sand sheets and low dunes intergrade with the fine-textured soils of the plateau. Sand sage Colorado Plateau shrubland (Figure 3) occurs on sand sheets and low dunes, where it is the most common vegetation, and on sandy plateaus; sand sage/black grama occurs only on sand sheets and low dunes. Together they contribute 6.8 percent total cover. Bigelow's sagebrush–Torrey's jointfir occurs on a variety of habitats: sandy plateau soils, often with sandstone cobble; the top of erosional bentonite badlands; and on basaltic uplands near Headquarters Mesa. It is the third most common shrubland type at the park (5.5% cover). Fourwing saltbush/galleta and snakeweed–(prickly pear)/galleta dwarf-shrubland are common on the plateau environments. Fourwing saltbush, which often has a steppe-like aspect, is the most common shrubland on the plateau (14.4% cover).

The three remaining shrublands occur on basalt uplands (cliffrose–crispleaf buckwheat shrubland), talus slopes (Mormon tea–three-leafed sumac talus shrubland), and erosional badlands (blue sage dwarf-shrubland, only at Chinde Mesa in Arizona). Mormon tea–three-leafed sumac talus shrubland characterizes all the talus slopes at the park.

Shrub-Herbaceous Vegetation

Shrub-herbaceous vegetation, often referred to as steppe, has few or no trees and more grasses than shrubs. Generally, shrub cover is more than half of the relative cover of the herbaceous layer. Shrub-herbaceous vegetation is a common vegetation type at the park, comprising more than 26.5 percent of

the land cover. It consists of four associations within four alliances (Table 3).

Giant sandreed desert wash shrub-herbaceous vegetation, which occurs in and adjacent to washes, is quite striking due to the height of the dominant grass, often more than a meter tall. On the plateau, New Mexico saltbush/alkali sacaton–(galleta) shrub-herbaceous vegetation is the most common association, but it is also found on erosional badlands and outwash plains. It is the most common vegetation type overall, with 24 percent cover. Rubber rabbitbrush/blue grama shrub-herbaceous and winterfat/blue grama dwarf-shrubland are found on the plateau, and often occur in a mosaic with grasslands. In addition, rubber rabbitbrush/blue grama shrub-herbaceous vegetation also occurs on sand sheets and low dunes.

Herbaceous Vegetation

Perennial grasses with no trees and few shrubs characterize herbaceous vegetation, or grasslands. Cover of forbs may be significant. Generally, shrub cover is less than half of the relative cover of the herbaceous layer. The grasslands at the park do characteristically have some shrubs present. It consists of nearly 6 percent of the land cover. We described six associations within four alliances (Table 4).

The most common grassland associations occur with a mix of blue grama, galleta, and alkali sacaton grasses either dominating or co-dominating the associations. They occur on the plateau and occasionally in sparse patches in erosional badlands. The six associations—black grama–galleta herbaceous, blue grama–galleta herbaceous, blue grama herbaceous vegetation (Figure 4), galleta–alkali sacaton herbaceous, alkali sacaton–blue grama herbaceous, and alkali sacaton southern Plains herbaceous vegetation—form a mosaic of common and widely occurring grasslands over 5.7 percent of the park.

Sparse and Forb Vegetation

Sparse vegetation (Table 5) is characteristic of erosional badlands, which occur in the northern Painted Desert wilderness area, at

Figure 3. Sand Sage Colorado Plateau Shrublands on a low dune east of Jim Camp Wash and south of Rainbow Forest Museum at Petrified Forest National Park, Arizona.

Figure 4. Blue grama herbaceous vegetation south of Agate House at Petrified Forest National Park, Arizona. The picture, which was taken in a drought year, shows the low vegetative growth response of the grama grasses resulting from little precipitation.

Figure 5. New Mexico saltbush badland sparse vegetation north of the Flattops at Petrified Forest National Park. Petrified wood stones and cobble cover the soil surface, which is common on erosional badlands.

the Tepees, around Blue Mesa, and in Rainbow, Jasper, and Crystal Forests. The sparse vegetation characteristically has between 1 and 10 percent vegetative cover. Areas with less than 1 percent vegetation cover are classified as barren.

Of the three sparse vegetation associations—New Mexico saltbush badland sparse (Figure 5), slenderleaf buckwheat sparse dwarf-shrubland, and Arizona siltbush sparse dwarf-shrubland)—the New Mexico saltbush badland sparse vegetation is the most common association in the erosional badlands (14.4%). This association also occurs on outwash plains. The slenderleaf buckwheat sparse shrubland typically occurs on the white sandstone cap often occurring on top of the bentonite hills.

Another sparsely vegetated substrate of the park, which is naturally highly disturbed, comprises unstabilized sand sheets and dune hummocks. Here we find that infestations of the invasive, nonnative, annual forb Russian thistle (*Salsola tragus*) have proliferated; this is the park special

Russian thistle sand dune vegetation. This forb is an annual that generally sprouts in the late spring and can dominate the open sandy areas during the summer and early fall. Although it seems to have invaded extensively throughout the park, it is only on the sand sheets and dunes that Russian thistle cover is the dominant species.

DISCUSSION

The vegetation of Petrified Forest is more complex and varied than one would think from a quick trip through the park. Many of the dominant plants have low stature and cannot be distinguished until the viewer is close. Yet on close inspection, there is never a single dominant plant species but rather a rich mosaic of grasslands, steppe, and shrubland types with different species dominating at different locations.

In our description of the vegetation associations, alliances, and park specials, we have emphasized the topographic and soil correlates of the vegetation distribution. Other factors that influence the expression of vegetation in the park are drought and invasive plants.

Precipitation in the park is biseasonal, with a winter precipitation period and a summer monsoon period. The grasses of the park, in particular, respond closely to this seasonality of precipitation, with some grass species showing the most growth in the spring warmup (cool season grasses) and others showing the most growth in response to the summer monsoons (warm season grasses). Climatic events that reduce precipitation during the winter, summer, or both seasons inhibit plant growth and reproduction and may kill plants. The drought in the southwestern United States in the early 2000s has greatly reduced the apparent cover of vegetation as plants responded with reduced vegetative growth and dieback. Climate change, especially change that brings warmer temperatures and decreases in precipitation and/or changes in the monsoon patterns, can be expected to change the characteristic plant distribution over time.

A study encompassing Headquarters Mesa, the Puerco River corridor, and parts of the southern part of the park revealed more than 25 invasive nonnative plants (Thomas et al. 2003b, 2004). The most prolific was Russian thistle, which occurred in more than three-fourths of the area sampled. Invasive plants increasingly threaten wildland ecosystems, not only interacting with the native plants and animals but possibly increasing the frequency and magnitude of fires. In the event of long drought, the invading species can magnify the effects of reduced water by sprouting earlier than the native plants and thus garnering much available soil moisture that would have been available for the native plants.

ACKNOWLEDGMENTS

K. Wright and L. Rodriquez collected field data in 1996 and 1997 and S. Till, L. Larsen, and J. Lambert collected field data in 2003. The authors thank A. Cully, J. Vogel, and three anonymous reviewers for thoughtful comments on the paper.

LITERATURE CITED

Driscoll, R. S., D. L. Merkel, D. L. Radloff, D. E. Synder, and J. S. Hagihara. 1984. An Ecological Land Classification Framework for the United States. U.S. Forest Service Misc. Publ. No. 1439. Washington, D.C.

Dufrêne, M., and P. Legendre. 1997. Species assemblages and indicator species: The need for a flexible asymmetrical approach. Ecological Monographs 67 (3): 345–366.

Federal Geographic Data Committee. 1997. Vegetation Classification Standard, FGDC-STD-005. Washington, D.C.

Grossman, D. H., D. Fabre-Langendoen, A. S. Weakley, M. Anderson, P. Bourgeron, R. Crawford, K. Goodin, R. Landal, K. Metzler, K. Patterson, M. Pyne, M. Reid, and L. Sneddon. 1998. International Classification of Ecological Communities: Terrestrial Vegetation of the United States. Volume 1. The Nature Conservancy, Arlington, Virginia.

McCune, B., and J. B. Grace. 2002. Analysis of Ecological Communities. MjM Software Design, Gleneden Beach, Oregon.

McCune, B., and M. J. Mefford. 1999. PC-ORD. Multivariate Analysis of Ecological Data, version 4. MjM Software Design, Gleneden Beach, Oregon.

NatureServe Explorer: An Online Encyclopedia of Life. 2003. Version 1.8 . NatureServe, Arlington, Virginia. Available at http://www.natureserve.org/explorer.

Thomas K. A., M. Hansen, and C. Seger. 2003a. Vegetation of Petrified Forest National Park, Arizona: Part I: Vegetation of Petrified Forest National Park, Arizona. U.S. Geological Survey Biological Resources Discipline.

Thomas, K. A., B. Dale, and R. Hunt. 2003b. Petrified Forest National Park Weed Inventory and Mapping Project. Annual Report FY 2002. U.S. Geological Survey Southwest Biological Science Center.

Thomas, K. A., R. Hunt, and M. Ostrowski. 2004. Petrified Forest National Park Weed Inventory and Mapping Project. Annual Report FY 2003. U.S. Geological Survey Southwest Biological Science Center.

UNESCO. 1973. International Classification and Mapping of Vegetation. Series 6, Ecology and Conservation. United Nations Educational, Scientific, and Cultural Organization, Paris.

UNDERSTORY VEGETATION RESPONSES TO MULTI-AGED GROUP SELECTION HARVESTING AND PRESCRIBED FIRE IN NORTHERN ARIZONA

John D. Bailey and Robert K. Speer

Ponderosa pine (*Pinus ponderosa*) forests in the semi-arid Southwest have become increasingly dense over the past century. Changes in land use practices brought by Euro-American settlers beginning in the late 1800s created these shifts in forest structure (Covington et al. 1994, 1997; Moore et al. 1999). The current high density of ponderosa pine forests allows little sunlight for understory vegetation (Naumburg and DeWald 1999). Extensive grazing of domestic livestock has further depleted a rich understory of grasses and forbs that once out-competed pine seedlings, and this depleted understory once drove the frequent surface fires that further prohibited extensive pine regeneration (Korb and Springer 2003). In addition to supporting a natural fire regime (Laughlin et al. 2004), healthy understory communities promoted a number of ecosystem functions including hydrology and soil stabilization (Korb and Springer 2003), net primary productivity, nutrient cycling, and forage for wildlife and domestic livestock (Harris and Covington 1983). Of particular interest in regards to harvesting, burning, and general soil disturbance is the introduction and spread of alien species (Crawford et al. 2001; Sieg et al. 2003; Korb et al. 2004). Alien species are of importance to ecosystem function and health because they alter successional pathways by out-competing native pioneer species, thereby altering ecosystem functions normally performed by native species (Fornwalt et al. 2003).

Our understory vegetation study was conducted on one of 13 sites in the national Fire and Fire Surrogate (FFS) Program, a large multi-disciplinary study that examines the effectiveness of mechanical harvesting and prescribed burning, alone and in combination, for reducing wildfire risk while improving forest health (Weatherspoon 2000). Our goal was to learn more about how perennial and annual understory plants respond to increasing intensities of management (burn only, harvest only, harvest & burn), with a focus on species richness and native and exotic ground cover.

STUDY AREAS AND METHODS

Relatively homogenous forested areas were selected for the three FFS Southwest Plateau replicate blocks, and were named after nearby land features: KA Hill, Rudd's Tank, and Powerline (Figure 1). Each of the three replicate sites was a stand of dense younger trees that regenerated from the scattering of presettlement individuals left over from turn-of-the-century logging. Each was divided randomly into the four assigned FFS treatments: control, burn only, harvest only, and harvest & burn. Data are presented throughout the paper in this order, consistent with an increasing level of management or disturbance intensity. Sizes of treatment stands ranged from 16 to 37 ha depending on natural topography and stand boundaries.

Our mechanical treatment, identical across all sites, was designed as a multi-aged group selection harvest with the goal of restoring the natural clumping of healthy, vigorous trees of all diameter classes (Bailey and Covington 2002). The target basal area

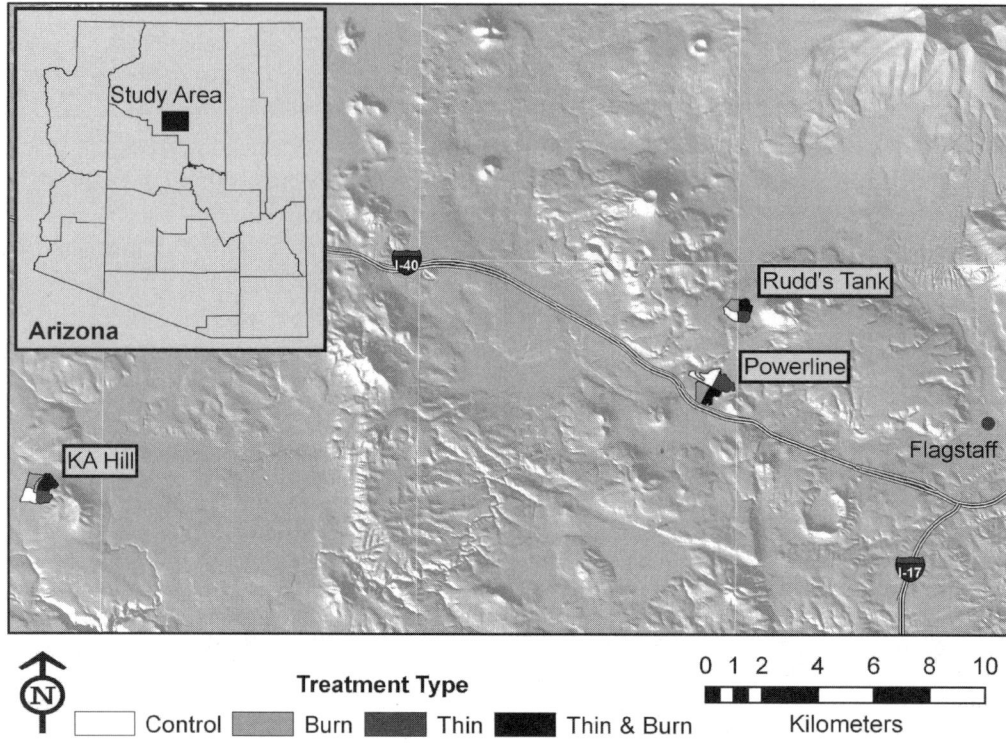

Figure 1. Locations of the multi-disciplinary Southwest Plateau Fire and Fire Surrogate Project sites, located in northern Arizona in the wildland-urban interface of Flagstaff, AZ.

of the prescription was 13 m²/ha, which removed an average of 52 percent of the basal area across all units using ground-based equipment; this occurred in the autumn of 2002. Slash was mechanically aggregated in 10–20 large piles per unit, distributed throughout the units; slash piles were later hauled away in the harvest-only treatment but burned on-site in late October and early November of 2003 in the harvest & burn treatment units. No sample plots fell in or near these large burn scars. Pile burning was augmented with a light surface pre-scribed fire between piles; similarly, burn-only treatments were implemented under the same fire weather conditions as a surface prescribed fire only in an otherwise unman-aged unit. Flame lengths were typically less than 50 cm with slow rate of spread.

Thirty-six gridpoints were permanently established with rebar markers in each of the 12 stands on a 50 m spacing. All grid points were located at least 50 m from the bounda-ry of the treatment to avoid edge effect. Ten of the 36 points in each treatment were randomly selected as permanent inventory points which were measured pre-treatment in 2001 and post-treatment in 2005. Inven-tories were timed to take place at the peak of productivity, which coincides with seasonal (June to September) monsoon rain. Both inventory years were similar for this period: in 2001, 17.7 cm of rain (89% of average) was recorded at the Flagstaff Airport weather station, compared with 17.2 cm (87% of average) in 2005 (Western Regional Climate Center, www.wrcc.dri.edu). This period was preceded by average winter/spring precipitation in 2001 but an abundant snow pack and spring rains in 2005.

Understory species were identified and their respective ground cover was recorded following a modified-Whittaker sampling design (Stohlgren et al. 1995). Unknown species were keyed and identified at the Weaver Herbarium at Northern Arizona University. Most species were identified and inventoried at the species level, but 8 of the 151 plant species found in this particular FFS study (*Allium* spp., *Arabis* spp., *Carex* spp., *Castilleja* spp., *Ipomoea* spp., *Opuntia* spp., *Oxalis* spp., and *Solidago* spp.) were identified only to the genus level, as keying non-flowering specimens to the species level was difficult or impossible. Species were categorized as native or alien as currently designated on the USDA Plants Database (2005). When the Plants Database yielded uncertain or conflicting nativity definitions, local sources were cross-checked (Buckley 2005).

In each 20 x 50 m modified-Whitaker plot, the number of species was counted over a total of 130 m^2, including ten 1 m^2 subplots, two 10 m^2 subplots, and one 100 m^2 subplot. Species were inventoried similarly, at the 1000 m^2 level during the post-treatment inventory. Species richness-(log)area curves were developed from the nested 1, 10, and 100 m^2 subplots and tested against the known 1000 m^2 levels. It was found that the linear regressions under-predicted the actual number of species by 19 percent on average; however, such species counts were under-predicted on some plots by as much as 49 percent. While under-predicting is acknowledged in the modified-Whittaker design (Stohlgren 1995), this extremely large difference in richness prompted us to use the 130 m^2 estimate in order to make a direct comparison between known inventories.

Percent cover of individual species was recorded for ten 1 m^2 non-contiguous subplots along the perimeter of the modified-Whitaker plot. Percent ground cover was inventoried by visually estimating the basal area of individual forb, graminoid, and shrub species. Templates were used as a visual guide to quantify cover of species as < 0.05, 0.1, 0.5, 1, 2, or 5 percent (for noticeably more than 2%). Above this 5 percent cover, a species was quantified into a cover class: 5–25, 25–50, 50–75, 75–95, and 95–100 percent, though > 5 percent cover of any single species was rare and > 25 percent was extremely rare. Such cover classes were used to avoid discrepancies in visual estimation between different field crews; 5–25 percent was converted to 15 percent for averaging and exerted little influence on the data base.

Treatments were first summarized at the plot level and analyzed as treatments within replicates to investigate treatment effects at a localized scale not influenced by preexisting site-to-site differences. Since two plots within the Powerline control area were accidentally burned during nearby burning, they were subsequently not included in this analysis, resulting in n = 8 for the Powerline control treatment and n = 10 for all other treatments. To compare across sites, analyses were conducted using a randomized complete block design ANOVA, resulting in n = 3 for each treatment. Plots within treatments were treated as subsamples. All test assumptions including normality and additivity were met. Treatment differences are pseudoreplicated at this level of analysis, given how treatments were implemented.

Comparisons of native and alien species richness and cover were made between the pre-treatment and post-treatment inventories on the permanent plots; two-tailed tests were used to detect either increases or decreases in this initial evaluation of treatment effects. Hypothesis testing of future inventories at these permanent plots can use single-tailed tests based on the increases or decreases observed in this study. Analysis of variance (ANOVA) was used to compare true differences among treatments both before and after treatment, and Tukey's Honestly Significant Difference (HSD) test was then used to group treatments into homogenous subsets that were significantly different from each other. Significance levels for ANOVA F-tests, Tukey HSD tests, and t-tests were set to $\alpha = 0.05$.

RESULTS

Treatments Within Replicate Sites

There were no significant pre-treatment differences in native species richness among treatment units at KA Hill or Powerline (p = 0.185 and 0.599, respectively), but there was a significant pre-existing difference in native richness among treatment units at Rudd's Tank ($p \leq 0.001$). After implementation, all treatment units within replicate sites were significantly different from each other ($p \leq 0.001$). No significant changes were observed over time in any of the three control treatments; the burn-only treatment at KA Hill decreased significantly in native species richness, while remaining unchanged in the burn-only treatment at Rudd's Tank, and increasing significantly at Powerline. The harvest-only and harvest & burn treatments yielded a slight increase in native species richness at KA Hill, and significant increases at Rudd's Tank and Powerline (Table 1).

There were no significant differences in pre-treatment conditions of percent ground covered by native species at Powerline (p = 0.634), but pre-treatment differences existed among the stands at KA Hill and Rudd's Tank (p = 0.050 and < 0.001, respectively). Post-treatment, there was no significant difference among stands at KA Hill (p = 0.096), but significant differences did exist among the stands at Rudd's Tank and Powerline (p = 0.001 and 0.007, respectively). No significant changes in native species ground cover were observed in the control treatments; native species ground cover increased significantly in the burn-only treatment at KA Hill, but no significant change was observed from burning only at Rudd's Tank or Powerline. Native species ground cover increased in all of the harvest only and harvest & burn treatments, and increased significantly at KA Hill and Powerline (Table 1).

Pre-treatment conditions show that there were no significant differences in alien species richness among treatments at KA Hill or Rudd's Tank (p = 0.693 and 0.124), but there was a significant difference among units pre-treatment at Powerline (p = 0.040). After treatments, all units within replicate sites were significantly different from each other (p = 0.002, < 0.001, and < 0.001 for KA Hill, Rudd's Tank, and Powerline, respectively). Alien species richness decreased in all of the control treatment areas, and decreased significantly at KA Hill. Alien species richness increased significantly in the burn-only treatment at Rudd's Tank and Powerline, but like native richness, alien richness decreased in the burn-only treatment at KA Hill. Harvest-only and harvest & burn treatments showed increases in alien species richness at KA Hill and significant increases at Rudd's Tank and Powerline (Table 1).

No significant differences in percent of ground covered by alien species existed among units within replicates before treatment (p = 0.162, 0.481, and 0.099 for KA Hill, Rudd's Tank, and Powerline, respectively), but all were significantly different 2 years after treatments were completed (Table 1). No significant changes were observed in alien species cover in any of the control treatments. The burn-only treatments at KA Hill and Rudd's Tank increased slightly in alien cover, and increased significantly at Powerline. Significant increases in alien species cover occurred in the harvest-only and harvest & burn treatments at Rudd's Tank and Powerline, and increased slightly in these treatments at KA Hill (Table 1).

Randomized Completely Blocked Design

There were no significant pre-treatment differences across the overall blocked design in native species richness among units (p = 0.302), and despite the observed increases in richness at each replicate, there were only marginal post-treatment differences (p = 0.142). Pre- and post-treatment levels of native species richness changed very little in the control treatments, but increasing levels of richness were observed from the burn-only, harvest-only, and harvest & burn treatments (Figure 2a). Native species ground cover levels were similar among treatment areas before treatments (p = 0.5214); after treatment, significant differences existed among treatment types (p = 0.029). Similar to the changes in native richness, very little

Table 1. Changes in understory vegetation at three northern Arizona study sites due to treatment at the replicate stand level. Values are reported as means (standard error), followed by paired t-test p-values, with significance α = 0.05. Significant changes between pre-treatment and post-treatment values are represented in bold (n = 10 per replicate site except for Powerline control where n = 8).

Treat	Native Richness Pre-Treat	Post-Treat	T-Test p	Native Cover Pre-Treat	Post-Treat	T-Test p	Alien Richness Pre-Treat	Post-Treat	T-Test p	Alien Cover Pre-Treat	Post-Treat	T-Test p
KA Hill												
Control	24.5 (0.7)	26.0 (0.7)	0.105	2.6 (0.6)	3.6 (0.6)	0.051	2.0 (0.4)	**1.3 (0.4)**	**0.004**	.008 (.003)	.020 (.012)	0.237
Burn Only	22.6 (0.6)	**20.0 (0.7)**	**0.004**	1.1 (0.2)	**2.4 (0.3)**	**0.002**	1.4 (0.2)	0.9 (0.2)	0.138	.004 (.001)	.011 (.010)	0.469
Harvest Only	23.9 (0.8)	26.3 (1.2)	0.135	1.2 (0.3)	**3.7 (0.7)**	**<0.001**	1.7 (0.5)	2.9 (0.7)	0.111	.003 (.001)	.012 (.005)	0.144
Harvest & Burn	25.0 (0.9)	26.3 (1.2)	0.152	1.5 (0.4)	**4.5 (0.7)**	**<0.001**	1.8 (0.3)	2.9 (0.3)	0.057	.007 (.002)	.235 (.165)	0.199
Rudd's Tank												
Control	24.6 (1.0)	24.3 (1.0)	0.656	1.5 (0.4)	1.5 (0.4)	0.858	1.2 (0.3)	1.0 (0.3)	0.591	.003 (.001)	.002 (.001)	0.193
Burn Only	34.1 (1.0)	34.0 (1.2)	0.932	2.5 (0.2)	3.2 (0.4)	0.094	2.0 (0.3)	**3.5 (0.4)**	**<0.001**	.007 (.002)	.012 (.003)	0.063
Harvest Only	27.7 (1.0)	**36.0 (1.2)**	**<0.001**	0.9 (0.1)	**3.1 (0.3)**	**<0.001**	1.4 (0.3)	**5.1 (0.5)**	**<0.001**	.004 (.002)	**.029 (.006)**	**0.004**
Harvest & Burn	33.6 (1.1) <0.001	**39.4 (1.5)** <0.001	**0.007**	3.5 (0.6)	6.4 (1.4)	0.065	2.3 (0.5)	**4.3 (0.5)**	**0.034**	.003 (.004)	**.014 (.023)**	**0.035**
Powerline												
Control	16.0 (2.0)	15.4 (2.0)	0.503	1.6 (0.7)	1.7 (0.7)	0.578	1.8 (0.7)	1.3 (0.7)	0.351	.005 (.003)	.010 (.007)	0.429
Burn Only	17.7 (1.4)	**25.1 (1.8)**	**0.001**	1.4 (0.3)	1.2 (0.5)	0.510	0.8 (0.4)	**4.7 (0.5)**	**<0.001**	.001 (.001)	**.018 (.006)**	**0.020**
Harvest Only	16.3 (1.7)	**25.1 (2.1)**	**0.004**	1.5 (0.5)	**3.4 (0.6)**	**<0.001**	1.2 (0.2)	**2.6 (0.8)**	**<0.001**	.008 (.005)	.487 (.160)	0.008
Harvest & Burn	18.4 (1.3)	**26.1 (1.0)**	**<0.001**	0.8 (0.4)	**4.4 (0.9)**	**0.002**	2.5 (0.4)	**7.4 (0.4)**	**<0.001**	.005 (.002)	**.558 (.096)**	**<0.001**

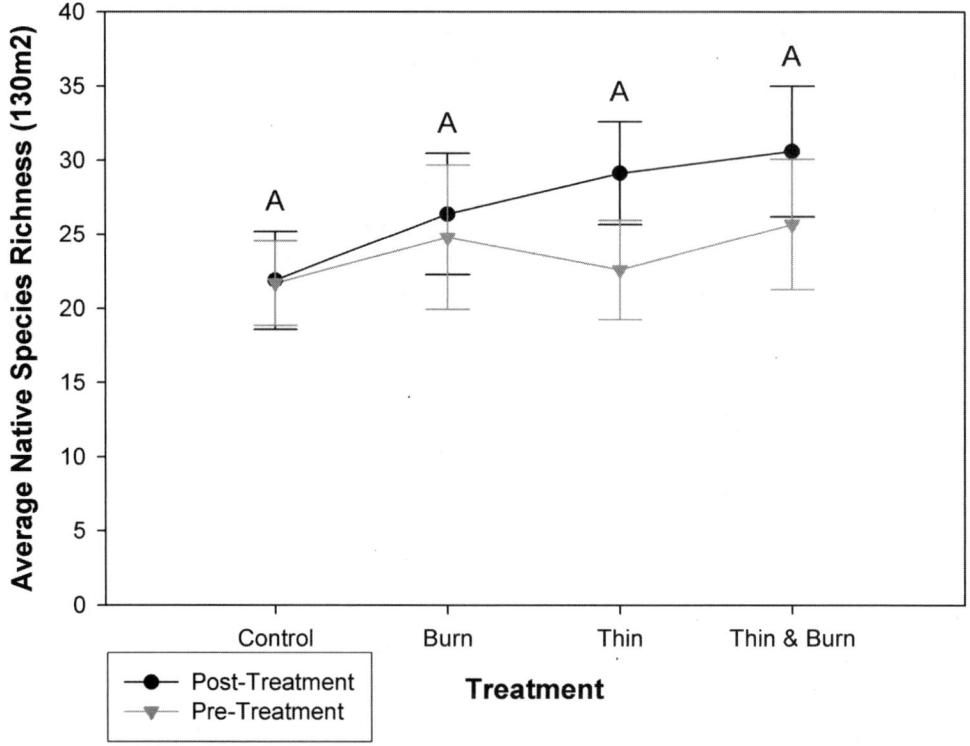

Figure 2. Mean native species richness (a) at the 130 m² scale and mean percent ground cover (b, see next page) across three northern Arizona study sites. Error bars represent standard error; superscript values Tukey's HSD homogenous subsets, * denotes significant increases between pre- and post-treatment values, with a significance of α = 0.05. No significant differences were detected among pre-treatments levels.

difference was detected in the control treatments, while the burn-only treatment increased slightly. Percent native species ground cover increased significantly from pre- to post-treatment in the harvest-only and harvest & burn treatments, and post-treatment levels of native ground cover were significantly higher in the harvest & burn than in other treatments (Figure 2b).

No pre-treatment differences in alien species richness existed among the treatment areas ($p = 0.228$), but differences among treatment became apparent post-treatment ($p = 0.021$). Alien species richness increased in all treated areas but decreased in the control area. The harvest-only and harvest &

burn treatments were significantly higher in alien species richness than the control (see Figure 3a). Similarly, there were no pre-treatment differences in alien species percent ground cover among treatments ($p = 0.486$), but there were significant differences among areas post-treatment ($p = 0.021$). Again, there was little change in alien species percent ground cover in control and burn-only areas, but the harvest-only and harvest & burn treatments were significantly higher in alien species ground cover (Figure 3b).

DISCUSSION

Understory vegetation response at the Southwest Plateau FFS site showed that

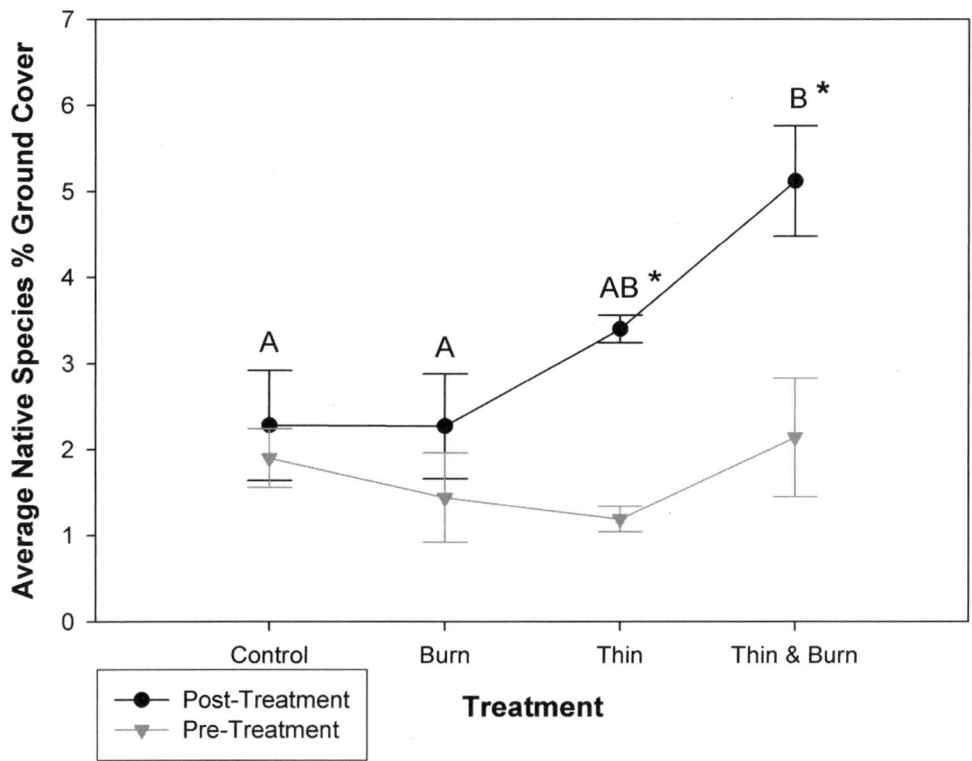

harvesting and burning, alone and in combination, increases both native and alien species richness and abundance. As management intensities increased (burn only, harvest only, harvest & burn), the understory responses increased. Uresk and Severson (1989) noted the same types of increases in understory production from a reduction in overstory density, relating understory production to overstory basal area of ponderosa pine. The post-treatment target basal area of 13 m^2/ha at the Southwest Plateau FFS sites falls within the 10–14 m^2/ha threshold pointed out by Uresk and Severson to maximize productivity of both the understory and overstory in Black Hills ponderosa pine forests.

A similar overstory-understory trend was also observed by Moore and Deiter (1992) in northern Arizona, relating overstory stand density index (SDI) to understory produc-

tion. SDI measures are best applied to even-aged stands, as the method relies on the quadratic mean diameter (QMD) of the stand. In the multi-aged group selection restoration treatments at our study sites, however, each diameter class can represent the QMD, which allows SDI to be computed for each diameter class and summed. Using this method, the pre-treatment SDIs of the harvested stands at our study sites were 481, 491, and 508 at KA Hill, Rudd's Tank, and Powerline, respectively. After harvesting, the SDI of these stands was 237, 264, and 202. This large drop in SDI at our study sites moved the stands from the low productivity break point of 500 SDI described by Moore and Deiter (1992) to well within the productive zone under an SDI of 500.

Areas that were mechanically treated in this FFS understory study showed significant but small increases in native species

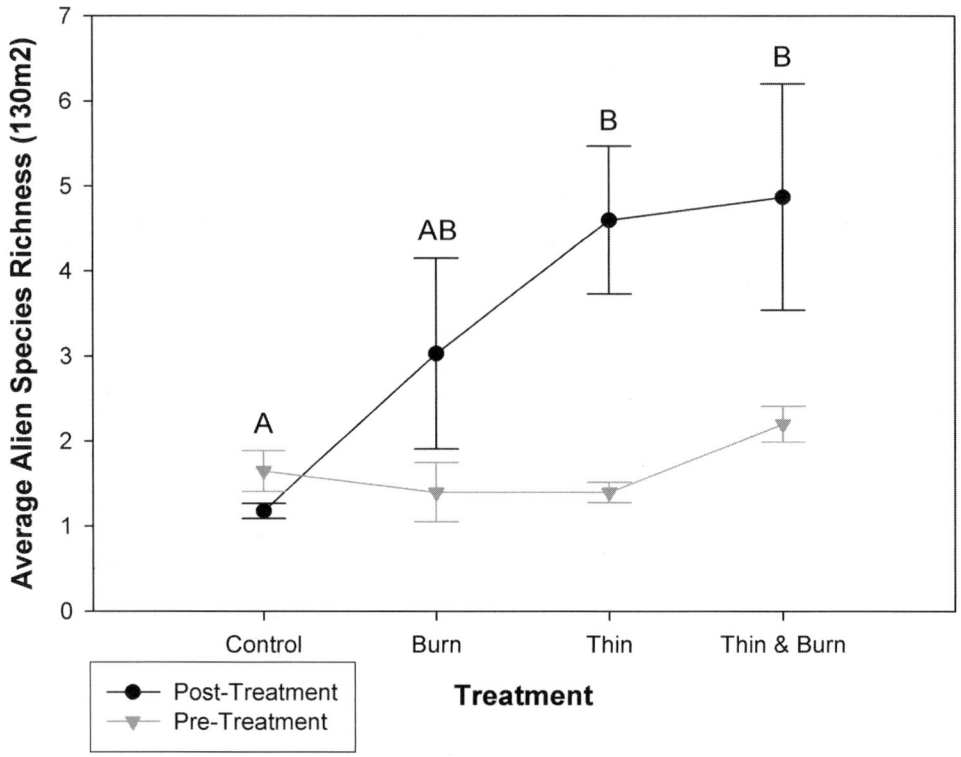

Figure 3. Mean alien species richness (a) at the 130 m² scale and mean percent ground cover (b, see next page) across three northern Arizona study sites. Error bars represent standard error; superscript values represent Tukey's HSD homogenous subsets with a significance of α = 0.05

ground cover (2–3%) and richness (~5 species, a 20% increase). Harvesting also resulted in smaller increases in alien species richness and ground cover that were significantly greater than that of the controls. The trend of increasing richness and abundance of alien and native species in harvest-only treatments was also found by Wienk et al. (2004), whose harvesting treatments reduced basal area to 12 m²/ha, and Griffis et al. (2001) who found an increase in graminoid production as management intensities increased, but also observed only a slight increase in richness between unmanaged stands and thinned stands. Griffis et al. (2001) acknowledged that their thinned stands were still of high basal area (16 to 23

m²/ha), a disturbance level which may not be intensive enough for a large understory response as described by the Deiter and Moore (1992) thresholds.

Both native and alien species richness and cover responded most strongly to the combination of harvesting and burning treatments, yielding levels significantly higher than in the controls, where alien cover and native richness and cover stayed relatively consistent during the 4 years of this study despite rainfall patterns. Harvesting and burning was found to increase understory production by Covington et al. (1997), and Abella and Covington (2004) found both native and alien species richness increased after harvesting and burning.

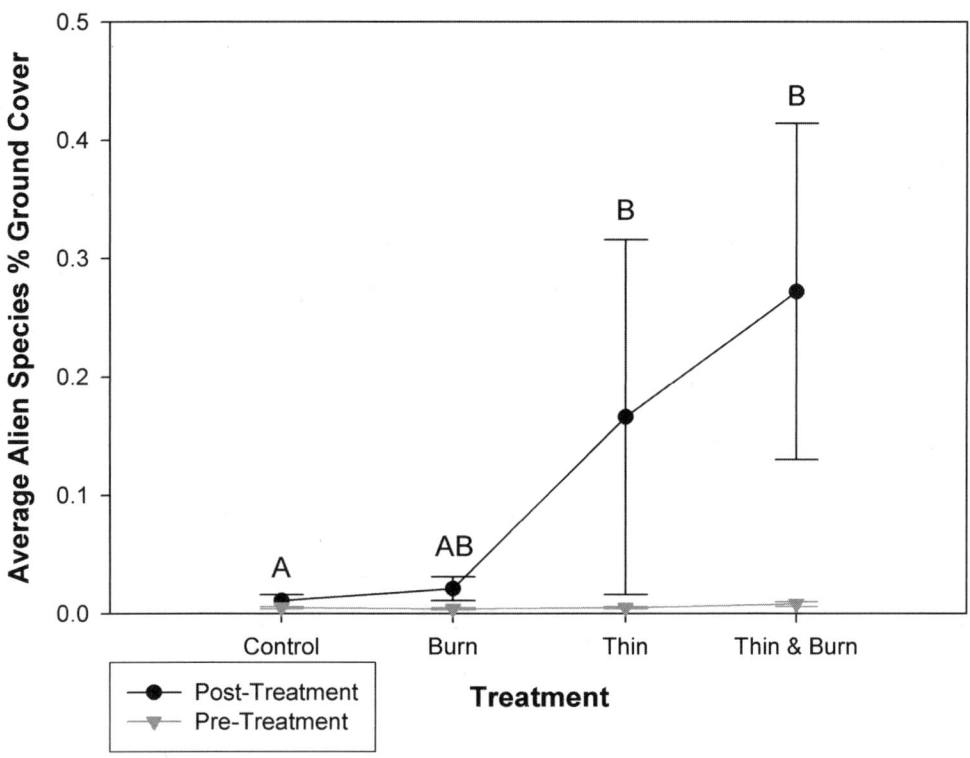

Wienk et al. (2004) found understory production to increase more with thinning and burning than with thinning or burning alone, and found a slight increase in species richness with thinning and burning as opposed to thinning or burning alone. Vose and White (1991) found species-specific responses to burning. Squirreltail (*Elymus elymoides*), the only true grass occurring at every inventory plot in our study, was shown by Vose and White (1991) to increase in biomass significantly in open "sawtimber" stands, similar to our harvested units, but not in "pole" stands such as our burn-only treatment.

Burning alone stimulated insignificant increases in native species richness and cover given only minor changes in overstory condition and relatively little site disturbance. Harris and Covington (1983) and

Laughlin et al. (2004) noted such an increase in understory production due to prescribed burning alone, and Andariese and Covington (1986) found that understory vegetation increased 2 years after the introduction of prescribed fire. In a similar study in the Black Hills ponderosa pine forests of South Dakota, Wienk et al. (2004) found that biomass and richness increased in burn-only treatments. The burn-only treatments in our study showed even smaller increases in alien species richness and cover, consistent with other similar studies treated only with fire (Crawford et al. 2001; Griffis et al. 2001; Wienk et al. 2004). Laughlin et al. (2004) did not find that exotic species increased after burning only, but they acknowledged that this may be due to the remote location of their study area on the Kaibab Plateau. This is unlike our study area in a wildland-urban interface, which inherently has more traffic and therefore vectors allowing non-native species to colonize.

Alien species richness surprisingly decreased (0.15) with time in our control areas, in contrast to the recent findings of Fornwalt et al. (2003) who found that non-native species have colonized both managed and undisturbed areas equally in ponderosa pine forests on the Colorado Front Range. Several recent studies acknowledge that management disturbance provides a vector for alien species to colonize (Crawford et al. 2001; Sieg 2003; Korb et al. 2004; Wienk et al. 2004), but there is little documentation of a decrease in alien species richness when left undisturbed. On the control plots, we observed 9 of the 15 total alien species in this study. Field bindweed (*Convolvulus arvensis*), lambsquarters (*Chenopodium album*), prickly lettuce (*Lactuca serriola*), and the common dandelion (*Taraxacum officinale*) all declined during the 4 years of this study across sites. This decrease in alien species richness and frequency may be due to species-specific natural germination cycles of annual and biennial plants relative to our limited 2-year sampling. While late summer rainfall was very similar between the 2 inventory years, late winter was much wetter in 2005, when 39.0 cm of precipitation (200% of average) fell between January and April, compared to 17.7 cm (91% of average) in the same time period in 2001. Similarly, the wetter spring that northern Arizona experienced during the post-treatment inventory may have allowed native cool season (C3) grass species to out-compete alien species. Cover of C3 grasses found on the control plots (*Elymus elymoides, Festuca arizonica*, and *Poa fendleriana*) averaged 0.19 percent in 2001, but 0.52 percent with the wetter winter in 2005.

FIRE AND FIRE SURROGATE NETWORK

Metlen et al. (2004) found the opposite general trends that we observed at a similar FFS study site in an eastern Oregon ponderosa pine–Douglas fir forest, where harvesting was conducted to reduce basal area from 18.5 to 10.3–14.4 m^2/ha. Both treatments involving prescribed fire showed a neutral richness response and reduced cover, while the harvest-only treatment saw a reduction in richness and little change in cover (Metlen et al. 2004). One possible explanation for this difference in findings is timing of their inventory, as responses are variable the first year after burning, appearing more strongly in the second year (Andariese and Covington 1986). Another difference in these modest responses may be due to the light fuels reduction thinning, which started as more productive, relatively open stands (18.5 m^2/ha, 364.4 stems/ha), when compared with the heavier release harvests of our study or the study area of Wienk et al., which started as less productive, denser stands (27.0 m^2/ha, 501.5 stems/ha and 60.0 m^2/ha, 2700 stems/ha, respectively). Comparison of the effects from similar restorative fuel-reduction treatments in different geographic areas is a broad goal of the Fire and Fire Surrogate Program. As more sites continue to report findings, further comparisons, such as this one, can be made to evaluate the ecosystem responses in specific regions.

CONCLUSIONS

Understory vegetation responses to the harvesting and burning treatments in our study were strong, and they suggest that more intensive management regimes yield higher native and alien vegetation cover and richness. This general trend has been documented repeatedly (e.g., Crawford et al. 2001; Griffis et al. 2001; Wienk et al. 2004; Laughlin et al. 2004). We will continue to monitor these permanent plots to document whether these trends continue over time or plateau or decline as time from disturbance passes.

Promoting a more robust understory may be a desired management objective for grazing and foraging, soil stabilization, nutrient cycling, influencing fire behavior, ecological restoration, or simply for aesthetics. This study showed that these goals can be achieved at different levels by changing intensities of management activities. However, the risk of invasive alien species colonizing after treatment should be weighed carefully, as any of the management activities presented here provided a vector for colonization in this landscape.

ACKNOWLEDGMENTS

This research was supported in part by funds provided by the USDA Forest Service, Rocky Mountain Research Station, through agreement 03-JV-11221615-290. Thanks to M. Hurteau, C. Sieg, and anonymous reviewers for their helpful feedback.

LITERATURE CITED

Abella, S. R., and W. W. Covington. 2004. Monitoring an Arizona ponderosa pine restoration: Sampling efficiency and multivariate analysis of understory vegetation. Restoration Ecology 12 (3): 359.

Andariese, S. W., and W. W. Covington. 1986. Changes in understory production for three prescribed burns of different ages in ponderosa pine. Forest Ecology and Management 14: 193–203.

Bailey, J. D., and W. W. Covington. 2002. Evaluating ponderosa pine regeneration rates following ecological restoration treatments in northern Arizona, USA. Forest Ecology and Management 155: 271–278.

Buckley, S., editor. 2005. Plants of Northern Arizona Forests, draft version. Ecological Restoration Institute, Northern Arizona University, Flagstaff.

Covington, W. W., and M. M. Moore. 1994. Southwestern ponderosa pine forest structure and resource conditions: Changes since Euro-American settlement. Journal of Forestry 92: 39–47.

Covington, W. W., P. Z. Fulé, M. M. Moore, S. C. Hart, T. E. Kolb, J. N. Mast, S. S. Sackett, and M. R. Wagner. 1997. Restoring ecosystem health in ponderosa pine forests of the Southwest. Journal of Forestry 95: 23–29.

Crawford, J. A., C. H. A. Wahren, S. Kyle, and W. H. Moir. 2001. Responses of exotic plant species to fires in *Pinus ponderosa* forests in northern Arizona. Journal of Vegetation Science 12: 261–268.

Fornwalt, P. J., M. R. Kaufmann, L. S. Huckaby, J. M. Stoker, and T. J. Stohlgren. 2003. Non-native plant invasions in managed and protected ponderosa pine/Douglas-fir forests of the Colorado Front Range. Forest Ecology and Management 177: 515–527.

Griffis, K. L., J. A. Crawford, M. R. Wagner, and W. H. Moir. 2001. Understory response to management treatments in northern Arizona ponderosa pine forests. Forest Ecology and Management 146 (1-3): 239–245.

Harris, G. R., and W. W. Covington. 1983. The effect of a prescribed fire on nutrient concentration and standing crop of understory vegetation in ponderosa pine. Canadian Journal of Forest Research 13(3): 501–507.

Korb, J. E., and J. D. Springer. 2003. Understory vegetation. In Ecological Restoration of Southwestern Ponderosa Pine Forests, edited by P. Friederici. Island Press, Washinton, D.C.

Korb, J. E., N. C. Johnson, and W. W. Covington. 2004. Slash pile burning effects on soil biotic and chemical properties and plant establishment: Recommendations for amelioration. Restoration Ecology 12(1): 52–62.

Laughlin, D. C., J. D. Bakker, M. T. Stoddard, M. L. Daniels, J. D. Springer, C. N. Gildar, A. M. Green, and W. W. Covington. 2004. Toward reference conditions: Wildfire effects on flora in an old-growth ponderosa pine forest. Forest Ecology and Management 199: 137–152.

Metlen, K. L., C. E. Fiedler, and A. Youngblood. 2004. Understory response to fuel reduction treatments in the Blue Mountains of Northeastern Oregon. Northwest Science 78 (3): 175–185.

Moore, M. M., and D. A. Deiter. 1992. Stand density index as a predictor of forage production in northern Arizona pine forests. Journal of Range Management 45 (3): 267–271.

Moore, M. M., W. W. Covington, and P. Z. Fulé. 1999. Reference conditions and ecological restoration: A southwestern ponderosa pine perspective. Ecological Applications 9(4): 1266.

Naumburg, E., and L. E. DeWald. 1999. Relationships between *Pinus ponderosa* forest structure, light characteristics, and understory graminoid species presence and abundance. Forest Ecology and Management 124 (2-3): 205–215.

Sieg, C. H., B. G. Phillips, and L. P. Moser. 2003. Exotic invasive plants. In Ecological Restoration of Southwestern Ponderosa Pine Forests, edited by P. Friederici. Island Press, Washington, D.C.

Stohlgren, T. J., M. B. Falkner, and L. D. Schell. 1995. A modified-Whittaker nested vegetation sampling method. Vegetatio 117: 113–121.

Uresk, D. W., and K. E. Severson. 1989. Understory-overstory relationships in ponderosa pine forests, Black Hills, South Dakota. Journal of Range Management 42: 230–208.

USDA, NRCS. 2005. The PLANTS Database, Version 3.5 (http://plants.usda.gov). Data compiled from various sources by Mark W. Skinner. National Plant Data Center, Baton Rouge, Louisiana.

Vose, J. M., and A. S. White. 1991. Biomass response mechanisms of understory species the first year after prescribed burning in an Arizona ponderosa-pine community. Forest Ecology and Management 40: 175–187.

Weatherspoon, C. P. 2000. A proposed long-term national study of the consequences of fire and fire-surrogate treatments. In Proceedings of the Joint Fire Science Conference and Workshop, edited by Neuenschwander and Ryan, pp. 117–126. University of Idaho, Moscow.

Wienk, C. L., C. H. Sieg, and G. R. McPherson. 2004. Evaluating the role of cutting treatments, fire and soil seed banks in an experimental framework in ponderosa pine forests of the Black Hills, South Dakota. Forest Ecology and Management 192 (2–3): 375–393.

FIVE YEARS OF VEGETATION CHANGE FOLLOWING HIGH-SEVERITY FIRE AND FIRE-FIGHTING ACTIVITIES IN GRAND CANYON NATIONAL PARK

Julie Crawford

Prior to the establishment of Grand Canyon National Park (GRCA) in 1919 and for some years thereafter, livestock grazing took place over the entire park, with the exception of concessionaire areas at Bright Angel Point on the North Rim and the Village area on the South Rim (Crosby 1923). This grazing, coupled with periods of deer overpopulation (Binkley et al. 2006) and decades of fire suppression (Fulé et al. 2002; Wolf and Mast 1998; White and Vankat 1993), has resulted in reduced herbaceous layers, fine fuels, and fire frequency. In the forest communities of the North Rim, this fire frequency reduction has led to invasions of small shade-tolerant and fire-intolerant *Abies concolor* (Gord. & Glend.) Lindl. ex Hildebr. (white fir), homogeneity of forest structure, and high fuel loads. Although at least some of the North Rim forests studied by White and Vankat (1993) and Fulé et al. (2003) were characterized by naturally high-severity fire regimes, the current structure of these forests has resulted in a greater likelihood of higher severity fire at larger than historic scales. This is a serious concern for park resource managers and it has led to increased prescription burns in these forests in recent decades.

The Outlet fuel-reduction prescription fire on the North Rim, which was ignited near Outlet Canyon on 24 April 2000, burned 465 acres (188 ha). However, this prescribed fire jumped its intended boundary, and the National Park Service classified the burn as the Outlet wildfire on 9 May 2000. The Outlet wildfire, referred to herein as the Outlet fire, which began near Bright Angel Spring, eventually burned 8776 acres (3552 ha) in GCNP (Figure 1). It spread in a northeast direction to burn 5109 acres (2068 ha) of Forest Service land, and continued to burn for 36 days until it was contained on 15 June. Since the park's inception, other than the Saddle Mountain fire of 1960, fire has been suppressed throughout the area burned by the Outlet fire.

The North Rim of GRCA is diverse elevationally (roughly 6000–9200 ft, or 1829–2804 m) and topographically (slope aspect, position, and inclination), supporting a variety of vegetation communities including pinyon-juniper woodland, ponderosa pine forest, montane grassland, mixed conifer forest, and spruce-fir forest (Rasmussen 1941; Merkle 1954). Although the Outlet fire burned in portions of all of these communities, the mixed conifer forest was the dominant community altered by the fire. Study transects fell within the mixed conifer forest and montane grassland communities.

The mixed conifer forest is characterized by *Pinus ponderosa* (ponderosa pine), *Pseudotsuga menziesii* (Mirbel) Franco (Douglas fir), *A. concolor*, *Picea pungens* Engelm. (blue spruce), *Pinus strobiformis* Engelm. (southwestern white pine), and *Populus tremuloides* Michx. (quaking aspen; Moir and Ludwig 1979). This is largely due to moisture availability, warm daytime temperatures during the growing season, and generally deep permeable soils capable of storing water (Vankat 2004). The mixed conifer forests of the North Rim occur between 8250 and 8700 ft (2515–2652 m) in elevation with montane

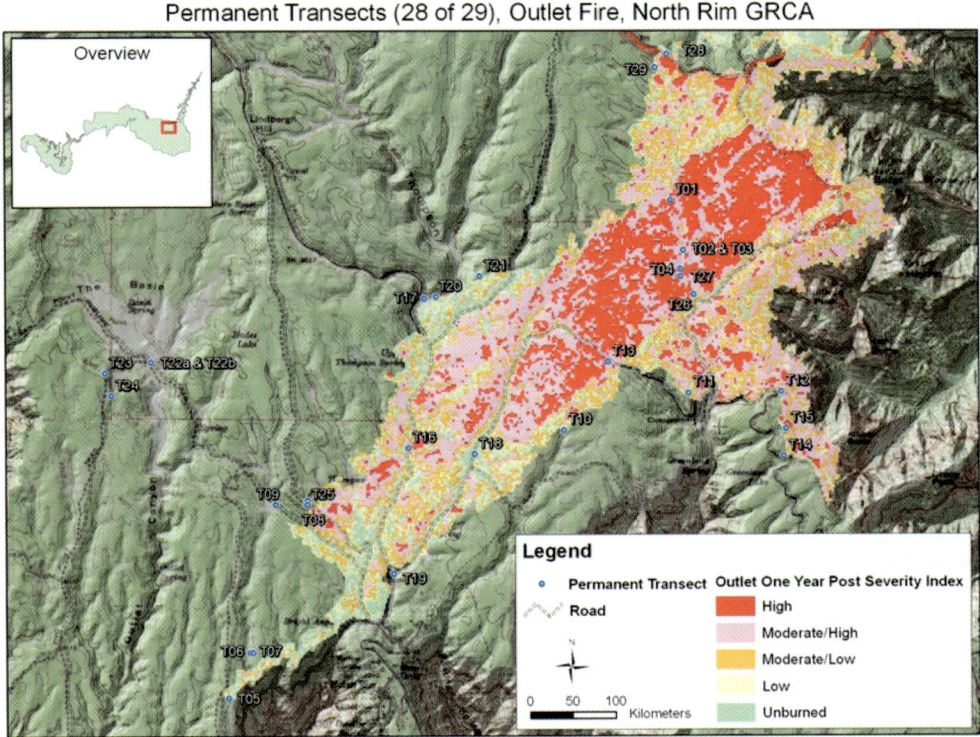

Figure 1. Location, severity, and boundaries of the 2000 Outlet fire, North Rim, GRCA. Plot locations of 28 of the 29 permanent point-line transects studied between 2000 and 2004 are shown in blue.

grassland occurring in level basins and valley bottoms throughout the forested areas (Merkle 1962; Rasmussen 1941). The climate of the North Rim consists of cold winters, warm to cool summers, and both winter and mid-summer precipitation (Vankat 2004).

Much of the Outlet fire resulted in high-severity impact—that is, stand-replacing fire where most if not all overstory trees were killed. This was especially the case in the mixed conifer forest of the Bright Angel Creek drainage, an area of complete overstory loss, where GRCA resource managers were most concerned with the possibility of erosion on the steep slopes. In addition, a great deal of fire-fighting equipment and many personnel were involved with fighting this fire. These activities impacted soil and vegetation directly through removal of vegetation and duff and by vegetation trampling and soil compaction. These types of disturbance have been shown to provide an avenue for exotic plant invasion and spread (Baskin 1998; Kotanen 1997; Fox and Fox 1986; Elton 1958)—an additional concern to GRCA resource managers. Although numerous studies have investigated the effects of fire-fighting foams and retardants on vegetation (see Adams and Simmons 1999), there is apparently only one study addressing vegetation response to mechanical fire-fighting activity (Caling and Adams 1999). Similarly, only Huisinga et al. (2005) have addressed vegetation response to high-severity fire in mixed conifer forests of the Southwest. The objectives of this study were to examine vegetation response to both high-severity burning in mixed conifer forest and to fire-fighting activity.

In order to assess vegetation change, in particular the response of exotic species, in areas at risk of severe erosion and in areas of fire-fighting activity, a system of permanent point-line vegetation transects was established. Because of the linear nature of handline and drop-point areas and for economy of time and resources, linear transects were chosen for this study. Burn transects were placed in the high-severity burn area of Bright Angel Creek; drop and dip transects were placed in the areas used to stage fire-fighting activity. Handline transect locations were chosen at random. These transects were established 3 months post fire and were measured on several occasions during the summers of 2001 and 2002, and one time per year in late July in 2003 and 2004. National Park Service reports for each year of this study are on file with the GRCA Science Center (Murov 2000; Eisenberg et al. 2001; Eisenberg and Watters 2002; Crawford and Whitchurch 2003; Crawford and Straka 2004).

Several drop-points (areas of fire-fighting equipment staging and fire-fighter drop-off and pick-up) and dip sites (locations of portable water storage tanks for use by helicopter dip buckets) were established during fire-fighting efforts. These sites, collectively referred to here as drop-points, were combined for analysis because only one transect crossed a dip site and it was destroyed by heavy equipment between the 2002 and 2003 visits. Fire-fighting efforts also included the construction of approximately 24 miles (38.6 km) of hand-dug fireline, or "handline." Constructing handline involved scraping the ground down to mineral soil (removing all organic and combustible materials) using hand tools and, following fire suppression, involved erosion control and contour resurfacing.

METHODS
Field

Twenty-nine line transects (7 high-severity burn, 13 drop-point, and 9 handline) were established in the summer of 2000, with initial measurements taken between September and October of that year. To reveal the full vegetation expression throughout the growing season, these transects were read four times from July to September in 2001 and 2002. During 2003 and 2004, transects were revisited and measured in late July following the onset of monsoons when most vegetation is identifiable.

UTM coordinates for transect start and end posts were recorded using Garmin GPS units (map datum NAD27) with compass bearings between permanent posts recorded (13° east declination). Transect photographs from both the start and end posts were initiated in 2001 and repeated during each year of the study. In all years following initial setup, the permanent posts of most transects were relocated using a combination of UTM coordinates, photographs, maps, and written narratives. The few transects with missing start or end posts were aligned using photographs.

Each transect was 25 m long, with the exception of transects 6 and 7, which were both 12.5 m long and in close proximity to one another; their data were combined for analysis. A tape measure was stretched the length of each transect and secured at both ends onto the start and end posts. Every 0.25 m, a 3/4-inch diameter stick was placed vertically on the left side of the tape. Whatever the stick was touching at that spot—biotic or abiotic (soil, rock, litter, course woody debris)—was recorded. One hundred points were read for each transect.

Unknown species were collected from areas near, but not on transects and pressed for later identification. Specimens were identified in the Deaver Herbarium of Northern Arizona University and deposited both in the Deaver Herbarium and at the GRCA Museum Collections Herbarium.

Analysis

Summary statistics were calculated for each year of study across lifeform group (graminoid, forb, shrub, tree seedling, ferns and fern allies) and across disturbance type (burn, handline, drop-point); comparative graphs and charts were also created. For comparison purposes, all summaries and statistics contained in this report relate to data collected in September–October of 2000, and July of 2001, 2002, 2003, and 2004.

Species diversity of each transect was calculated ($\log_{10}$) using the Shannon index (H'), taking into account both abundance (richness) and evenness (Magguran 1988) and species richness (total number of species per sample; Moore and Chapman 1986). Differences in diversity and richness between disturbance types and years were analyzed with repeated measures MANOVA using a Greenhouse-Geiser adjustment

to account for autocorrelation. If treatment effects were significant, treatment differences within years were tested using one-way ANOVA and Tukey's post-hoc tests.

PC-Ord Indicator Species Analysis was used to calculate an importance value and associated p-value (Monte Carlo test) for each species in the first and last years of the study. These values were used to identify the most important species that characterized the three disturbance types immediately following the fire and most recently, in 2004. Indicator values were calculated by combining relative abundance (proportional abundance of a particular species in a particular group relative to the abundance of that species in all groups) and relative frequency (proportional frequency of the species in each group) following the method of Dufrene and Legendre (1997).

Using the program PC-Ord, nonmetric multidimensional scaling ordination was used to look for trends; this method created a spatial representation of species composition and cover data from our 29 transects over all years of study. We used SYSTAT to create 5 percent confidence ellipses around years and treatments. These ellipses serve as a visual aid to help distinguish the center of the three groups of points. Using the software PRIMER, the non-parametric ANOSIM procedure of Clarke (1993) was used to compare floristic differences among the three disturbance types and between the years 2000 and 2004.

RESULTS

More than 100 species of vascular plants were recorded from transects over the 5 years of study (Table 1). Eight exotic species were found in the transects; four are considered by GRCA biologists to be a high priority for management, based on their present level of impact and their innate ability to become invasive, but with low feasibility of control: smooth brome (*Bromus inermis* Leyss.), cheatgrass (*Bromus tectorum*), Kentucky bluegrass (*Poa pratensis* L.), and common sheep sorrel (*Rumex acetosella* L.). The remaining four species are lower in priority as they have less impact on park

Table 1. List of all plants found on 29 transects of the Outlet fire during 2000–2004. Nomenclature follows NRCS (2005).

Scientific Name	Common Name	Family	Nativity	Life Form
Abies concolor (Gord. & Glend) Lindl. ex Hildebr.	White fir	Pinaceae	Native	Tree
Achillea millefolium L.	Western yarrow	Asteraceae	Native	Herbaceous
Achnatherum nelsonii (Scribn.) Barkworth	Columbia needlegrass	Poaceae	Native	Graminoid
Agoseris glauca (Pursh) Raf.	Pale agoseris	Asteraceae	Native	Herbaceous
Agropyron spp.	Wheatgrass	Poaceae	Native	Graminoid
Agrostis scabra Willd.	Rough bentgrass	Poaceae	Native	Graminoid
Antennaria marginata Greene	Whitemargin pussytoes	Asteraceae	Native	Herbaceous
Antennaria parvifolia Nutt.	Rocky Mountain pussytoes	Asteraceae	Native	Herbaceous
Antennaria rosulata Rydb.	Kaibab pussytoes	Asteraceae	Native	Herbaceous
Arabis drummundii Gray	Drummond's rockcress	Brassicaceae	Native	Herbaceous
Arenaria fendleri Gray	Fendler's sandwort	Caryophyllaceae	Native	Herbaceous
Arenaria lanuginosa (Michx.) Rohrb.	Spreading sandwort	Caryophyllaceae	Native	Herbaceous
Artemisia campestris L.	Common sagewort	Asteraceae	Native	Herbaceous
Artemisia dracunculus L.	False tarragon	Asteraceae	Native	Herbaceous
Artemisia filifolia Torr.	Sand sagebrush	Asteraceae	Native	Herbaceous
Artemisia ludovisiana Nutt.	Cudweed sagewort	Asteraceae	Native	Herbaceous
Astragalus humistratus Gray	Groundcover milkvetch	Fabaceae	Native	Herbaceous
Blepharoneuron tricholepis (Torr.) Nash	Pine dropseed	Poaceae	Native	Graminoid
Bouteloua gracilis (Willd. ex Kunth) Lag. ex Griffiths	Blue grama	Poaceae	Native	Graminoid
Bromus anomalus Rupr. ex Fourn.	Nodding brome	Poaceae	Native	Graminoid
Bromus ciliatus L.	Fringed brome	Poaceae	Native	Graminoid
Bromus inermis Leyss.	Smooth brome	Poaceae	Exotic	Graminoid
Bromus tectorum L.	Cheatgrass	Poaceae	Exotic	Graminoid
Campanula parryi Gray	Parry's bellflower	Campanulaceae	Native	Herbaceous
Carex bolanderi Olney	Bolander's sedge	Cyperaceae	Native	Graminoid
Carex geyeri Boott	Elk sedge	Cyperaceae	Native	Graminoid
Carex rossii Boott	Ross's sedge	Cyperaceae	Native	Graminoid
Castilleja miniata Dougl. ex Hook.	Giant red Indian paintbrush	Scrophulariaceae	Native	Herbaceous
Ceanothus fendleri Gray	Fendler's ceanothus	Rhamnaceae	Native	Shrub
Chenopodium fremontii S. Wats.	Fremont's goosefoot	Chenopodiaceae	Native	Herbaceous
Cirsium wheeleri (Gray) Petrak	Wheeler's thistle	Asteraceae	Native	Herbaceous
Corydalis aurea Willd.	Golden corylalis	Fumariaceae	Native	Herbaceous
Cryptantha setosissima (Gray) Payson	Bristly catseye	Boraginaceae	Native	Herbaceous
Draba sp.	Draba	Brassicaceae	Native	Herbaceous
Elymus elymoides (Raf.) Swezey	Squirreltail	Poaceae	Native	Graminoid
Epilobium angustifolium L.	Fireweed	Onagraceae	Native	Herbaceous

Table 1 (continued)

Scientific Name	Common Name	Family	Nativity	Life Form
Equisetum sp.	Horsetail	Equisetaceae	Native	Fern/allie
Erigeron flagellaris Gray	Trailing fleabane	Asteraceae	Native	Herbaceous
Erigeron formosissimus Greene	Beautiful fleabane	Asteraceae	Native	Herbaceous
Erigeron spp.	Fleabane	Asteraceae	Native	Herbaceous
Eriogonum racemosum Nutt.	Redroot buckwheat	Polygonaceae	Native	Herbaceous
Eriogonum umbellatum Torr.	Sulfer buckwheat	Polygonaceae	Native	Herbaceous
Festuca arizonica Vasey	Arizona fescue	Poaceae	Native	Graminoid
Festuca ovina L.	Sheep fescue	Poaceae	Native	Graminoid
Fragaria virginiana Duchesne	Virginia strawberry	Rosaceae	Native	Herbaceous
Gayophytum diffusum Torr. & Gray	Bigflower Groundsmoke	Onagraceae	Native	Herbaceous
Gentianella tenella (Rottb.) Boerner	Dane's dwarf gentian	Gentianaceae	Native	Herbaceous
Geranium richardsonii Fisch. & Trautv.	Richardson's geranium	Geraniaceae	Native	Herbaceous
Hackellia sp.	Hackellia	Boraginaceae	Native	Herbaceous
Helianthella quinquenervis (Hook.) Gray	Nodding dwarf-sunflower	Asteraceae	Native	Herbaceous
Heliomeris multiflora Nutt.	Showy goldeneye	Asteraceae	Native	Herbaceous
Heterotheca villosa (Pursh) Shinners	Hairy goldaster	Asteraceae	Native	Herbaceous
Hieracium fendleri Schultz-Bip.	Yellow hawkweed	Asteraceae	Native	Herbaceous
Hymenoxys cooperi (Gray) Cockerell	Cooper's rubberweed	Asteraceae	Native	Herbaceous
Hymenoxys subintegra Cockerell	Arizona rubberweed	Asteraceae	Native	Herbaceous
Koeleria macrantha (Ledeb.) J.A. Schultes	June grass	Poaceae	Native	Graminoid
Lactuca tatarica (L.) C.A. Mey.	Blue lettuce	Asteraceae	Native	Herbaceous
Ligusticum porteri Coult. & Rose	Porter's wild lovage	Apiaceae	Native	Herbaceous
Linum lewisii Pursh	Blue flax	Linaceae	Native	Herbaceous
Lotus wrightii (Gray) Greene	Wright's deervetch	Fabaceae	Native	Herbaceous
Lupinus sp.	Lupine	Fabaceae	Native	Herbaceous
Machaeranthera bigelovii var. *mucronata* (Greene) B.L. Turner	Bigelow's tansyaster	Asteraceae	Native	Herbaceous
Machaeranthera canescens (Pursh) Gray	Hoary aster	Asteraceae	Native	Herbaceous
Mahonia repens (Lindl.) G. Don	Oregon grape	Berberidaceae	Native	Shrub
Medicago lupulina L.	Black medic	Fabaceae	Exotic	Herbaceous
Mertensia franciscana Heller	Franciscan bluebells	Boraginaceae	Native	Herbaceous
Muhlenbergia montana (Nutt.) A.S. Hitchc.	Mountain muhly	Poaceae	Native	Graminoid
Oreochrysum parryi (Gray) Rydb.	Parry's goldenrod	Asteraceae	Native	Herbaceous
Orthocarpus luteus Nutt.	Yellow owlclover	Scrophulariaceae	Native	Herbaceous
Packera multilobata (Torr. & Gray ex Gray) W.A. Weber & A. Löve	Lobeleaf groundsel	Asteraceae	Native	Herbaceous

Table 1 (continued)

Scientific Name	Common Name	Family	Nativity	Life Form
Pascopyrum smithii (Rydb.) A. Löve	Western wheatgrass	Poaceae	Native	Graminoid
Pedicularis centranthera Gray	Dwarf lousewort	Scrophulariaceae	Native	Herbaceous
Penstemon barbatus (Cav.) Roth	Beardlip penstemon	Scrophulariaceae	Native	Herbaceous
Penstemon virgatus Gray	Upright blue beardtongue	Scrophulariaceae	Native	Herbaceous
Phacelia egena (Greene ex Brand) Greene ex J.T. Howell	Kaweah River Scorpion-weed	Scrophulariaceae	Native	Herbaceous
Phlox austromontana Coville	Mountain phlox	Polemoniaceae	Native	Herbaceous
Picea engelmannii Parry ex Engelm	Engelmann spruce	Pinaceae	Native	Tree
Poa fendleriana (Steud.) Vasey	Mutton grass	Poaceae	Native	Graminoid
Poa pratensis L.	Kentucky bluegrass	Poaceae	Exotic	Graminoid
Polygonum aviculare L.	Prostrate knotweed	Polygonaceae	Exotic	Herbaceous
Polygonum douglasii Greene	Douglas' knotweed	Polygonaceae	Native	Herbaceous
Populus tremuloides Michx.	Quaking aspen	Salicaceae	Native	Tree
Potentilla crinita Gray	Bearded cinquefoil	Rosaceae	Native	Herbaceous
Potentilla hippiana Lehm.	Horse cinquefoil	Rosaceae	Native	Herbaceous
Pseudocymopterus montanus (Gray) Coult. & Rose	False springparsley	Apiaceae	Native	Herbaceous
Pseudostellaria jamesiana (Torr.) W.A. Weber & R.L. Hartman	Sticky starwort	Caryophyllaceae	Native	Herbaceous
Pseudotsuga menziesii (Mirbel) Franco	Douglas fir	Pinaceae	Native	Tree
Pteridium aquilinum (L.) Kuhn	Bracken Fern	Dennstaedtiaceae	Native	Fern/allie
Ranunculus cardiophyllus Hook.	Heartleaf buttercup	Ranunculaceae	Native	Herbaceous
Robinia neomexicana Gray	New Mexico locust	Fabaceae	Native	Shrub
Rubus idaeus L.	Western red raspberry	Rosaceae	Native	Shrub
Rumex acetosella L.	Common sheep sorrel	Polygonaceae	Exotic	Herbaceous
Silene scouleri Hook.	Scouler catchfly	Caryophyllaceae	Native	Herbaceous
Solidago canadensis L.	Canadian goldenrod	Asteraceae	Native	Herbaceous
Solidago multiradiata Ait.	Rocky Mountain goldenrod	Asteraceae	Native	Herbaceous
Solidago nana Nutt.	Baby goldenrod	Asteraceae	Native	Herbaceous
Solidago velutina DC	Sparse goldenrod	Asteraceae	Native	Herbaceous
Solidago wrightii Gray	Wright's goldenrod	Asteraceae	Native	Herbaceous
Stipa spp.	Needle and thread	Poaceae	Native	Graminoid
Taraxacum officinale G.H. Weber ex Wiggers	Common dandelion	Asteraceae	Exotic	Herbaceous
Thalictrum fendleri Engelm. ex Gray	Fendler's meadowrue	Ranunculaceae	Native	Herbaceous
Thinopyrum intermedium (Host) Barkworth & D.R. Dewey	Intermediate Wheatgrass	Poaceae	Exotic	Graminoid
Tragopogon dubius Scop.	Goatsbeard	Asteraceae	Exotic	Herbaceous

resources: common dandelion (*Taraxacum officinale* G.H. Weber ex Wiggers), goatsbeard (*Tragopogon dubius* Scop.), black medic (*Medicago lupulina* L.), and prostrate knotweed (*Polygonum aviculare* L.). Relative cover by lifeform group and disturbance type is shown in Figure 2. Transects in all three disturbance types and in all years of study were dominated by graminoids, with Bolander's sedge (*Carex bolanderi* Olney) dominant in both burn and handline areas and the exotics Kentucky bluegrass, wheatgrass (*Agropyron* spp.), and smooth brome dominant in drop-points.

Species diversity was highest in drop-point transects in 2001 (0.83), followed closely by burn transects in 2002 (0.81; Figure 3). In burn and drop-point transects, diversity increased sharply in the first few years, attesting to a spike in ruderal species presence, and they have decreased steadily thereafter; trends in species richness are similar. While there was an initial increase in mean diversity in handline transects, there has been a stabilizing trend in the last 2 years (Figure 3). Diversity and richness differed among treatments over time (repeated measures MANOVA, F = 13.35, $p <$

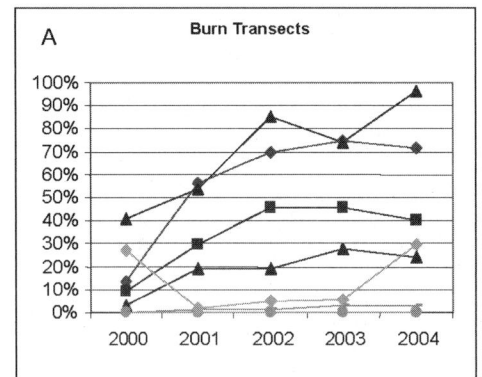

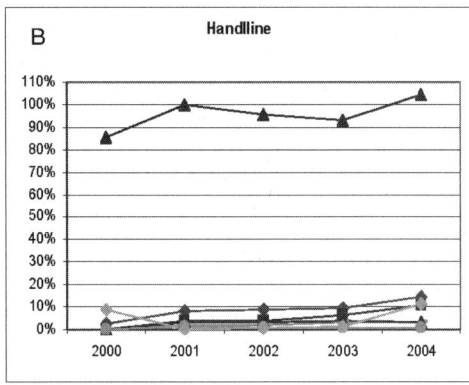

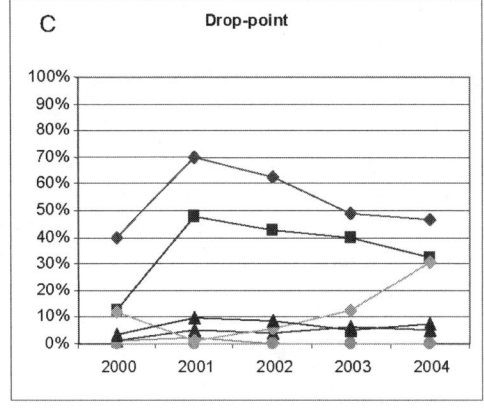

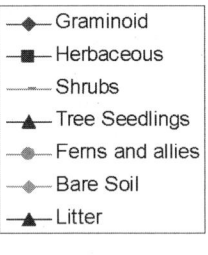

Figure 2. Relative cover of burn, handline, and drop-point transects, grouped by lifeform and the dominant abiotic factors bare soil and litter, during 5 years (2000–2004) of post-fire vegetation transect measurement, Outlet fire, North Rim, GRCA.

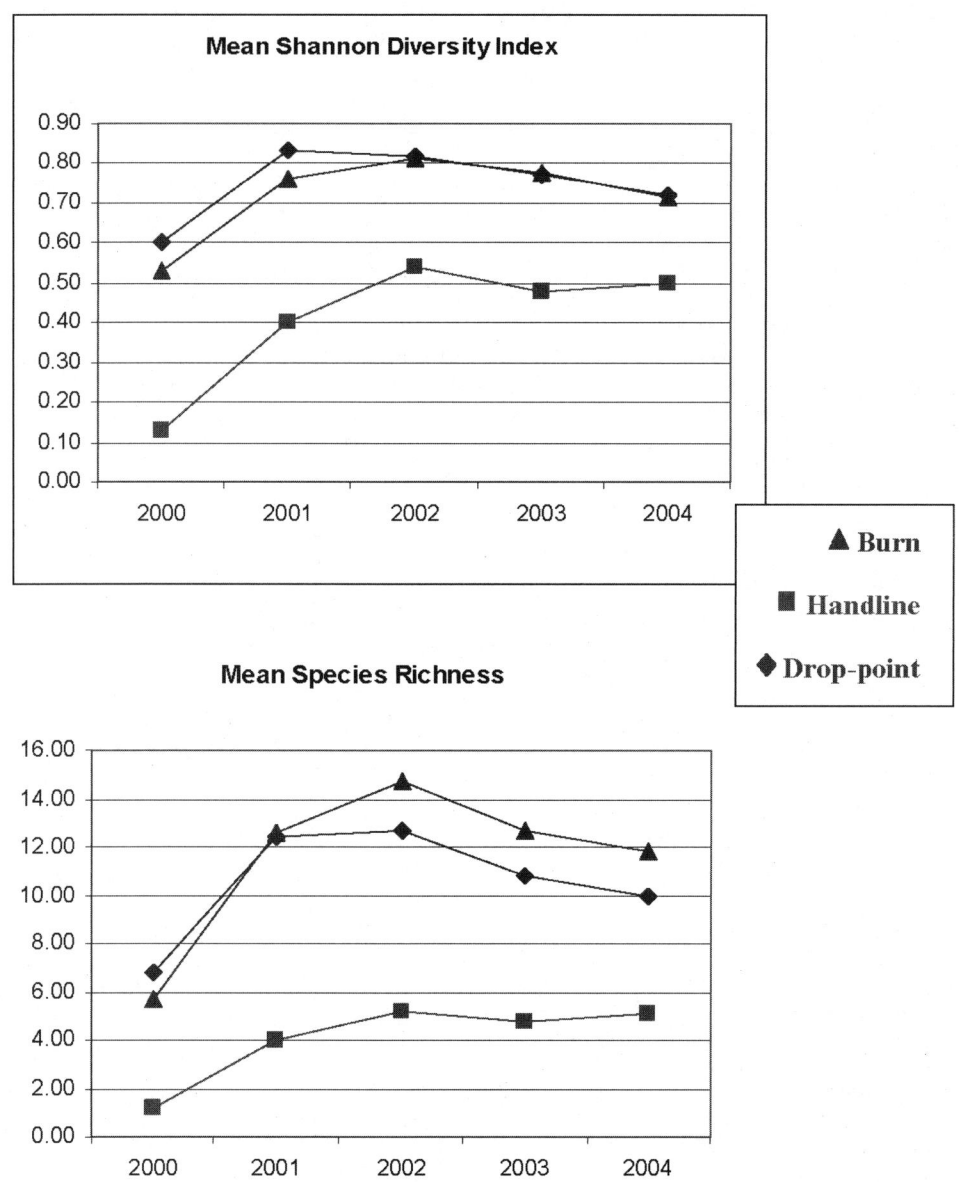

Figure 3. Mean Shannon diversity index and mean species richness by disturbance type over time from 29 transects in the Outlet fire, North Rim, GRCA. The handline transects had significantly lower mean richness and mean diversity than the drop-point or burn transects in all years of study.

0.001 and F = 17.98, $p < 0.001$, respectively). The handline transects had significantly lower richness and diversity than the drop point or burn transects in all years of study (Figure 3).

Species composition and importance changed over time, as reflected by the indicator values and significance of Table 2. Four of the top indicator species of burn transects in 2000 remain indicators in 2004— quaking aspen (*Populus tremuloides*), sedge (*Carex* spp.), common yarrow (*Achillea millefolium* L.), and Virginia strawberry (*Fragaria virginiana* Duchesne). The exotic cheatgrass, although still very low in cover, was not an indicator in 2000, but became an indicator of the high-severity burn transects by 2004. Drop-point transects remain dominated by exotic and native weedy species

such as dandelion and fleabane (*Erigeron* spp.). Although not an indicator in 2000, the exotic Kentucky bluegrass had become the primary indicator of drop-point transects by 2004. In contrast, the additional drop-point transect indicators of 2000, wheatgrass and smooth brome, dropped from the list of species of highest importance by 2004. Indicator species in handlines varied between 2000 and 2004 but only sedge was significant in either year; this is likely due to a lack of species abundance and diversity.

Ordination revealed a clear difference in floristic composition between the three disturbance types (see Figure 4). This visual difference was verified by the ANOSIM procedure of Clark (global R = 0.274; $p < 0.001$). Ordination also revealed a separation and gradual shift of floristic composition

Table 2. Indicator species analysis results for 2000 and 2004, sorted by disturbance type, for the Outlet fire on the North Rim of the Grand Canyon. *Carex* spp. was predominantly *Carex bolanderi*, with some *C. rossii*, and *C. occidentalis*. *Erigeron* spp. consisted of *E. divergens*, *E. flagellaris*, *E. formosissimus*, and *E. speciosus*. *Solidago* spp. consisted of *S. canadensis*, *S. multiradiata*, *S. nana*, *S. missouriensis*, *S. velutina*, and *Oreochrysum parryi*. Indicator values and significant *p*-values are in bold.

Plant	Burn	Drop-point	Handline	*p*-value
2000				
Populus tremuloides Michx.	**67**	3	1	**0.001**
Carex spp.	**52**	17	12	**0.042**
Solidago spp.	**35**	3	0	0.113
Achillea millefolium L.	**32**	3	4	0.156
Fragaria virginiana Duchesne	**29**	0	0	0.119
Geranium richardsonii Fisch. & Trautv.	**29**	0	0	0.119
Taraxacum officinale G.H. Weber ex Wiggers	0	**54**	0	**0.014**
Agropyron spp.	1	**47**	1	0.063
Bromus inermis Leyss.	0	**38**	0	0.059
Erigeron spp.	3	**35**	0	0.135
Carex spp.	52	17	**12**	**0.042**
Festuca ovina L.	0	2	**11**	0.575
Bromus ciliatus L.	0	5	**6**	0.866
2004				
Carex spp.	**76**	11	6	**0.001**
Populus tremuloides Michx.	**73**	6	6	**0.003**
Achillea millefolium L.	**65**	1	1	**0.007**
Fragaria virginiana Duchesne	**57**	0	0	**0.002**
Bromus tectorum L.	**43**	0	0	**0.014**
Poa pratensis L.	0	**52**	2	**0.017**
Elymus elymoides (Raf.) Swezey	13	**46**	1	0.066
Taraxacum officinale G.H. Weber ex Wiggers	0	**45**	0	**0.019**

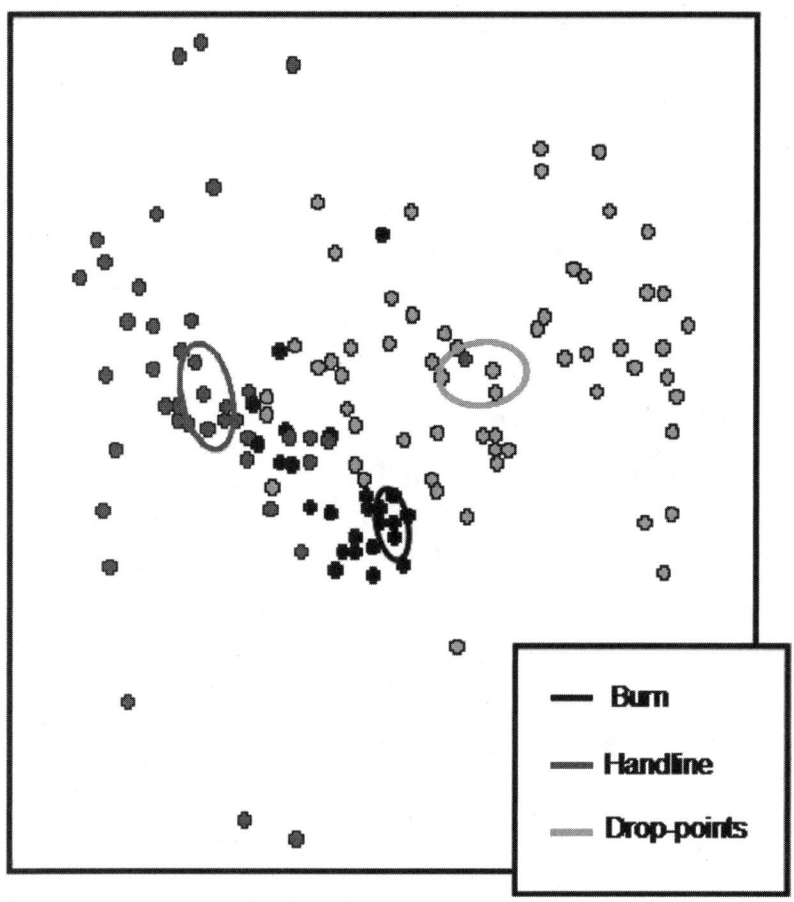

Figure 4. Non-metric multidimensional scaling ordination reveals a clear separation in species composition and abundance between the disturbance types of burn, handline, and drop-points on the Outlet fire, North Rim, GRCA. This ordination represents data from 2000–2004 combined.

between years, as shown in the 5 percent confidence ellipses of Figure 5. This result was also verified by the ANOSIM procedure, which found significant differences between floristic composition between 2000 and 2004 (global R = 0.12; p = 0.01).

Results of ANOSIM reveal statistically significant differences in floristic composition and cover between handline and burn in both 2000 and 2004 (p = 0.01 and p = 0.06, respectively) and between handline and drop-point in 2000 and 2004 (p = 0.03 and

p = 0.02). However, there was no statistical difference in the floristic composition and cover between burn and drop-point transects in either year (p = 2.18 and p = 0.28).

Burn transects were statistically different (p = 0.01) between years, with greatest recovery occurring in the first year. Following a ruderal pulse which decreased by the third year, burn transects were dominated primarily by two rhizomatous native species, Bolander's sedge and quaking aspen. Few exotic species were present in the burn tran-

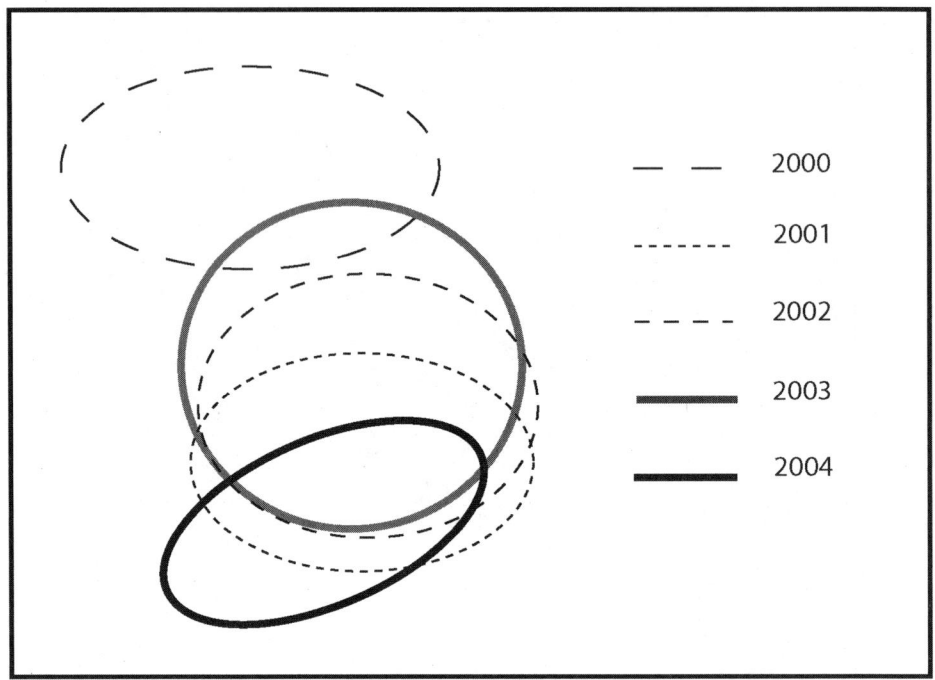

Figure 5. Non-metric multidimensional scaling ordination by year, showing only the 5 percent confidence ellipses, reveals a separation in species composition and abundance between transects over time. This ordination represents all three disturbance types combined. The arrow shows the direction of the change from 2000 to 2004.

sects and all had relatively low cover. Drop-point transects were not statistically different ($p = 1.98$) between years. These transects contained the highest percentage of exotic species in all years, but also supported many native species. There was no statistical difference ($p = 0.18$) between the handline transects of 2000 and 2004, showing no significant vegetation recovery in these areas.

DISCUSSION
High-Severity Burn

After an initial post-fire ruderal vegetation pulse, diversity in the post-burn transects has been steadily decreasing. Understory vegetation has become more homogeneous, with the dominance of the two native rhizomatous species Bolander's sedge and quaking aspen (see also Crawford 2005). These two species, in combination with

mechanically dropped trees and brush on the steepest slopes, may have aided in the prevention of post-fire erosion. The lack of species diversity, and hence habitat diversity in this area, however, might have detrimental impacts on other trophic levels in both the short and long term.

As quaking aspen is seral in conifer habitat types, its abundance throughout much of the interior West may result from historic wildfires, after which it can dominate for many decades. Though historic stand-replacing fires were uncommon on the North Rim (usually related to drought years), and fires that did occur were small and topographically limited, North Rim patches of aspen are likely the result of historic high-severity fires (Vankat 2004). The large scale of modern stand-replacing fires, such as the Outlet fire, will perpetuate the homogeneity of mixed conifer forests and thus the probability of future large-scale disturbance (White and Vankat 1993).

Similar to other North Rim vegetation studies (Huisinga et al. 2005; Laughlin et al. 2004; Gildar et al. 2004), exotic species were low both in cover and richness in our study. This is likely due to low levels of historic disturbance relative to the other conifer forests of northern Arizona that experienced large exotic plant populations following high-severity fire (Crawford et al. in review; Crawford et al. 2001; Griffis et al. 2000). Cheatgrass is an important post-fire species and though its cover was low in all years of study, its abundance has increased since 2000 and the species became an indicator of high-severity burn transects in 2004.

Drop-point

The montane grassland of the North Rim is a natural feature not caused by fire, but maintained by fire and climatic conditions (Rasmussen 1941; Moore and Huffman 2004). These communities were heavily grazed historically by livestock and both historically and currently by native ungulates and small mammals. This high level of disturbance affects species composition by reducing some species while encouraging others. The original floristic composition of the North Rim montane grassland and the history behind the introduction of exotic species such as Kentucky bluegrass and smooth brome to these areas for livestock grazing are unknown to the author. Although it has been reported that areas containing Kentucky bluegrass appear to recover in a few years following grazing reduction (Dick-Peddie 1993; Wolters 1996), both Kentucky bluegrass and smooth brome are prominent members in many of the montane grassland communities of the North Rim today (personal observation). Both grasses reproduce by both seed and rhizome, giving them a competitive advantage. In addition, graminoids have been shown to be the most resistant plants to trampling (Cole 1995). Bates (1935) stated that cryptophytes (plants with resting buds lying just beneath the surface of the ground)

such as Kentucky bluegrass are the most resistant life form to trampling.

Drop-point transects are roadside pull-outs, primarily in montane grassland communities, that are also used by park visitors; therefore, it is difficult to determine if the native ruderal and exotic vegetation dominant in these transects is a direct result of fire-fighting activity or simply a measure of the continued disturbance from visitor use and native herbivore activity. While this is possible, it is interesting to note that in our study, similar to burn transects, there was an initial spike in species diversity and richness followed by a decreasing trend (Figure 3). This pattern implies that a disturbance event in 2000 did occur—that is, a spike in ruderal species followed by a rapid decline. Without pre-fire or exclosure data, it is impossible to determine the direct cause of the greater richness and cover of exotic species in the drop-point transects.

Handline

It is clear that handline transects have experienced little, if any recovery after 5 years. Species diversity in these transects is significantly lower than in the other disturbance types. As noted, handline creation involves removing all vegetation, duff, and soil until a mineral substrate is exposed, thus removing all propagules near the surface. The uppermost soil layer contains a higher proportion of propagule richness and abundance than deeper soil layers (Rydgren and Hestmark 1997; Warr et al. 1994). In addition, seedbanks of forests that experience infrequent disturbance, such as the long-unburned mixed conifer forest on the North Rim, are generally smaller than systems with frequent disturbance (Warr et al. 1994). In a study of ponderosa pine forests on the North Rim, Gildar et al. (2004) reported that "the buried viable seed bank was depauperate relative to the above-ground plant community, suggesting that seed rain is a more important source of propagules than the soil seed bank." The understory vegetation of unburned areas in the vicinity of the handlines is sparse and offers little opportunity for seed rain (personal observation).

Caling and Adams (1999) reported that bulldozer trails created to prevent fire spread in a small area in Victoria, Australia had numerous negative ecological effects. They found an increase in habitat fragmentation and associated edge effects, an increase in soil compaction due to bulldozer construction and subsequent trampling during fire-suppression operations, and a significantly higher ($p < 0.05$) exotic species presence in bulldozer trails than in areas of wildfire. It is probable that if native ruderal species do not fill the early successional niche created by the handlines of the Outlet fire, eventually exotic species might. Even though richness and cover of exotic plants was low overall, these species do exist in the area and many, such as cheatgrass and dandelion, have mechanisms for long-distance dispersal. The current post-fire site preparation methods of contouring soil surfaces for erosion control are simply not enough to facilitate natural regeneration on handlines, at least in the short term.

CONCLUSIONS

This study brought to light the need to further investigate the effects of high-severity fire on understory vegetation recovery in mixed conifer forests of the Southwest as well as the effects of fire-fighting activity on vegetation in general. Concerning the damage associated with fire control activities, Boura (1996) recommends a "sympathetic use of hand tools, limited vehicle movements off road, and use of natural or existing control lines when ever possible." In addition to these recommendations, GRCA managers should (1) require mitigation for fire fighters and their equipment to eliminate the spread of exotic plants, (2) continue and expand this study to investigate vegetation response at additional sites of fire and fire-fighting activity, including experiments on the effectiveness and efficiency of restorative seeding using locally collected native species, and (3) encourage the local collection and storage of native seed for post-disturbance management within the park.

The Outlet fire burned more than 13,000 acres (5261 ha) of mostly mixed conifer forest on the North Rim of Grand Canyon National Park and the Kaibab National Forest in 2000. Post-fire monitoring began that summer and continued through 2004. This study examined post-fire vegetation change in relation to three types of disturbed sites: high-severity burned areas, fire-fighting staging areas, and fire-fighting handlines. Indicator species analysis, nonmetric multidimensional scaling, and ANOSIM were employed to determine indicator species and trends among disturbance types and across years. There were statistically significant differences in floristic composition, cover, and diversity over time and among disturbance types. Burned sites had the highest vegetation cover in the first year and cover remained high through 2004. Diversity in the burned areas decreased following dieback of the initial ruderal invasion and by 2004 became largely floristically homogeneous with high cover of two native rhizomatous species. Few exotic species were present in high-severity burn transects, although by 2004 cheatgrass (*Bromus tectorum* L.) had become an indicator species. Staging areas used in fire-fighting contained the greatest number of exotic species in all years of study, but this may be related to continued use of these roadside areas by park visitors. Areas of handlines showed no statistically significant differences between 2000 and 2004, indicating that no vegetation recovery has occurred. Current methods of site rehabilitation after handline construction would need modification to improve vegetation recovery. Continued monitoring is essential for understanding long-term changes in these high-elevation forests.

ACKNOWLEDGMENTS

I wish to acknowledge N. Juarez-Cummings for her promotion of the continuation of this project; C. Bliss, D. Reese, K. Straka, and M. Zylo for their invaluable assistance in the field; and B. Parker for diligence with data entry. I would also like to acknowledge the numerous people who worked on this project 2000–2003, including I. Clausen, B. Eisenberg, K. Fawcett, F. Hays, L. Makarick, M. Murov, B. Salter, A. Sokolowsky, K.

Watters, and J. White. Thanks also go out to Daniel Laughlin of the Ecological Restoration Institute for guidance in statistical analysis. Funding was provided by the Intermountain Regional Office of the National Park Service–Burned Area Rehabilitation, the Grand Canyon National Park Foundation, and the Cooperative Conservation Initiative.

LITERATURE CITED

Adams, R., and D. Simmons. 1999. Ecological effects of fire fighting foams and retardants. Australian Bushfire Conference Proceedings, Albury, July 1999.

Baskin, Y. 1998. Winners and losers in a changing world. Bioscience 48(10): 788–792.

Bates, G. H. 1935. The vegetation of footpaths, sidewalks, cart-tracks and gateways. The Journal of Ecology 23(2):470–487.

Binkley, D., M. M. Moore, W. H. Romme, and P. M. Brown. 2006. Was Aldo Leopold right about the Kaibab deer herd? Ecosystems 9(2): 227–241.

Boura, J. 1996. Reconciling fire protection and conservation issues at the urban-forest interface. In Fire and Biodiversity: The effects and effectiveness of fire management. Biodiversity Series 8. Biodiversity Unit, Victorian National Parks Association, Canberra.

Caling, T. M., and R. Adams. 1999. Ecological impacts of fire suppression operations in a small vegetation remnant. Australian Bushfire Conference Proceedings, Albury, July 1999.

Clarke, K. R. 1993. Non-parametric multivariate analyses of changes in community structure. Australian Journal of Ecology 18: 117–143.

Cole, D. N. 1995. Experimental trampling of vegetation. II. Predictors of resistance and resilience. The Journal of Applied Ecology 32(1): 215–224.

Crawford, J. 2005. The 2004 Outlet fire relevé vegetation survey. Report to the National Park Service.

Crawford, J., and K. Straka. 2004. The Outlet fire of 2000; Results from five years of vegetation monitoring, North Rim, Grand Canyon National Park. Report to the National Park Service.

Crawford, J., and G. Whitchurch. 2003. Four Years of vegetation change following the Outlet fire, North Rim Grand Canyon National Park: Effects of fire and fire-fighting activities. Report to the National Park Service.

Crawford, J., C.-H. Wahren, W. Moir, and S. Kyle. 2001. Response of exotic plant species to fires in *Pinus ponderosa* forests in northern Arizona. Journal of Vegetation Science 12: 261–268.

Crawford, J., W. Moir, and C. Sieg. In review. Five years of vegetation change following wildfire in a northern Arizona *Pinus ponderosa* forest. Forest Ecology and Management.

Crosby, W. 1923. Superintendent's annual report, Grand Canyon National Park.

Dick-Peddie, W. A. 1993. New Mexico Vegetation: Past, Present, and Future. University of New Mexico Press, Albuquerque.

Dufrene, M., and P. Legendre. 1997. Species assemblages and indicator species: The need for a flexible asymmetrical approach. Ecological Monographs 67: 345–366.

Eisenberg, B., and K. Watters. 2002. Final report 2002 revegetation and exotic species surveys in the Outlet fire Grand Canyon National Park and Kaibab National Forest, Arizona. Report to the National Park Service.

Eisenberg, B., J. White, B. Salter, and K. Fawcett. 2001. Report 2001 revegetation and exotic species surveys in the Outlet fire, Grand Canyon National Park and Kaibab National Forest, Arizona. Report to the National Park Service.

Elton, C. 1958. The Ecology of Invasions by Plants and Animals. Methuen, London.

Fox, M., and B. Fox. 1986. The susceptibility of natural communities to invasion. In Ecology of Biological Invasions, edited by R. H. Groves and J. J. Burdon, pp. 57–66. Cambridge University Press.

Fulé, P., W. Covington, M. Moore, T. Heinlein, and A. Waltz. 2002. Natural variability in forests of the Grand Canyon, USA. Journal of Biogeography 29: 31–47.

Fulé, P., J. Crouse, T. Heinlein, M. Moore, W. Covington, and G. Verkamp. 2003. Mixed-severity fire regime in a high-elevation forest of Grand Canyon, Arizona, USA. Landscape Ecology 18: 465–486.

Gildar, C. N., P. Z. Fulé, and W. W. Covington, W.W. 2004. Plant community variability in ponderosa pine forest has implications for reference conditions. Natural Areas Journal 24: 101–111.

Griffis, K., J. Crawford, M. Wagner, and W. Moir. 2000. Understory response to management treatments in northern Arizona ponderosa pine forests. Forest Ecology and Management 146: 239–245.

Huisinga, K. D., D. C. Laughlin, P. Z. Fulé, J. D. Springer, and C. M. McGlone. 2005. Effects of an intense prescribed fire on ground-flora in a mixed conifer forest. Journal of the Torrey Botanical Society 132(4): 590–601.

Kotanen, P. 1997. Effects of experimental soil disturbance on revegetation by natives and exotics in coastal California meadows. The Journal of Applied Ecology 34(3): 631–644.

Laughlin, D., J. Bakker, M. Stoddard, M. Daniels, J. Springer, C. Gildar, A. Green, and W. Covington. 2004. Toward reference conditions: Wildfire effects on flora in an old-growth ponderosa pine forest. Forest Ecology and Management 199: 137–152.

Magurran, A. 1988. Ecological Diversity and its Measurement. Princeton University Press, Princeton, New Jersey.

Merkle, J. 1954. An analysis of the spruce-fir community on the Kaibab Plateau, Arizona. Ecology 35(3): 316–322.

Merkle, J. 1962. Plant communities of the Grand Canyon area, Arizona. Ecology 43(4): 698–711.

Moir, W., and J. Ludwig. 1979. A classification of spruce-fir and mixed conifer habitat types of Arizona and New Mexico. USDA Forest Service Research Paper RM-207. Rocky Mountain Forest and Range Experiment Station, Fort Collins, Colorado.

Moore, M. M., and D. W. Huffman. 2004. Tree encroachment on meadows of the North Rim, Grand Canyon National Park, Arizona, USA. Arctic, Antarctic, and Alpine Research 36(4): 474–483.

Moore, P., and S. Chapman. 1986. Methods in Plant Ecology. Blackwell, Oxford.

Murov, M. 2000. First report revegetation and exotic species surveys outlet fire area Grand Canyon National Park and Kaibab National Forest, Arizona. Report to the National Park Service.

NRCS. 2005. The Plants Database, version 3.5 (http://plants.usda.gov). Data compiled from various sources by Mark W. Skinner. National Plant Data Center, Baton Rouge, Louisiana.

Rasmussen, I. 1941. Biotic communities of Kaibab Plateau, Arizona. Ecological Monographs 11(3): 229–275.

Rydgren, K., and G. Hestmark. 1997. The soil propagule bank in a boreal old-growth spruce forest: Changes with depth and relationship to aboveground vegetation. Canadian Journal of Botany 75: 121–128.

Vankat, J. 2004. Montane and subalpine terrestrial ecosystems of the southern Colorado Plateau—Literature review and conceptual models, rev. ed. In Vital Signs Monitoring Plan for the Southern Colorado Plateau Network: Phase II Report (Supplement I), edited by L. Thomas, C. Lauver, M. Hendrie, N. Tancreto, J. Whittier, J. Atkins, M. Miller, and A. Cully, pp. 1–100. NPS Southern Colorado Plateau Network, Flagstaff, Arizona (http://www1.nature.nps.gov/im/units/scpn/phase2.htm).

Warr, S., M. Kent, and K. Thompson. 1994. Seed bank composition and variability in five woodlands in south-west England. Journal of Biogeography 21: 151–168.

White, M., and J. Vankat. 1993. Middle and high elevation coniferous forest communities of the North Rim region of Grand Canyon National Park, Arizona, USA. Vegetation 109: 161–174.

Wolf, J., and J. Mast. 1998. Fire history of mixed-conifer forests on the North Rim, Grand Canyon National Park, Arizona. Physical Geography 19: 1–14.

Wolters, G. L. 1996. Elk effects on Bandelier National Monument meadows and grasslands. In Fire Effects in Southwestern Forests: Proceedings of the Second La Mesa Fire Symposium, technical edit by C. D. Allen, pp. 196–205. Los Alamos, New Mexico, 29–31 March 1994. RM-GTR-286, USDA Forest Service, Rocky Mountain Forest and Range Experiment Station, Fort Collins, Colorado.

NATURAL VARIATION IN DIVERSITY AND INVASION PATTERNS OF THE GRAND STAIRCASE–ESCALANTE NATIONAL MONUMENT, UTAH

Alycia W. Crall, Thomas J. Stohlgren, Paul Evangelista, and Deb Guenther

Various theories have been proposed to explain patterns of species richness using measures of productivity (Palmer 1994; Waide et al. 1999; Mittelbach et al. 2001). A predominant theory has generalized this relationship to a hump-shaped/unimodal curve, with species richness increasing and then decreasing as productivity increases (Grime 1973a, 1979; Huston 1979; Tilman 1982; Rosenzweig 1992; Huston 1994; Grace 1999). However, some authors have suggested that surveys of species richness conducted over limited productivity ranges are less likely to detect a hump-shaped relationship than studies conducted over a broad productivity range (Begon et al. 1990; Rosenzweig 1992, 1995; Huston 1994; Grace 1999). Therefore, data are clearly lacking to establish only one relationship between native species richness and productivity.

To add to this complexity, the effects that non-native species invasions have on productivity, and consequently on native species richness, are even less understood. It is not known which factors make a vegetation type vulnerable to plant invasion. A long-held theory of invasion states that disturbed, species-poor communities are more susceptible to invasion by non-natives due to a lack of biotic resistance from such factors as competition or predation (Elton 1958; Simberloff 1986). More diverse vegetation types are also more likely to utilize available resources more completely, making it difficult for new plant species to establish (Crawley 1987; but see Stohlgren et al. 1998b). However, this theory has been challenged recently as new research has found a higher risk of invasion into highly diverse vegetation types with intermediate levels of disturbance, such as tall grass prairies, wet meadows, and riparian zones (Robinson et al. 1995; Planty-Tabacchi et al. 1996; Wiser et al. 1996; Stohlgren et al. 1999, 2001).

There are various mechanisms that can make species-rich vegetation types more susceptible to invasion than species-poor vegetation types. Species richness tends to be low in stressful environments as a result of few species being able to survive under harsh conditions (Grime 1973a, 1973b). If species-poor vegetation types are a result of limited resources, then non-natives are also unlikely to establish and succeed in those areas and would more likely be found in areas of greater species richness and resources (Stohlgren et al. 1998b, 1999).

Natural and anthropogenic disturbances are also correlated with the vulnerability of habitats to invasion (Fox and Fox 1986; Hobbs 1989; Hobbs and Huenneke 1992). As niche space in a vegetation type becomes available through disturbance, the establishment of a non-native species may be possible because of open space and increased nutrient availability (Robinson et al. 1995). However, establishment of non-native species into these areas may still be limited by dispersal or seed availability (Rosentreter 1994).

All of these theories are confounded by studies being conducted at multiple spatial and temporal scales (Levine and D'Antonio 1999; Stohlgren 2002). Several multi-scale observational studies have shown both a negative and a positive relationship between native and non-native species richness at small spatial scales (Brown and Peet 2003; Fridley et al. 2004), whereas a positive relationship was seen at larger spatial scales in most cases (Stohlgren et al. 1998b, 1999). This may be a consequence of differences in primary controls on diversity. At smaller spatial scales (plant neighborhoods), native and non-native species richness may be negatively correlated because of competitive exclusion. At larger spatial scales, the effects of competition might be reduced or reversed because most competitors have similar habitat requirements, resulting in coexistence (Levine and D'Antonio 1999). Nevertheless, differences at multiple scales may make it difficult for researchers to develop broad generalizations related to non-native species invasions.

In addition, research findings are dependent on the vegetation type's stage of invasion at a certain point in time (i.e., temporal scale). Because vegetation surveys only record one point in time, their findings detect current native and non-native species richness, which may have changed since initial invasion and may be different in the future (Levine and D'Antonio 1999). Positive relationships between native and non-native species richness may occur only in the early stages of invasion. Later in the invasion process, certain non-native invaders have the capability to drastically alter an ecosystem (Vitousek et al. 1987; D'Antonio 2000). In such cases, native species richness is likely to be reduced as a result of the non-native species' ability to gain dominance under these new conditions. Stohlgren et al. (1999) found a negative correlation between native and non-native species richness in some heavily invaded vegetation types (possible long invasion history) at the 1 m^2 scale, but less invaded vegetation types (possible short invasion history) showed a positive relationship between the two variables at the same scale.

As with native species richness theory, contradictory findings among small-scale experiments, mathematical models, and large-scale observational studies have resulted in the inability to develop a general ecological theory of invasion (Levine and D'Antonio 1999; Stohlgren 2002). Thus, it remains unclear as to what role productivity and disturbance may play in determining native and non-native species richness.

The objectives of this study were to evaluate the relationships between native and non-native plant species richness and cover in the Grand Staircase–Escalante National Monument and to provide some insight into how these relationships might be affected by productivity and disturbance across vegetation types at different spatial scales. In addition, we determined where non-native species have successfully established and gained dominance within the monument to help guide and direct future management efforts.

We developed four hypotheses. (1) The common unimodal relationship between species richness and total cover (a productivity surrogate) would be found for native and non-native species when looking across all vegetation types because they span the range of low to high productivity. However, this relationship should only show a monotonic increase for xeric and a monotonic decrease for mesic vegetation types because they only represent a portion of the unimodal curve. (2) Native and non-native species richness and cover would be greatest in the mesic vegetation types compared to the xeric vegetation types because of greater resource availability. (3) Non-native species richness and cover would be positively correlated with native species richness and cover within and across vegetation types at the plot scale because non-native plant species are known to invade and coexist in areas of high species richness at this spatial scale. However, the reverse will be found at smaller spatial scales due to competitive interactions. (4) Disturbance would increase non-native species richness and cover because disturbance is known to facilitate the establishment and potential dominance of non-native plant species.

METHODS
Study Area

The Grand Staircase–Escalante National Monument, located in southern Utah, ranges in elevation from 1372 to 2530 m (Grand Staircase–Escalante National Monument 2000). The climate of the region is generally temperate and arid with the average annual precipitation approaching 250 mm. Mean summer temperatures range from 16 to 32° C, and winter temperatures range from –9 to 4° C (National Climatic Data Center 2003).

Field Sampling

From March 1998 through March 2001, 309 modified-Whittaker vegetation plots were established within the study area using a stratified random sampling design (Figure 1; Stohlgren et al. 1995, 1998a). Plots were randomly located within 15 basic vegetation types (identified using remotely sensed information) that encompassed the range of moist/dry and disturbed/undisturbed habitats (Tables 1 and 2). Because productivity is correlated with various climatic factors, especially water in arid environments (Leith and Whittaker 1975), an analysis of variance (ANOVA) was performed ($p < 0.05$) to compare climatic data over the 4 years that the vegetation was sampled. Yearly means were calculated for all 4 years at five weather stations adjacent to the monument (Bryce Canyon National Park, Escalante, Kanab, and Panguitch in Utah, and Page in Arizona).

Each modified-Whittaker plot captured an area of 1000 m² (20 x 50 m) and contained nested subplots of 1 m², 10 m², and 100 m² (Stohlgren et al. 1995, 1998a). The plots were placed parallel to the environmental gradient of the selected vegetation type to capture habitat heterogeneity. In the 1 m² subplots, percent absolute foliar cover of each species, species height, and percent cover of bare ground, microbiotic crust, rock, litter, duff, water, and dung were recorded. Presence of species not recorded in the 1 m² subplots were noted in the 10 m² and 100 m² subplots and the entire 1000 m² plot (Stohlgren et al. 1995). Plants were sampled during peak biomass production of most species, and species that could not be identified in the field were collected for later identification.

Species Richness and Productivity Relationships

We conducted several analyses to test our first hypothesis. Plant species were first identified as native or non-native according to the Natural Resource Conservation Service's Plants database (USDA 2003). Species whose origin could not be classified were excluded from the analysis. A surrogate for productivity (i.e., total cover, TC) was used to address this relationship:

$$TC = \sum_{i=1}^{s} c_i ,$$

where c_i is the absolute percent cover for each species, and s is the number of species in a subplot.

Many researchers have used surrogates for production and productivity to provide estimates of these parameters using non-destructive means rather than more labor-intensive methods such as clipping (Rosenzweig 1968; Lieth 1973; Bond 1983). Volume (i.e., species cover multiplied by height) is commonly used as a surrogate for production, but it may not always be well correlated with productivity (Barbour et al. 1987). For example, forested vegetation types will typically have much greater volume than grassland vegetation types, but may not be more productive (Barbour et al. 1987). Using cover alone as a surrogate adjusts for this sort of bias by only estimating the current year's growth (i.e., canopy cover) versus many years' growth (i.e., tree height). Given our need to compare very different growth forms, we concluded that cover would be the most appropriate surrogate for productivity across multiple vegetation types, and we used it in all analyses. However, the limitations of this surrogate should be noted because other analyses using more accurate measures of productivity could produce somewhat different results.

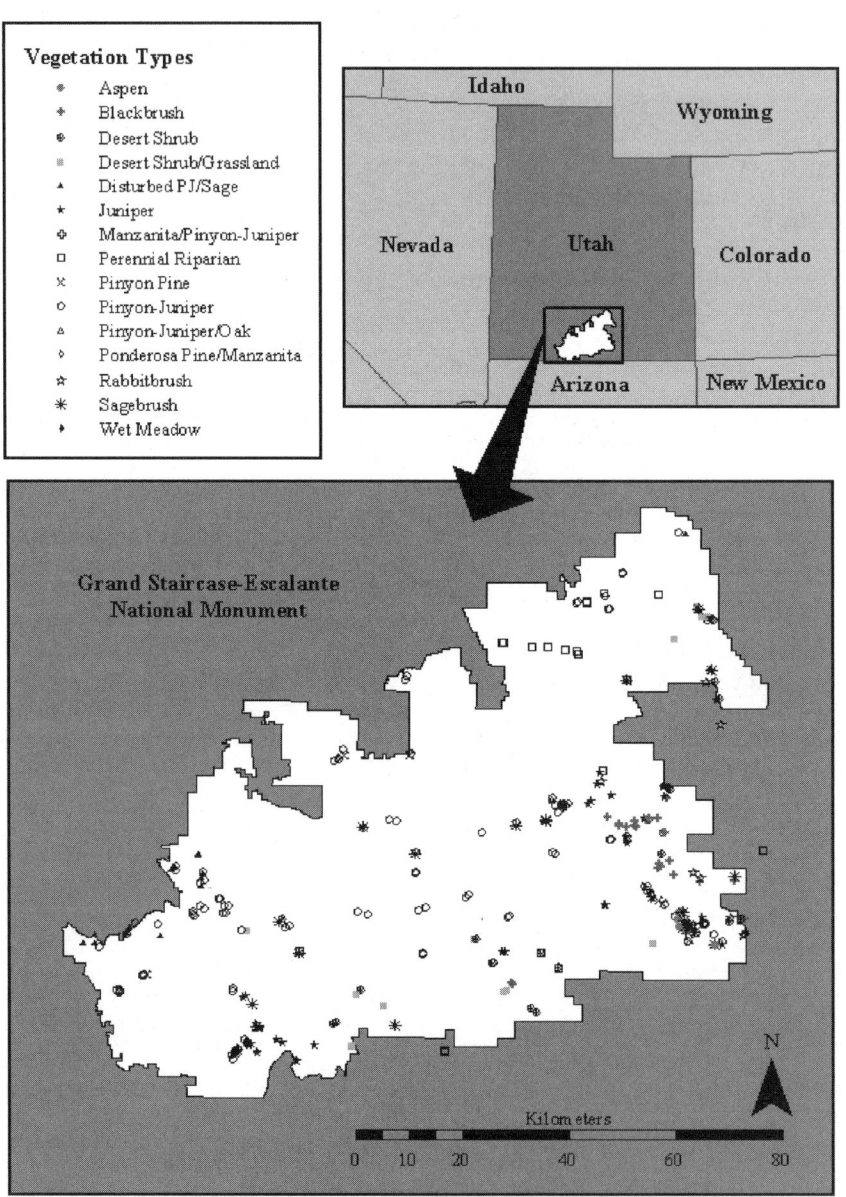

Figure 1. Plot locations by vegetation type within Grand Staircase–Escalante National Monument.

Table 1. Mean native and non-native species richness and cover (standard errors in parentheses) in 1 m² subplots averaged for each plot within Grand Staircase–Escalante National Monument, Utah, and their correlations (r).

Vegetation Type	Acronym	Sample Size (1 m²)	Average Species Richness			Average Cover		
			Native	Non-Native	r*	Native	Non-Native	r*
Desert Shrub	DSH	25	4.6 (0.4)	0.8 (0.1)	-0.34	17.3 (2.2)	2.2 (0.7)	-0.09
Blackbrush	BLB	26	4.2 (0.3)	0.6 (0.1)	0.07	29.0 (2.0)	2.2 (0.7)	0.07
Desert Shrub/Grassland	DSG	15	5.9 (0.5)	0.7 (0.1)	0.09	22.5 (4.5)	1.9 (0.5)	0.17
Sagebrush	SGB	26	4.7 (0.3)	0.7 (0.1)	0.004	26.0 (2.7)	3.8 (0.9)	0.40*
Juniper	JNP	33	4.4 (0.3)	0.6 (0.1)	-0.50*	20.0 (2.1)	3.8 (1.2)	0.58*
Pinyon-Juniper	PIJ	82	4.1 (0.2)	0.2 (0.04)	0.53*	27.9 (1.6)	0.7 (0.2)	0.33*
Disturbed Pinyon-Juniper	DPJ	35	3.1 (0.3)	1.7 (0.2)	-0.55*	15.2 (2.4)	11.9 (1.7)	-0.56*
Pinyon Pine	PIP	6	3.6 (0.6)	0.3 (0.2)	0.88	31.9 (7.0)	0.8 (0.6)	-0.59
Pinyon-Juniper/Oak	PJO	15	4.9 (0.4)	0.2 (0.1)	0.04	42.4 (5.1)	0.5 (0.3)	0.24
Pinyon-Juniper/Manzanita	PJM	5	7.0 (1.1)	0.02 (0.02)	-0.83	38.4 (7.0)	0.01 (0.01)	-0.98*
Ponderosa Pine/Manzanita	PPM	5	4.5 (0.7)	0.3 (0.2)	0.71	43.0 (9.6)	1.2 (0.9)	0.76
Rabbitbrush	RBB	6	3.4 (0.6)	0.8 (0.2)	0.64	29.7 (5.1)	5.5 (3.3)	0.86*
Aspen	ASP	6	5.3 (0.5)	1.3 (0.2)	-0.83*	57.8 (6.1)	8.7 (2.5)	-0.31
Wet Meadow	WTM	3	5.3 (0.8)	2.3 (0.1)	0.99*	37.1 (9.6)	27.5 (5.3)	-0.48
Perennial Riparian	PRI	18	3.6 (0.4)	1.1 (0.1)	0.18	39.3 (5.2)	10.6 (2.0)	0.07

* indicates r values that are significant ($p < 0.05$).

Table 2. Mean native and non-native species richness and cover (standard errors in parentheses) in 1000 m² subplots within Grand Staircase–Escalante National Monument, Utah, and their correlations (r).

Vegetation Type	Acronym	Sample Size (1 m²)	Species Richness			Average Cover		
			Native	Non-Native	r^*	Native	Non-Native	r^*
Desert Shrub	DSH	25	23.8 (1.5)	2.0 (0.3)	-0.72*	17.3 (2.2)	2.2 (0.7)	-0.09
Blackbrush	BLB	26	21.7 (1.0)	1.2 (0.2)	0.004	29.0 (2.0)	2.2 (0.7)	0.07
Desert Shrub/Grassland	DSG	15	28.1 (1.3)	1.5 (0.3)	0.04	22.5 (4.5)	1.9 (0.5)	0.17
Sagebrush	SGB	26	24.3 (1.6)	1.8 (0.3)	-0.14	26.0 (2.7)	3.8 (0.9)	0.40*
Juniper	JNP	33	26.4 (1.4)	1.7 (0.3)	-0.37*	20.0 (2.1)	3.8 (1.2)	0.58*
Pinyon-Juniper	PIJ	82	27.2 (0.7)	0.9 (0.1)	-0.27	27.9 (1.6)	0.7 (0.2)	0.33*
Disturbed Pinyon-Juniper	DPJ	35	17.9 (1.3)	3.9 (0.4)	-0.43*	15.2 (2.4)	11.9 (1.7)	-0.56*
Pinyon Pine	PIP	6	25.3 (3.0)	0.7 (0.4)	0.84*	31.9 (7.0)	0.8 (0.6)	-0.59
Pinyon-Juniper/Oak	PJO	15	33.4 (2.3)	1.0 (0.4)	0.75*	42.4 (5.1)	0.5 (0.3)	0.24
Pinyon-Juniper/Manzanita	PJM	5	37.6 (1.5)	0.2 (0.2)	0.25	38.4 (7.0)	0.01 (0.01)	-0.98*
Ponderosa Pine/Manzanita	PPM	5	29.6 (2.2)	0.6 (0.2)	0.38	43.0 (9.6)	1.2 (0.9)	0.76
Rabbitbrush	RBB	6	24.9 (1.5)	3.0 (0.8)	0.27	29.7 (5.1)	5.5 (3.3)	0.86*
Aspen	ASP	6	33.3 (3.1)	4.7 (1.0)	0.14	57.8 (6.1)	8.7 (2.5)	-0.31
Wet Meadow	WTM	3	33.7 (5.0)	6.7 (0.9)	0.99*	37.1 (9.6)	27.5 (5.3)	-0.48
Perennial Riparian	PRI	18	27.0 (2.6)	5.9 (0.5)	0.14	39.3 (5.2)	10.6 (2.0)	0.07

* indicates r values that are significant ($p < 0.05$).

The large spatial extent of the study area and nested design of the modified-Whittaker plots could result in spatial dependency in the data at either the plot or subplot scale. Therefore, we tested for spatial autocorrelation for each variable (i.e., total species, total cover, native species, native cover, non-native species, non-native cover) within and among plots using Moran's I (Cliff and Ord 1981). Because these tests were significant ($p < 0.05$), subplot data were averaged within each plot and both linear and quadratic spatial autoregressive models were fit to the averaged subplot data and original plot data (Upton and Fingleton 1985; Cressie 1993). Spatial autoregressive models have the form of a regular regression model with the addition of a random effect (λ) that models the spatial structure of the residuals. The residuals for the models were inspected for normality and homogeneity of variance to make sure all spatial dependency had been removed from the data. The linear and quadratic spatial autoregressive models were then compared using Akaike's Information Corrected Criteria (AICC), and the model with the lowest AICC was chosen to be most representative of the relationship between the two variables (i.e., species richness and total cover; Hurvich and Tsai 1989).

Species data from the 1 m^2 subplots were used to model the average number of total, native, and non-native plant species per subplot (i.e., species richness) within each vegetation type to total species cover. Cumulative species data from the 1000 m^2 plots were also used to model the relationship between the number of species per plot and the average total species cover. Thus, cover values were identical at both scales. These models included all species, only native species, and only non-native species in the data set. Because the disturbed pinyon-juniper vegetation type represents unnatural disturbance within this system, it was excluded from this analysis. Subplots and plots were then divided into xeric (i.e., desert shrub, blackbrush, desert shrub/grassland, sagebrush, juniper, pinyon-juniper, pinyon pine, pinyon-juniper/oak, pinyon-juniper/manzanita, and ponderosa pine/

manzanita) and mesic (i.e., rabbitbrush, aspen, wet meadow, and perennial riparian) vegetation types to look at differences in the relationship of species richness to total species cover separately under dry and wet conditions. Log$_{10}$ transformations ($x + 1$) were used for all data that did not meet the models' assumptions. Transformed regression lines were then backtransformed to display the original, nontransformed data to aid in analysis and interpretation.

Invasion of Different Vegetation Types

We conducted various analyses to test our hypotheses related to relationships of native to non-native species richness and native to non-native species cover. We used analysis of variance to compare the means of these variables among the 15 vegetation types. To test for differences between mesic and xeric vegetation types, t-tests were performed to find significant differences ($p < 0.05$) between native and non-native species richness and native and non-native cover. If variances were unequal, Satterthwaite's modified t-test was used (Satterthwaite 1946). To determine the relationship of native and non-native species richness and cover to each other, either a linear spatial autoregressive model or a common linear regression was used depending on the significance ($p < 0.05$) of the spatial autoregressive parameter (λ) in the model. This was done for each vegetation type at both scales and for all vegetation types, excluding the disturbed pinyon-juniper plots. We also ran regressions to see if species richness at the subplot scale could predict species richness at the plot scale. This was done for native, non-native, and total species richness.

Disturbed Vs. Undisturbed Pinyon-Juniper Plots

The disturbed pinyon-juniper vegetation type included plots that had been chained, sprayed with herbicides, burned, or seeded, or a combination of these practices, within the past 10–40 years. Such disturbances are atypical for the monument. To test our hypothesis related to disturbance, simple t-tests were performed to find significant differences ($p < 0.05$) between native and non-native species richness and native and non-native species cover in the disturbed pinyon-juniper vegetation type and the pinyon-juniper vegetation type. If variances were unequal, Satterthwaite's (1946) modified t-test method was used.

RESULTS

Species Richness and Productivity Relationships

Values for mean monthly temperature, precipitation, and snowfall across years showed no significant difference ($p < 0.05$) for the Bryce Canyon, Escalante, and Page weather stations. The only significant differences across years were in the mean monthly temperature data from Panguitch and the mean monthly snowfall data from Kanab and Panguitch. Thus, any differences in temperature or precipitation over the 4 years should not have seriously compromised comparisons of species richness and cover across vegetation types.

At the 1 m^2 subplot scale, relationships between total and native species richness and total cover were similar across all vegetation types, with species richness increasing gradually as total cover values increased (Figure 2a, b). Total cover explained little of the variation in total ($R^2 = 0.16$; Figure 2a) and native species richness ($R^2 = 0.13$; Figure 2b) in this analysis. Non-native species richness showed a similar pattern to total and native species richness, but total cover explained twice as much variation ($R^2 = 0.33$; Figure 2c). This pattern was also seen in the xeric vegetation types at this scale, with total, native, and non-native species having R^2 values of 0.14, 0.14, and 0.33, respectively.

Relationships between total, native, and non-native species richness and total cover were similar at the 1000 m^2 plot scale across all vegetation types (Figure 3). For total species, species richness remained relatively constant at lower total cover values, and showed a strong increase as total cover values increased (Figure 3a). The same pattern was seen for native species, but the increase at high total cover values was only

slight (Figure 3b). For non-native species, species richness remained relatively constant as total cover increased (Figure 3c). The R^2 values were similar at this scale, ranging from 0.16 to 0.29 (Figure 3). Similar patterns were seen in the xeric vegetation types, with total, native, and non-native species having R^2 values of 0.16, 0.17, and 0.22, respectively. Within the mesic vegetation types at the plot scale, an increase in species richness was found as total species cover increased. Little variation was explained here as well, but R^2 values increased for total and native species and decreased for non-native species when compared to the relationship across all vegetation types and xeric vegetation types ($R^2 = 0.23_{total}$, 0.20_{native}, $0.15_{non-native}$).

In addition, total cover explained little of the variation in total and native species richness in the mesic vegetation types at the subplot scale (see Figure 4; $R^2 = 0.36_{total}$, 0.34_{native}, $0.13_{non-native}$). The R^2 values were twice as great for total and native species and less for non-native species when compared to the relationship found across vegetation types and within xeric vegetation types. The shape of the regression also differed for total, native, and non-native species richness for the mesic vegetation types, with species richness increasing at lower total cover values and decreasing as total cover values increased, producing a slight hump-shaped/unimodal pattern (Figure 4).

Invasion of Different Vegetation Types

At the subplot scale, native species richness ranged from 3.1 ± 0.3 for the disturbed pinyon-juniper vegetation type to 7.0 ± 1.1 for the pinyon-juniper/manzanita vegetation type (Table 1). The same vegetation types had the lowest and highest native species richness values at the plot scale, ranging from 17.9 ± 1.3 to 37.6 ± 1.5 (Table 2). The pinyon-juniper/manzanita and the wet meadow vegetation types had the lowest and highest values for non-native species richness at both scales (Tables 1 and 2). Non-native species richness values ranged from 0.02 ± 0.02 to 2.3 ± 0.1 at the subplot scale and 0.2 ± 0.2 to 6.7 ± 0.9 at the plot scale.

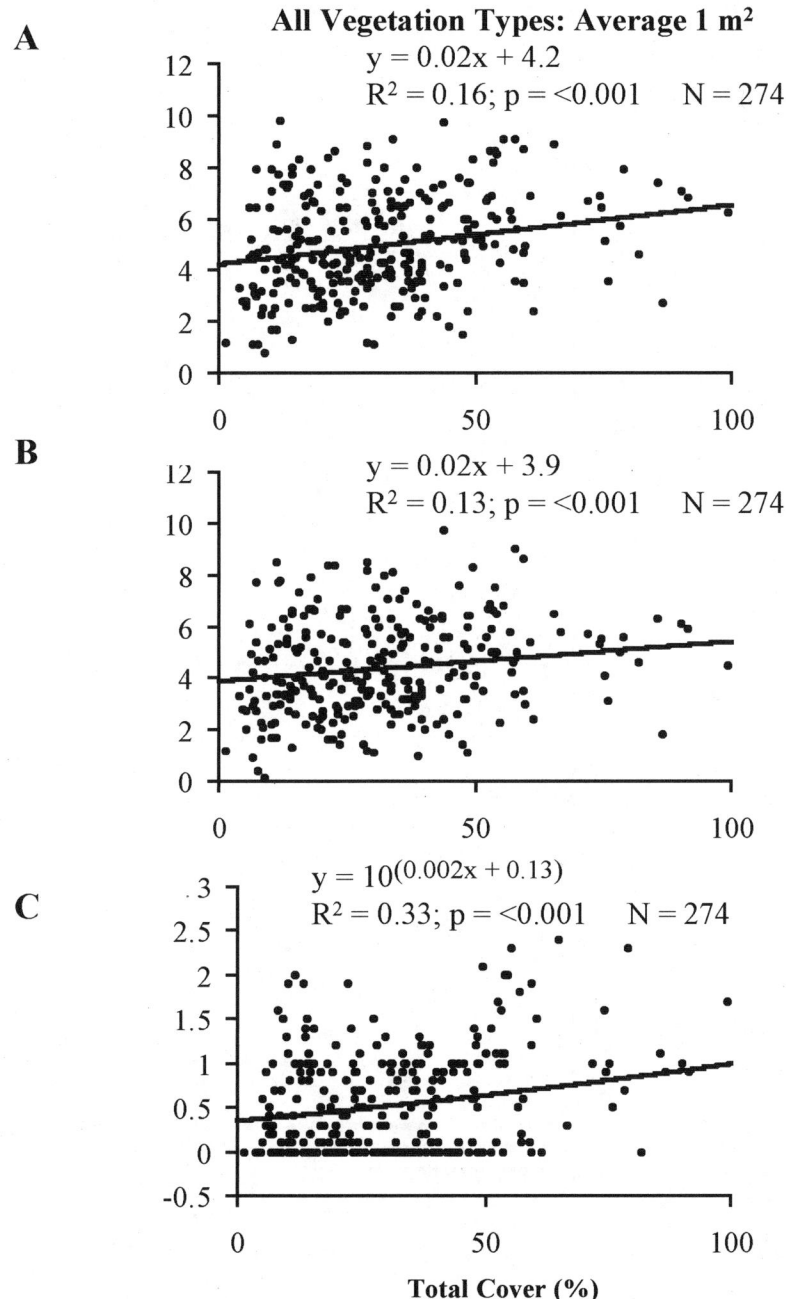

Figure 2. Relationship between total (A), native (B), and non-native (C) plant species richness and total cover averaged for 1 m² subplots within all vegetation types, excluding the disturbed pinyon-juniper vegetation type.

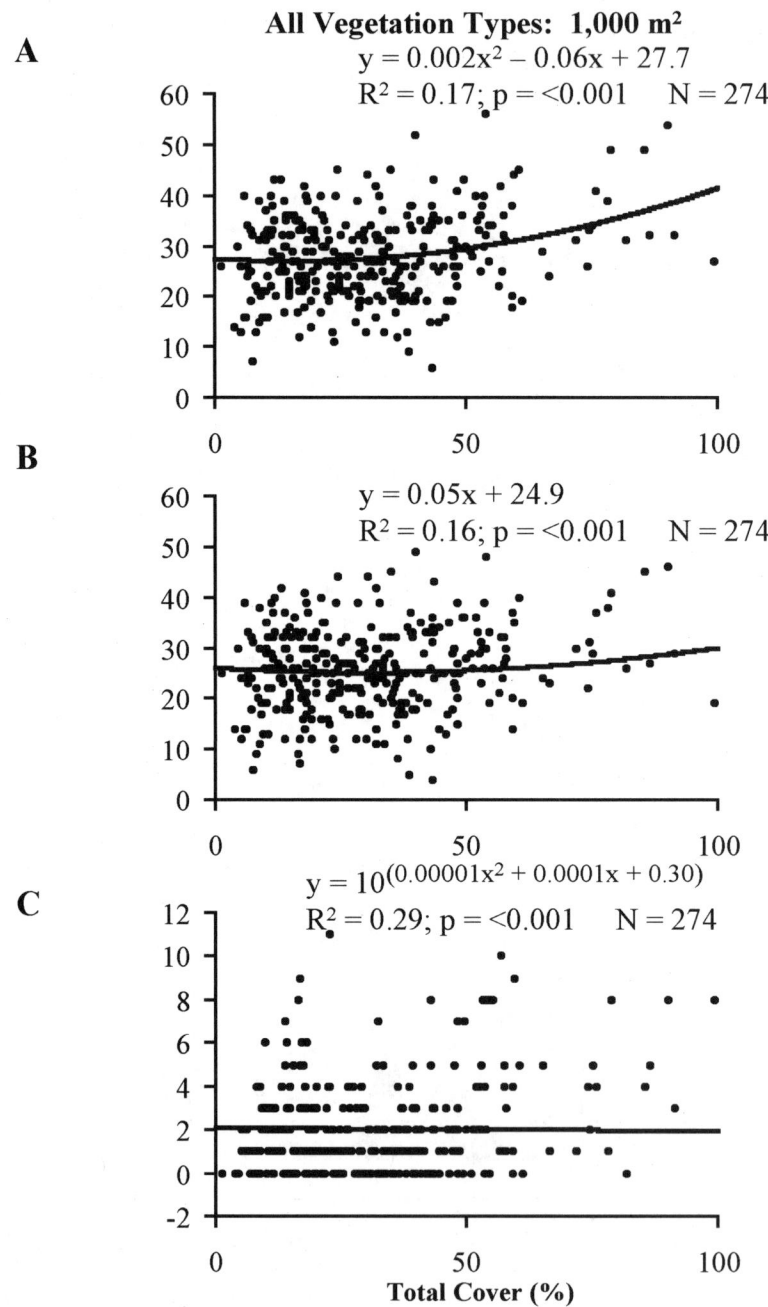

Figure 3. Relationship between total (A), native (B), and non-native (C) plant species richness and total species cover in 1000 m² plots for all vegetation types, excluding the disturbed pinyon-juniper vegetation type.

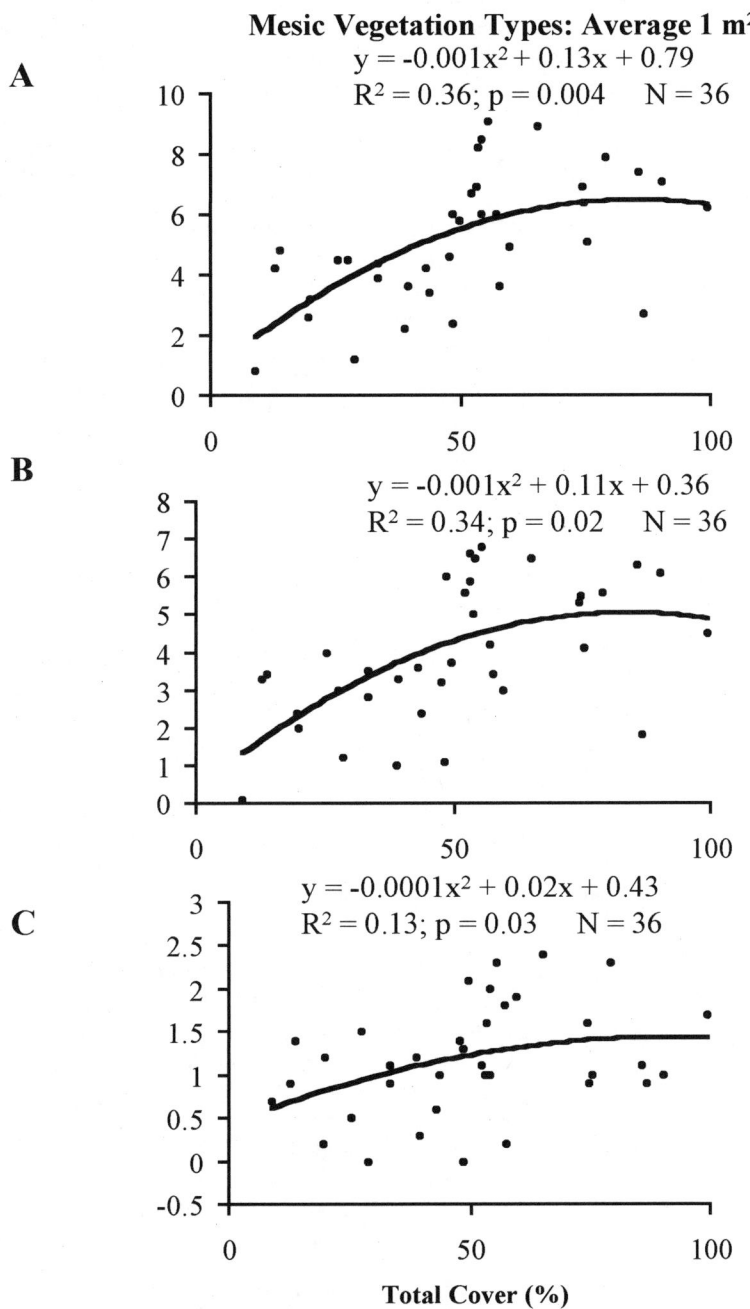

Figure 4. Relationship between total (A), native (B), and non-native (C) plant species richness and total cover averaged for 1 m² subplots within mesic vegetation types.

The t-test comparisons of xeric to mesic vegetation types yielded unsurprising results. Mean native cover was significantly higher ($p < 0.05$) in the mesic vegetation types ($40.3 \pm 3.5\%$) than the xeric vegetation types ($27.6 \pm 1.0\%$), with the aspen vegetation type having the greatest cover for native species (Tables 1 and 2). Non-native species cover was also significantly higher in the mesic vegetation types ($11.0 \pm 1.7\%$) compared to the xeric vegetation types ($1.9 \pm 0.3\%$), with the wet meadow vegetation type having the greatest cover for non-native species. Native species richness was not significantly different for the xeric and mesic vegetation types at the plot scale, but native species richness was significantly greater in the xeric vegetation types at the subplot scale. Non-native species richness was significantly greater within the mesic vegetation types when compared to the xeric vegetation types at both scales (Tables 1 and 2).

In comparing native and non-native species richness and cover among vegetation types, two interesting findings emerged. The wet meadow vegetation type had significantly higher non-native cover when compared to all the other vegetation types ($p < 0.05$; Tables 1 and 2). The disturbed pinyon-juniper vegetation type had significantly lower native species richness than all the other vegetation types at the plot scale.

Only a few of the correlations of native to non-native species richness in each vegetation type were significant. The wet meadow showed a significant linear increase in non-native species richness to increases in native species richness at both scales. The disturbed pinyon-juniper and juniper vegetation types showed a significant, negative correlation between native and non-native species richness at both scales (Tables 1 and 2; Figure 5).

However, some of the significant relationships were scale dependent. Although the rabbitbrush and pinyon-juniper vegetation types showed a significant, positive relationship between native and non-native species richness at the subplot scale, these relationships were not significant at the plot scale. The aspen vegetation type demon-strated a significant, negative correlation at the 1 m^2 subplot scale but a nonsignificant, positive correlation at the plot scale (Tables 1 and 2; Figure 5). The pinyon pine and pinyon-juniper/oak vegetation types showed a significant, positive relationship at the plot scale and the desert shrub vegetation type showed a significant, negative relationship at the same scale. These relationships were not significant for all these vegetation types at the subplot scale.

Four of the nine (sagebrush, juniper, pinyon-juniper, and rabbitbrush) vegetation types that showed a positive correlation between native and non-native species cover were significant. The disturbed pinyon-juniper and the pinyon-juniper/manzanita vegetation types had native and non-native cover values that were significantly, negatively correlated.

Across vegetation types, excluding the disturbed pinyon-juniper plots, a significant, positive correlation was found between native and non-native species richness at the subplot ($r = 0.55$) and plot ($r = 0.49$) scales (Figure 5). Native and non-native species cover values were also significantly, positively correlated ($r = 0.54$; Figure 5). A significant, positive correlation was also found between species richness at the plot and subplot scales. Values of r for native, non-native, and total species richness were 0.78, 0.90, and 0.70, respectively.

Disturbed Vs. Undisturbed Pinyon-Juniper Plots

The t-tests comparing native and non-native species richness and cover between the undisturbed and disturbed plots at both scales were all significant ($p < 0.05$). Native species richness was significantly higher in undisturbed (4.1 ± 0.2) compared to disturbed plots (3.1 ± 0.3) at the subplot scale. These results were also found at the plot scale for undisturbed (27.2 ± 0.7) and disturbed plots (17.9 ± 1.3). Non-native species richness was significantly higher in disturbed plots at the subplot (1.7 ± 0.2 vs. 0.2 ± 0.04) and plot scales (0.9 ± 0.1 vs. 3.9 ± 0.4).

For native cover, undisturbed plots in the pinyon-juniper vegetation type ($27.9 \pm 1.6\%$)

A

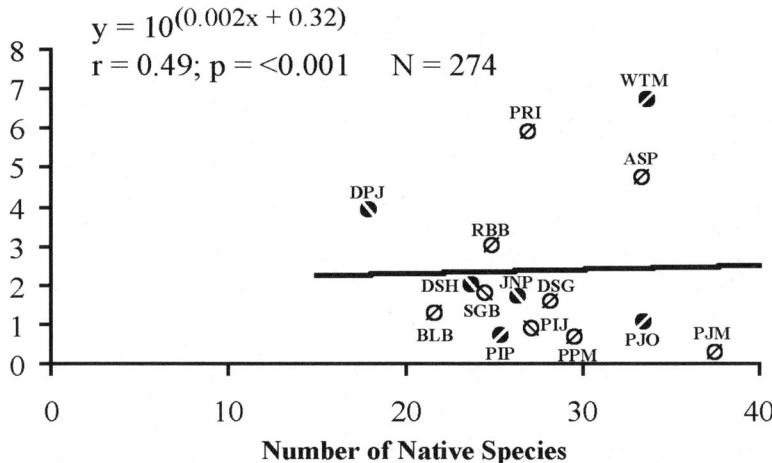

B

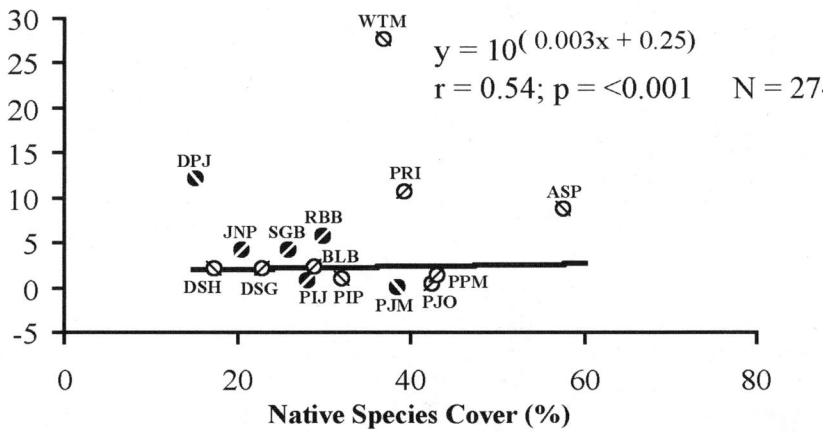

Figure 5. Mean native and non-native species richness (A) and cover (B) for plots within each vegetation type (acronyms given in Tables 1 and 2). Regressions across plots exclude the disturbed pinyon-juniper vegetation type. Regressions within vegetation types show positive (circle with right slash) or negative (circle with left slash) correlations with cross bars and are darkened if significant ($p < 0.05$).

had a significantly higher mean than the disturbed plots (15.2 ± 2.4%). For non-native cover, the opposite was found, with the disturbed plots (11.9 ± 1.7%) having a significantly higher cover than the undisturbed plots (0.7 ± 0.2%).

DISCUSSION

Productivity Surrogate Vs. Species Richness

Our results did not fully support our first hypothesis. The hump-shaped/unimodal pattern was not the most common richness-productivity relationship found for the plots and subplots across all vegetation types. However, combinations of relationships were observed, with variations occurring at different scales and among different vegetation types. Such variations in the relationship between species richness and total cover suggests that one pattern between the two variables is not representative of this landscape.

Regressions across all vegetation types at both scales showed an increase in species richness as total cover increased. It may be argued that the lack of a declining phase in the relationship here is a result of the lack of a productivity gradient that extends from regions of extremely low to regions of extremely high productivity (Begon et al. 1990; Rosenzweig 1992, 1995; Huston 1994; Grace 1999). However, this argument detracts from the fact that most terrestrial biomes have low productivity (Whittaker and Likens 1975), suggesting that the hump-shaped pattern may not be applicable to a majority of natural landscapes. This is further emphasized by the monotonic increase in total cover for the xeric vegetation types at both scales because most of the vegetation types (88%) fell into the xeric category.

Although we hypothesized that a linear decrease would be found for the mesic vegetation types, it was the only analysis that resulted in a slight hump-shaped curve. This relationship was only found at the 1 m² subplot scale, indicating that the effects of competition and resource homogeneity may be slightly more powerful at neighborhood scales in the most productive habitats.

Although R^2 values ranged between 0.13 and 0.36 for the correlations of total, native, and non-native species to total cover, other predictors of species richness should also be examined. Productivity (or some surrogate), in conjunction with other variables, could be used in multivariate analyses to better explain patterns of species richness. While conducting a comprehensive review of the literature pertaining to the relationship between species richness and productivity, Grace (1999) found that multivariate analyses that include biomass along with other factors that control species richness (i.e., time lags, species composition, plant morphology, plant density, soil microbial effects) are more accurate in explaining variations in species richness. He recommended the use of a new model that uses disturbance, vegetation type biomass, colonization, the species pool, and spatial heterogeneity to account for most of the variability in species richness. From these results, it appears that foliar cover in isolation could be used as a fairly reliable indicator of total, native, or non-native species richness, but more detailed field measurements to quantify resource availability and productivity may be needed to better predict patterns of species richness across vegetation types at multiple spatial scales.

Invasion of Vegetation Types In the Monument

In general, species richness was highly variable across all the vegetation types in the monument. The wet meadow vegetation type and the aspen vegetation type both had high numbers of native and non-native species, supporting the hypothesis that non-native species tend to invade areas of high native species richness. These are locally rare, mesic vegetation types of high native diversity that have been shown to be vulnerable to non-native species invasions in past studies (Stohlgren et al. 1997, 1998b, 1999). Because grazing and other disturbances are commonplace in the monument, there may be enough frequent, small-scale disturbances to provide available resources (i.e., light, water, nutrients) to both native and non-native species to allow them to coexist (Robinson et al. 1995; Stohlgren et al. 1998b, 1999). Such an observation coincides with the intermediate disturbance hypothesis (Connell 1978; Huston 1979). If areas with the greatest species richness have intermediate levels of disturbance, it would be probable that favorable microsites with available resources are continuously made available for non-native species to invade through the removal of other species.

It may be surprising that the perennial riparian vegetation type did not have high native and non-native species richness values similar to that of the aspen and wet meadow vegetation types. This may be a result of the characteristics of the non-native species that have invaded these areas (e.g., Vitousek et al. 1987; D'Antonio 2000). That is, the perennial riparian habitat has had a long invasion history that may have provided enough time for the non-native invaders to alter the characteristics of this ecosystem, resulting in losses to species richness. Most of the non-native cover in this vegetation type is represented by three invasive non-native plant species: saltcedar (*Tamarix ramosissima*), Russian olive (*Elaeagnus angustifolia*), and cheatgrass (*Bromus tectorum*). With time, Russian olive is capable of producing dense thickets that displace the dominant riparian trees throughout its range, thereby limiting native species richness (Pearce and Smith 2001). *Tamarix* spp. have also been shown to cause substantial reduction in native riparian vegetation, especially in areas where it has reached high densities (Di Tomaso 1998). In addition, cheatgrass has been shown to displace the seedlings of many native perennial grasses and shrubs (Aguirre and Johnson 1991).

Not all the vegetation types supported our hypothesis that non-native species would be found in areas of high native species richness. Native species richness was greatest in the pinyon-juniper/manzanita vegetation type, which had the lowest non-native species richness and cover per plot. A similar situation was found in the pinyon-juniper/oak vegetation type, which had the

third-highest number of native species richness per plot but low non-native species richness. It may be that in some habitats a few species may be able to monopolize light, water, and nitrogen, reducing resource availability to potential non-native invaders (Stohlgren et al. 1999).

In addition, non-natives may be unable to establish in species-rich vegetation types due to the small-scale heterogeneity. Twelve endemic species are known to occur in the pinyon-juniper/manzanita and pinyon-juniper/oak vegetation types. More specifically, on average, 2.5 endemic species per 1000 m^2 plot occur in the pinyon-juniper/manzanita vegetation type (greater than all other vegetation types) and 1.1 endemic species per plot occur in the pinyon-juniper/oak vegetation type (Stohlgren et al. 2005). This may be evidence of the existence of highly specialized microhabitats that are not easily invaded by a generalist non-native species (Stohlgren et al. 2001). The monument has not been isolated from disturbance and it has a large non-native seed source, so it is unlikely that these vegetation types have low non-native species richness because of a short invasion history.

Our hypothesis related to differences in native and non-native species richness between the xeric and mesic vegetation types was supported by our results for non-native species, but not for native species. Non-native species richness was significantly greater in the mesic vegetation types, but this difference was not significant at the plot scale, and the xeric vegetation types had greater native species richness at the subplot scale. This may be a result of native species being more adapted to the harsh, arid environment found within the xeric plots. Because water is one of the most limiting resources in arid environments, it is likely that the increased availability of water in the mesic vegetation types is better able to support greater numbers of non-native species especially if these species are not adapted to drier, more stressful conditions.

Native and non-native cover both increased significantly in the mesic vegetation types compared to the xeric vegetation types. Biomass and productivity are also known to increase with moisture availability (Rosenzweig 1968; Lieth 1973), and it may even be considered the environmental factor most correlated with productivity in arid environments (Ludwig 1986). Therefore, it would be reasonable to expect an increase in mean native and non-native cover in the more mesic vegetation types of the monument.

Relationships at Multiple Scales

We hypothesized that non-native species richness and cover would increase with native species richness and cover within and across vegetation types at the plot scale because of increasing evidence that non-native species tend to invade areas of high native species richness at larger spatial scales (Stohlgren et al. 1998b) and across vegetation types (Stohlgren et al. 1999). For this study, non-native species richness and cover were significantly, positively correlated to native species richness and cover, supporting this theory and our hypothesis.

Looking within vegetation types at the plot scale, a significant, positive correlation between native and non-native species richness was only found for three vegetation types, while three other vegetation types showed a significant, negative correlation. Similarly, it was not always found that native species richness was a strong predictor of non-native species richness within each vegetation type. The R^2 values ranged from .14 to .98 at the plot scale for the regressions that were significant. Therefore, depending on the vegetation type, a little or almost all of the variation in non-native species richness could be explained by native species richness. These inconsistencies were also found at the subplot scale. Considering the complexity of the monument's landscape, it is not surprising that patterns among these variables were not consistent among vegetation types. This suggests that generalizations regarding the relationships between native and non-native species should not be made. As with total cover as a predictor, it might be necessary to use multivariate analyses to

better predict overall patterns of species richness within some of these vegetation types.

Because cover values were averaged across $1\ m^2$ subplots, the dependence of these relationships on scale could not be examined. However, several significant relationships were apparent between native and non-native cover. These relationships were found for both mesic and xeric vegetation types. Only two of the six vegetation types that demonstrated a significant relationship were negative. Again, our hypothesis was only partially supported by our results, and a generalization could not be made about relationships of native to non-native cover within vegetation types.

An interesting finding in this analysis was that small-scale species richness was a good predictor of large-scale species richness. This is an important finding for land managers who do not have the resources to conduct large-scale sampling efforts similar to that conducted for this study. Instead, smaller scale studies of species richness may be adequate to determine species richness at larger scales.

The Role of Disturbance

Our data supported other findings that disturbance facilitates invasion (Fox and Fox 1986; Hobbs 1989; Hobbs and Huenneke 1992). We found an increase in non-native species richness and non-native cover with disturbance. Because native species richness and native cover also decrease within the disturbed vegetation type, it may be possible that the disturbance is opening up niche space for non-native invasions through the removal of native species (D'Antonio et al. 2001). Once non-native species have become established, they also have the potential to modify the habitat in a way that allows other non-native species to invade, further increasing non-native species richness and reducing native species richness (Vitousek et al. 1987). It is important to note that some of these differences might be exaggerated considering that some of the heavily disturbed plots within this vegetation type were re-seeded with non-native species. However, seeding is a common restoration practice and its effects should be recognized.

The role of atypical disturbance in this system is also made evident by the fact that native species richness was significantly lower in this vegetation type compared to all other vegetation types. This is not surprising considering that such unnatural, heavy disturbances are likely to remove many native species or make conditions unfavorable for their growth and reproduction. In addition, non-native species richness and cover were significantly, negatively correlated to native species richness and cover within this vegetation type. Thus, these types of disturbances have the potential to reverse patterns of plant species richness and cover if carried out on large areas of the landscape.

Over time, the establishment of non-native species through heavy, large-scale disturbance may prove to have negative consequences for plant species richness in the monument. Research has shown that invasions that tend to have the most dramatic impacts are those in which the non-native is functionally dissimilar from the native species (D'Antonio 2000). For instance, disturbance regimes may be created that the native species are unable to tolerate as a result of not evolving with such disturbances. This leads to native species reductions and the proliferation of non-native species in a positive feedback cycle (D'Antonio 2000).

Management Implications

Armed with knowledge about which vegetation types have high non-native richness and cover, managers within the monument and other public lands will be better able to target areas at high risk to invasion. In the case of the monument, areas of high productivity and disturbance should receive highest priority for invasive species control. Reducing disturbance in these invaded communities may help restore more natural conditions to these systems, but careful manipulation of these environmental controls will be needed to minimize impacts to native species.

Disturbance was shown to drastically decrease native species richness and cover and increase non-native species richness and cover. Because the disturbances represented by this vegetation type are unnatural to the monument, these findings suggest that practices such as chaining, spraying herbicides, burning, and seeding need to be limited within its boundaries. If such disturbances are allowed to persist and increase in size, many of the relationships between species richness and productivity at larger spatial scales have the potential to be reversed.

ACKNOWLEDGMENTS

We would like to thank Grand Staircase–Escalante National Monument and the Bureau of Land Management for their funding of this research, and the U.S. Geological Survey for logistical support. Various crews conducted the field work that provided the data for this analysis. R. Reich, P. Chapman, S. Boyden, and S. Merrill provided help with statistical analyses. G. Newman, C. Jarnevich, and S. Boyden provided helpful comments and guidance for earlier drafts of this manuscript. To all we are thankful.

LITERATURE CITED

Aguirre, L., and D. A. Johnson. 1991. Influence of temperature and cheatgrass competition on seedling development of two bunchgrasses. Journal of Range Management 44: 347–354.

Barbour, M. G., J. H. Burk, and W. D. Pitts. 1987. Terrestrial Plant Ecology, 2nd ed. Benjamin/Cummings, Menlo Park, California.

Begon, M., J. L. Harper, and C. R. Townsend. 1990. Ecology. Blackwell Scientific, Boston, Massachusetts.

Bond, W. 1983. On alpha diversity and the richness of the Cape Flora: A study in southern Cape fynbos. In Mediterranean-Type Ecosystems: The Role of Nutrients, edited by F. J. Kruger, D. T. Mitchell, and J. U. M. Jarvis, pp. 337–356. Springer-Verlag, Berlin.

Brown, R. L., and R. K. Peet. 2003. Diversity and invasibility of southern Appalachian plant communities. Ecology 84: 32–39.

Cliff, A. D., and J. K. Ord. 1981. Spatial Processes: Models and Applications. Pion, London.

Connell, J. H. 1978. Diversity in tropical rain forests and coral reefs. Science 199: 1302–1310.

Crawley, M. J. 1987. What makes a community invasible? In Colonization, succession, and stability, edited by M. J. Crawley and P. J. Edwards, pp. 429–451. Blackwell Scientific, Oxford.

Cressie, N. A. C. 1993. Statistics for Spatial Data. John Wiley and Sons, New York.

D'Antonio, C. M. 2000. Fire, plant invasions, and global changes. In Invasive Species in a Changing World, edited by H. A. Mooney and R. J. Hobbs, pp. 65–93. Island Press, Washington, D.C.

D'Antonio, C. M., R. F. Hughes, and P. M. Vitousek. 2001. Factors influencing dynamics of two invasive C4 grasses in seasonally dry Hawaiian woodlands. Ecology 82: 89–104.

Di Tomaso, J. M. 1998. Impact, biology, and ecology of saltcedar (Tamarix spp.) in the southwestern United States. Weed Technology 12: 326–336.

Elton, C. S. 1958. The Ecology of Invasions by Animals and Plants. Methuen, London.

Fox, M. D., and B. J. Fox. 1986. The susceptibility of natural communities to invasion. In Ecology of Biological Invasions: An Australian Perspective, edited by R. H. Groves and J. J. Burdon, pp. 57–66. Australian Academy of Sciences, Canberra.

Fridley, J. D., R. L. Brown, and J. E. Bruno. 2004. Null models of exotic invasion and scale-dependent patterns of native and exotic species richness. Ecology 85: 3215–3222.

Grace, J. B. 1999. The factors controlling species density in herbaceous plant communities: An assessment. Perspectives in Plant Ecology, Evolution, and Systematics 2: 1–28.

Grand Staircase–Escalante National Monument. 2000. Grand Staircase–Escalante National Monument: Approved Management Plan Record of Decision. BLM/UT/PT-99/020+1610. Bureau of Land Management, Cedar City, Utah.

Grime, J. P. 1973a. Competitive exclusion in herbaceous vegetation. Nature 242: 344–347.

Grime, J. P. 1973b. Control of species diversity in herbaceous vegetation. Journal of Environmental Management 1: 151–167.

Grime, J. P. 1979. Plant Strategies and Vegetation Processes. John Wiley and Sons, New York.

Hobbs, R. J. 1989. The nature and effects of disturbance relative to invasions. In Biological Invasions: A Global Perspective, edited by J. A. Drake, H. A. Mooney, F. di Castri, R. H. Groves, F. J. Kruger, M. Rejmanek, and M. Williamson, pp. 389–405. . Wiley and Sons, New York.

Hobbs, R. J., and L. F. Huenneke. 1992. Disturbance, diversity, and invasion: Implications for conservation. Conservation Biology 6: 324–337.

Hurvich, C. M., and C.-L. Tsai. 1989. Regression and time series model selection in small samples. Biometrika 76: 297–307.

Huston, M. A. 1979. A general hypothesis of species diversity. American Naturalist 113: 81–101.

Huston, M. A. 1994. Biological Diversity: The Coexistence of Species in Changing Landscapes. Cambridge University Press, Cambridge, UK.

Levine, J. M., and C. M. D'Antonio. 1999. Elton revisited: A review of evidence linking diversity and invasibility. Oikos 87: 1–11.

Leith, H. 1973. Primary production: Terrestrial ecosystems. Human Ecology 1: 303–332.

Leith, H. and R.L. Whittaker. 1975. Primary Productivity of the Biosphere. Springer-Verlag, New York.

Ludwig, J. A. 1986. Primary production variability in desert ecosystems. In Pattern and Process in Desert Ecosystems, edited by W. G. Whitford, pp. 5–17. University of New Mexico Press, Albuquerque.

Mittelbach, G. G., C. F. Steiner, S. M. Scheiner, K. L. Gross, H. L. Reynolds, R. B. Waide, M. R. Willig, S. I. Dodson, and L. Gough. 2001. What is the observed relationship between species richness and productivity? Ecology 82: 2381–2396.

National Climatic Data Center. 2003. National Climatic Data Center. Available at http://www.ncdc.noaa/gov/oa/ncdc.html.

Palmer, M. W. 1994. Variation in species richness: Towards a unification of hypotheses. Folia Geobot. Phytotax., Praha 29: 511–530.

Pearce, C. M., and D. G. Smith. 2001. Plains cottonwood's last stand: Can it survive invasion of Russian olive onto the Milk River, Montana floodplain? Environmental Management 28: 623–637.

Planty-Tabacchi, A., E. Tabacchi, R. J. Naiman, C. Deferrari, and H. Decamps. 1996. Invasibility of species-rich communities in riparian zones. Conservation Biology 10: 598–607.

Robinson, G. R., J. F. Quinn, and M. L. Stanton. 1995. Invasibility of experimental habitat islands in a California winter annual grassland. Ecology 76: 786–794.

Rosentreter, R. 1994. Displacement of rare plants by exotic grasses. In Proceedings, Ecology and Management of Annual Rangelands, edited by S. B. Monsen and S. G. Kitchen, pp. 170–175. Intermountain Research Station General Technical Report 313. USDA-USFS, Ogden, Utah.

Rosenzweig, M. L. 1968. Net primary productivity of terrestrial communities: Prediction from climatological data. American Naturalist 102: 67–74.

Rosenzweig, M. L. 1992. Species diversity gradients: We know more and less than we thought. Journal of Mammalogy 73: 715–730.

Rosenzweig, M. L. 1995. Species Diversity in Space and Time. Cambridge University Press, Cambridge, UK.

Satterthwaite, F.W. 1946. An approximate distribution of estimates of variance components. Biometrics Bulletin 2: 110–114.

Simberloff, D. 1986. Introduced insects: A biogeographic and systematic perspective. In Ecology of Biological Invasions of North America and Hawaii, edited by H. A. Mooney and J. A. Drake, pp. 3–26. Springer-Verlag, New York.

Stohlgren, T. J. 2002. Beyond theories of plant invasions: Lessons from natural landscapes. Comments on Theoretical Biology 7: 355–379.

Stohlgren, T. J., M. B. Falkner, and L. D. Schell. 1995. A modified-Whittaker nested vegetation sampling method. Vegetatio 117: 113–121.

Stohlgren, T. J., G. W. Chong, M. A. Kalkhan, and L. D. Schell. 1997. Rapid assessment of plant diversity patterns: A methodology for landscapes. Ecological Monitoring & Assessment 48: 25–43.

Stohlgren, T. J., K. A. Bull, and Y. Otsuki. 1998a. Comparison of rangeland sampling techniques in the Central Grasslands. Journal of Range Management 51: 164–172.

Stohlgren, T. J., K. A. Bull, Y. Otsuki, C. A. Villa, and M. Lee. 1998b. Riparian zones as havens for exotic plant species in the central grasslands. Plant Ecology 138: 113–125.

Stohlgren, T. J., D. Binkley, G. W. Chong, M. A. Kalkhan, L. D. Schell, K. A. Bull, Y. Otsuki, G. Newman, M. Bashkin, and Y. Son. 1999. Exotic plant species invade hot spots of native plant diversity. Ecological Monographs 69: 25–46.

Stohlgren, T. J., Y. Otsuki, C. A. Villa, M. Lee, and J. Belnap. 2001. Patterns of plant invasions: A case example in native species hotspots and rare habitats. Biological Invasions 3: 37–50.

Stohlgren, T. J., D. A. Guenther, P. Evangelista, and N. Alley. 2005. Patterns of plant species richness, rarity, endemism, and uniqueness in an arid landscape. Ecological Applications 15: 715–725.

Tilman, D. 1982. Resource Competition and Community Structure. Princeton University Press, Princeton, New Jersey.

Upton, G. L. G., and B. Fingleton. 1985. Spatial Data Analysis by Example. Vol. 1. Point Pattern and Quantitative Data. John Wiley, Chichester, England.

USDA. 2003. The PLANTS Database. National Plant Data Center, Baton Rouge, Louisiana.

Vitousek, P. M., L. R. Walker, L. D. Whiteaker, D. Mueller-Dombois, and P. A. Matson. 1987. Biological invasion by *Myrica faya* alters ecosystem development in Hawaii. Science 238: 802–804.

Waide, R. B., M. R. Willig, C. F. Steiner, G. G. Mittelbach, L. Gough, S. I. Dodson, G. P. Juday, and R. R. Parmenter. 1999. The relationship between productivity and species richness. Annual Review of Ecology and Systematics 30: 257–300.

Whittaker, R. H., and G. E. Likens. 1975. The biosphere and man. In Primary Productivity of the Biosphere, edited by H. Lieth and R. H. Whittaker, pp. 305–328. Springer-Verlag, New York.

Wiser, S. K., R. B. Allen, P. W. Clinton, and K. H. Platt. 1996. Invasibility of species-poor forest by a perennial herb over 25 years. Bulletin of the Ecological Society of America 77: 488.

RESTORING ROADS IN AN ARIZONA PONDEROSA PINE FOREST: EVALUATING THE EFFICACY OF FUNGAL INOCULUM AND MULCH AMENDMENTS

Joseph M. Trudeau

In the southern Rocky Mountains area, nearly half of all ponderosa pine forest lies within 0.25 miles of a road (Southern Rockies Ecosystem Project 2000). The impacts of roads and trails in these forests are of particular concern to ecological restorationists (Covington 2003). The effects of roads on ecosystems are numerous and have been well documented. In a review of these effects, Trombulak and Frissell (2000) concluded that roads affect terrestrial and aquatic communities in the following ways: increased animal mortality from construction and vehicle collisions, modification of animal behavior, alteration of the physical and chemical environment, spread of exotic species, and increased alteration and use of natural communities by humans. In addition, roads in ponderosa pine forests can alter natural fire regimes by interrupting surface fuels and enabling recreationists to start fires in otherwise unreachable areas (Covington 2003). Friederici (2003:22) has suggested that future restoration projects in ponderosa pine ecosystems may focus on more "broad-based and holistic restoration that includes road and trail closures," but for there to be effective improvements in habitat connectivity, fuel continuity, and other measures, regionally specific information regarding techniques and methods is needed (Elseroad et al. 2003).

Road removal is increasingly being used as a method for restoring pre-disturbance hydrology, ecosystem processes, and habitat continuity. Protocols have been established that guide the process of road ripping, which uses bulldozers to drag rock rippers or subsoilers through the road and break the compacted soil into aggregates of varying sizes. However, little research has been done to assess the effectiveness of these procedures in restoring critical ecosystem processes or attributes (Elseroad et al. 2003; Switalski et al. 2004). Luce (1997), in a study of the effects of ripping on infiltration capacity, concluded that ripping and subsoiling provide temporary and marginal improvements in hydrologic and ecological function. Elseroad (2001) confirmed these conclusions in a northern Arizona ponderosa pine forest. Ripping alone cannot be considered an effective restoration technique. The Society for Ecological Restoration (SER) considers a restored ecosystem to be one that is self-sustaining in structure and function, resilient to normal ranges of stress and disturbance, and able to interact with contiguous ecosystems (SER 2002). A road that is only ripped and then left to recover on its own may not meet these criteria within reasonable time periods. More effective procedures that restore the biological, physical, and chemical processes and properties of the disturbed soils and adjacent environment are needed in order to complement and complete holistic restoration of Southwest ponderosa pine forests.

One possible method for increasing the effectiveness of road restoration is the use of fungal inoculum to assist plant and microbial communities to achieve pre-disturbance conditions. Road construction removes organic soil layers and leaves a surface that is primarily mineral soil, which lacks the symbiotic and other fungi (Harvey et al. 1979) necessary to complete essential processes in the soil food web, such as nutrient cycling and plant community support. Several studies have documented that the fungal propagules that may be necessary for restoration are lacking in the damaged and compacted soils of dirt forest roads. Amaranthus and Trappe (1993), for example, observed that topsoil in a Pacific Northwest Douglas fir forest was a reservoir of fungal spores and other propagules of organisms important for decomposition, nutrient cycling, and mycorrhiza formation; after fire-induced erosion, the lack of this topsoil greatly impaired rehabilitation. Morman and Reeves (1979) documented a lack of arbuscular mycorrhizal (AM) inoculum in ripped road surfaces in northwestern Colorado, which they suggest significantly impairs the restoration of stable ecosystems on disturbed lands. Also, Reeves et al. (1979) found that more than 99 percent of plant cover in a natural site was mycorrhizal, but less than 1 percent of plant cover on a roadbed relied on mycorrhizae.

There has been limited exploration of the possibilities of using fungal inoculum in ecosystem restoration. Johnson (1998) examined the effects of AM inoculation on the invasive, non-mycotrophic Russian thistle (*Salsola kali*), and the perennial, mycotrophic switchgrass (*Panicum virgatum*) in an abandoned mine setting. Johnson found that plots with mycorrhizal inoculum resisted invasion by the exotic weed and supported growth of the perennial native grass. Despite the positive effects of inoculation, Johnson (1998) concluded that additions of local soil organic material may be more cost effective than the expensive process of inoculation. In another study, Stamets (2005) used fungal inoculum in road restoration in the Pacific Northwest, where wood chips were inoculated with saprophytic and mycorrhizal fungi to accelerate decomposition and provide the improved conditions offered through mycorrhizal associations. Advantages of this technique included sediment containment, moisture enhancement,

habitat recovery, soil structural improvements, and aesthetic enhancement of the road (Stamets 2005).

To date, no studies have been published that evaluate the use of fungal inoculum in road restoration in ponderosa pine forests. The purpose of my study was thus twofold: to evaluate the effect of commercially available mycorrhizal inoculum and mulch amendments on plant establishment on a ripped and seeded road in a northern Arizona ponderosa pine forest, and to evaluate the ability of commercially available saprophytic fungal inoculum to colonize ponderosa pine mulch.

METHODS
Experimental Design

This research was conducted at the Centennial Forest (CF) near Flagstaff, Arizona. The CF is a research, education, and demonstration forest co-managed by the Arizona State Lands Department and Northern Arizona University. Three temporary logging roads at the CF (35° 11.6' N, 111° 45.3' W) 10 km (6 miles) west of Flagstaff, Arizona were selected for study (Figure 1). The roads were built to facilitate a 2003 restoration thinning project and were then closed and ripped by September of that year. The roads were bladed to push topsoil aside, and no outside materials were incorporated. The ponderosa pine–Gambel oak (*Pinus ponderosa–Quercus gambelii*) forest lies at approximately 2250 m (7400 ft) in elevation. Roads were flat (0% slope) and soils were clayey-skeletal montmorillonitic Typic Eutroboralfs (Miller et al. 1995). Mean annual precipitation for the area is 21.5 inches with most occurring in winter and late summer, measured at the Flagstaff WSO AP Station. During the year of the road construction and ripping (2003) the area received below-normal precipitation (17.9 inches), but 2004 was above normal (23.6 inches), as was 2005 (24.0 inches; Western Regional Climate Center, http:// www.wrcc.dri.edu/index.html, retrieved on 4 April 2007).

A section of each road was selected that was straight for 55 m, allowing for the installation of five experimental blocks located

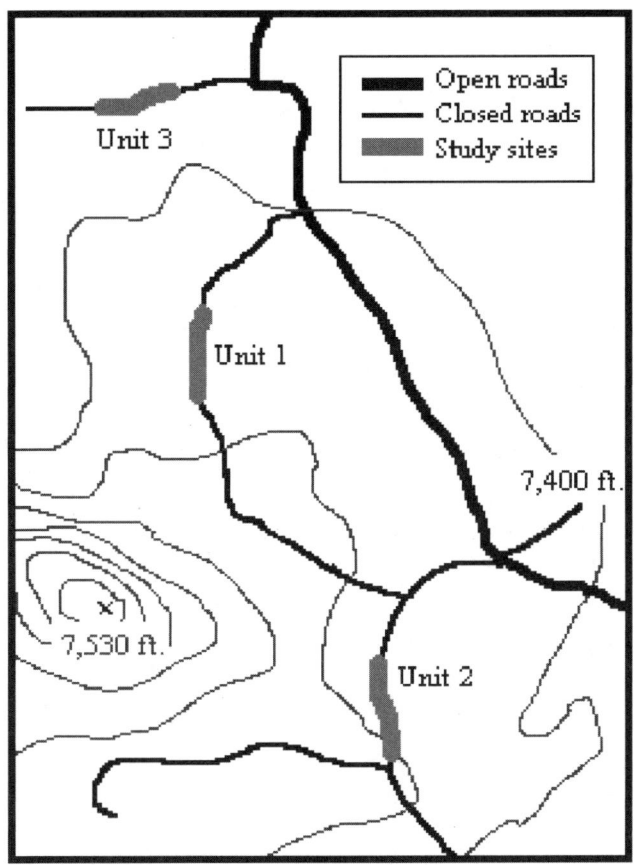

Figure 1. Study units at the Centennial Forest.

end-to-end in succession. Each block was 11 m long and 2 m wide, divided into four 1.5 x 2 m plots with 1 x 2 m buffers between each plot, resulting in 2 m buffers between blocks (Figure 2). Four treatments were assigned systematically to enhance ease of relocation and measurement in subsequent years. Treatments included a control (CON), addition of mycorrhizal inoculum (MYC), addition of mycorrhizal and saprophytic inoculum and mulch (SAP), and addition of only mulch (MUL). All plots were seeded with a native seed mix consisting of 56 percent (by weight) bottlebrush squirrel-tail (*Elymus elymoides*), 20 percent Arizona fescue (*Festuca arizonica*), 16 percent spike

muhly (*Muhlenbergia wrightii*), and 8 percent silvery lupine (*Lupinus argenteus*), added at a rate of 9.75 g/m^2. These same species and rates were used by Elseroad et al. (2003). Mulch (ground ponderosa pine slash) was applied to a depth of 10 cm after seeding.

Fungal Inoculum

The mycorrhizal inoculum was a commercially available (Mycogrow Micronized Endo-Ecto Seed Mix, purchased from Fungi Perfecti in Kamilche Point, Washington; www.fungi.com) mixture of three arbuscular mycorrhizal fungi: *Glomus aggregatum, G. intraradices,* and *G. mosseae.* Mycorrhizal fungi of the genus *Glomus* are the most

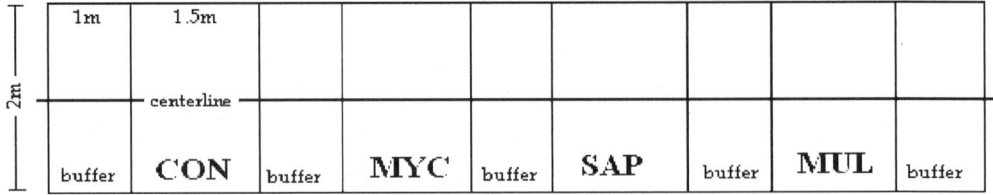

Figure 2. Block design with plot layout and dimensions.

important fungus partners forming mycorrhizae (Hudler 1998), and *G. mosseae* has been documented in Arizona with Arizona fescue (Molina et al. 1978). Also in the mix are the ectomycorrhizal fungi *Pisolithus tinctorus* and four species of *Rhizopogon*. The product, sold in powder form, is intended to be mixed with seeds as they are dispersed into the soil. Seeds in the MYC and SAP treatments were mixed with 7.5 g of the powder per plot, then lightly raked to mix with the upper layer of the soil.

The saprophytic inoculum was the commercially available "dowel spawn" of the species *Hypholoma capnoides*, also purchased from Fungi Perfecti. These 1 x 2 cm alder wood dowels have been aggressively colonized by the fungus, as evidenced by masses of hyphae growing within and among the dowels. After spreading out 5 cm of mulch, 2000 ml of dowel spawn were evenly distributed over the mulch, and the remaining 5 cm of mulch was added on top. All the amendments were made over 3 days, on 60 total plots, in mid-August of 2004.

Although using native species in ecological restoration is important, determining if a fungus is native can be challenging. *Hypholoma capnoides* has never been collected in close proximity to the study site, but it has been collected and positively identified at several locations in the region. In Colorado, *H. capnoides* was observed at the Colorado Mycological Society's autumn 2001 mushroom fair (http://www.capsandstems.com/id.htm, retrieved 30 June 2004), and in New Mexico at the 1997 New Mexico Mycological Society's Foray near Chama, New Mexico (http://www.mycowest.org/fungi/s1997.ht

m, retrieved 5 October 2005). In Arizona, *H. capnoides* has been documented in ponderosa pine–Douglas fir–aspen forests (*P. ponderosa–Pseudotsuga menziesii–Populus tremuloides*) on the North Rim of Grand Canyon National Park (GCNP), and in ponderosa pine–Douglas fir forests in the Chiricahua Mountains in the southeastern part of the state (Jack States, personal communication 23 September 2004). In a study of belowground fungal sporocarp production on the Kaibab (North Rim, GCNP) and Coconino (region of present study) Plateaus, States and Gaud (1997) found a 78 percent coefficient of similarity for species that contributed more than 1 g to annual biomass, suggesting that fungal species encountered on the North Rim, such as *H. capnoides*, may also be found in the region of the present study. Given the ubiquitous nature of fungi, and the unreliable fruiting both spatially and temporally (Arnolds 1988), it is possible that *H. capnoides* is native to the Flagstaff area but this has yet to be documented.

Vegetation Field Sampling

Data were collected during two 2-week sessions, in mid-October of 2004 (2 months after applying amendments) and in mid-October of 2005 (14 months after application). Early October marked the end of monsoon season in both years, although the autumn and early winter of 2004 maintained steady precipitation (Table 1). Vegetation sampling in 2004 consisted of identification of species present on plots and a measurement of abundance by counting individuals. Grasses, unless developed enough to identify to the species level, were categorized

Table 1. Precipitation records for Flagstaff, Arizona, 2003–2005 (in inches). Data from Western Regional Climate Center (http://www.wrcc.dri.edu/index.html).

	2003	2004	2005
Jan	0.14	0.76	6.58
Feb	2.75	1.06	4.19
Mar	1.13	0.74	2.43
Apr	0.44	1.81	2.15
May	0.73	0.00	0.08
Jun	0.04	0.02	0.40
Jul	3.40	1.47	2.51
Aug	3.03	4.71	3.41
Sep	2.62	1.76	0.46
Oct	0.14	3.51	1.59
Nov	2.51	3.10	0.20
Dec	0.92	4.67	0.01
Mean	17.85	23.61	24.01

simply as grass seedlings. Vegetation sampling in 2005 consisted of identifying all plants to species level, measuring abundance by counting individuals, and estimating percent areal cover at 1 percent intervals, with less than 1 percent being tallied as 0.5 percent. All grasses were developed enough to be identified correctly, so the grass seedling category was not necessary.

Fungal Field Sampling

Fungal data were collected at the same time as the vegetation sampling. The saprophytic community was sampled by direct observation of fungal hyphae spreading from the inoculated dowels into the pine mulch and forming connections with pieces of the wood. For the 2004 measurement, a 20 x 30 cm subplot (2% of total plot) was randomly selected to be sampled within each SAP plot. Mulch was carefully removed from the selected subplot until dowels were encountered, and mulch in the vicinity of the dowel was then examined for colonization. If no dowels were encountered, another subplot was selected. Subplots were thoroughly examined to ensure that all dowels were accounted for. Any observation of colonization was counted as a success, no matter how small the hyphal strand.

For the 2005 measurement, five randomly selected subplots were examined (10% sample). Mulch was removed from the subplot until dowels were encountered or it was determined to be lacking inoculum. If a subplot was devoid of dowels, it was not replaced but was counted as absent of inoculum. Also recorded was the presence or absence of resident mycelia. Early in the measurement process, it was observed that many resident fungal hyphae were not directly connected to inoculum points and appeared to come from the soil or mulch. The first road measured was not quantified for fungi because the observation came after plots had been disturbed, but the following two roads were measured for any effects the inoculum may have had on resident soil fungi. In addition to sampling five subplots in the SAP plots, five subplots were also sampled in MUL plots. Two morphological types were encountered: a white filamentous hyphae that emerged from rocks and pieces of wood that were imbedded in the road surface prior to any amendments, and a dull-white to creamy yellow, cottony hyphae that formed at times dense networks within the woodchip matrix. If either was encountered and it was determined to not be associated with any dowel, presence of resident mycelia was recorded. Resident mycelium varied in the density of its colonization of the mulch and the breadth of coverage. This was quantified by rating the degree of colonization as weak, medium, or strong. A weak score was recorded for resident mycelium that occupied less than 5 cm^3 of the 20 x 30 cm subplot, a medium score was recorded for resident mycelium that occupied 5–10 cm^3 of the subplot, and a strong score was recorded for anything that exceeded 10 cm^3 of the subplot.

Statistical Analysis

Vegetation data, which were distributed normally, were initially analyzed using a blocked ANOVA ($\alpha = 0.05$) to determine the effects of treatment on the following variables: grass seedling abundance, species richness, and total species abundance. Following a significant overall test, means were compared by pairwise Tukey-Kramer post hoc tests to determine differences between plots with mulch (SAP and MUL) and those without mulch (CON and MYC).

Similarly, a blocked ANOVA ($\alpha = 0.05$) was used to determine any effects of the presence or density of resident mycelia, and depth of mulch on survival of inoculum. Mycorrhizal fungi were not measured in the field or through any form of bioassay. The effects of mycorrhizae on plants were inferred through analysis of the vegetation data.

RESULTS

Vegetation Response

Two months after application, in 2004, treatments adequately explained the variation in grass seedling abundance and species richness (ANOVA $P < 0.0001$). Post-hoc analysis showed that the difference was between pairs; CON and MYC plots had significantly higher abundance and richness than SAP and MUL plots (Figures 3 and 4; Tukey-Kramer $P < 0.05$).

In 2005, 14 months after application, variation in richness and abundance was again explained by treatment (ANOVA $P < 0.0001$), and again the difference resided between pairs; post-hoc analysis showed that CON plots were not different from MYC plots and SAP were not different from MUL, but SAP and MUL plots had significantly lower species richness and total species abundance than CON and MYC (Tukey-Kramer $P < 0.05$). There was a mean increase in species richness of two species per plot for all treatments from 2004 to 2005. The mean abundance per plot of bottlebrush squirrel-tail, Arizona fescue, silvery lupine,

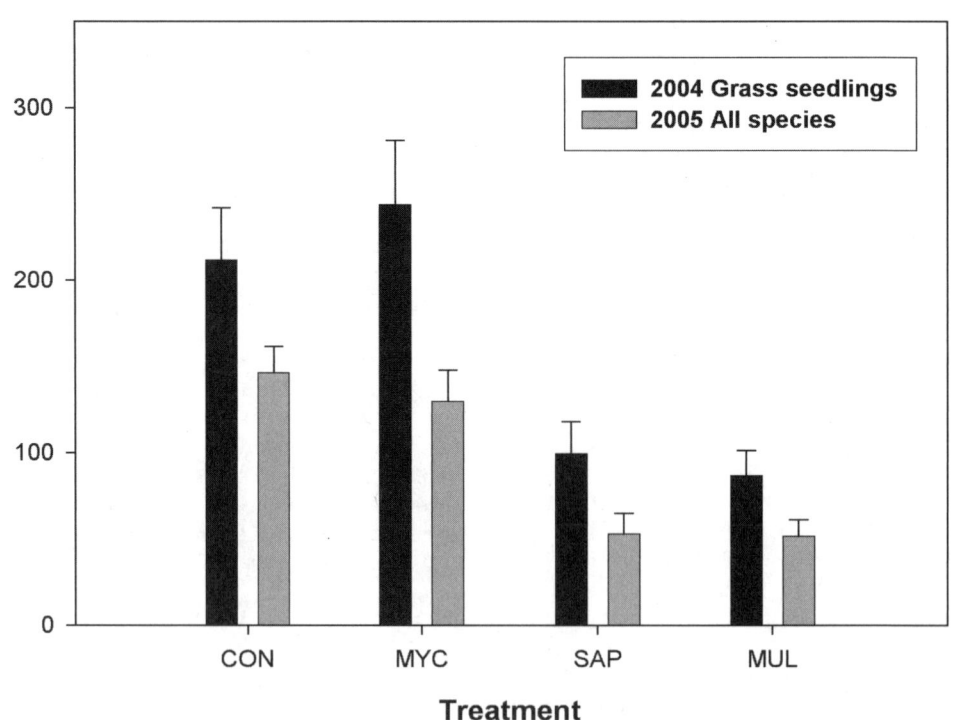

Figure 3. Grass seedling abundance in 2004 and all-species abundance in 2005, with standard errors.

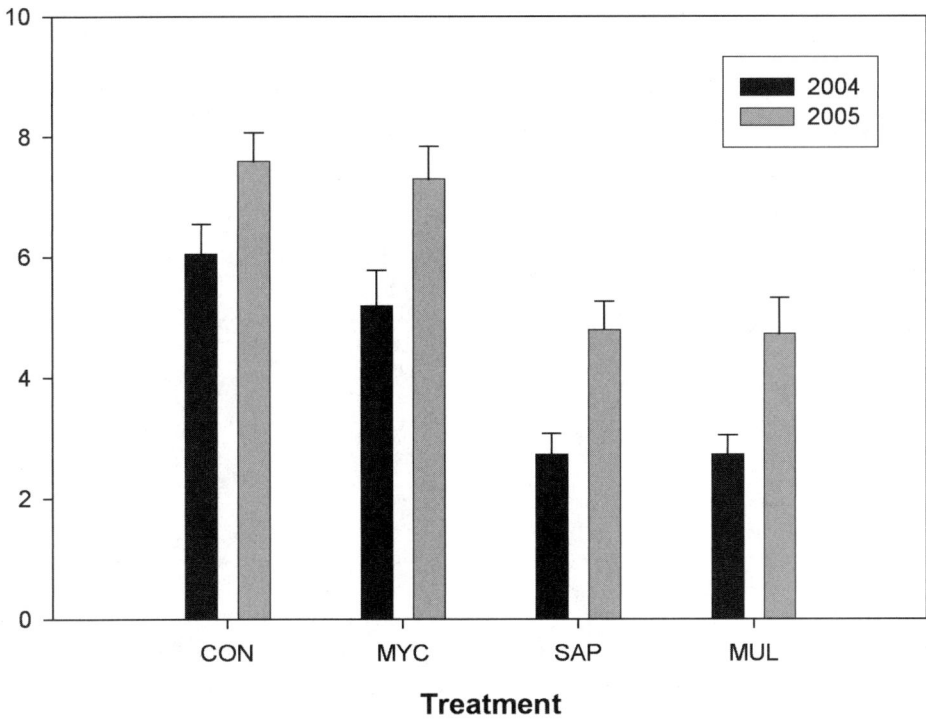

Figure 4. Species richness in 2004 and 2005, with standard errors.

and spike muhly was 42.3, 3.7, 1.5, and 0.2 plants per plot, respectively, and they were found in 55, 25, 29, and 2 of the 60 total plots. One ponderosa pine seedling was encountered in a control plot.

Although mulched plots (SAP and MUL) suppressed the establishment of most species, the substrate was conducive to the establishment of Gambel oak seedlings (Figure 5). Two seedlings were found to be growing from acorns, confirming that they were not sprouts from former trees. More oaks were found on SAP and MUL plots than on CON and MYC, and oaks established especially well on one of the roads. Treatment explained some variability in oaks (R^2 0.18, ANOVA $P = 0.0085$), indicating that mulch may provide conditions suitable for oak establishment. However, of the 111 oak seedlings counted, 100 were on

one road, suggesting that conditions there had a strong effect on recruitment. There were significantly more oaks on SAP plots than other treatments (Figure 5; Tukey-Kramer $P < 0.05$).

Efficacy of Saprophytic Inoculation

In 2004, 70 percent of inspected dowels showed successful survival and colonization, but by 2005 only 34 percent of subplots showed signs of success. In 2004 the inoculum on one of the roads was vigorously colonized by a resident soil fungus, which appeared similar to members of the genus *Trichoderma*, which often excrete substances capable of killing mycelia of other fungi (Ingold and Hudson 1993). By 2005 there was zero survival of *H. capnoides* on this road, which was also the first to be measured, so there was no quantification of the

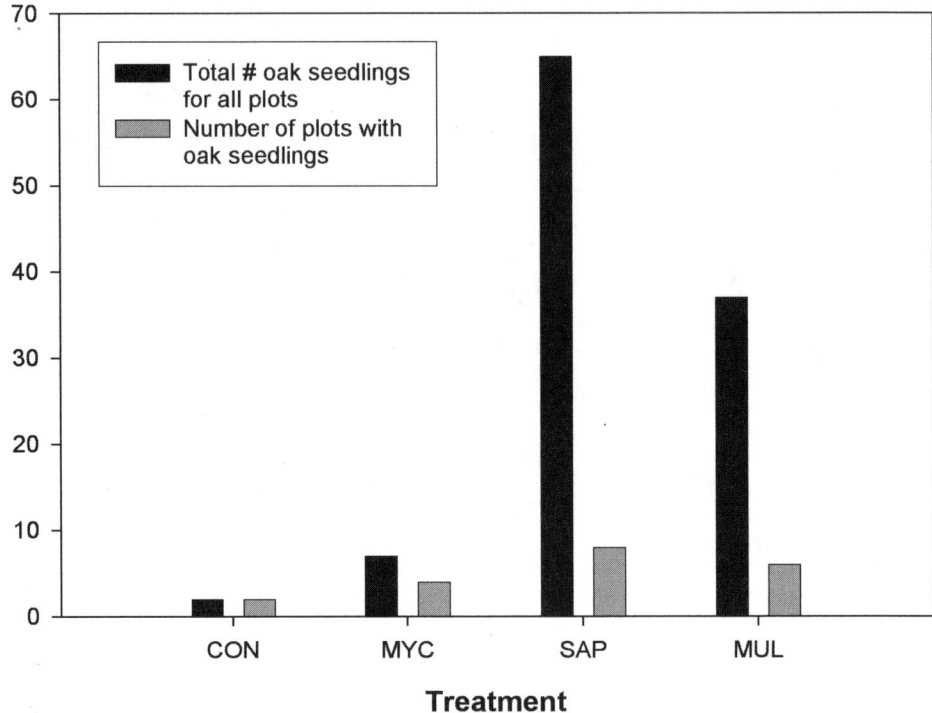

Figure 5. Abundance of Gambel oak (*Quercus gambelii*) seedlings across all plots per treatment and number of plots with established oak seedlings.

resident soil fungi. The following two units were examined more critically for resident soil fungi. Sixty-eight percent of sampled subplots that had been inoculated (SAP plots) were colonized by resident fungi, and 60 percent of MUL plots were colonized. There were no relationships between success of inoculation and presence of resident fungi or density of resident colonization.

DISCUSSION

The results of this study provide little support for the use of commercial fungal inoculum on ripped roads in Arizona ponderosa pine forests. The mycorrhizal inoculum had no significant effect on plant abundance and richness or on the establishment of seeded species in plots with or without mulch. The initial response of a greater abundance of grass seedlings on MYC plots proved to be an insignificant variation that reversed for all species the following year. The negative findings are surprising in light of higher plant survival (St. John 1996) and species diversity (Amaranthus and Trappe 1993) at other sites inoculated with mycorrhizae. The insignificant effects observed in the present study may be due to the method of application or environmental factors. St. John (1996, 1998) suggested that inoculum placed on the soil surface will be less functional than inoculum that is buried; in my study the inoculum was lightly raked into the soil. It may therefore not have been tilled deep enough into the soil to protect it from predation, UV exposure, or desiccation. Future trials with powdered inoculum would benefit from adding it at the time of ripping to increase mixing with deeper soils. As plant roots tend to be most dense near the surface, inoculating any deeper than several inches is not advised.

The hot and dry climate may have been an unfavorable environment for successful germination and formation of mycorrhizae, despite the wet autumn of 2004. The mix was purchased from a distributor of fungal products in the Pacific Northwest, where the climate is damper and more conducive to robust fungal development. The fungi in the mix may have been adapted to those conditions and not the arid Southwest, with its long periods of low or no precipitation as happened in May and June of both 2004 and 2005. Johnson (1998), who studied the reclamation of taconite mine tailings, used soil from a nearby location as a source of fungal inoculum and reported enhanced performance of a native perennial grass. That method may be more effective than using a generic mix of species in powder form, and it may prevent the introduction of exotic species (Bagley 1999).

Elseroad et al. (2003) found that bottle-brush squirrel-tail established in the highest density on a ripped road in an Arizona ponderosa pine forest, followed by Arizona fescue, silvery lupine, and spike muhly, which is consistent with the results of my study. Seeding with silvery lupine has not proven to be effective on ripped roads in Arizona, but future trials would be useful as it is important to facilitate the establishment of nitrogen-fixing species in degraded soils (Bradshaw 1987). Also, silvery lupine seeds were not scarified prior to use in my study, which may explain the poor response.

Most of the plants present across all three road units were volunteer species, including Douglas' knotweed (*Polygonum douglasii*), annual muhly (*Muhlenbergia minutissima*), Wright's deervetch (*Lotus wrightii*), Rusby's milkvetch (*Astragalus rusbyi*), and King's lupine (*Lupinus kingii*), which Elseroad et al. (2003) also encountered frequently. In addition, spreading fleabane (*Erigeron divergens*), spreading groundsmoke (*Gayophytum diffusum*), Fremont's goosefoot (*Chenopodium*

fremontii), and Navajo cinquefoil (*Potentilla subviscosa*) were commonly observed, suggesting that local seed sources were abundant enough to affect revegetation.

Organic amendments have variable effects on road revegetation (Switalski et al. 2004), but there are reports of long-term positive effects of woodchips and wood mulch. Cotts et al. (1991) reported suppression of plant community development the first year after application of woodchips, but during the second year development rebounded to normal levels. Petersen et al. (2004) found that woodchips spread over hydroseeded native grasses at Bryce Canyon National Park were as effective as any other treatment at promoting revegetation. Additionally, the woodchips offered erosion protection not offered by other amendments, and they can mitigate the surface sealing that is common after roads are ripped and left to recover on their own (Bradley 1997; Elseroad 2001). Paschke et al. (2000) observed that a wood-shaving-based fiber cloth alone did not significantly negatively affect plant establishment, and when used in conjunction with fertilizer, it had significant positive effects. Wood mulch had a strong inhibitory effect on plant growth in the present study, which is unfortunate in terms of establishing a diverse plant community. Possible reasons for the suppressed plant abundance and richness are cooler soil temperatures from the insulating effect, limited light reaching young plants, and physical barriers to small seedlings (Petersen et al. 2004). These results are different from those of Elseroad et al. (2003), who found that mulch 2 cm deep did not affect total plant density or cover in an Arizona ponderosa pine forest approximately 16 km (10 miles) north of the present study. The mulch in my study was applied to a depth of approximately 10 cm, which could explain this inconsistency. However, she observed that mulched plots had increased establishment of Arizona fescue, which I did not observe.

Although mulch inhibited most plant species, it did seem to provide suitable habitat for the establishment of Gambel oak. In terms of permanently restoring roads, oaks may be highly effective plants. As they mature, they will become natural barriers to vehicle passage, they will contribute to organic soil creation with their annual leaf fall, their roots will be effective at loosening the compacted soil, and trees offer microhabitat values that grasses and forbs do not. A possible mechanism for the introduction of acorns to the mulch may be caching by birds, squirrels, or other mammals. That most oaks were found on one road implies that there may be more seed trees in that area, or that there may be limited available substrates suitable for caching by wildlife, or a combination of the two.

There has been limited research into the inoculation of wood waste with fungi. Stamets (2005) found that saprophytic inoculum rapidly colonized woodchips and straw in the moist climate of Tahuya State Park in Washington, near the location from which the inoculum was cultured. Within 3 years, extensive mycelium had spread throughout the woodchips and enhanced decomposition as well as contributed to sediment retention by filtering runoff (Stamets 2005). However, introducing saprophytes into a substrate may not be necessary. Mayfield et al. (1990) studied the saprophytic fungal response to silvicultural site preparations and found that 40 species of saprophytic fungi had colonized logging residue after 2 years. In the case of roads, where there are nearby sources of inoculum such as adjacent forest, there is likely to be adequate inoculum potential (Jack States, personal communication 23 September 2004). The majority of mulched plots (both SAP and MUL) in my study became colonized by fungi that already existed in the soil or nearby, suggesting that inoculation with a saprophyte appears unnecessary. That the mulch was colonized by local fungi is a testament to the ubiquitous nature and effective life history of Nature's most important degrader of wood and that there may still have been viable spores in the road, which was used for less than a year. Soil fungal communities are highly spatially heterogeneous (Smith et al. 2004; Stendell et al. 1999), which may explain why one unit was rapidly colonized by an aggressive antagonist and the other two units showed more survival, albeit decreasing. In addition, that there were no relationships between survival of inoculum, presence of resident fungi, and degree of resident colonization suggests that not only did the saprophytic inoculum have little success in colonizing the mulch, but it also had little effect on attracting, resisting, or facilitating resident fungi.

Future studies with fungal inoculum in different regions should consider locating or developing inoculum that is adapted to the environmental conditions of that region. Until aggressive regional strains of mycorrhizal or saprophytic fungi can be cultured and produced for distribution, as is the case in the Pacific Northwest, these methods will remain less effective on the Colorado Plateau at restoring roads than spreading local soils and mulch and then allowing local fungi to colonize naturally. Application of woodchips or mulch has proven effective in maintaining or increasing vegetation cover while reducing erosion (Stamets 2005; Petersen et al. 2004; Bradley 1997; Cotts et al. 1991). Mycorrhizal inoculum is typically used in the form of local soil (Johnson 1998) and that may be the most effective method in Arizona ponderosa pine forests. Bradshaw (1987) suggested that speed of attainment, cheapness, reliability, and stability should be considered in land reclamation projects. The cost of purchasing inoculum is not consistent with these considerations at this time. However, spreading mulch from logging residue can provide important habitat for oak regeneration, which has positive long-term implications for road restoration, is inexpensive, has long-term benefits (Cotts et al. 1991; Petersen 2004), and promotes soil stability. In addition, mulching waste wood may have less of a negative effect on forest ecosystems than burning slash piles, which has proven deleterious effects (Tiedemann et al. 2000; Korb et al. 2004).

ACKNOWLEDGMENTS

I would like to thank the numerous students, staff, and faculty at the Ecological Restoration Institute at Northern Arizona University, who are too many to mention, for the generous provision of help in project implementation, statistical consultation, funding, and reviews of earlier drafts of this manuscript. I would also like to thank J. Bell for insights into fungal ecology, D. Sumerlin and P. Stamets at Fungi Perfecti for financial and intellectual assistance, and J. J. Smith and the Centennial Forest for permitting the research. Most of all, I thank people of the Ecological Restoration Institute for allowing me the freedom of exploration for this little-understood realm of restoration ecology.

LITERATURE CITED

Amaranthus, M. P., and J. M. Trappe. 1993. Effects of erosion on ecto- and VA-mycorrhizal inoculum potential in soil following forest fire in southwest Oregon. Plant and Soil 150(1): 41–49.

Arnolds, E. 1988. The changing macromycete flora of the Netherlands. Transactions of the British Mycological Society 90: 391–406.

Bagley, S. 1999. Desert road removal: Creative restoration techniques. Road Rip-porter 4(4): 12–13.

Bradley, K. 1997. An evaluation of two techniques for the utilization of logging residues: Organic mulch for abandoned road revegetation and accelerated decomposition in small chipped piles. Master's thesis, University of Montana, Missoula.

Bradshaw, A. D. 1987. The reclamation of derelict land and the ecology of ecosystems. In Restoration Ecology: A Synthetic Approach to Ecological Research, edited by W. R. Jordan, M. E. Gilpin, and J. D. Aber, pp. 53–74. Cambridge University Press, Cambridge.

Cotts, N. R., E. F. Redente, and R. Schiller. 1991. Restoration methods for abandoned roads at lower elevations in Grand Teton National Park, Wyoming. Arid Soil Research and Rehabilitation 5: 235–249.

Covington, W. W. 2003. The evolutionary and historical context. In Ecological Restoration of Ponderosa Pine Forests, edited by P. Friederici, pp. 26–47. Island Press, Washington, D.C.

Elseroad, A. 2001. Forest roads in northern Arizona: Recovery after closure and revegetation techniques. Master's thesis, Northern Arizona University, Flagstaff.

Elseroad, A. C., P. Z. Fulé, and W. W. Covington. 2003. Forest road revegetation: Effects of seeding and soil amendments. Ecological Restoration 21(3): 180–185.

Friederici, P. 2003. The "Flagstaff Model." In Ecological Restoration of Ponderosa Pine Forests, edited by P. Friederici, pp. 7–25. Island Press, Washington, D.C.

Harvey, A. E., M. J. Larsen, and M. F. Jurgensen. 1979. Comparative distribution of ectomycorrhizae in soils of three western Montana forest habitat types. Forest Science 25(2): 350–358.

Hudler, G. W. 1998. Magical Mushrooms, Mischievous Molds. Princeton University Press, Princeton.

Ingold, C. T., and H. J. Hudson. 1993. The Biology of Fungi, 6th ed. Chapman & Hall, London.

Johnson, N. C. 1998. Responses of Salsola kali and Panicum virgatum to mycorrhizal fungi, phosphorous and soil organic matter: Implications for reclamation. Journal of Applied Ecology 35: 86–94.

Korb, J. E., N.C. Johnson, and W. W. Covington. 2004. Slash pile burning effects on soil biotic and chemical properties and plant establishment: Recommendations for amelioration. Restoration Ecology 12(1): 52–62.

Luce, C. H. 1997. Effectiveness of road ripping in restoring infiltration capacity of forest roads. Restoration Ecology 5(3): 265–270.

Mayfield, J. E., M. B. Edwards, and W. V. Dashek. 1990. Relationship of macrofungal population to silvicultural treatments in a recently harvested pine forest. Forest Ecology and Management 31: 109–119.

Miller, G., N. Ambos, P. Boness, D. Reyhar, G. Robertson, K. Scalzone, R. Steinke, and T. Subirge. 1995. Terrestrial ecosystem survey of the Coconino National Forest. USDA Forest Service, Southwestern Region.

Molina, R. J., J. M. Trappe, and G. S. Strickler. 1978. Mycorrhizal fungi associated with Festuca in the western United States and Canada. Canadian Journal of Botany 56: 1691–1695.

Morman, T., and F. B. Reeves. 1979. The role of endomycorrhizae in revegetation practices in the semi-arid West. II. A bioassay to determine the effect of land disturbance on endomycorrhizal populations. American Journal of Botany 66(1): 14–18.

Paschke, M. W., C. DeLeo, and E. F. Redente. 2000. Revegetation of roadcut slopes in Mesa Verde National Park, U.S.A. Restoration Ecology 8(3): 276–282.

Petersen, S. L., B. A. Roundy, and R. M. Bryant. 2004. Revegetation methods for high-elevation roadsides at Bryce Canyon National Park, Utah. Restoration Ecology 12(2): 248–257.

Reeves, F. B, D. Wagner, T. Moorman, and J. Kiel. 1979. The role of endomycorrhizae in revegetation practices in the semi-arid West. I. A comparison of incidence of mycorrhizae in severely disturbed vs. natural environments. American Journal of Botany 66(1) 1–13.

Smith, J. E., D. McKay, C. G. Niwa, W. G. Thies, G. Brenner, and J. W. Spatafora. 2004. Short-term effects of seasonal prescribed burning on the ectomycorrhizal fungal community and fine root biomass in ponderosa pine stands in the Blue Mountains of Oregon. Canadian Journal of Forest Research 34: 2477–2491.

Society for Ecological Restoration Science & Policy Working Group. 2002. The SER Primer on Ecological Restoration (www.ser.org/).

Southern Rockies Ecosystem Project. 2000. The State of the Southern Rockies Ecoregion. Southern Rockies Ecosystem Project, Nederland, Colorado.

Stamets, P. 2005. Mycelium Running. Ten Speed Press, Berkeley.

States, J. S., and W. S. Gaud. 1997. Ecology of hypogeous fungi associated with ponderosa pine. I. Patterns of distribution and sporocarp production in some Arizona Forests. Mycologia 89(5): 712–721.

Stendell, E. R., T. R. Horton, and T. D. Bruns. 1999. Early effects of prescribed fire on the structure of the ectomycorrhizal fungus community in a Sierra Nevada ponderosa pine forest. Mycological Research 103(10): 1353–1359.

St. John, T. 1996. Mycorrhizal inoculation: Advice for growers and restorationists. Hortus West 7(2): 1–4.

St. John, T. 1998. Mycorrhizal inoculation in habitat restoration. Land and Water (Sep–Oct) 1998: 17–19.

Switalski, T. A., J. A. Bissonette, T. H. DeLuca, C. H. Luce, and M. A. Madej. 2004. Benefits and impacts of road removal. Frontiers in Ecology and the Environment 2(1): 21–28.

Tiedemann, A. R., J. O. Klemmedson, and E. L. Bull. 2000. Solution of forest health problems with prescribed fire: Are forest productivity and wildlife at risk? Forest Ecology and Management 127: 1–18.

Trombulak, S. C., and C. A. Frissell. 2000. Review of ecological effects of roads on terrestrial and aquatic communities. Conservation Biology 14(1): 18–30.

GAINING INSIGHTS FROM THE PAST

MODELING FUTURE PLANT DISTRIBUTIONS ON THE COLORADO PLATEAU: AN EXAMPLE USING *PINUS EDULIS*

Kenneth L. Cole, Kirsten E. Ironside, Samantha T. Arundel, Philip Duffy, and John Shaw

The recent mortality of some plant species in the U.S. Southwest has been attributed to the ongoing drought conditions over the last decade. This mortality has been especially acute in populations of *Pinus edulis* (Colorado pinyon pine; hereafter abbreviated as pinyon), a widespread and highly visible species (Shaw 2006; Shaw et al. 2005; Mueller et al. 2005). These recent mortality events may be similar to changes expected to occur in the future because of global climate warming (Breshears et al. 2005). That is, the consequences of periodic episodes of low precipitation can be exacerbated by higher temperatures, which increase drought stress. Here we demonstrate new techniques for modeling the effect of future climates on plant species using pinyon as an example. Using the techniques described here, similar results could be generated for any plant species or for any climatically controlled environmental process.

MODELING CHANGE USING CLIMATIC ENVELOPES

As climates have changed in the past, the geographic distributions of plant species have shifted as well. Future climate change will also result in the mortality of some plant populations as the suitable climate for these species shifts to new regions. This shifting of climate and populations has occurred naturally through time, though sometimes at notably faster rates. If the rate of the climate shift exceeds the migrational capabilities of the species, it must lag the climate shift, eventually catching up through migrational processes such as seed dispersal, maturation, succession, and equilibration within their new plant assemblages.

Through modeling it is possible, given any new climate, to predict where a species should be eliminated and where it could survive. Many models assume that a species is already present across its new potential range, waiting for the new climate to allow its seeds to sprout. In our model, we instead incorporate data on the species' past and ongoing rates of migration in order to determine which of these new potential areas the species will likely expand into, and which will remain beyond its range during that period.

We developed our computer models using the "individualistic" species approach to vegetation change. Each species is treated as an individual entity and its climate envelope is modeled using its geographic range during the late twentieth century. This approach integrates the variability within the current population including regional genetic variants and local microhabitats. The model also incorporates all of the indirect physical and biological processes influencing the range that are modulated by climate. Processes such as fire frequency, soil microfauna, and arthropod occurrences are represented in the model to the extent that they are controlled by climate and have influenced the species' range during the twentieth century. For example, if the species is susceptible to bark beetle infestations as warmer temperatures allow for increased beetle populations (Williams and Liebhold 2002), then this temperature limit should be expressed somewhere as a geographic limit to the species range. But the model could not predict the effects of new beetle species recently introduced from other continents.

The incorporation of cross-correlations between these multiple limiting factors is maximized by analyzing the entire twentieth century range of the species (Jackson and Overpeck 2000). But this climate envelope model cannot predict the effect of new or unique combinations of biological and physical variables not evident near the species range. Most important, the physiological effects of higher CO_2 concentrations (Waterhouse et al. 2004) are not incorporated because the twentieth-century species range did not contain CO_2 levels as high as current or predicted future values. The results of FACE (Free Air CO_2 Enrichment) experiments and/or observations of ongoing landscape changes from permanent plots can eventually be used to incorporate this effect.

RECENT SHIFTS OF PINYON

The rate of plant migrations is determined by several factors, the most obvious of which is long-distance seed dispersal. But to be effective, these long-distance events must be frequent because seeds are rarely successful in colonizing new habitats, especially those already occupied by other species. If a seedling does survive, that individual must mature to provide a local seed source. Then this local process of further establishment and further maturation will involve additional time lags for late successional species such as pinyon. Parameterizing models of these processes requires values from observations, but the relationships between frequency and distance for long-distance dispersals cannot be measured experimentally but only inferred (Cain et al. 2000; Davis et al. 1986) because as the distance of these dispersals becomes greater, their frequency rapidly decreases below a measurable threshold. As a result, some studies retreat toward anecdotes of extreme dispersal events without information about the frequency of such events. A more promising approach toward estimating plant migra-

tions might be to observe and measure ongoing and past plant migrations, rather than to model their constituent processes.

Over the last century pinyon has been observed encroaching into grazed grasslands adjacent to existing woodlands (Rosenstock and van Riper 2001). Some researchers have observed that pinyon encroachment lags behind a more rapid juniper encroachment (Blackburn and Tueller 1970), as pinyon seedlings often become established under juniper "nurse" trees (Chambers 2001). Unfortunately, there seem to be few estimates on this rate of pinyon encroachment by distance. Instead, the rate is usually estimated as the number of years it takes for a forest to develop in a grazed or burned, or a chained area that is already adjacent to an existing forest. As a result, these estimates may be more useful as a measure of successional processes that follow the local availability of seed.

While studying woodland recovery after three stand-replacing fires near the northern limit of pinyon at Dutch John in Utah, Goodrich and Barber (1999) estimated that despite already having a rich seed source nearby, these small areas (13–82 ha in size) would take more than 200 years to return to the pre-burn mature successional pinyon woodland stage. A similar study in north-central Arizona estimated that 215 years may "underestimate" the time required for development of a mature pinyon forest (Tress and Klopatek 1987). From these estimates, it would seem that mature pinyon woodland requires a significant amount of time for development, well after it comes within the distance of abundant seed dispersal.

In order to study the issue on a much broader scale of time and space, records of paleo-migrations can be obtained from fossil pine needles in packrat middens. These records encompass the entire migration and successional cycle from mature woodland phase to mature woodland phase.

PAST CLIMATE WARMING

During the span of history most accessible for paleoecological study in the western

United States—the last 20,000 years—the most extreme climate warming occurred between about 11,700 to 11,500 years ago (Overpeck et al. 2003). This sudden temperature warming occurred at the end of the cold Younger Dryas period during the transition between the last ice age and the Holocene. During this time, temperatures increased more than 4° C in less than 200 years in the western United States (Barron et al. 2003; Cole and Arundel 2005). Uncertainties in the dating of these records probably underestimate the rapidity of this abrupt warming, shown to have happened in less than 100 years by better-dated chronologies (Severinghaus et al. 1998; Grootes and Stuiver 1997). Following this abrupt climate warming, temperatures then continued to increase more slowly, possibly reaching a maximum several thousand years later (Figure 1).

The rate and magnitude of the climate warming that occurred 11,600 years ago is remarkably similar to the warming that is expected to occur in the western United States over the next 100 years (Overpeck et al. 2003). General circulation models (GCMs) project that temperatures should warm 2–5° C over this period, depending upon many variables (Solomon et al. 2007).

Reconstructions of past temperatures are showing increasing overlap from region to region, especially in isotopic records from sea and ice cores (Barron et al. 2003; Hendy et al. 2002; Hughen et al. 2000; Grootes and Stuiver 1997). These records portray large temperature fluctuations throughout the last portions of the ice age, culminating in the rapid early Holocene warming, but then temperatures flattened out, depicting the Holocene climate as relatively stable (see Figure 1).

In contrast to sea and ice core temperature records (e.g., Grootes and Stuiver 1997), terrestrial Holocene paleoenvironmental records show much more variability and much less agreement with each other. Although broad patterns are evident on large spatial and temporal scales, there is very little correspondence in Holocene chronologies of smaller regions such as the Colorado Plateau (Thompson et al. 1993). Changes in precipitation seasonality due to changing solar insolation complicates terrestrial records in areas with a biseasonal precipitation regime such as the southwestern United States. But biological proxy data are further affected by the complexity of ecological dynamics following times of extreme change (Cole 1985). Because of this lack of firm conclusions on Holocene climates on the Colorado Plateau, we refer to the observed migrations across the plateau as primarily a delayed (lagging) response to the climate warming at the start of the Holocene (Davis 1989). Although no two climate events can be exactly the same, these past migration rates can at least serve as a first-order estimate for the possible rate of future migration.

PAST PINYON MIGRATIONS

The past migrations of plant species on the Colorado Plateau can be retrospectively observed by analyzing plant macrofossils found in packrat middens (Betancourt et al. 1990). Individual fossil pine needles can be radiocarbon dated using accelerator mass spectrometry (AMS), which produces much more precise results (Van Devender et al. 1985) compared to conventional dating. Since each fossil pine needle probably grew within 30 m of the fossil midden (K. Cole, unpublished data), this produces a highly reliable record of the species' occurrence at a specific time and location.

A consistent application of the same criteria of AMS-dated pinyon pine needles across time should produce comparable information for calculating past migration rates. Because pine needles are not likely to be recorded until it is a dominant tree on the landscape, it is an extremely conservative record. The recorded migration rate integrates not only the geographic problem of seed dispersal into the new areas, but also the other local ecological processes of succession, maturation, and eventual dominance of pinyon in the new community.

Needles identified as *Pinus edulis* are first verified in the eastern Grand Canyon during the Younger Dryas period (Cole and Arundel 2005; Cole et al., in review). Following

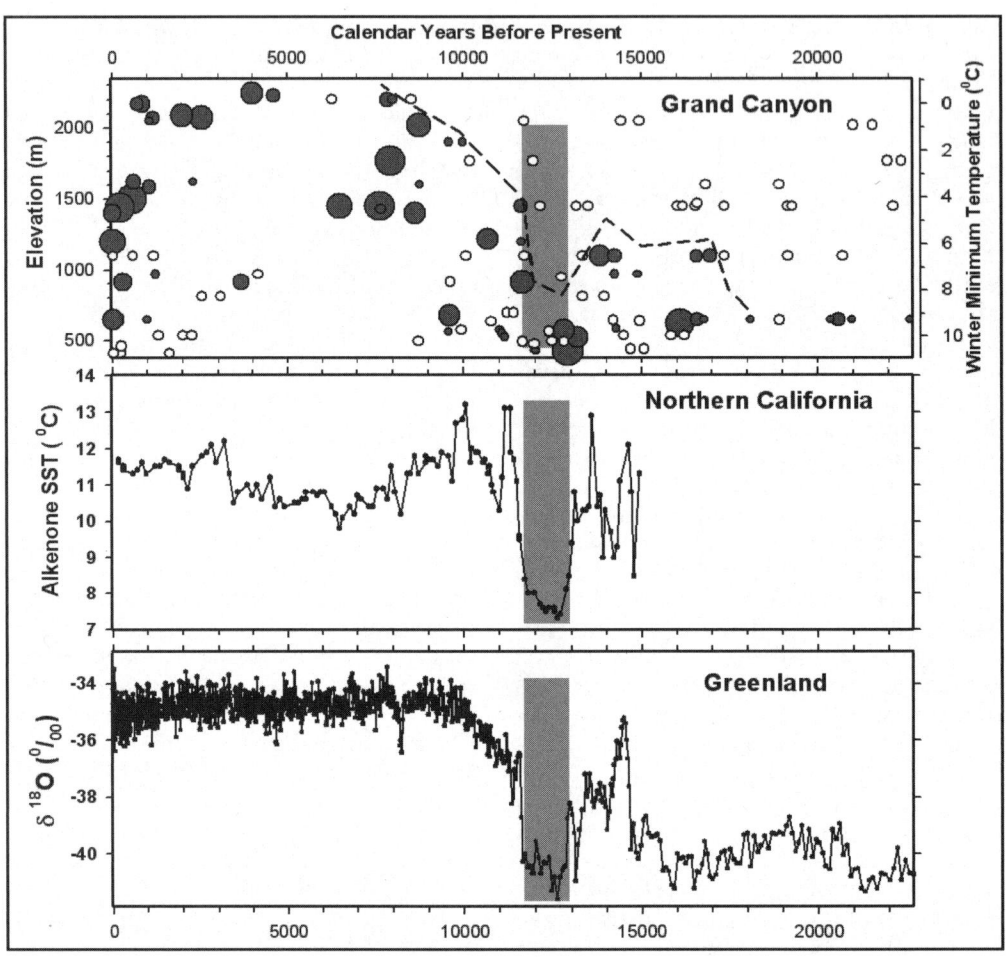

Figure 1. Indicators of temperature over the last 22,000 years illustrating the sharp warming that occurred at the end of the Younger Dryas interval (shaded zone). Top Panel: Departure of winter minimum temperature below current values (right scale), as indicated by the upper elevational limit (left scale) of Utah agave (*Agave utahensis*) in the Grand Canyon, Arizona. This upper limit is estimated as shown by the dashed line. Gray circles are scaled to represent the concentration of Utah agave fossils from packrat middens. Small open circles are packrat middens lacking Utah agave (modified from Cole and Arundel 2005). Middle Panel: Mean annual sea surface temperature estimates derived from alkenone chemistry from a sea core (ODP-1019) taken off the coast of northern California (modified from Barron et al. 2003). Bottom Panel: Oxygen isotopic record from the GRIP2 ice core from Greenland (modified from Grootes and Stuiver 1997).

the warming at the end of the Younger Dryas, these needles are then found in middens farther and farther to the north (Figure 2), suggesting a fairly continuous northward migration starting at about 11,600 years ago in the eastern Grand Canyon (Cole et al., in review) and ending in the Dutch John Mountains on the Utah-Wyoming border between 800 and 1000 years ago (Gray et al. 2006). Throughout the Holocene this northward migration from northern Arizona to southernmost Wyoming averaged 43 m/yr in latitudinal distance. Although its overall progression seems constant over the entire period, migration must have been characterized by episodic periods of relative stability interspersed with periods of rapid local invasion during decades of favorable climate, such as that described by Gray et al. (2006).

Some tree species are thought to have responded to past major warming more rapidly than others, especially in regions with low climatic and topographic variability. King and Herstram (1997) calculated migration rates of 180 and 156 m/yr for American beech (*Fagus grandifolia*) and eastern hemlock (*Tsuga canadensis*) in the deciduous forests of the eastern United States. Yansa (2006) estimated the early Holocene northward migration of white spruce (*Picea glauca*) in the northern Great Plains to have averaged 300 m/yr. Considering the magnitude of climate change expected in the near future, it is well to keep in mind that what has been called "the paradox of rapid plant migration" is only 100 to 500 m/yr (Clark et al. 1998). Even applying the most rapid migration scenario imaginable, complete species responses to climate change expected over the next 100 years could still take over 1000 years.

METHODS

In order to determine how climate controls pinyon distribution, we began with its current distribution and climate. Creating a model detailed enough to predict local changes on the landscape requires information at an extremely high resolution. We have thus integrated detailed information from sources such as the USDA's Forest Inventory and Analysis program (Gillespie 1999), which has sampled roughly every 5 km in a systematic grid to create a highly detailed map of current tree distributions. Another important component for creating detailed models is having high-resolution information about twentieth-century climates. Using the relationships between elevation, slope aspect, and climate, monthly climate surfaces for North America were created at a ~1 km scale using data from the Global Historical Climatology Network. Species distribution and climate maps are available from our website (http://www.usgs.nau.edu/global_change).

The GCM results applied to this study use an intermediate CO_2 scenario and a high-resolution application of NCAR's CCM3 model (Duffy et al. 2003; Govindasamy et al. 2003). The results from this simulation project a warming on the Colorado Plateau of about 3.5° C (summer maximum temperature) and 2–3° C (winter minimum temperature) between now and the time when the global CO_2 reaches twice its preindustrial concentration (2 x CO_2), roughly estimated to be the year 2100.

We next employed a program called ClimLim (Arundel 2005) to analyze geographic relationships between climate and species ranges. Because all climate variables change across space, the geographic range of any plant will encompass a range of each variable. But some of these climate variables have little influence in controlling the plant species distribution; they are merely coincidental to its geographic range. In order to determine which climate variables are the most important, ClimLim ranks each climate variable based on how well it controls the spatial distribution of the species (Arundel 2005). While some species, such as agave, are highly influenced by cold temperatures (Cole and Arundel 2005), others, such as pinyon, are highly influenced by the seasonality of precipitation (Figure 3; Cole and Arundel 2007), although temperature variables are still important.

Once the climate variables with the greatest geographic control of a species are identified, they are combined into a spatial

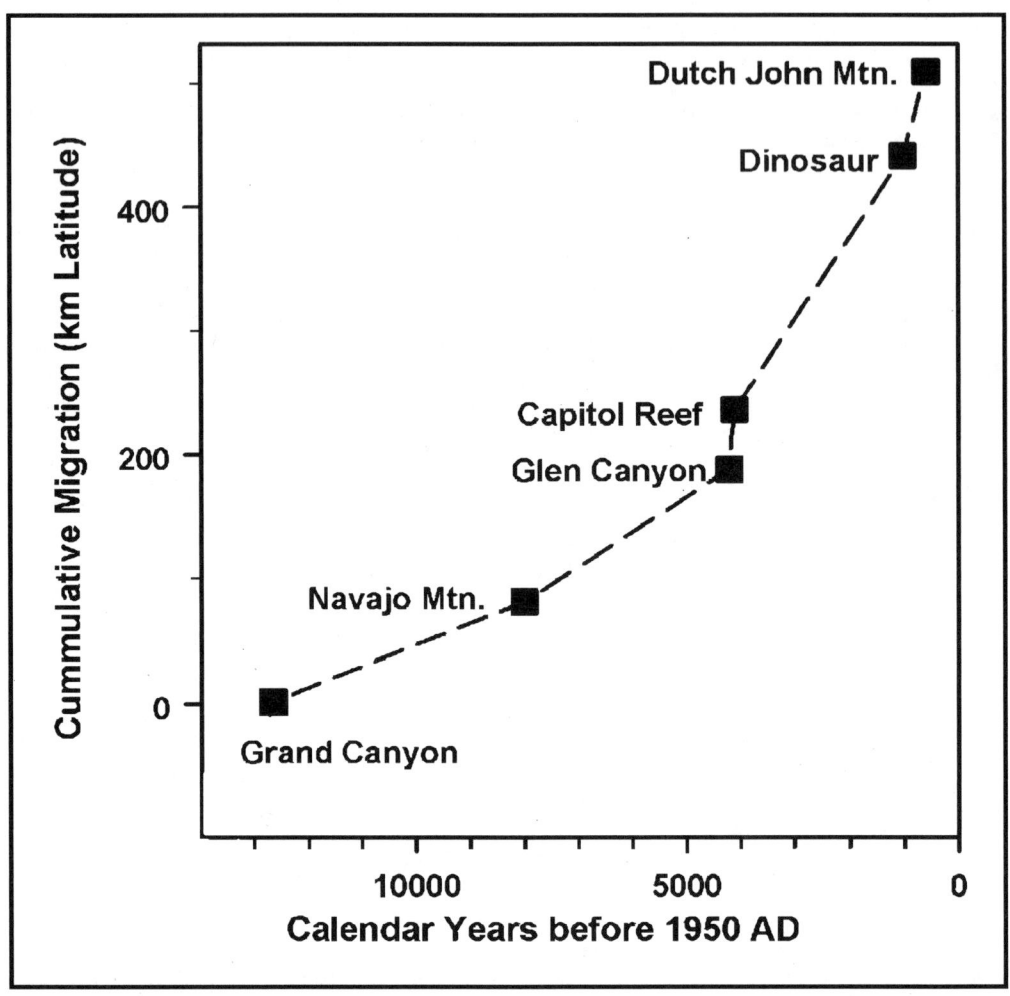

Figure 2. Northward migration of pinyon from north-central Arizona toward southern Wyoming since the last major climate warming 11,600 years ago. Names on the chart refer to packrat midden localities where the first arrival of pinyon has been documented using radiocarbon ages directly on needles (modified from Cole et al., in review).

model identifying the geographic areas that are optimal for species growth. This spatial model, called a probability surface, is generated using a multiple quadratic logistic regression of the most important variables. To create the probability surfaces for pinyon (Figure 4), variables representing January, May, June, August, and October precipitation (Figure 3) were combined with January, March, and May minimum temperatures and June maximum temperatures.

To predict future areas of suitable climate for pinyon occupancy, we use results from general circulation models (GCMs). Most GCMs have poor resolution (approximately 310 km) because they are simulating complex atmospheric and oceanic processes for the entire globe. The GCM data that we used for predicting impacts of climate change on pinyon were produced through a special

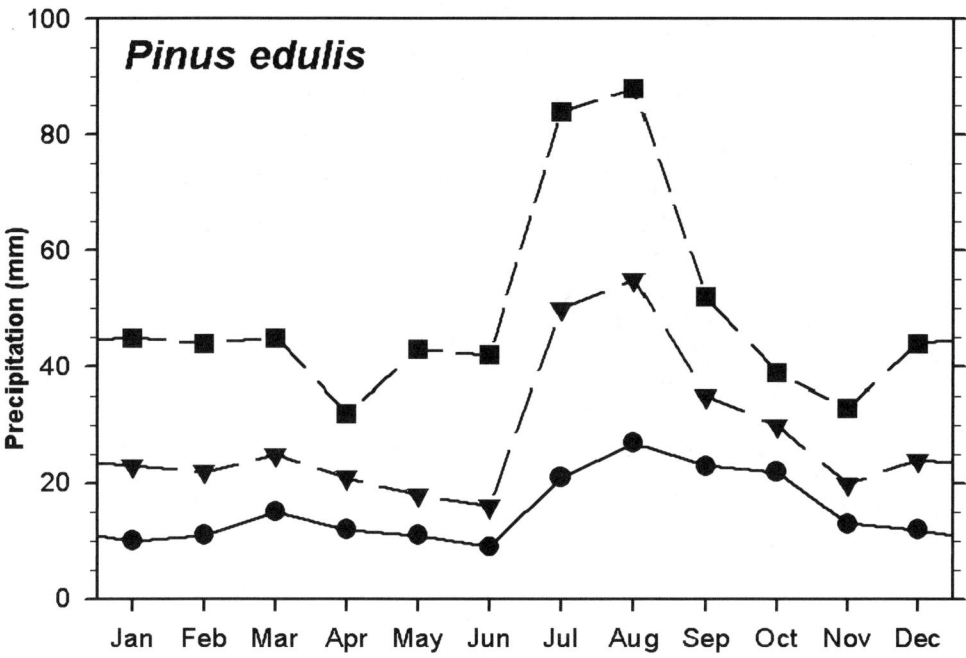

Figure 3. Precipitation seasonality over the range of pinyon. The spread between the 233,636 values representing each km² within the species range is shown by month. The bottom line (solid with circles) displays the 5th percentile, the middle line the 50th (median) percentile (dashed with triangles), and the top line the 95th percentile (long dashes with squares) of all values (modified from Cole et al., in press).

high-resolution run of the National Center for Atmospheric Research (NCAR) Community Climate Model version 3.6.6 (CCM3; Duffy et al. 2003; Govindasamy et al. 2003). Though this CCM3 data set has high spatial resolution (approximately 75 km) relative to other GCM data, this scale is still coarse for determining local changes in climate. Because of topographic diversity, climates are variable across the Colorado Plateau, and these results must be "downscaled" to show a detail level relevant to land managers. To downscale the CCM3 results, the difference between the present climate and the simulated 2 x CO_2 climate was calculated within each modeled 75 km grid square and these difference values were then applied to each 1 km grid within that larger grid using our

downscaled monthly climate surfaces for North America. This technique, sometimes called the "delta-change method" (Hay et al. 2000), takes advantage of the Colorado Plateau's high topographic diversity and low atmospheric humidity which cause temperature and precipitation to be highly correlated with local elevations.

The model predicting suitable climate for pinyon was then applied to the future climate surfaces to predict areas where it would have suitable climate in the future, approximately around 2100 AD (see Figure 4B). Information from past pinyon movements was then applied to approximate areas within its new climatic range into which it could potentially expand over such a short period.

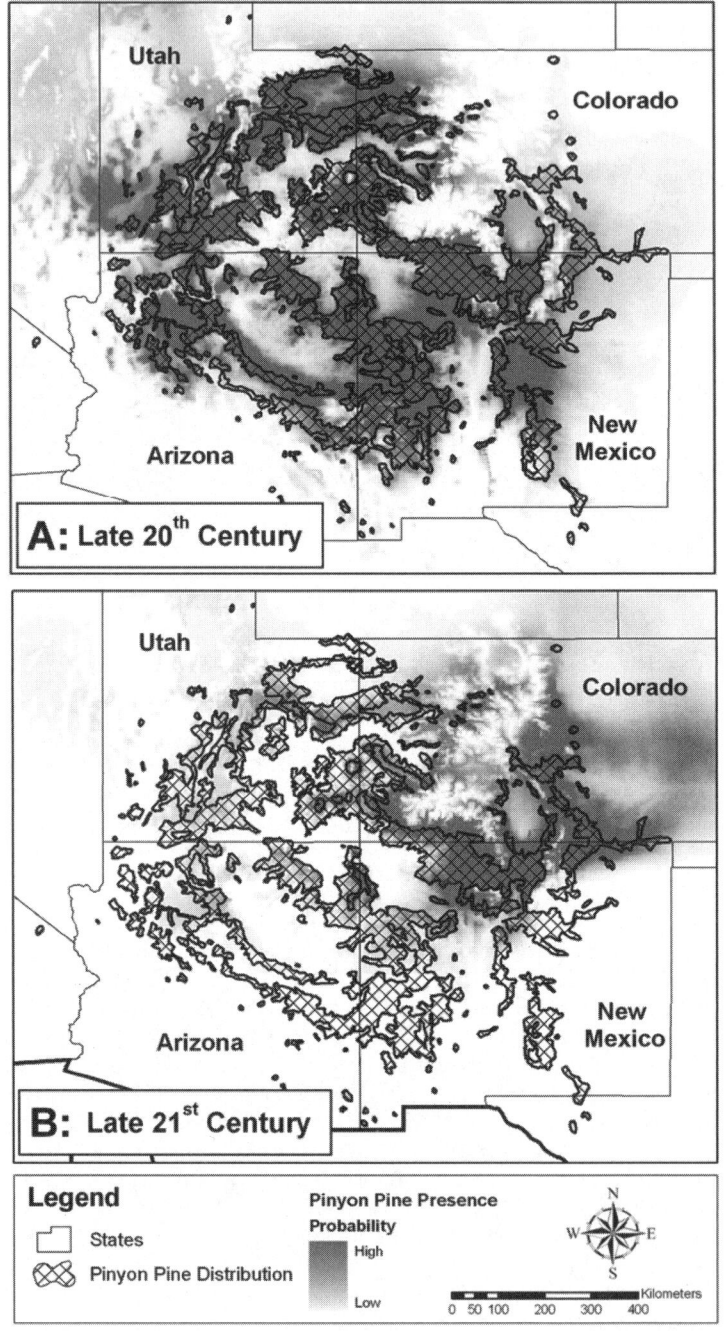

Figure 4. The current distribution of pinyon contrasted with the climate model prediction of its probability of occurring under a (a) late twentieth century climate and (b) late twenty-first century climate ($2 \times CO_2$) as generated using the CCM3 model.

RESULTS AND DISCUSSION

Our model (Figure 4A) reveals a close alignment between the present range of pinyon and model predictions. By modeling with the 2 x CO_2 probability surface resulting from the future GCM results, we see a proposed movement of pinyon northward (Figure 4B); the results suggest a profound shift in the potential range of pinyon over about the next 100 years. Most of pinyon's current range in Arizona and Utah will become increasingly inhospitable, but it should be able to expand northeastward and to higher elevations in Colorado and northernmost New Mexico. It is also predicted to thrive at higher elevations in the Chuska, Abajo, and La Sal Mountains, and in the Rocky Mountains (Figure 5).

Our model also predicts that pinyon would expand rapidly along the front range of the Colorado Rocky Mountains, assuming a generous future migration rate of 100 m per year over the next 100 years (= 10 km) outward from existing stands. This rate is more than twice the 43 m/yr observed in fossil records (Figure 2) and is less than half the time estimated for pinyon woodland to mature where seeds are already available (Goodrich and Barber 1999; Tress and Klopatek 1987). Despite the expansion rate provided by our model, the pinyon woodland is only estimated to fill in a small percentage of its new potential range during the next 100 years (Figure 5).

Although the results shown in Figure 5 may seem extreme, they are actually more moderate than the only other published projections for this species (Thompson et al. 1998:19). Thompson's model, applying very different climate modeling methods, different GCM inputs, and coarser-resolution 15 km grid cells, predicted the elimination of pinyon from almost all of the Colorado Plateau except the highest parts of the Mogollon Rim, and projected its future range as mostly in Oregon and Wyoming.

Our results need to be further verified through the application of climate data from multiple GCMs. The modeling of multiple species will further allow more complex predictions of climate effects on plant associations to test possible plant interactions and the individualistic model of species response. Preliminary data on a major associate of pinyon in Arizona—one-seed juniper (*Juniperus monosperma*)—suggest that it will respond to these climatic changes very differently. Whereas the model portrays pinyon as persisting at the highest elevations, the same climate scenario suggests that *Juniperus monosperma* will continue its ongoing expansion into the lower elevation grasslands of northern Arizona (Ironside 2006).

Verification of the results of our model will eventually come as pinyon mortality occurs where climatic conditions are predicted to become unfavorable and establishment occurs where climatic conditions are predicted to become more favorable. The predictions of our model can be downscaled to local landscape scales, allowing land managers to monitor these predicted changes in specific plots. Because model predictions are based on quantitative probability scales, the fate of different growth classes may be best predicted at finer levels on that scale. For example, seedlings are usually far more sensitive to climate than adults.

Loss of pinyon on the Colorado Plateau due to climate change has already begun, especially in areas most affected by rising temperature. Recent droughts are causing increased pine mortality due to the higher temperatures that occur during the drought (Breshears et al. 2005). The low mortality along the Colorado front range and high mortality in Arizona and at low elevations in Utah are consistent with the CCM3 model's predictions. The recent high mortality observed in northern New Mexico was not predicted by our model, probably because the CCM3 scenario predicts increased summer monsoon precipitation for northern New Mexico, but recent summers (2000–2005) have instead been notably dry. This disparity suggests that either 2 x CO_2-like precipitation scenarios are not yet evident in New Mexico or the CCM3 model is overestimating this variable. Different GCMs have greater correlation for temperature estimates

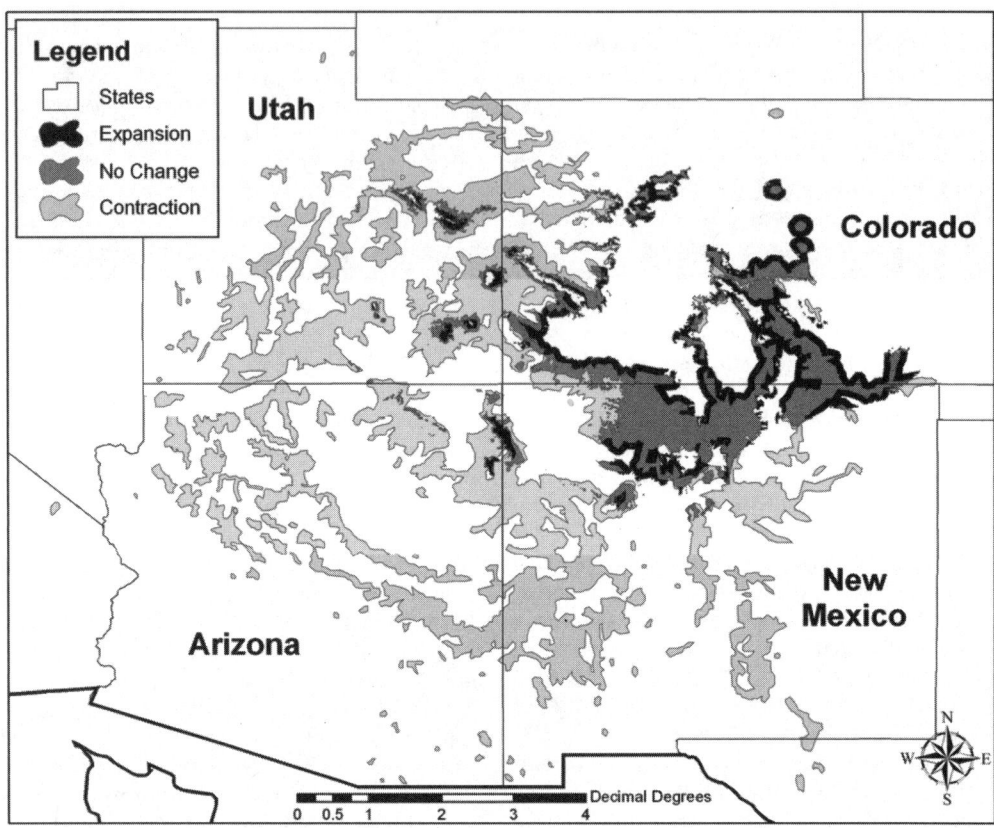

Figure 5. Areas that should experience a contraction (mortality) of pinyon over the next 100 years (light gray) are contrasted with areas where it should persist (dark gray) and areas where it should expand (black), assuming an average migration rate of 100 m/yr over the next 100 years.

than for precipitation (Coquard et al. 2004). Comparison with other GCMs will help further explore this issue.

CONCLUSIONS

Predicting the effects of future climate change on a plant species requires knowledge of the plant's current distribution, climate tolerances, and migratory response to change, as well as the geography of future climates. Using new modeling techniques and applying them to Colorado pinyon pine (*Pinus edulis*), we have found that the climatic envelope occupied by each species can be modeled through a geostatistical analysis of the twentieth-century climates over its current range. Our model incorporates all of the climate-modulated physical and biological variables occurring near the continental range of the species during the twentieth century. We developed models of future potential geographic ranges by applying this climatic envelope to future climate predictions from general circulation model (GCM) results. Finally, in order to distinguish between this future potential climate range and the species' likely future range, we applied a spatial model of the species' observed migration rate in response to past and ongoing climate warming. Through the compilation of spatially detailed data for our twentieth-century climate model, the GCM modeling, and the species distribution data, our results are projected to a landscape grid scale of ~1 km^2. This detailed projection allows application of the results to individual land management areas as well as specific predictions to assist monitoring.

Our modeling results suggest that over the next 100 years the range of *Pinus edulis* will continue to profoundly contract in Arizona, Utah, and southern New Mexico, but will expand in Colorado and northernmost New Mexico. The results from this one GCM scenario imply a large magnitude of change for this species and delineate useful areas to focus future monitoring efforts.

REFERENCES CITED

Arundel, S. T. 2005. Using spatial models to establish climatic limiters of plant species' distributions. Ecological Modeling 182: 159–181.

Barron, J. A., L. Huesser, T. Herbert, and M. Lyle. 2003. High-resolution climatic evolution of coastal northern California during the past 16,000 years. Paleoceanography 18: 1020–1035.

Betancourt, J., T. Van Devender, and P. S. Martin, editors. 1990. Fossil packrat middens: The last 40,000 years of biotic change in the arid West. University of Arizona Press, Tucson.

Blackburn, W. H., and P. T. Tueller. 1970. Pinyon and juniper invasion in black sagebrush communities in east-central Nevada. Ecology 51: 841–848.

Breshears, D. D., N. S. Cobb, P. M. Rich, K. P. Price, C. D. Allen, R. G. Balice, W. H. Romme, J. H. Kastens, M. L. Floyd, J. Belnap, J. J. Anderson, O. B. Myers, and C. W Meyer. 2005. Regional vegetation die-off in response to a global change type drought. Proceedings of the National Academy of Science 102: 15144–15148.

Cain, M. L., B. G. Milligan, and A. E. Strand. 2000. Long-distance seed dispersal in plant populations. American Journal of Botany 87: 1217–1227.

Chambers, J. C. 2001. *Pinus monophylla* establishment in an expanding *Pinus-Juniperus* woodland: Environmental conditions, facilitation, and interacting factors. Journal of Vegetation Science 12: 27–40.

Clark J. S., C. Fastie, G. Hurtt, S. T. Jackson, C. Johnson, G. King, M. Lewis, J. Lynch, S. Pacala, I. C. Prentice, E. W. Schupp, T. Webb III, and P. Wyckoff. 1998. Reid's Paradox of rapid plant migration. BioScience 48: 13–24.

Cole, K. L. 1985. Past rates of change, species richness, and a model of vegetational inertia in the Grand Canyon, Arizona. American Naturalist 125: 289–303.

Cole, K. L., and S. T. Arundel. 2005. Carbon-13 isotopes from fossil packrat pellets and elevational movements of Utah agave reveal Younger Dryas cold period in Grand Canyon, Arizona. Geology 33: 713–716.

Cole, K. L., and S. T. Aundel. 2007. Modeling the climatic requirements for Southwestern plant species. In Proceedings of the Twenty-First Annual Pacific Climate Workshop, edited by S. Starratt, P. Cornelius, and J. Joelson Jr., pp. 31–39. Technical Report 77, State of California Interagency Ecological Program for the San Francisco Estuary.

Cole, K. L., J. Cannella, J. Fisher, P. Koehler, S. Arundel, J. Mead, and K. Ironside. In review. The biogeographic histories of *Pinus edulis* and *P. monophylla* over the last 40,000 years. Journal of Biogeography.

Cole, K. L., J. Fisher, S. Arundel, J. Cannella, and S. Swift. In press. Geographic and climatic limits of needle types of one and two-needled pinyon pines. Journal of Biogeography.

Coquard, J., P. B. Duffy, K. E. Taylor, and J. P. Iorio. 2004. Present and future surface climate in the western USA as simulated by 15 global climate models. Climate Dynamics 23: 455–472.

Davis, M. A. 1989. Insights from paleoecology on global change. Bulletin of the Ecological Society of America 70: 222–228.

Davis, M. A., K. D. Woods, S. L. Webb, and R. P. Futyma. 1986. Dispersal versus climate: Expansion of *Fagus* and *Tsuga* into the upper Great Lakes Region. Vegetatio 67: 93–103.

Duffy, P. B., B. Govindasamy, J. P. Iorio, J. Milovich, K. R. Sperber, K. E. Taylor, M. F. Wehner, and S. L. Thompson. 2003. High resolution simulations of global climate, part 1: Present climate. Climate Dynamics 21: 371–390.

Gillespie, A. J. R. 1999. Rationale for a national annual forest inventory program. Journal of Forestry 97: 16–20.

Goodrich, S., and B. Barber. 1999. Return interval for pinyon-juniper following fire in the Green River corridor, Near Dutch John, Utah. In Proceedings: Ecology and Management of Pinyon-Juniper Communities Within the Interior West, pp. 391–393. USDA forest Service, RMRS-P-9.

Govindasamy, B., P. B. Duffy, and J. Conquard. 2003. High resolution simulations of global climate, part 2: Effects of increased greenhouse gases. Climate Dynamics 21: 391–404.

Gray, S. T., J. L. Betancourt, S. T. Jackson, and R. Eddy. 2006. Role of multidecadal climate variability in a range extension of pinyon pine. Ecology 87: 1124–1130.

Grootes, P. M., and M. Stuiver. 1997. Oxygen 18/16 variability in Greenland snow and ice with 103- to 105-year time resolution. Journal of Geophysical Research 102: 26,455–26,470.

Hay, L. E., R. L. Wilby, and G. Leavesley. 2000. A comparison of delta change and downscaled GCM scenarios for the three mountainous basins in the United States. Journal of the American Water Resources Association 36: 387–397.

Hendy, I. L., J. P. Kennett, E. B. Roark, and B. L. Ingram. 2002. Apparent synchroneity of submillennial scale climate events between Greenland and Santa Barbara Basin, California, from 30–10 ka. Quaternary Science Reviews 21: 1167–1184.

Hughen, K., J. Southon, S. Lehman, and J. Overpeck. 2000. Synchronous radiocarbon and climate shifts during the last deglaciation. Science 290: 1951–1954.

Ironside, K. 2006. Climate change research in national parks; paleoecology, policy, and modeling the future. Master's thesis, Northern Arizona University, Flagstaff.

Jackson, S. T., and J. T. Overpeck. 2000. Responses of plant populations and communities to environmental changes of the late Quaternary. Paleobiology 26 (Supplement 4): 194–220.

King, G. A., and A. A. Herstram. 1997. Holocene tree migration rates objectively determined from fossil pollen data. In Past and Future Rapid Environmental Changes: The Spatial and Evolutionary Responses of Terrestrial Biota, edited by B. Huntley, W. Cramer, A. V. Morgan, H. C. Prentice, and J. R. M. Allen, pp. 91–101 . Springer-Verlag, Berlin, Germany.

Mueller, R. C., C. M. Scudder, M. E. Porter, R. T. Trotter, C. A. Gehring, and T. G. Whitham. 2005. Differential tree mortality in response to severe drought: Evidence for long-term vegetation shifts. Journal of Ecology 93: 1085–1093.

Overpeck, J. T., C. Whitlock, and B. Huntley. 2003. Terrestrial biosphere dynamics in the climate system: Past and future. In Paleoclimate, Global Change, and the Future, edited by K. D. Alverson, R. S. Bradley, and T. Pedersen, pp. 81–103. Springer, Berlin.

Rosenstock, S. S., and C. van Riper III. 2001. Breeding bird response to juniper woodland expansion in Northern Arizona grasslands. Journal of Range Management 54: 226–232.

Severinghaus, J. P., T. Sowers, E. Brook, R. Alley, and M. Bender. 1998. Timing of abrupt climate change at the end of the Younger Dryas interval from thermally fractionated gasses in polar ice. Nature 391: 141–146.

Shaw, J. D. 2006. Population-wide changes in pinyon-juniper woodlands caused by drought in the American Southwest: Effects on structure, composition, and distribution. In Patterns and Processes in Forest Landscapes, Consequences of Human Management, edited by R. Lafortezza and G. Sanesi, pp. 117–124. Proceedings of the 4th IUFRO Working Party 8.01.03, 26–29 September 2006. Locorotondo, Bari, Italy.

Shaw, J. D., B. E. Steed, and L. T. DeBlander. 2005. Forest Inventory Analysis (FIA) Annual inventory answers to the Question: What is happening to pinyon-juniper woodlands? Journal of Forestry 280–285.

Solomon, S., D. Qin, M. Manning, Z. Chen, M. Marquis, K. V. Averyt, M. Tignor, and H. L. Miller, editors. 2007. Climate Change 2007: The Physical Science Basis. Contribution of Working Group I to the Fourth Assessment Report of the Intergovernmental Panel on Climate Change. Cambridge University Press, UK.

Tress, J. A., and J. K. Klopatek. 1987. Successional changes in community structure of pinyon-juniper woodlands in north-central Arizona. In Proceedings: Pinyon-Juniper Conference, pp. 80–85. USDA Forest Service, INT-215.

Thompson, R. S., C. Whitlock, P. J. Bartlein, S. P. Harrison, and W. G. Spaulding. 1993. Climatic changes in the western United States since 18,000 yr B.P. In Global Climates Since the Last Glacial Maximum, edited by H. Wright, J. Kutzbach, T. Webb, W. Ruddiman, F. A. Street-Perrott, and P. Bartlein. University of Minnesota Press, Minneapolis.

Thompson, R. S., S. E. Hostetler, P. J. Bartlein, and K. H. Anderson. 1998. A strategy for assessing potential future changes in climate, hydrology, and vegetation in the western United States. USGS Circular 1153 (http://pubs.usgs.gov/circ/1998/c1153/c1153_4.htm).

Van Devender, T. R., P. Martin, R. Thompson, K. Cole, A. J. T. Jull, A. Long, L. Toolin, and D. Donahue. 1985. Fossil packrat middens and the tamdem accelerator mass spectrometer. Nature 317: 610–613.

Waterhouse, J. S., V. R. Switsur, A. C. Barker, A. H. C. Carter, D. L. Hemming, N. J. Loader and I. Robertson. 2004. Northern European trees show a progressively diminishing response to increasing atmospheric carbon dioxide concentrations. Quaternary Science Review 23: 803–810.

Williams, D. W., and A. M. Liebhold. 2002. Climate change and the outbreak ranges of two North American bark beetles. Agricultural and Forest Entomology 4: 87–99.

Yansa, C. H. 2006. The timing and nature of Late Quaternary vegetation changes in the northern Great Plains, USA and Canada: A reassessment of the spruce phase. Quaternary Science Reviews 25: 263–281.

THE ROLE OF AEOLIAN SEDIMENT IN THE PRESERVATION OF ARCHAEOLOGICAL SITES ALONG THE COLORADO RIVER IN THE GRAND CANYON

Amy E. Draut and David M. Rubin

Since the closure of Glen Canyon Dam in 1963, major changes have occurred in the natural hydrologic and sedimentary regime in the Colorado River corridor through the Grand Canyon (e.g., Andrews 1986; Webb et al. 1999; Topping et al. 2003). The dam has reduced the supply of sediment at the upstream boundary of Grand Canyon National Park (the head of Marble Canyon) by about 95 percent (Topping et al. 2000a), and the Paria River is now the only major supplier of sediment to this system (Figure 1). The other substantial post-dam supplier of sediment is the Little Colorado River, which enters the Colorado 98 km downstream from the Paria River and supplies an additional 10–15 percent of the pre-dam sediment load (Topping et al. 2000a). Operation of the dam smoothes seasonal variation in river discharge; daily discharge typically fluctuates over a much greater range than in the pre-dam state (Topping et al. 2003). Regulation of the river has important implications for the storage and redistribution of sediment. Without regular floods, the relatively small sediment load that the river carries cannot be deposited at the higher elevations in the channel that received sediment regularly before dam closure. The channel of the pre-dam river accumulated sand when discharges were below about 9000 cfs (Topping et al. 2000a). In the absence of flows below 5000 cfs, which are not permitted under the 1996 Record of Decision signed by the Secretary of the Interior, the sediment-storage capability in the main channel is greatly reduced (Topping et al. 2000a, 2003). Most sediment supplied by tributaries below the dam is exported from the canyon on time scales of weeks to months under present flow operations (Topping et al. 2000a, 2000b; Rubin et al. 2002; Wright et al. 2005).

Numerous studies have identified the physical and biological consequences of the altered hydrograph and diminished sediment content of the Colorado River. With respect to sediment, the size and number of subaerial sand deposits have decreased systemwide over the past four decades, punctuated by episodic, localized aggradation during the 1983 flood of 97,000 cfs, the 1996 and 2004 controlled-flood experiments (45,000 and 41,000 cfs, respectively), and sediment input from occasional tributary floods (Beus et al. 1985; Schmidt and Graf 1987; Budhu and Gobin 1994; Kearsley et al. 1994; Kaplinski et al. 1995; Schmidt and Leschin 1995; Wiele et al. 1996; Hazel et al. 1999; Schmidt et al. 2004; Topping et al. 2006). Encroachment of riparian vegetation on sandbars has also contributed significantly to the loss of open sand area along the river (e.g., Turner and Karpiscak 1980).

When the Bureau of Reclamation sponsored the creation of the Glen Canyon Environmental Studies (GCES) research initiative in 1982, research objectives included physical and biological resources but not cultural resources (Fairley et al. 1994; Fairley 2003). At that time it was widely believed that, because few archaeological remains were preserved within the annual flood zone, cultural features would not be greatly affected by dam operations. However, more recent studies indicate that alterations in the flow and sediment load of the Colorado River by

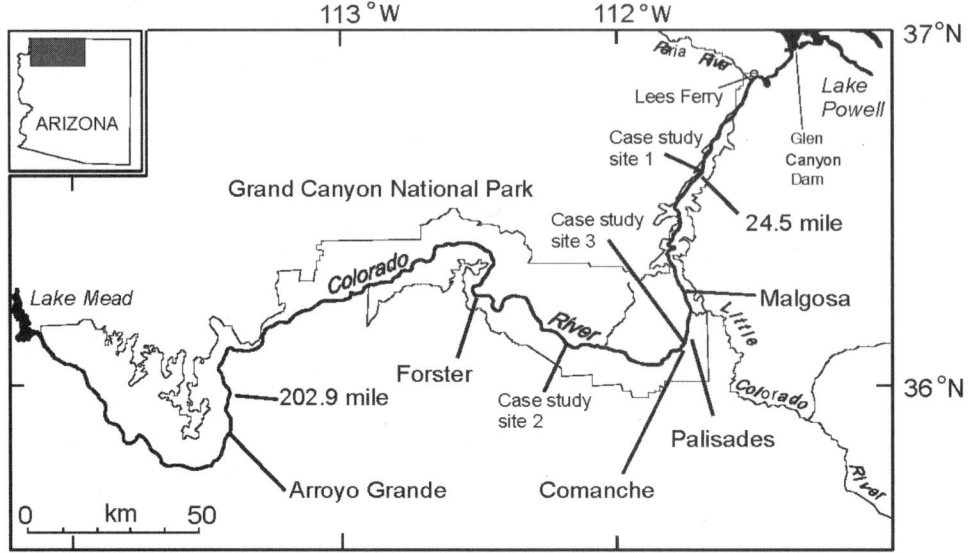

Figure 1. Location map showing the Colorado River through Grand Canyon, with study sites indicated. Stratigraphy and geomorphology were studied in detail at Palisades, Comanche, and Arroyo Grande. Weather stations collected data at 24.5-Mile, Malgosa, Palisades, Comanche, Forster, and 202.9-Mile. Locations of case study sites 1–3 are indicated.

Glen Canyon Dam may be affecting archaeological sites within the river corridor, even above the annual flood limit (Hereford et al. 1993; Yeatts 1996, 1997; Thompson and Potochnik 2000; Draut et al. 2005).

Of the ~500 documented archaeological sites in the river corridor between Glen Canyon Dam and Separation Canyon (255 river miles), more than 330 are considered to be within the Area of Potential Effect (APE) of dam operations designated by the NPS (Fairley et al. 1994; U.S. Department of the Interior 1996; Fairley 2005). Many of the sites are located in or on poorly consolidated sedimentary deposits that are actively eroding due to gully incision, aeolian deflation, and, to a lesser degree, visitor impact (e.g., Leap et al. 2000; Fairley 2003; Pederson et al. 2003; Balsom et al. 2005). Gully incision and the base level to which small, ephemeral drainage systems respond were first proposed to be linked to dam operations by Hereford et al. (1993), who documented the

erosion of pre-dam fluvial terrace deposits, and associated archaeological sites, by gullies that formed in the mid-1970s when rainfall was unusually high. These observations led to their hypothesis that rapid incision was caused by the lowering of the local base level at the mouths of ephemeral drainages to meet the new, post-dam elevation of high-flow sediment deposition.

Thompson and Potochnik (2000) modified the base-level concept to include the restorative effects of fluvial deposition in the mouths of gullies and arroyos (raising the local base level) and the potential for aggradation of pre-dam fluvial terraces by aeolian redistribution of flood sand. Neal et al. (2000) suggested that aeolian deposition in incipient gullies is "one of the strongest restorative forces operating at archaeological sites." Repeated high-resolution mapping conducted after the 1996 controlled-flood experiment confirmed that floods can deposit sediment in arroyo mouths, and also suggested that those deposits can be a source for wind-blown sand that accumulates at higher elevation, causing sediment accretion above the flood-stage elevation (Yeatts 1997; Hazel et al. 2000).

INVESTIGATING THE ROLE OF AEOLIAN SEDIMENT TRANSPORT

In May of 2003, reconnaissance work was performed by personnel from the USGS, Hopi and Hualapai Tribes, GeoArch Inc., and the Western Area Power Administration (WAPA), led by NPS archaeologists J. R. Balsom, J. L. Dierker, and L. M. Leap. This group of scientists visited 38 archaeological sites where aeolian sediment was believed to be important as the material either underlying the site or forming a protective cover. This initial survey, while noting that the selected sites were likely not typical of all sites in the river corridor, determined that the majority of the sediment beneath or above these particular archaeological sites had been transported by wind from fluvial deposits at the river margins (Draut et al. 2005). Some of the sites were built on or buried by sediment directly deposited by large floods of the Colorado River. It was apparent that aeolian sand transport to archaeological sites was limited largely by sediment supply and vegetation rather than by wind. Reductions in the amount of exposed sand on sandbars and increased submergence of sandbars by daily flow fluctuations can be expected to reduce aeolian sand transport to nearby dune fields that contain archaeological sites (e.g., Ash and Wasson 1983; Buckley 1987; Namikas and Sherman 1995; Wiggs et al. 2004). Reduced aeolian transport, in turn, would facilitate the growth of existing arroyos and the establishment of new ones (Thompson and Potochnik 2000). The degradation of archaeological sites by arroyo incision was apparent at many of the sites visited in May 2003; repeat photography at those sites by the NPS over the past decade supports this observation (e.g., Leap et al. 2000). Some archaeological sites are affected by small drainages that might be repaired with moderately increased aeolian sand deposition. In some locations, the filling of small gullies (< 1 m wide) by aeolian sand would aid preservation of archaeological features that were stabilized by checkdam construction. Still other sites are threatened by large drainages (meters wide and deep) that could not reasonably be expected to fill with aeolian sediment given the current local geomorphology and sand supply.

Observations made during this reconnaissance work in May 2003 formed the basis for more detailed sedimentary and geomorphic studies in 2004, and for the establishment of weather stations that monitored wind, rain, and aeolian sand transport between late 2003 and early 2006. We discuss the results of these studies, present criteria used to assess the sensitivity of particular areas to dam operations (with respect to aeolian sand), and consider three case studies of the assessment process. Dam operation has caused a reduction in sandbar size, reducing the supply of sand available for transport from these upwind sources to some archaeological sites; the reduced aeolian transport is likely responsible for gullying and deflation at some sites. Cultural sites at which aeolian sand is derived from river-level sandbars could potentially benefit from new sand deposited on the upwind sandbars during sediment-rich controlled floods. We demonstrate that the restoration potential for cultural sites in aeolian deposits can be increased by dam operations that maximize the exposed fluvial sandbar area from April through early June, when aeolian sediment transport is greatest.

METHODS

This study combined the analysis of prehistoric sedimentary and geomorphic environments with that of modern aeolian processes and rainfall to assess the sensitivity of specific areas of the river corridor and associated cultural sites to dam operations. Site-specific evaluation is very important because the relative importance of fluvial, aeolian, and local sediment sources, as well as wind and precipitation patterns, differs widely among sites in the same region of the river corridor.

Sedimentary and Geomorphic Analyses

Understanding the local pre-dam sedimentary and geomorphic setting is an important precursor to any future assessments of dam-operation effects in any specific area. Field investigations in the Palisades, Comanche, and Arroyo Grande areas of the river corridor (Figure 1) during May of 2004 assessed the relative importance of various depositional processes in these archaeologically significant areas (cf. Hereford 1996). As shown in aerial photographs, each of these locations has experienced a reduction of exposed sand area since the 1960s (Grams and Schmidt 1999; Draut and Rubin 2007).

Prehistoric sedimentary environments were interpreted by examining sedimentary profiles in pits and arroyo walls; no cultural artifacts were exposed or collected. Sedimentary characteristics were described and unit thicknesses were measured in detail (Draut et al. 2005). Within vertical exposures, small-scale sedimentary structures (textures that form from aggregates of sediment grains during or shortly after deposition) were used where possible to infer the depositional environment (e.g., Hunter 1977; Rubin 1987). Structures characteristic of fluvial and aeolian environments could commonly be identified by their dimensions, grain-size sorting, and spatial orientation of bedding; although some of the same sedimentary structures occur in both aeolian and fluvial deposits, differences in appearance often allow their depositional settings to be differentiated.

Monitoring Wind, Aeolian Sediment Transport, and Precipitation

Stratigraphic and geomorphic investigations, which identify the local importance of aeolian sedimentation in the past, were complemented by measurements of modern weather and sand transport to provide a more complete picture of the natural and anthropogenic processes affecting archaeological sites. Instrument stations were installed temporarily in the river corridor to monitor wind, precipitation, and aeolian sand transport in the vicinity of cultural sites. Data collected at these stations constitute the only continuous weather record from the river corridor between 2003 and 2006, with the exception of wind and rainfall measurements made at Phantom Ranch (near river mile 88) by the NPS. At three sites (24.5-Mile, Malgosa, and Palisades; Figure 1), data were collected between November 2003 and January 2006. Two instrument stations were used at each of these three sites: one at lower elevation near the river, and one farther from the river at higher elevation within a dune field. One instrument station at each of three additional sites (Comanche, Forster, and 202.9-Mile; Figure 1) operated between April 2004 and January 2006. These stations were deployed in areas known to contain archaeological sites but not near enough to any sites that equipment could damage cultural features. All study sites had experienced a reduction in exposed fluvial sandbar area since Glen Canyon Dam operations began (based on aerial photographs; Draut and Rubin 2007) but had experienced new deposition on sandbars during the 1996 experimental flood, suggesting potential sensitivity of associated aeolian deposits to dam operations. The timing of the November 2004 controlled flood, which occurred approximately halfway through this study, allowed the effects of that experiment on aeolian transport rates to be evaluated specifically (see below).

Detailed specifications of the instrument deployments have been presented by Draut and Rubin (2005; see Figure 2). Digital measurements of mean wind velocity, maximum gust velocity, mean wind direction, and total precipitation were recorded at each station (using spinning-cup anemometers mounted on tripods and tipping-bucket rain gages, respectively) with a 4-minute sampling interval using battery-powered data loggers. Aeolian sand flux was monitored using passive-sampling Big Spring Number Eight (BSNE) sand traps (Fryrear 1986; Stout and Fryrear 1989) that were emptied every 4–8 weeks. Four traps were mounted on a vertical pole at each station at heights of 1.0, 0.7, 0.4, and 0.1 m; these traps have vanes that orient them into the wind. The total sand

Figure 2. The instrument station in use at the mouth of Forster Canyon from April 2004 until January 2006. Four sand traps are set up on a vertical pole at the left side of the photo. A tripod with two anemometers is in the center; the data logger is attached to this tripod also but is not visible. A rain gage is shown to the right of the tripod. Equipment was camouflaged with paint and dead branches to reduce visual impact to canyon visitors.

mass collected from each trap during every site visit was weighed; mass-transport rates were calculated for each interval between visits by integration from 0 to 1 m of a combined power-law and exponential function fitted to the mass versus height data (Sterk and Raats 1996). The error in wind direction is estimated at no more than 10 degrees. For the magnitude component of wind velocity, analyses include error of 0.5 m s^{-1} for velocities < 17 m s^{-1} and ± 3% for velocities between 17 and 30 m s^{-1}. The accuracy of sand-flux measurements is dictated by sand-trap efficiency as calibrated in wind-tunnel studies (see Draut and Rubin 2007 for complete discussion of accuracy and precision in weather measurements).

RESULTS

Sedimentary analyses indicate that Colorado River flooding, aeolian dune development and migration, and slope-wash and debris-flow deposition have affected archaeological sites in the Palisades, Comanche, and Arroyo Grande study areas to varying degrees (see also Hereford et al. 1996). The dominant substrate underlying prehistoric cultural sites at Palisades and Arroyo Grande, where Holocene fluvial terraces had previously been recognized (Hereford 1996; Hereford et al. 1996, 2000), consists of fluvial deposits. The number and thickness of fluvial deposits generally decrease with distance from the river (Draut et al. 2005). At Palisades, six of nine archaeological sites were built on Colorado River flood deposits and the remaining three sites were built on distal debris-flow sediment. At Comanche, five of the six archaeological sites were built on, and are partially buried by, aeolian sediment; the sixth and largest site is on a terrace that consists of fluvial, aeolian, and slope-wash deposits. The most extensive of four archaeological sites at Arroyo Grande, which includes more than 20 features, was established on a terrace surface and within alluvium of the terrace that consists primarily of interbedded fluvial and locally derived slope-wash sediment.

Wind velocities and sand transport were greatest during the spring windy season (April through early June), with maximum winds locally more than 25 m/s and transport rates locally as high as ~5 kg/cm/day (Draut and Rubin 2005, 2006). At all weather stations, rates of sand transport in the spring were 5–15 times higher than in other seasons. Although the dominant wind direction varied substantially with location, in Marble Canyon (between Lees Ferry and river mile 62) the greatest sediment-transport potential was directed upstream during the windy season. Precipitation records showed that the same storm event could induce rainfall with great spatial variability—a common occurrence in the arid Southwest. Cumulative rainfall also varied substantially with location throughout the river corridor, presumably due to local topography; the Malgosa site (river mile 57.5) consistently received substantially more rain (commonly twice as much, per rainfall event) than the Palisades site only 8 miles downstream.

DISCUSSION

Dunes are present on terraces at Palisades, Comanche, and Arroyo Grande; however, aeolian sediment was inferred to be more important as a protective cover than as the substrate on which most cultural sites were originally built. Sites in all three areas are affected today by gully incision, wind, slope-wash, and occasional debris flows. Gully development and aeolian deflation have exposed and eroded cultural sites; deflation exposes artifacts to erosion during rainfall, and dune migration causes downslope movement of artifacts. The deposition of Colorado River flood sediment, which episodically affected cultural areas at Palisades, Comanche, and Arroyo Grande in the past, no longer reaches the elevations of those archaeological sites under current dam operations, even during the controlled high flows of 1996 and 2004. Many cultural sites have been preserved until the present due to burial by flood deposits and dunes combined with a lack of severe gully erosion. Slope-wash and debris-flow sediment contributes a smaller volume at these three sites than the cover provided by thicker fluvial and aeolian strata. However, the exposures of interbedded aeolian and local sediment studied in other areas of the river corridor indicate that a thin horizon of poorly sorted slope-wash or debris-flow sediment can form a resistant cap that protects the thicker, more erodible aeolian sediment beneath.

Wind velocities and aeolian sand transport rates were consistently greater at the higher-elevation weather stations compared to those measured at river level. This is attributed not only to the effects of topography but also to the presence of vegetation near the river, which reduces wind velocity and consequently the potential for sediment entrainment (Ash and Wasson 1983; Buckley 1987; Bauer et al. 1996). Sand transport may also be lower near the river because interstitial moisture in sandbars limits the entrainment of sand by wind (e.g., Wiggs et al. 2004); under normal dam operations, sandbars inundated by the river do not have time to dry thoroughly after the flow recedes before being inundated again the following day. Dune fields with abundant biological soil crust had sand-transport rates an order of magnitude lower than those measured on more "active" dune fields, where soil crust is rare (cf. Goossens 2004).

Modern vs. Relict Fluvial Sourced Aeolian Sand Deposits

Wind direction and magnitude indicate the potential for sediment transport from fluvial sandbars to dune fields (Draut and Rubin

2005, 2006). Our findings show that aeolian deposits in the Grand Canyon along the river corridor fall into two categories: modern fluvial sourced (MFS) and relict fluvial sourced (RFS). The two categories are distinguished by the relative locations of aeolian deposits, modern fluvial sandbars (defined as those at or below the 45,000 cfs stage), and the dominant wind direction causing local sediment transport. MFS aeolian deposits formed as the wind transported sand inland from a river-level sandbar, creating a dune field directly downwind (Figure 3). Dune fields at 24.5-Mile, Malgosa, and Forster are classified as MFS deposits (Draut and Rubin 2007).

RFS aeolian deposits formed as the wind eroded and redeposited sediment that comprised extensive pre-dam flood deposits, generating aeolian dunes essentially "in

place" over parts of terraces left by floods higher than any post-dam floods have been (up to and exceeding 200,000 cfs). The most significant sediment source of RFS deposits is not modern, active sandbars but the large pre-dam flood deposits (see also Burke et al. 2003). Dunes at Palisades, Comanche, 202.9-Mile, and Arroyo Grande are RFS deposits (Draut and Rubin 2007). Because the formation of RFS aeolian deposits was linked inextricably with deposition of fluvial sediment during large, sediment-rich, pre-dam floods, it is unlikely that substantial new sediment will be supplied to RFS dune fields without similar large, sediment-rich floods in the future.

MFS aeolian deposits tend to be smaller and to occur more commonly in the Grand Canyon than RFS deposits. Aeolian deposits of both types are known to contain archaeo-

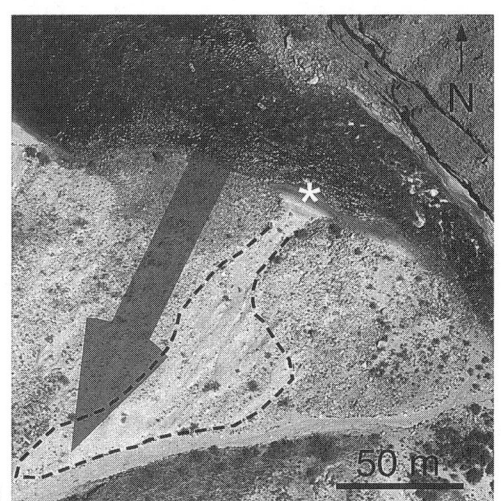
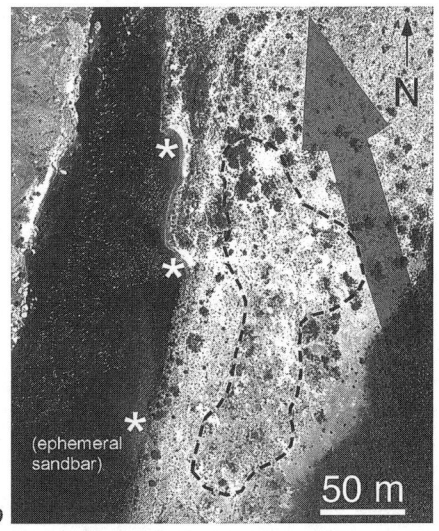

Figure 3(a). Modern fluvial sourced (MFS) aeolian dune field at the mouth of Forster Canyon (river mile 123.0). The wind blows consistently from the NNE (transparent arrow; Draut and Rubin 2005, 2006); net sediment transport is toward the SSW. Substantial wind velocities (> 20 m/s during the spring windy season) transport sediment away from a river-level sandbar (asterisk) to form a dune field (dashed outline) downwind. (3b). Relict fluvial sourced (RFS) aeolian deposits, such as this one at Palisades (river mile ~66) formed as the wind reworked sediment deposited by large pre-dam floods, generating aeolian dunes essentially "in place" over parts of fluvial terraces. The dominant wind direction at Palisades is from the SSE (transparent arrow). Modern fluvial sand deposits (asterisks) at Palisades are not directly upwind of the aeolian dune field (dashed outline), so little sediment from those sandbars reaches the dunes.

logical material. Both MFS and RFS dune fields are sensitive to Glen Canyon Dam operations, but for different reasons. MFS deposits, directly downwind of fluvial sandbars, receive a quantity of wind-blown sand that is a function of wind velocity, exposed dry fluvial sandbar area, and sediment-trapping vegetation. Because fluvial sandbar area and riparian vegetation both respond to dam operations, changes to either can affect the delivery of aeolian sand to MFS dune fields downwind. Because the primary sediment source for RFS dunes was high-elevation pre-dam flood deposits, no substantial new deposition will occur on RFS dune fields as long as the lack of large sediment-rich floods continues. As vegetation and soil crust grow on RFS dunes over time, aeolian sand transport decreases such that these areas become increasingly unable to compensate for precipitation-induced gully erosion (Draut and Rubin 2007). Large gullies and arroyos can develop in pre-dam terraces and associated RFS dune fields as a result.

Evaluating Site Sensitivity to Dam Operations

Draut et al. (2005) proposed a series of steps that can be used to determine the role of aeolian sediment in preserving an archaeological site, and to assess the effects of dam operations on aeolian sedimentation there. Here we review those site-evaluation criteria and present three examples of their use. These guidelines apply specifically to aeolian sedimentation and to MFS aeolian deposits, though our approach could be modified to address a wider range of processes that affect archaeological features.

An essential first step in evaluating factors that contribute to cultural-site preservation is to determine the depositional context of sediment on which the site was built, and the depositional context of sediment that has buried and protected the site. Depositional settings are interpreted by examining geomorphic features, sedimentary structures on the land surface, and subsurface sedimentary structures viewed in vertical section. Field investigations can be supplemented by

laboratory analysis of sediment particle-size distribution (e.g., Rubin et al. 1998; Burke et al. 2003; Draut et al. in press) and by radiocarbon dating of materials that contain organic matter or charcoal (whether cultural or non-cultural). Radiocarbon dating (not used during our study) provides information on the age of cultural occupation and can also be used to quantify recurrence intervals of floods and debris flows in the study area.

If aeolian deposition has buried and protected a given cultural site, it is necessary to establish whether some of the aeolian sediment covering the site has been lost. Recognizing whether sand cover has been lost by wind deflation or gully incision is an important step in assessing the risk of cultural-site degradation (or the degree to which degradation has already occurred). If aeolian sediment has not been lost at a particular site, it may not be at immediate risk of artifact loss. Loss of aeolian sediment can be documented most accurately using repeated high-resolution mapping to measure deflation of a land surface or the dimensions of growing gullies. If quantitative surveys are not available, loss of sediment can be qualitatively identified using repeated ground-based photography and geomorphic evidence of deflation. Evidence for deflation includes winnowed lag deposits on the land surface, exhumation of plant roots, and pedestal development (although erosion from rain may also leave small-scale pedestals).

Deflation of aeolian dunes represents a net loss of sediment from an area that had formerly experienced net deposition (creating the dune forms). This could arise from a change in the balance between sediment supply and the sediment-transport capacity of local winds. If one assumes that local wind conditions have not changed significantly and that a decrease in sediment supply is more likely, the next step is to identify the source of the aeolian sediment that has buried the archaeological site in question. Source areas for aeolian sand are best identified by measuring wind conditions. The longer the time interval over which wind data are gathered, the more accurate the resulting sediment-transport predictions will be. Records of wind speed and direction, such as those obtained during this study, can be used to generate vector sums that demonstrate the potential for local aeolian sediment transport (Draut and Rubin 2005, 2006). If long-term weather monitoring is not feasible, recent wind conditions can be inferred from the orientation of ripples, sand shadows (which form in the lee of obstacles), and dune morphology.

After local wind conditions have been used to identify a sand source, it must be asked whether characteristics of that sand source have changed and consequently decreased sand supply to downwind MFS deposits and archaeological sites. Aerial photography, high-resolution mapping, and oblique, ground-based photography can effectively document changes in source-area characteristics that might affect the ability of that sandbar to serve as a sand source for MFS dunes (e.g., Schmidt et al. 2004; Draut and Rubin 2007). Important changes in sediment-source characteristics include a decrease in exposed sand area caused by sandbar erosion, vegetation growth, increased moisture content of the sandbar (more frequent inundation than in the past), greater surface exposure of rocks, driftwood, or other obstacles that inhibit wind, or the development of biological soil crust on the land surface.

If the supply of aeolian sediment to a particular site has declined, it is necessary to judge whether renewed deposition of aeolian sand would substantially enhance protection of the archaeological site. Sites at which increased aeolian deposition would substantially enhance preservation of cultural features are those where the greatest degradation occurs by deflation, or from incision by gullies small enough to be healed by wind-blown sand (< 1 m in width and depth, judging by exposures of filled gullies observed during this study and by Thompson and Potochnik 2000). Aeolian sedimentation is unlikely to prevent the loss of archaeological features that are threatened more immediately by the incision of a major arroyo or side-canyon channel than by deflation.

The sensitivity of a particular MFS aeolian deposit (and any associated archaeological sites) to dam operations can be assessed by observing how closely the condition of the aeolian deposit is linked to the condition of nearby fluvial sandbars, which in turn are greatly affected by the dam-controlled flow and sediment content of the river (e.g., Wiele et al. 1996; Webb et al. 1999; Schmidt et al. 2004; Topping et al. 2006). At cultural sites that could be better preserved by additional aeolian sand, there are several ways to accomplish this goal. At some sites, erosion control by checkdams can keep gullies small enough to be partially or fully healed by windblown sand (Pederson et al. 2003). Localized mitigation efforts may substantially restore site conditions in such cases (Leap and Coder 1995; Balsom et al. 2005). In many areas, however, and especially if a larger-scale solution is desired, restoration of cultural sites that rely on aeolian sediment cover can be achieved only by restoration of aeolian sand sources. Because Colorado River sand deposits constitute the largest source of sediment that could be mobilized by wind, increasing the area of exposed sandbar is the most effective way to enhance the potential for transport to and deposition on MFS aeolian dune fields and associated archaeological sites. Rebuilding sandbars by controlled flooding under sediment-enriched conditions thus holds the most promise for rebuilding aeolian deposits and protecting cultural sites above the flood level (Yeatts 1997; Fairley 2005; Draut and Rubin 2006; Topping et al. 2006).

Three Case Studies

This study identified archaeological sites where erosion is linked to dam operations, sites at which erosion is unrelated to dam operations, and some sites where the relationship between dam operations and site degradation remains unclear (Draut and Rubin 2007). Case studies of each situation are presented here. Detailed locations of the archaeological sites are confidential; they are referred to as Sites 1–3.

Site 1 displays considerable sensitivity to dam operations. Cultural artifacts at Site 1 include a roasting feature, potsherds, and the remains of a possible shelter. Based on geomorphic observations and a shallow test pit dug into sediment near this site in May of 2003, it was determined that aeolian sediment forms both the substrate beneath the site and the cover that has kept the site largely intact. Located near the landward, upper-elevation end of an aeolian dune field, this site has experienced loss of aeolian sediment manifested by downslope movement of artifacts as the land surface has deflated (based on more than 10 years of monitoring by the NPS). Although the dune field shows evidence of active sand transport in the vicinity of the site, a large area of the dune field ~10 m away has undergone pronounced deflation and soil-crust growth, indicating little or no currently active sand deposition. The source of the aeolian sediment that partially covers Site 1 has been identified as a fluvial sandbar located about 80 m directly upwind of the site. This source was identified based on wind conditions measured at a weather station that operated nearby for more than a year. Oblique historical photographs show a major decrease in the area and volume of that river-level sandbar between the 1920s and 1970s (Turner and Karpiscak 1980), a trend confirmed by aerial photographs from the 1960s and 1990s; this suggests that loss of the sandbar since dam closure is very probably linked to the deflation and limited sand transport in the dune field ~80 m downwind. Site 1 is presently intact enough that it could be better preserved with additional aeolian sediment cover—the site and its surroundings are not affected by gully incision, visitor trails, or any other degradational processes.

The apparent sensitivity of Site 1 to dam operations implies that rebuilding the fluvial sandbar could enhance wind-blown sediment delivery to the dune field, helping to cover artifacts. This sandbar did receive substantial new sediment during the 1996 flood experiment and again during the November 2004 flood experiment (Draut and Rubin 2006). Aeolian sand-transport rates showed greater sand flux during the spring 2005 (post-flood) windy season compared to spring 2004 (pre-flood; Draut and Rubin 2006). This provides an example of dam operations having been used to rebuild a sandbar from which sand transport to an archaeologically significant MFS aeolian deposit was then enhanced.

Site 2 shares many characteristics with Site 1; both contain cultural features that were built on and buried by aeolian sediment. A natural exposure reveals debris-flow sediment underlying the aeolian dune field in the area of Site 2. Advanced deflation and well-developed biological soil crust are apparent throughout the upper dune field. Although wind was not measured directly near Site 2, dune morphology and the orientation of sand shadows in the lower, more active part of the dune field, which was photographed during four visits over multiple seasons, indicate that the source of the aeolian sediment is a fluvial sandbar ~70 m away from Site 2. Comparing aerial photographs taken in the 1960s with those from the 1990s, we noted a pronounced decrease in the open sand area of the sandbar coupled with increased vegetation on the dunes. Thus far, evidence suggests that the deflation and reduced sand mobility on this MFS dune field are linked to post-dam shrinkage of the river-level sandbar. However, although the general condition of the dune field likely responds to dam operations, rapid degradation of Site 2 cannot be linked to Glen Canyon Dam because other factors unrelated to wind-blown sand affect the site on shorter time scales. A large tributary channel has cut into the aeolian sediment at the edge of the dune field. Cultural features from Site 2 (Figure 4) erode more substantially from episodic tributary floods and debris flows than from processes that could be reversed with more wind-blown sand (aeolian deflation or formation of small gullies).

In contrast to Sites 1 and 2, where cultural site sensitivity to dam operations can be either confirmed or refuted, respectively, at Site 3 the effect of dam operations remains uncertain. Like the first two cases, Site 3 was built on and buried by aeolian sediment; this

Figure 4. NPS archaeologist Jennifer Dierker stands by the edge of a major tributary channel that has eroded into a cultural site. A roasting feature that was built on and buried by aeolian sediment is exposed in the cutbank; the aeolian sediment is underlain by the poorly sorted sediment of a debris-flow deposit.

site includes several roasting features in a large, sparsely vegetated dune field. There is evidence for loss of the aeolian sediment that previously covered Site 3, through deflation, slumping, and downslope movement of artifacts. Much of this movement is associated with dune migration, based on observations of dune morphology and slip-face orientations. Minor soil crust is present in some areas, but most of the dune field appears subject to active sediment transport. The dominant local wind direction, measured at an instrument station near Site 3 for more than a year, indicates that the most likely source of sediment for this dune field is a large, pre-dam fluvial terrace ~150 m upwind. The present morphology of the terrace suggests substantial loss of sediment

from this source area: the terrace is incised by an arroyo network 1–2 m deep and up to 10 m wide, and its surface has been colonized by abundant vegetation and biological soil crust. These observations are consistent with the fact that post-dam floods have not deposited sediment on or near this terrace.

Although conditions in the aeolian dunes may be related to lack of sediment deposition on the upwind fluvial terrace, it is not clear whether Site 3 would be better preserved if deposition from a very large, sediment-rich flood were to restore sand transport from the terrace to the dune field. Aeolian processes (dune migration) are responsible for the disturbance of artifacts at Site 3, but dune migration occurs naturally in active dune fields. In a natural system, continued dune migration would be expected to eventually cover Site 3. Over time scales of years to decades, migrating dunes would expose and cover any given part of the dune field repeatedly, and downslope movement of artifacts would be expected as the sand shifts under them. Dunes near this site would need to be surveyed repeatedly over decades to determine whether dune migration will continue as it would in a natural system with plentiful sediment, or whether the dune field will show signs of exacerbated sediment-supply limitation in the future (increase in vegetation, soil crust, or winnowed lag deposits; decreased mobility of dune forms). The latter situation might indicate that deprivation of sediment caused by dam operations had prolonged the exposure of these artifacts, thereby increasing the risk of their damage and loss.

Effects of the November 2004 High-Flow Experiment on Aeolian Sand Transport

Sandbar restoration was the primary goal of the controlled-flood experiment in November 2004, the second experiment of its kind. The first, done in March 1996 under conditions of sediment depletion in the Colorado River channel, temporarily increased the surface area and volume of higher-elevation parts of sandbars in the Grand Canyon (Rubin et al. 1998, 2002; Hazel et al. 1999; Topping et al. 2006). Much of the sediment deposited on the upper parts of the sandbars, however, was supplied by the lower parts of sandbars upstream, and the surface area of most new higher-elevation deposits decreased substantially within a year (Hazel et al. 1999). To ensure a greater supply of sand than was available in 1996, the 2004 controlled flood took place after substantial sand had recently been supplied to upper Marble Canyon by the Paria River and other tributaries. Conditions during the 2004 high flow therefore represented sediment enrichment in the river channel relative to the 1996 flood (Topping et al. 2006).

The design of the 2004 flood experiment included a 60-hour steady flow of 41,000 cfs released from Glen Canyon Dam in late November. The area and the volume of sandbars in Marble Canyon above about river mile 40 were significantly greater than before this high flow; approximately half of the surveyed sand deposits were much larger than they had been immediately after the 1996 flood (Hazel et al. 2005; Topping et al. 2006). Localized deposition and erosion were documented below river mile ~40, and a consistent pattern of sandbar aggradation could not be demonstrated, implying that in future floods more sand must be present in the river channel to achieve deposition on sandbars throughout the canyon (Topping et al. 2006).

In addition to the sandbar surveys that followed the 2004 flood experiment (Hazel et al. 2005), fluvial sand deposits at our aeolian study sites were photographed immediately before and after the flood (November and December of 2004) and in March, May, and September of 2005. At all six locations where weather stations were present, major deposition of new sand occurred as a result of the November 2004 high flow (Draut and Rubin 2005). Because these sandbars are the sources from which wind transports sediment to MFS aeolian deposits, substantial deposition on the sandbars implies that sediment-rich floods have excellent potential for replenishing sand on MFS dunes and preserving associated archaeological sites. However, if maximizing windblown sediment is one of the management goals of a controlled flood, flows that occur between the flood and the next windy season must be managed to maximize dry sandbar area during the windy season (April–June). Although the November 2004 flood deposited new sand on many sandbars (and also promoted aeolian sand entrainment by decreasing the roughness of the land surface—covering vegetation, rocks, and driftwood), much of the sand was removed by high daily flow fluctuations (5000–20,000 cfs) between January and March 2005, before the windy season began (Figures 5 and 6).

At sites where flood sand was entirely removed before the 2005 windy season began, windy season aeolian sand-transport rates in 2005 were comparable to or lower

Figure 5. Photographs taken at 24.5-Mile before (a, b) and after (c, d) the November 2004 flood experiment and (e, f) after several months of high daily flow fluctuations. In b, d, and f, arrows mark boulders that appear in these three pictures, for reference. The pre-flood photos (a, b) were taken on 17 November 2004, at a discharge of 8000 cfs. Post-flood photos (c, d) were taken on 4 December 2004, at a discharge of 8000 cfs. Note that the new flood deposit both increased the area of open sand near the river and decreased the roughness of the land surface by covering vegetation, rocks, and driftwood. Photos that followed high flow fluctuations (e, f) were taken on 8 March 2005, at a discharge of approximately 8000 cfs. New sand deposited by the November flood experiment covered vegetation, driftwood, and rocks to a thickness of up to ~1 m (d), but this had been substantially eroded by March 2005 (Draut and Rubin 2005).

Figure 6. Photographs taken on river right at Malgosa just before (a) and after (b) the 2004 flood and (c) after several months of 5000–20,000 cfs daily flow fluctuations. The pre-flood photo (a) was taken on 17 November 2004, at a discharge of 8000 cfs. The post-flood photo (b) was taken on 9 December 2004, at a discharge of 8000 cfs. The photo that followed high flow fluctuations (c) was taken on 13 March 2005, at a discharge of 8000 cfs. New sand deposited here during the flood was ~2 m thick (a person standing on top of the deposit is visible for scale in the background of b). By March 2005 the flood deposit had been almost entirely removed, and the sandbar in c appears nearly identical to its pre-flood state (Draut and Rubin 2005).

than those measured during the 2004 windy season, given similar wind conditions. At 24.5-Mile, where approximately half of the flood-deposited sand remained by March 2005 (Figure 5), the windy season sand flux near river level was approximately double that measured in 2004 (Figure 7) and wind-blown sand from the flood deposit was observed to heal small gullies (Draut and Rubin 2006). However, no similar increase in sand flux relative to the previous year was measured at the upper-elevation station in the dune field at 24.5-Mile (Figure 8). Note that unusually wet weather may have inhibited windblown-sand transport in the spring of 2005 (Draut and Rubin 2006), but sand transport near river level at 24.5-Mile was still much greater than the year before even with the unusually wet conditions. The dominant WSW wind direction at 24.5-Mile moves sand from the area of the lower station toward the upper station (Draut and Rubin 2005, 2006). Although the influence of the new flood sediment was apparent near the river (measured at the lower station, Figure 7), its effects were not apparent in the upper part of the dune field (measured at the upper station, Figure 8).

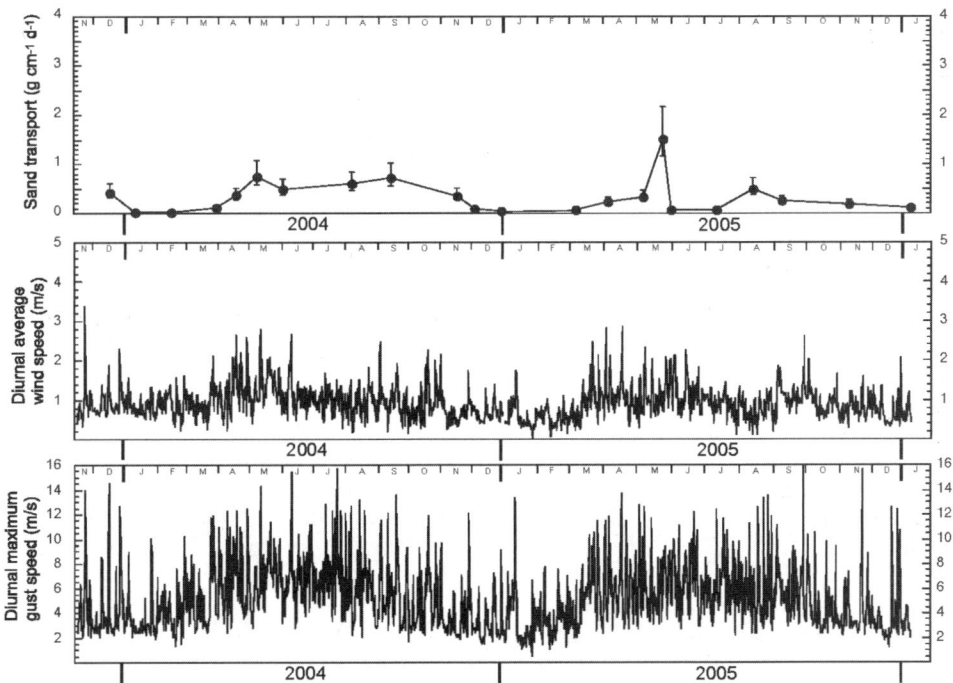

Figure 7. Wind and aeolian sediment-transport data for the lower of the two stations deployed at 24.5-Mile, near river level, from November 2003 through December 2005 (Draut and Rubin 2005, 2006). Sand transport is reported in grams per day transported between ground surface and 1 m (the elevation of the uppermost sand trap), normalized to a width of 1 cm to yield g/cm/d; sand mass collected from the four traps at each visit has been integrated over 1 m and divided by the number of days since the traps had last been emptied to obtain these values. Wind data from an anemometer at a height of 2.0 m are presented as diurnal averaged wind speed and diurnal maximum gust speed using daytime (0600–1800 hrs) and nighttime (1800–0600 hrs) averages of the data points collected at 4-minute intervals.

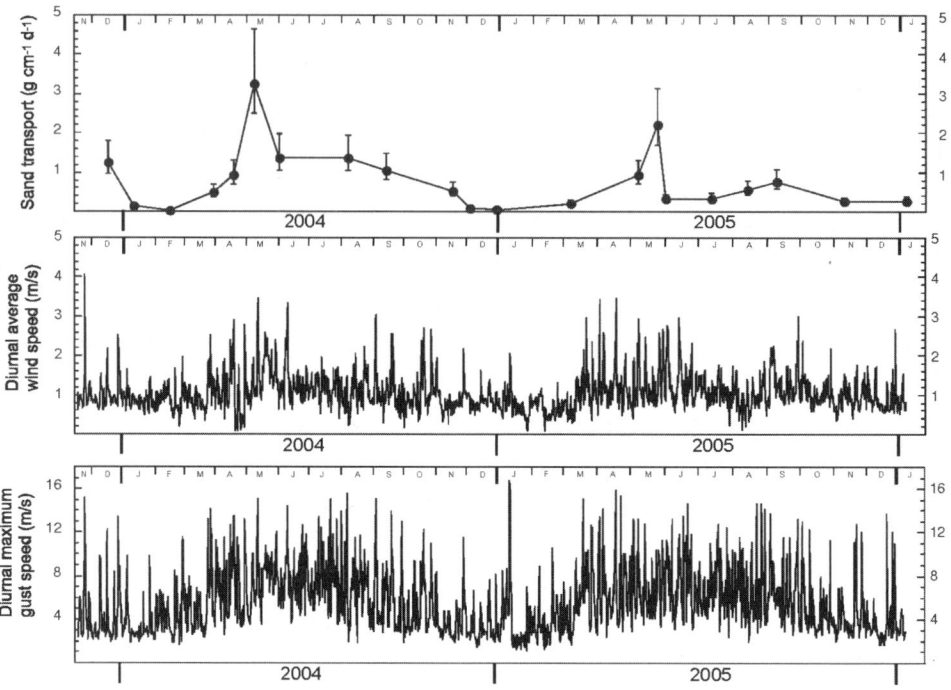

Figure 8. Wind and aeolian sediment-transport data for the upper of the two stations deployed at 24.5-Mile, at the high-elevation end of a dune field, from November 2003 through January 2006 (Draut and Rubin 2005, 2006).

In many areas of the river corridor, new sand would need to reach the highest parts of aeolian dune fields in order to benefit archaeologically significant locations. Because the sand flux measured at the upper station at 24.5-Mile exceeded sand flux at the lower station even with the new flood sand present, net flux into this dune field was still negative, indicating loss of sediment by deflation. Given wind conditions similar to those measured in 2005, sediment flux at the lower station would need to be about 50 percent higher than in 2005 to equal the sand flux at the upper station; such a situation would mean the dune field had no net gain or loss of sand. We speculate that if the sandbar at 24.5-Mile had still been as large in the spring of 2005 windy season as it was

immediately after the November 2004 flood (Figure 5), sand flux into the dune field might have balanced or exceeded flux out of the dune field, resulting in a net sediment gain in the dune field.

CONCLUSIONS

Past and present sedimentary processes can be evaluated along with modern wind and sand-transport rates to assess the sensitivity to dam operations of specific areas of the Colorado River corridor in Grand Canyon and associated cultural sites. Field study should be as detailed and as quantitative as possible, and should be conducted over the longest possible time scales to determine the effect, if any, of dam operations on a particular archaeological site.

Aeolian deposits in the river corridor fall broadly into two categories: (1) modern fluvial sourced (MFS) deposits, which form as the wind transports sand inland from < 45,000 cfs stage sandbars, creating aeolian dunes directly downwind, and (2) relict fluvial sourced (RFS) deposits, which formed as wind eroded and redistributed sediment of extensive pre-dam fluvial terraces. Archaeological material is known to occur in aeolian deposits of both types.

Some archaeological sites in MFS dunes have been negatively affected by the loss of aeolian sand caused by decreased sand supply on upwind sandbars, a process attributable to dam operations. These same sites could benefit from aeolian redistribution of new sand deposited on fluvial sandbars by sediment-rich controlled floods. The November 2004 high flow resulted in major deposition of new sand in many areas that are sediment sources for MFS aeolian deposits; wind reworking of 2004 flood sand was also observed to heal small gullies locally. The potential of the 2004 flood to supply new sand to MFS aeolian deposits and their associated archaeological sites was greatly reduced, however, by 3 months of high daily flow fluctuations that removed much of the new sand before the start of the first post-flood windy season in April 2005. The restoration potential for cultural sites in aeolian deposits can be maximized by using dam operations (controlled floods and post-flood flows) that maximize the exposed sand area on fluvial sandbars from April through early June, when wind-borne sediment transport is greatest.

ACKNOWLEDGMENTS

This study was supported by funding from the U.S. Bureau of Reclamation through the Grand Canyon Monitoring and Research Center and is a cooperative effort with the National Park Service. We thank our collaborators: N. Andrews, J. Balsom, M. Barger, J. Dierker, H. Fairley, C. Fritzinger, R. Griffiths, J. Hazel, R. Hunter, L. Jackson, M. Kaplinski, K. Killoy, L. Leap, T. Melis, F. Nials, T. Sabol, E. Todd, D. Topping, R. Tusso, N. Voichick, and M. Yeatts. We also thank the many river guides who have assisted with this study. Reviews by R. Hereford, P. Barnard, J. Lovich, D. Topping, and two anonymous reviewers improved this manuscript substantially.

LITERATURE CITED

Andrews, E. D. 1986. The Colorado River: A perspective from Lees Ferry, Arizona. In Surface Water Hydrology, edited by M. G. Wolman and H. C. Riggs, pp. 304–310. Geological Society of America, Decade of North American Geology O-1.

Ash, J. E., and R. J. Wasson. 1983. Vegetation and sand mobility in the Australian desert dunefield. Zeitschrift für Geomorphologie Suppl. Bd. 45: 7–25.

Balsom, J. R., J. G. Ellis, A. Horn, and L. M. Leap. 2005. Using cultural resources as part of the plan: Grand Canyon management and implications for resource preservation. In The Colorado Plateau II: Biophysical, Socioeconomic, and Cultural Research, edited by C. van Riper III and D. J. Mattson, pp. 367–377. University of Arizona Press, Tucson.

Bauer, B. O., R. G. D. Davidson-Arnott, K. F. Nordstrom, J. Ollerhead, and N. L. Jackson. 1996. Indeterminacy in aeolian sediment transport across beaches. Journal of Coastal Research 12: 641–653.

Beus, S. S., S. W. Carothers, and C. C. Avery. 1985. Topographic changes in fluvial terrace deposits used as campsite beaches along the Colorado River in Grand Canyon. Journal of the Arizona-Nevada Academy of Science 20: 111–120.

Buckley, R. 1987. The effect of sparse vegetation on the transport of dune sand by wind. Nature 325: 426–428.

Budhu, M., and R. Gobin. 1994. Instability of sandbars in Grand Canyon. Journal of Hydraulic Engineering 120: 918–933.

Burke, K. J., H. C. Fairley, R. Hereford, and K. S. Thompson. 2003. Holocene terraces, sand dunes, and debris fans along the Colorado River in Grand Canyon. In Grand Canyon Geology, edited by S. S. Beus and M. Morales, pp. 352–370. Oxford University Press, New York.

Draut, A. E., and D. M. Rubin. 2005. Measurements of wind, aeolian sand transport, and precipitation in the Colorado River corridor, Grand Canyon, Arizona—November 2003 to December 2004. U.S. Geological Survey Open-File Report 2005-1309. Available at http://pubs.usgs.gov/of/2005/1309/.

Draut, A. E., and D. M. Rubin. 2006. Measurements of wind, aeolian sand transport, and precipitation in the Colorado River corridor, Grand Canyon, Arizona—January 2005 to January 2006. U.S. Geological Survey Open-File Report 2006-1188. Available at http://pubs.usgs.gov/of/2006/1188/.

Draut, A. E., and D. M. Rubin. 2007. The role of aeolian sediment in the preservation of archaeological sites in the Colorado River corridor, Grand Canyon, Arizona—Final report on research activities, 2003–2006. U.S. Geological Survey Open-File Report 2007-1001. Available at http://pubs.usgs.gov/of/2007/1001/.

Draut, A. E., D. M. Rubin, J. L. Dierker, H. C. Fairley, R. E. Griffiths, J. E. Hazel Jr., R. E. Hunter, K. Kohl, L. M. Leap, F. L. Nials, D. J. Topping, and M. Yeatts. 2005. Sedimentology and stratigraphy of the Palisades, Lower Comanche, and Arroyo Grande areas of the Colorado River corridor, Grand Canyon, Arizona. U.S. Geological Survey Scientific Investigations Report 2005-5072. Available at http://pubs.usgs.gov/sir/2005/5072/.

Draut, A. E., D. M. Rubin, J. L. Dierker, H. C. Fairley, R. E. Griffiths, J. E. Hazel Jr., R. E. Hunter, K. Kohl, L. M. Leap, F. L. Nials, D. J. Topping, and M. Yeatts. Application of sedimentary-structure interpretation to geoarchaeology in the Colorado River corridor, Grand Canyon, Arizona. Geomorphology, in press.

Fairley, H. C. 2003. Changing River: Time, Culture, and the Transformation of Landscape in the Grand Canyon—A Regional Research Design for the Study of Cultural Resources Along the Colorado River in Lower Glen Canyon and Grand Canyon National Park, Arizona. Prepared for the U.S. Geological Survey, Grand Canyon Monitoring and Research Center, Flagstaff, Arizona. Technical Series 79, Statistical Research, Inc.

Fairley, H. C. 2005. Cultural resources in the Colorado River corridor. In The State of the Colorado River Ecosystem in Grand Canyon, edited by S. P. Gloss, J. E. Lovich, T. S. and Melis, pp. 177–192. U.S. Geological Survey Circular 1282.

Fairley, H. C., P. W. Bungart, C. M. Coder, J. Huffman, T. L. Samples, and J. R. Balsom. 1994. The Grand Canyon River Corridor Survey Project: Archaeological Survey Along the Colorado River Between Glen Canyon Dam and Separation Canyon. Cooperative Agreement No. 9AA-40-07920, Grand Canyon National Park. Prepared in cooperation with the Bureau of Reclamation, Glen Canyon Environmental Studies, Flagstaff, Arizona.

Fryrear, D. W. 1986. A field dust sampler. Journal of Soil and Water Conservation 41: 117–119.

Goossens, D. 2004. Effect of soil crusting on the emission and transport of wind-eroded sediment: Field measurements on loamy sandy soil. Geomorphology 58: 145–160.

Grams, P. E., and J. C. Schmidt. 1999. Integration of photographic and topographic data to develop temporally and spatially rich records of sand bar change in the Point Hansbrough and Little Colorado River confluence study reaches. Report submitted to Grand Canyon Monitoring and Research Center.

Hazel, J. E. Jr., M. Kaplinski, R. Parnell, M. Manone, and A. Dale. 1999. Topographic and bathymetric changes at thirty-three long-term study sites. In The Controlled Flood in Grand Canyon, edited by R. H. Webb, J. C. Schmidt,

G. R. Marzolf, and R. A. Valdez, pp. 161–183. Washington, D. C., American Geophysical Union: Geophysical Monograph 110.

Hazel, J. E., M. Kaplinski, M. Manone, and R. Parnell. 2000. Monitoring arroyo erosion of pre-dam river terraces in the Colorado River ecosystem, 1996–1999, Grand Canyon National Park, Arizona. Northern Arizona University, Department of Geology. Draft final report to the Grand Canyon Monitoring and Research Center, cooperative agreement CA 1425-98-FC-40-22630.

Hazel, J. E. Jr., M. Kaplinski, R. Parnell, J. C. Schmidt, and D. J. Topping. 2005. A tale of two floods: Comparing sandbar responses to the 1996 and 2004 high-volume experimental flows on the Colorado River in Grand Canyon. Colorado River Ecosystem Science Symposium, Tempe, Arizona, October 25–27.

Hereford, R. 1996. Surficial geology and geomorphology of the Palisades Creek area, Grand Canyon National Park, Arizona. U.S. Geological Survey Miscellaneous Investigations Series Map I-2449, scale 1:2000 (with discussion).

Hereford, R., H. C. Fairley, K. S. Thompson, and J. R. Balsom. 1993. Surficial geology, geomorphology, and erosion of archeological sites along the Colorado River, eastern Grand Canyon, Grand Canyon National Park, Arizona. U.S. Geological Survey Open-File Report 93-517.

Hereford, R., K. S. Thompson, K. J. Burke, and H. C. Fairley. 1996. Tributary debris fans and the late Holocene alluvial chronology of the Colorado River, eastern Grand Canyon, Arizona. Geological Society of America Bulletin 108: 3–19.

Hereford, R., K. J. Burke, and K. S. Thompson. 2000. Quaternary geology and geomorphology of the Granite Park area, Grand Canyon, Arizona. U.S. Geological Survey Geologic Investigations Series I-2662, scale 1:2,000.

Hunter, R. E. 1977. Basic types of stratification in small eolian dunes. Sedimentology 24: 361–387.

Kaplinski, M., J. E. Hazel Jr., and S. S. Beus. 1995. Monitoring the effects of interim flows from Glen Canyon Dam on sand bars in the Colorado River corridor, Grand Canyon National Park, Arizona. Final report to Glen Canyon Environmental Studies, Northern Arizona University, Flagstaff.

Kearsley, L. H., J. C. Schmidt, and K. D. Warren. 1994. Effects of Glen Canyon Dam on Colorado River sand deposits used as campsites in Grand Canyon National Park, USA. Regulated Rivers: Research and Management 9: 137–149.

Leap, L. M., and C. M. Coder. 1995. Erosion control project at Palisades delta along the Colorado River corridor, Grand Canyon National Park. Report repared for the Bureau of Reclamation, Salt Lake City, Utah. Grand Canyon National Park, River Corridor Monitoring Project, Report 29.

Leap, L. M., J. L. Kunde, D. C. Hubbard, N. B. Andrews, C. E. Downum, A. R. Miller, and J. Balsom. 2000. Grand Canyon Monitoring Project 1992–1999: Synthesis and annual report FY99. Report prepared by Grand Canyon National Park and Northern Arizona University, submitted to the Bureau of Reclamation, Upper Colorado Region, Salt Lake City, Utah, Grand Canyon National Park, River Corridor Monitoring Project, Report 66.

Namikas, S. L., and D. J. Sherman. 1995. A review of the effects of surface moisture content on aeolian sand transport. In Desert Aeolian Processes, edited by V. Tchakerian, pp. 269–293. Chapman and Hall, London.

Neal, L. A., D. Gilpin, L. Jonas, and J. H. Ballagh. 2000. Cultural resources data synthesis within the Colorado River corridor, Grand Canyon National Park and Glen Canyon National Recreation Area, Arizona. SWCA, Inc. Cultural Resources Report 98-85 submitted to Grand Canyon Monitoring and Research Center.

Pederson, J. L., P. A. Peterson, W. W. Macfarlane, F. M. Gonzales, and K. Kohl. 2003. Mitigation, monitoring, and geomorphology related to gully erosion of cultural sites in Grand Canyon. Utah State University, Department of Geology, Logan. Report submitted to Grand Canyon Monitoring and Research Center.

Rubin, D. M. 1987. Cross-bedding, bedforms, and paleocurrents. Society of Economic Paleontologists and Mineralogists: Concepts in Sedimentology and Paleontology Volume 1.

Rubin, D. M., J. M. Nelson, and D. J. Topping. 1998. Relation of inversely graded deposits to suspended-sediment grain-size evolution during the 1996 flood experiment in Grand Canyon. Geology 26: 99–102.

Rubin, D. M., D. J. Topping, J. C. Schmidt, J. Hazel, M. Kaplinski, and T. S. Melis. 2002. Recent sediment studies refute Glen Canyon Dam hypothesis. EOS, Transactions of the American Geophysical Union 83: 277–278.

Schmidt, J. C., and J. B. Graf. 1987. Aggradation and degradation of alluvial sand deposits, 1965 to 1986, Colorado River, Grand Canyon National Park, Arizona. U.S. Geological Survey Open-File Report 87-555.

Schmidt, J. C., and M. F. Leschin. 1995. Geomorphology of post-Glen Canyon Dam fine-grained alluvial deposits of the Colorado River in the Point Hansbrough and Little Colorado River confluence study reaches in Grand Canyon National Park, Arizona. Report prepared for the U.S. Bureau of Reclamation, Glen Canyon Environmental Studies.

Schmidt, J. C., D. J. Topping, P. E. Grams, and J. E. Hazel. 2004. System-wide changes in the distribution of fine sediment in the Colorado River corridor between Glen Canyon Dam and Bright Angel Creek, Arizona. Final report submitted to Grand Canyon Monitoring and Research Center by the Department of Aquatic, Watershed, and Earth Resources, Utah State University, Logan.

Sterk, G., and P. A. C. Raats. 1996. Comparison of models describing the vertical distribution of wind-eroded sediment. Soil Science Society of America Journal 60: 1914–1919.

Stout, J. E., and D. W. Fryrear. 1989. Performance of a windblown-particle sampler. Transactions of the American Society of Agricultural Engineers 32: 2041–2045.

Thompson, K. S., and A. R. Potochnik. 2000. Development of a geomorphic model to predict erosion of pre-dam Colorado River terraces containing archaeological resources. SWCA, Inc. Cultural Resources Report 99-257 submitted to Grand Canyon Monitoring and Research Center.

Topping, D. J., D. M. Rubin, and L. E. Vierra Jr. 2000a. Colorado River sediment transport 1. Natural sediment supply limitation and the influence of Glen Canyon Dam. Water Resources Research 36: 515–542.

Topping, D. J., D. M. Rubin, J. M. Nelson, P. J. Kinzel III, and I. C. Corson. 2000b. Colorado River sediment transport 2. Systematic bed-elevation and grain-size effects of sand supply limitation. Water Resources Research 36: 543–570.

Topping, D. J., J. C. Schmidt, and L. E. Vierra Jr. 2003. Computation and analysis of the instantaneous-discharge record for the Colorado River at Lees Ferry, Arizona—May 8, 1921 through September 30, 2000. U.S. Geological Survey Professional Paper 1677.

Topping, D. J., D. M. Rubin, J. C. Schmidt, J. E. Hazel Jr., T. S. Melis, S. A. Wright, M. Kaplinski, A. E. Draut, and M. J. Breedlove. 2006. Comparison of sediment-transport and bar-response results from the 1996 and 2004 controlled-flood experiments on the Colorado River in Grand Canyon. Proceedings of the 8th Federal Interagency Sediment Conference, Reno, Nevada, April 2006. ISBN 0-9779007-1-1. [CD-ROM]

Turner, R. M., and M. M. Karpiscak. 1980. Recent vegetation changes along the Colorado River between Glen Canyon Dam and Lake Mead, Arizona. U.S. Geological Survey Professional Paper 1132.

U.S. Department of the Interior. 1996. Record of Decision, Operation of Glen Canyon Dam. Final Environmental Impact Statement. Secretary of the Interior, Washington, D.C.

Yeatts, M. 1996. High elevation sand deposition and retention from the 1996 spike flow—An assessment for cultural resources stabilization. In Mitigation and Monitoring of Cultural Resources in Response to the Experimental Habitat Building Flow in Glen and Grand Canyons, Spring 1996, edited by J. R. Balsom and S. Larralde, pp. 124–158. Report submitted to the Bureau of Reclamation (Grand Canyon Monitoring and Research Center), Flagstaff, Arizona.

Yeatts, M. 1997. High elevation sand retention following the 1996 spike flow. Report to the Grand Canyon Monitoring and Research Center, Flagstaff, Arizona.

Webb, R. H., D. L. Wegner, E. D. Andrews, R. A. Valdez, and D. T. Patten. 1999. Downstream effects of Glen Canyon Dam on the Colorado River in Grand Canyon: A review. In The Controlled Flood in Grand Canyon, edited by R. H. Webb, J. C. Schmidt, G. R. Marzolf, and R. A. Valdez, pp. 1–21. American Geophysical Union, Geophysical Monograph 110, Washington, D.C.

Wiele, S. M., J. B. Graf, and J. D. Smith. 1996. Sand deposition in the Colorado River in the Grand Canyon from flooding of the Little Colorado River. Water Resources Research 32: 3579–3596.

Wiggs, G. F. S., A. J. Baird, and R. J. Atherton. 2004. The dynamic effects of moisture on the entrainment and transport of sand by wind. Geomorphology 59: 13–30.

Wright, S. A., T. S. Melis, D. J. Topping, and D. M. Rubin. 2005. Influence of Glen Canyon Dam operations on downstream sand resources of the Colorado River in Grand Canyon. In The State of the Colorado River Ecosystem in the Grand Canyon, edited by S. P. Gloss, J. E. Lovich, and T. S. Melis, pp. 17–31. U.S. Geological Survey Circular 1282.

Synthesis

INTEGRATING RESEARCH INTO BIOLOGICAL AND CULTURAL RESOURCES MANAGEMENT ON THE COLORADO PLATEAU— A SYNTHESIS

Charles van Riper III

In his book review of the Sixth Colorado Plateau Biennial Conference Proceedings: "The Colorado Plateau: Cultural, Biological and Physical Research," Fleming (2006) closes with the following statement: "A synthesis chapter by the editors on the issue of human impacts would have been a welcome addition to this potentially useful book." The following synthesis chapter provides that addition to this book, summarizing and integrating material from all previous chapters and also utilizing information from previous books published in the Colorado Plateau series. In addition to providing a synthesis, another goal of this chapter is to demonstrate how aspects of research have been used to enhance biological and cultural resource management.

This book is the eighth (and the third volume published by the University of Arizona Press) in a series of compilations that focus on the Colorado Plateau. These books highlight the integration of research into resource management efforts, as related to cultural, natural, and physical resources within the biogeographic province. The mix of chapters addresses management issues from many diverse resources and from disparate regions of the Colorado Plateau, specifically focusing on aspects of vegetation and wildlife research, combined with a series of chapters explaining integrative and collaborative tools that can be used to better manage natural and cultural resources on smaller and larger scales.

The 20 previous chapters in this book were selected from scientific presentations at the Eighth Biennial Conference of Research on the Colorado Plateau. The conference was held 7–10 November 2005 in Flagstaff, Arizona, hosted by the U.S. Geological Survey Southwest Biological Science Center's Colorado Plateau Research Station and the Center for Sustainable Environments at Northern Arizona University. The meeting theme revolved around research, inventory, and monitoring of lands over the Colorado Plateau, with a focus on the integration of research into resources management actions. Material presented in this book represents original research that has not been previously published, and every chapter contains a section on how that research can be best implemented by managers. Each paper selected for publication has been anonymously peer reviewed by at least two scientists from that specific research discipline. These contributed scientific studies each constitute a separate chapter, with the material subdivided into sections: Collaborating to Achieve Conservation, Assessing Large-scale Land-use Issues, Addressing Wildlife Issues, Addressing Vegetation Issues, Gaining Insights from the Past, and Synthesis.

The scientific works published in this Biennial Conference Series contribute significantly to presenting peer-reviewed results of collaborative efforts among scientists and land managers. The U.S. Geological Survey Southwest Biological Science Center's staff, university, and other partner agency scientists have worked closely with Colorado Plateau land managers from a variety of state and federal agencies, as well as from the private sector to achieve remarkable management results. Many of the protocols and techniques presently being utilized in land management units over the Colorado Plateau are a result of previous collaborative works published in this series of books. It has been clearly demonstrated that, because of similarities across the Colorado Plateau, techniques that work in one management unit can be applicable to many other areas. This is due primarily to the similarity of habitat and climatological conditions over the Colorado Plateau.

COLLABORATING TO ACHIEVE CONSERVATION

Collaboration is a critical component of almost every successful large-scale conservation effort. The opening chapter by Tilt et al. highlights the importance of collaborative efforts, a theme that is carried throughout this section of the book. The authors structure this opening chapter under the 11 lessons that they feel are critical for successful collaboration to occur: (1) Understand what collaboration is and is not, (2) Recognize challenge and time involved, (3) Exhaust traditional approaches (ripeness), (4) Build a common vision (passion for place, a community of purpose), (5) Create an open, inclusive, and transparent process, (6) Ensure stakeholders are representative of the community, (7) Provide facilitation and process, (8) Develop a common factual base, (9) Secure operational funding, (10) Achieve and communicate results, and (11) Meet or exceed applicable laws and be accountable. They illustrate aspects of successful partnerships with a series of stories from throughout the western United States, focusing on the successes of community-based collaboratives (CBCs). The one CBC that they feel has created a benchmark, and is the present standard for the Colorado Plateau, is the Diablo Trust. A background on this collaborative can be found in previous chapters of the Colorado Plateau book series (e.g., Sisk et al. 1999; Loeser et al. 2001). Initially founded in 1993 by two ranches, the Bar-T-Bar and Flying M, the Diablo Trust CBC was created to link private and public values

under one holistic goal: "to create sustainable rangeland management that maintains the tradition of working ranches and provides for economic viability while managing for ecosystem health." The focus area of this CBC is east of Flagstaff, Arizona, encompassing checker-boarded private and state lands, augmented with U.S. Forest Service summer grazing allotments. Collaborators of the Diablo Trust now include local ranchers, state and federal agencies, scientists, environmentalists, and other interested stakeholders.

Working with researchers at Northern Arizona University and Prescott College, and with the many products that have been produced in this biennial conference series (van Riper 1995; van Riper and Deshler 1997; van Riper and Stuart 1999; van Riper et al. 2001; van Riper and Cole 2004; van Riper and Mattson 2005), the Diablo Trust is one of the premier examples of a CBC that incorporates research and monitoring into rangeland conservation. One aspect that has led to the success of the Diablo Trust is the recognition that good land stewardship incorporates the integration of research into monitoring projects. Working with scientists in a collaborative environment has helped the Diablo Trust develop appropriate research questions that are relevant to the ranchers, while addressing perceived conflicts among stakeholders and the outside public (Sisk et al. 1999). In addition, integrated collaborative research and sound monitoring protocols can generate clear measures of effectiveness and progress from which to evaluate the success of any collaboration (Muñoz-Erickson and Aguilar-Gonzalez 2003).

As Tilt et al. clearly point out, the Diablo Trust CBC has yielded several benefits to scientists. This collaboration provides scientists with a landscape of resources at multiple scales that allow their studies to go from small plots to whole landscapes. In addition, the ability to collaborate with the people who manage the land results in more meaningful, insightful, and useful science (Sisk and Palumbo 2005). It must be recognized by scientists that, in order to continue this fruitful relationship among stakeholders, they must be willing to invest significant time into the collaborative process and anticipate a multi-decadal relationship. On the other hand, stakeholders and land owners must share a goal of sustaining research and monitoring over long time periods in order to generate information that is relevant to an ecological system that is often typified by slow responses interrupted by periodic bouts of dramatic change.

Although collaborations have their challenges and critics, Tilt et al. present outcomes and experiences of numerous CBC efforts, reaffirming that a collaborative process can be successful in developing long-term solutions for natural resource issues. Drawing from the experiences of these collaborations, the authors highlight several criteria that have contributed to the CBC success. First, all collaborative processes face issues where the problem and its solution are poorly understood, where there are few scientific data and little understanding of what that information means, and where personnel and financial resources are few or nonexistent. Next, they point out that conflicting values confuse the process and innovation is often viewed as risky and expensive. Not only must collaborations bring together a diverse and representative group of stakeholders, but they must also recognize the amount of time, effort, and funding that is necessary for creating and sustaining a successful collaborative process. Finally, the authors point out the importance of gaining the trust of the stakeholders and outside interest groups, which is accomplished by maintaining an open and transparent process that incorporates research and monitoring protocols that will properly evaluate the CBC goals. It is evident from the findings presented in this chapter that the spectacular natural resources of the Colorado Plateau will continue to serve as a focal draw for new CBCs being formed in the region, thus ultimately influencing all facets of collaborative efforts from which management policies will be developed.

The collaborative theme in this first portion of the book is continued in Chapter 2, where Turner et al. present the results of an effort between researchers and resource managers at Mesa Verde National Park. This collaborative modeling approach, called FRAME, details impacts of forest restoration policy at the park. The overall strategy of the FRAME Project was to combine the principles of collaboration with the adaptive capabilities of the U.S. Geological Survey modular modeling system (MMS), in order to develop a transportable, collaborative modeling approach to adaptive, multi-objective natural resource management. The group first collaboratively identified key system components, critical pathways, and associated conceptual models of pinyon-juniper ecosystem dynamics. They found that the recent invasion and rapid spread of cheatgrass in the park had the potential to significantly alter the fire regime by increasing fire frequency and impacting long-term vegetation successional patterns. This concern led the authors to focus on cheatgrass for their first modeling simulations. They modified the SIMPPLLE landscape model to capture key ecosystem components and dynamics of the conceptual models, which were then further refined through an iterative process in which project scientific experts helped define probabilities.

Model results presented by Turner et al. indicated the potential for frequent re-burning in the park, at intervals as short as a few years. These simulations suggested a fire rotation of approximately 45 years for the park as a whole, a dramatic change from the historic fire rotation that has previously been measured in centuries. The authors argue that such a disturbance regime would be far outside the historical range of variability for the ecosystem, and would likely lead to a substantial reduction and even local extirpation of many native plant species. They also showed that the projected changes in Mesa Verde's fire regime would bring an increased risk of significant debris-flow events, with the potential for substantial damage to water and cultural resources. The FRAME case study at Mesa Verde National Park provided an ideal opportunity to implement and refine the principles and components of a collaborative modeling approach. By coupling the principles of collaboration with integrated modeling approaches, the authors developed a collaborative modeling framework to facilitate adaptive, multi-objective resource management that would be applicable across a wide range of ecosystems.

We now see, across the Colorado Plateau, trends in public and private lands management toward integrated science approaches, with co-management of public lands, adaptive management in the face of uncertainty, and public engagement in land-use decision making, developed primarily in response to a greater appreciation of the inherent complexity and uncertainty in natural systems. We are also seeing an increased public scrutiny of decisions on public lands. The authors have developed their FRAME collaborative modeling approach to address the challenges faced by natural resource managers, and to provide those managers with mechanisms to effectively link integrated science to natural resource management needs. The FRAME approach can also readily be adapted to engage the public in participatory natural resource management efforts, and the authors demonstrate that this collaborative process could easily be applied to most management units over the entire Colorado Plateau.

ADDRESSING LARGE-SCALE LAND-USE ISSUES

At this point, the book departs from the arena of citizen-based collaboratives and moves into large-scale land management issues. The next three chapters focus on aspects of the GAP program, a computer-based Geographic Information System (GIS) tool initially developed by J. Michael Scott (Scott et al. 1993), a U.S. Geological Survey scientist at the University of Idaho. A number of research studies utilizing GAP have been published in chapters of previous books within the Colorado Plateau series (see especially the volumes of van Riper et al. 2001; van Riper and Cole 2004; van Riper and Mattson 2005). The GAP program provides information on ecosystem representa-tion by creating digital maps of conservation networks, providing an account of the representation of elements of biodiversity within a region (Crist and Scott 1999). Gap analysis uses the distribution of vegetation types and vertebrate species as indicators of biodiversity. Digital map overlays in GIS are used to identify individual species, species-rich areas, and vegetation types that are absent or underrepresented in existing management areas (Scott et al. 1993). These products are used to develop conservation strategies and to predict contributions of new management areas for biodiversity maintenance at landscape scales (Scott et al. 1991).

In Chapter 3, Ernst and Prior-McGee argue that the conservation of biological diversity is important for the maintenance of naturally functioning ecosystems, and to ensure preservation of species and communities as well as functional diversity of plant and animal populations. The Colorado Plateau, which is perhaps one of the most diverse ecoregions in North America, is characterized by unique geology and land-form features that create an environment that results in high endemism. Ernst and Prior-McGee demonstrate that the vulnerability and conservation of these unique Colorado Plateau resources can be adequately evaluated using the Southwest Regional Gap Analysis Project (SWReGAP) stewardship data set, and that the information provided can effectively assess general patterns of biodiversity protection within this ecoregion. They point out that this evaluation is important because federal agencies and tribal land stewards manage the majority (more than 75%) of the Colorado Plateau (the Bureau of Land Management manages 31% of the ecoregion, the National Park Service 7%, and the Forest Service 4% of the ecoregion), and maintenance of biodiversity with federal land management is easier to accomplish when compared to working with a mosaic of land-ownership patterns.

In examining degrees of protection, Ernst and Prior-McGee identified land management categorization schemes relative to the

purported degree of management for biodiversity maintenance for each managed
area. They listed four biodiversity management status categories as defined by Scott et
al. (1993), Edwards et al. (1994), and Crist et
al. (2000):

Status 1: An area having permanent protection from conversion of natural land cover
and mandated management plan in operation to maintain a natural state within which
disturbance events (of natural type, frequency, intensity, and legacy) are allowed to proceed without interference or are mimicked
through management.

Status 2: An area having permanent protection from conversion of natural land cover
and a mandated management plan in operation to maintain a primarily natural state,
but which may receive uses or management
practices that degrade the quality of existing
natural communities, including suppression
of natural disturbance.

Status 3: An area having permanent protection from conversion of natural land cover
for the majority of the area, but subject to
extractive uses of either a broad, low-
intensity type (e.g. logging) or localized type
(e.g. mining). It also confers protection to
federally listed endangered and threatened
species throughout the area.

Status 4: There are no known public or
private institutional mandates or legally
recognized easements or deed restrictions
held by the managing entity to prevent
conversion of natural habitat types to
anthropogenic habitat types. The area generally allows conversion to unnatural land
cover throughout.

The authors found that approximately 5
percent of the Colorado Plateau ecoregion
has permanent protection (GAP Status 1)
from conversion of natural land cover to anthropogenic land cover types. The National
Park Service manages 90 percent of the
Status 1 lands, with the largest parcels
including Grand Canyon, Canyonlands,
Zion, and Arches National Parks. The BLM
manages 7 percent of the Status 1 lands,
with most occurring as small and isolated

parcels in the form of administratively
designated Areas of Critical Environmental
Concern and Outstanding Natural Areas.
The U.S. Forest Service manages 3 percent of
the Status 1 lands, the largest being the
Kanab Creek Wilderness, and The Nature
Conservancy manages 1 percent. On the
Colorado Plateau, the state of Arizona man-
ages 63 percent of the Status 1 lands and
Utah manages 33 percent.

Status 2 lands constitute 12 percent of the
Colorado Plateau with 85 percent of those
lands managed by the BLM. About 60 percent of the Colorado Plateau is managed as
Status 3 lands (primarily multiple-use
lands), with 57 percent being tribal lands
and 33 percent BLM. Additionally, 23
percent of the Colorado Plateau is managed
as Status 4 lands, which are primarily
privately owned lands (62%) with no known
mandates that limit natural land cover
conversion to anthropogenic land uses. The
information provided by this chapter will be
effective in identifying land areas that are
presently being managed for biodiversity
and their levels of protection over the Colorado Plateau. By helping managers identify
locations of conservation lands, and the
stewards of those lands, the GAP program
allows the managers to better place their
parcels into a regional perspective. When
the land stewardship data are combined
with information on vegetation and species
richness, land stewards can then evaluate
how well their areas are contributing to
protecting biodiversity over the Colorado
Plateau.

The interface between conservation and
aspects of resource management, utilizing
the GAP program, is further developed in
Chapter 4 by Langs et al. These authors
build upon the framework of the previous
chapter, providing the reader with the first
mapping of natural land cover across the
Colorado Plateau using ecological systems.
They conducted a gap analysis for the
Colorado Plateau ecoregion through a
geospatial union of key environmental and
management data layers using ESRI ArcGIS
Desktop 9 software and Spatial Analyst extension. Their input data consisted of three

spatial databases developed by SWReGAP: land cover, land stewardship, and biodiversity management status categories. The stated goal of this chapter was to provide land managers and policy makers with information needed to make better-informed decisions when identifying priority areas for conservation.

Langs et al. documented 77 different land cover types that occur within the Colorado Plateau ecoregion, 62 of which they determined to be ecological systems. Only 7 of the 62 ecological systems had more than 5 percent of their mapped distribution within the Colorado Plateau ecoregion, which when combined represent approximately 75 percent of the total area. They also found that 48 of the ecological systems had 1 percent or less distribution within this ecoregion and 40 of these have 5 percent or less of their regional distribution on the Colorado Plateau. These ecological systems have either naturally restricted ranges, or although common, are considered peripheral to the Colorado Plateau.

The seven most abundant ecological systems on the Colorado Plateau were the Colorado Plateau Pinyon-Juniper Woodland (23% of the ecoregion), Inter-Mountain Basins Semi-Desert Shrub-Steppe (11%), Colorado Plateau Mixed Bedrock Canyon & Tableland (11%), Inter-Mountain Basins Semi-Desert Grassland (8%), Inter-Mountain Basins Mixed Salt Desert Scrub (7%), Colorado Plateau Blackbrush–Mormon tea Shrubland (7%), and Inter-Mountain Basins Big Sagebrush Shrubland (6%). There were five ecological systems they considered "nearly endemic" to the Colorado Plateau: Southern Colorado Plateau Sand Shrubland (99.7% of its mapped distribution falls the ecoregion), Colorado Plateau Blackbrush-Mormon-tea Shrubland (99.6%), Colorado Plateau Mixed Bedrock Canyon and Tableland (86%), Inter-Mountain Basins Mat Saltbush Shrubland (82%), and Inter-Mountain Basins Shale Badland (82%).

The use of conservation thresholds allowed Langs et al. to identify ecological systems with low representation in Status 1 and 2 lands (explained by Ernst and Prior-

McGee in the previous chapter). Langs et al. also identified six ecological systems with minimal protection within the ecoregion. On the other hand, they point out that there are many ecological systems within the Colorado Plateau that are either barren, sparsely vegetated, or have open-canopied scrubby vegetation (West and Young 2000). These systems occur on soils that are easily erodible such as sand sheets, dunes, and shale badlands. Wind and water degradation of the soil leads to degradation of the vegetation supported in these substrates. The presence of cryptogamic crusts plays an important role for many of these systems by facilitating the infiltration of water, increasing fertility, and reducing erosion of the soil (Belnap et al. 2001). The authors also point out that drought and increasing temperatures pose a near-future threat to the ecological systems of the Colorado Plateau.

In carrying through the earlier CBC theme of this book, Langs et al. point out that conservation at ecoregional (larger) scales requires the involvement of multiple partners and cooperative management among diverse land stewards. Partnerships with federal land management agencies, tribal entities, private land owners, academic institutions, and non-government organizations all play a vital role for ensuring successful, long-term conservation on the Colorado Plateau (Tuhy et al. 2002). One collaboration highlighted by the authors as an example of a CBC partnership that includes the Colorado Plateau is the Utah Partners for Conservation & Development. This CBC is composed of state, federal, and natural resource agencies, universities, county and local governments, private land owners, conservation organizations, and other vested stakeholders, who are working cooperatively to manage and restore rangelands in Utah (Utah Division of Wildlife Resources 2005). By having CBC groups utilizing regional GAP data, and by analyzing land-cover over the entire Colorado Plateau, land managers can strike a balance between biodiversity management with anthropogenic impacts, including development potential.

The next chapter serves as a summary of the large-scale land management section, as well as a transition into the Biological portion of this book. Boykin et al. incorporate protocols from the GAP program to create GIS models that map the distributions of wildlife over the Colorado Plateau. The authors use seven foundation GIS layers, ranging from dominant overstory vegetation through slope and aspect, to tree density and basal area to develop their models. From a survey of 40 academic institutions, they also provide a list of sensitive species that is incorporated into their model as a separate data layer. They then developed habitat models from literature reviews for each species using specific associations of available GIS environmental variables. Specific variables that they used included land cover, elevation, slope, aspect, distance to hydrological features, landform (after Manis et al. 2001), soils, and mountains. Models were constrained to the known range of the species using state, regional, and national references. Range data were converted to sub-basin watershed units (8-digit hydrologic units) using the National Hydrography Dataset (Boykin et al. 2006; see http://nhd.usgs.gov/).

Throughout Chapter 5 Boykin et al. point out that GAP analysis for vertebrate species is a process of intersecting habitat models with a data set of land stewardship that identifies levels of long-term conservation management. They provide spatial habitat models for 817 vertebrate species for the region comprising Arizona, New Mexico, Colorado, Utah, and Nevada, finding that total species richness was highest in areas of the Colorado and San Juan River drainages. They also demonstrate that patterns of richness vary among different vertebrate groups and subgroups, for example between the herpetofauna, bats, and large mammals. The information in this chapter was collected from the entire biogeographic province of the Colorado Plateau, from along the Mogollon Rim in the south to the White Mountains in eastern Arizona, up to the Green River in northern Utah, and west to the Mojave Desert, so land managers over

the entire biogeographic region will find this information useful.

The authors then provide an in-depth example of how their habitat model might be used, taking 19 amphibians with predicted habitat on the Colorado Plateau (51% of all amphibians modeled in the region), 341 birds (78%), 143 mammals (67%), and 78 reptiles (60%). They provide full results for the vertebrate models, including references, habitat data, modeling process, and textual and spatial models. The authors also calculated total species richness from the SWRe GAP data for the Colorado Plateau, and found an average of 354–390 animal species per drainage sub-basin. Species richness was higher in the eastern portion of the Colorado Plateau, with animals associated with the San Juan Mountains and the San Juan River, and on the western side of the plateau along the Colorado and Virgin Rivers. Compared to the entire SWReGAP region, Boykin et al. found that species richness on the Colorado Plateau was intermediate, with higher richness than more northern areas but lower richness than southern Arizona, much of New Mexico, and the Colorado Rocky Mountain Front Range. The information contained in Chapter 5 provides baseline information for conservation of animals over the entire Colorado Plateau, particularly when combined with other current inventory efforts. This chapter provides another useful tool for managers to better assess large-scale land-use issues over the Colorado Plateau, and this sound scientific tool can be used to enhance our understanding of vertebrate distributions on the plateau, and within the context of those species' habitats throughout southwestern North America.

ADDRESSING WILDLIFE ISSUES

This section of the book brings a focus to wildlife issues, and to addressing management concerns within this group of organisms over the Colorado Plateau. Chapter 6 serves as an introduction to wildlife issues, providing a historical account of the pronghorn antelope in Arizona, with a focus on Anderson Mesa. No area in Arizona is more frequently associated with pronghorn than the Anderson Mesa Game Management Unit. More than 25 percent of all of the pronghorn in the "Millennium" edition of the Arizona Wildlife Trophy Book came from this unit or from areas restocked with animals from Anderson Mesa (Lewis 2000). Three of the top five pronghorn trophies in the Boone and Crockett Club's North American Record Book are from Coconino County, where Anderson Mesa is located (Byers and Bettas 1999). Anderson Mesa was the site of Arizona's first legal pronghorn hunt and has been a focal point for pronghorn studies since the early 1930s. Pronghorn studies on Anderson Mesa have ranged from developing survey and capture methodologies (Wilkins and Welles 1944; Edwards 1947; Wallmo 1951), to determining seasonal food habits (Gay 1984), to evaluating reproductive performance (Erling 1956a, 1956b) to evaluating the effects of coyote control and other factors on fawn recruitment (Arrington 1947; Arrington and Edwards 1951; Neff and Woolsey 1980; Neff et al. 1985).

Brown provides a detailed history of the increases and declines in pronghorn recruitment rates and of the population sizes on Anderson Mesa. He documents that in the 1970s, declining pronghorn numbers resulted in an intensive study to determine if aerial gunning of coyotes could improve pronghorn numbers on Anderson Mesa. Although aerial gunning was expensive and politically unpopular, Brown concluded that these studies indicated pronghorn fawn recruitment could be improved by applying such control practices, as did Neff and Woolsey (1980) and Neff et al. (1985). When pronghorn recruitment and population numbers again declined in the 1990s, however, coyote reduction efforts were no longer deemed an effective solution, and since that time pronghorn recruitment on Anderson Mesa has been below herd maintenance levels (Yoakum 2003). Since the early 1900s pronghorn populations on Anderson Mesa have declined several times and then rebounded, demonstrating that the species is highly adaptable. But whether pronghorn on

Anderson Mesa can again attain their former numbers is problematic, and Brown argues that these animals are now subsisting on a declining forage base due to excessive elk and livestock use. He concludes this chapter by saying that sportsmen, ranchers, and the general public will have to press management agencies to reduce ungulate pressures and improve forage quality if mean annual pronghorn recruitment rates are to again exceed maintenance levels.

In Chapter 7, pronghorn antelope home range and the effects of Interstate 40 and the Burlington-Northern and Santa Fe railroad is examined at Petrified Forest National Park. The impact of transportation corridors on pronghorn in northern Arizona was first identified in the Colorado Plateau book series (Ockenfels at al. 1997), and later further detailed by van Riper and Ockenfels (1998) and then Bright and van Riper (2000). Hart et al. build upon these earlier studies by establishing an experimental study that examines the potential, non-lethal effects of transportation corridors on the basic ecology of pronghorn. The authors looked specifically at the one pronghorn herd that Ockenfels at al. (1997) documented as isolated, under comparatively unique conditions where Interstate 40 and the BNSF railroad constituted near impenetrable barriers. Hart et al. designed a manipulative study where fences were modified in an attempt to see if pronghorn would expand their home range, with the hope that by removing fences the confined animals would move across the railroad tracks and mix genetically with other pronghorn in the park.

After 2 years of manipulative studies, the authors' efforts were not successful in changing the movement patterns of the targeted pronghorn herd. Even after fence modification, they found consistent pronghorn avoidance of the I-40 freeway and the railroad, diminishing the odds that exploratory behavior would result in chance crossings. The isolated pronghorn at the park appear likely to remain so for the foreseeable future given the frequency of the train traffic and its inherent disturbance, as well as the other potential physical and psychological deterrents associated with the right-of-way. The authors state that it may ultimately be necessary to use overpasses or underpasses to enable pronghorn to negotiate these two transportation corridor barriers. However, they still believe that efforts to modify the right-of-way to enhance the potential for pronghorn crossings, such as those employed in this study, may have merit if the scope of the effort can be expanded both spatially and temporally. Given the high costs associated with creating structures to span or tunnel beneath the railroad, the authors recommend further investigation of the potential to enhance direct crossings of the right-of-way before more complicated and costly measures are pursued.

In Chapter 8 Wakeling and Riddering examine bighorn sheep being released in habitats based on a priority ranking system, and the possibility of differential mountain lion predation on those sheep. The authors agree that it seemed logical to assume that releases in lower-quality habitats would have lower survival and higher cause-specific mortality, but their analyses failed to support that assumption. They argue that a possible reason for the lack of a relationship to survival and the presently utilized priority ranking system is that habitat quality must fall below a critical threshold before bighorn survival is directly affected. The authors also point out that all past bighorn sheep releases have occurred primarily within the range of suitable habitats that are all above this critical threshold. Cunningham's (1989) original speculation that habitat must score higher than 50 to be suitable must be incorrect, because they found that bighorn sheep were capable of sustaining themselves in habitats that score as low as 40.

An alternate explanation that the authors explore in this chapter is that habitat quality scores that the Arizona Game and Fish Department presently use to identify bighorn sheep translocation sites are not good predictors of true bighorn habitat suitability. This latter rationale, however, seems unlikely because several studies have tested the suitability ranking method and found that these techniques are fairly reliable in selecting suitable bighorn habitats (e.g., Wakeling and Miller 1990). Most of the habitats where bighorn sheep releases have occurred scored over 40 points using the Cunningham-Brown criteria, and the highest-quality habitat in this evaluation received a score of 55. Moreover, habitats that received a numerical score in excess of 34 were not correlated with survival or mortality. Further, Wakeling and Riddering question the use of translocated animals as an effective surrogate for survival of resident bighorn sheep populations. The chapter also provides an analysis of data that examines the question "Do increases in mountain lion predation cause declines in bighorn sheep numbers?" Based on measured survival rates for translocated bighorn sheep populations, the authors point out that their analysis on survival and habitat quality did not support the hypothesis that translocations into lower-ranking habitats influenced mountain lion predation on, or survival of, bighorn sheep.

The next two chapters cover management aspects of avian resources on the Colorado Plateau. Chapter 9 by John Spence and Chapter 10 by Hurteau and her collaborators deal with birds in Grand Canyon National Park and on U.S. Forest Service lands in northern Arizona, respectively. Spence, who works as a resource manager in Glen Canyon National Recreation Area, has spent many years monitoring bird activities along the Colorado River. In Chapter 9 he analyzes 5 years of breeding bird survey data from the Colorado River between Glen Canyon Dam and upper Lake Mead. The principal emphasis of that program was to develop a baseline data set on relative abundance of riparian species, in order to develop a standardized methodology to monitor birds in riparian vegetation, and to examine aspects of statistical power in those data. Spence uses data from selected species to illustrate the relationships between relative abundance measures, abundance variability over time, and statistical power. He provides a quantitative model of avian community structure along the Colorado River in two national parks, and a power analysis related to the ability to accurately count birds, given differing avian guild assemblages and existing habitat structures associated with the Colorado River corridor.

Avian communities along the Colorado River have changed substantially since completion of Glen Canyon Dam in 1963, as predam vegetation along the river consisted of a thin riparian strip controlled primarily by spring flooding (Carothers and Brown 1991). Extensive stands of riparian habitat have become established on silt terraces where the Colorado River drains into Lake Mead. These recent habitat modifications have caused changes in the avian community (Brown et al. 1987). The monitoring program that Spence details in this chapter was established to provide data necessary to adaptively manage dam operations in order to minimize impacts to selected resources (National Resource Council 1999). He points out that various monitoring programs had been established as part of the environmental impact studies since 1982, under management of the Glen Canyon Environmental Studies Program (National Resource Council 1987, 1996), and a number of these studies have appeared as chapters in this Colorado Plateau book series (e.g., Felley and Sogge 1997). The avifauna, principally riparian breeding birds, bald eagles (*Haliaeetus leucocephalus*), and the endangered southwestern willow flycatcher (*Empidonax traillii extimus*), have been an integral part of past and ongoing monitoring studies along the river corridor (van Riper and Sogge 2004).

Spence points out that many birds are considered to be good indicators of ecosystem change because of their quick response. Such changes could be caused by climatic variation, invasion of the ecosystem by a new exotic species, recreational-based disturbances, changes in prey-base, differing management practices, or some combination of these factors. Due to the strong tendency of passerine birds to exhibit pronounced habitat selection (Hilden 1965; Cody 1985), Spence suggests that birds can be a useful group of organisms for monitoring habitat

effects in a dynamic system such as the Colorado River. Two of the major forcing variables presently controlling the Colorado River riparian system are quantity and timing of dam releases, so it is likely that most breeding birds are responding to changes in vegetation rather than to fluctuating river flows. By monitoring avian populations, changes in other components of the riparian ecosystem may be detected, and management practices can be developed to address any potential problems.

The principal goal of the study that Spence details in Chapter 9 was to determine whether a long-term monitoring program with adequate statistical power could be developed to detect trends in the riparian breeding bird community along the Colorado River. Power analysis is a necessary and important tool in the establishment of any monitoring program and it is particularly critical in the case of endangered species monitoring, as the failure to detect a decline may have disastrous consequences (Taylor and Gerrodette 1993). Most natural wildlife populations vary from year to year in abundance and this variation can result from numerous complex and interacting factors. In this study, Spence found that the power to detect change in less than 10 years only existed for a few species, such as Lucy's warbler and Bewick's wren, and only for very large effect sizes. Hence, long-term commitment of substantial financial and human resources would be needed to detect statistically defensible trends in the riparian bird community along the Colorado River. Such long-term commitments of time and resources are still rare in bird monitoring programs.

Spence also points out that it is important to understand that bird abundance within the study area is affected by numerous other variables outside the Colorado River corridor. The most important among these are non-breeding habitat changes and climate variability, both of which strongly influence bird survivorship. Glen Canyon Dam operations affect birds primarily through impacts on breeding habitat, but under normal operations these impacts are likely to be fairly minor compared with climate and habitat changes outside the Colorado River corridor. The major impacts of dam operations are the planned or unplanned floods, including those in 1983 and 1996. These floods can potentially scour out much of the riparian vegetation along the river corridor. Past flooding, particularly that in 1983, may explain many of the present differences found in the breeding bird communities between the mid-1980s and Spence's study, as the extent of riparian vegetation was much reduced after the 1983 event (Spence 2004; Holmes et al. 2005).

In summary, Spence demonstrates that in more temporally variable avian species it is often difficult to detect subtle long-term trends because of the natural variability in bird populations. His power analysis provides a measure of how well the monitoring program could detect a trend through variability in the monitoring data. In the absence of an estimate of the power of a monitoring program, resource managers and scientists cannot always know if change in a population (or species of interest) is statistically significant. Furthermore, without adequate power, managers may not be able to detect a significant change in a rare species that may be of management importance. This study used the approach of "prospective" power analysis (cf. Steidl et al. 1997), in which preliminary baseline data on population numbers and variability are gathered over a period of time and then, in turn, used to design an effective long-term monitoring program, examining factors like sample size considerations, sampling protocols, and duration of data collection. Although this model of predicted bird occurrence was developed along the Colorado River corridor, the technique would be widely applicable to other areas over the Colorado Plateau.

In the second bird chapter, Hurteau et al. document that in the past century forest management practices have significantly altered the function and structure of ecosystems in the Southwest that are dominated by ponderosa pine (*Pinus ponderosa*). There have been significant portions of earlier books in the Colorado Plateau series devoted to this subject (e.g., Garret et al. 1997; Garret and Soulen 1999; Bailey et al. 2001), all pointing out that fire suppression, grazing, and logging have resulted in a dense, closed-canopy forest with an increased susceptibility to stand-replacing wildfire. Mechanical thinning and prescribed fire, which are important tools in fuel reduction treatments, can mitigate the threat of stand-replacing wildfire, but the effects of these practices on wildlife communities are poorly understood. The authors explain that the Fire and Fire Surrogates (FFS) Program is a national study that seeks to quantify the effects of prescribed fire and mechanical thinning on numerous response variables, including wildlife. On three FFS Southwestern Plateau sites in northern Arizona, Hurteau et al. examined the short-term (3-year) avian community response to experimental thinning and prescribed fire treatments. For a suite of focal species selected from the total avian community, they evaluated changes in abundance and density resulting from different treatment types. Their results suggest that patterns in ranked avian abundance among treatments were significantly correlated, and that overall community structure was generally not affected by fuel reduction treatments, regardless of treatment type.

Among a suite of focal species, the authors found that response to specific treatment types was more variable. For example, western bluebird (*Sialia mexicana*) and dark-eyed junco (*Junco hyemalis*) densities increased in response to thinning and fire, alone and in combination. Mountain chickadee (*Poecile gambeli*) density decreased dramatically in all treatment types in the post-treatment period, while the pygmy nuthatch (*Sitta pygmaea*) exhibited a decrease in density during this same period. They also found that the yellow-rumped warbler (*Dendroica coronata*) exhibited a negative response to the thin-only treatment but a positive response to the burn-only treatment. Considering the wide spectrum of avian community attributes that the authors measured, their results provide essential baseline information for project-level planning. Moreover, their research provides evidence that perhaps the avian community may not be responding to fuel reduction treatments as previously believed. The authors' overarching conclusion was a recommendation that forest managers implement a mosaic of treatment types so as to best preserve avian habitats in ponderosa pine forests on the Colorado Plateau.

The next two chapters focus on other groups of vertebrates on the Colorado Plateau—herpetofauna and native fish. In Chapter 11 Trevor Persons and his colleagues provide a summary of the present status and historic changes of all reptiles and amphibians (herpetofauna) in national parks over the southern Colorado Plateau. This chapter provides a much-needed summary of numerous years of a biological inventory that has been undertaken at National Park Service sites. Although many national park areas on the Colorado Plateau were created primarily to protect remarkable deposits of cultural resources, the authors point out that these parks also preserve a diverse assemblage of vertebrate species. The inventory work outlined in this chapter began in 2000 when the National Park Service (NPS) initiated a nationwide program to inventory vertebrates and vascular plants within the parks. As part of this new inventory effort, 265 National Park units (e.g. parks, monuments, recreation areas, historic sites) were identified as having significant natural resources, and these were divided into 32 groups or "networks" based on geographical proximity and similar habitat types. The many NPS areas on the Colorado Plateau of Utah, northern Arizona, northwestern New Mexico, and western Colorado were divided into Northern and Southern Colorado Plateau networks. In this chapter Persons et al. summarize the results of their amphibian and reptile inventories at 19 parks within the Southern Colorado Plateau Inventory and Monitoring Network. They synthesize distribution and habitat information for all amphibian and reptile species across that network; the primary goal of their complete species inventory is to document at least 90 percent of the species present at each park. To evaluate their progress toward that goal,

they provide an estimated level of inventory completeness for each park. This chapter also provides an exhaustive checklist of all possible herpetofauna found over the southern Colorado Plateau, with an estimated level of inventory completeness for each species. The authors close the chapter with a list of considerations for future inventory work.

Chapter 12 details a unique application of inventory techniques, documenting fish species assemblages along drainages that have been disturbed by fire. Jonathan Long examines the potential impact that a forest fire in KP and Grant Creeks, within the White Mountain Apache Reservation, had on native Apache trout. These streams extend into mixed conifer forests where mixed-severity wildfires such as the KP and Steeple fires are typical. The two drainages are similar in geology, topography, and vegetation, and are similar to the majority of streams planned for recovery of Apache trout. Fish extirpations have been reported from streams in drier, lower-elevation forest types where wildfires have been more severe; however, long-term fire history studies suggest that high-severity wildfires do occur in high-elevation forest types during extended dry periods.

Long, with his colleagues at the U.S. Forest Service Rocky Mountain Research Station, sampled fish populations and habitat conditions at seven 50-m long sampling sites in KP and Grant Creeks in June 2004 after the KP fire was contained. Fish populations were resampled at six of the sites one year later. They also attempted to relocate sites that had been previously sampled for fish in September of 1995 by the Arizona Game and Fish Department, as part of their General Aquatic Wildlife Surveys program. Fish populations were sampled using backpack-mounted electro-shocking gear. Each reach was blocked off with nets to prevent fish from escaping during sampling, and each was sampled three times using the depletion method. They found that trout populations persisted following the mixed-severity wildfires in KP and Grant Creeks. These findings indicate that evacuation of populations, which is now standard procedure, may not be necessary at the higher elevations when a watershed is not severely burned. While many factors can influence the likelihood of fish persistence, Long suggests that burn severity can be determined through the use of satellite imagery, and until more confirmatory studies are conducted, the satellite imagery metric may help managers to quickly evaluate whether to evacuate Apache trout populations that are threatened by wildfire.

The final chapter that addresses wildlife issues deals with managing invertebrates within caves on the Colorado Plateau. Wynne et al. point out that cave environments are among the most fragile and understudied ecosystems on earth. From what scant information they can find, only limited research seems to have been conducted on caves in Grand Canyon National Park and over the southern Colorado Plateau. The authors reviewed all available literature and park cave trip reports, representing nine studies of 15 caves at Grand Canyon National Park. Chapter 13 lists approximately 37 cave-dwelling invertebrates that are known to occur in Grand Canyon caves (3 troglobites, 6 trogloxenes, 14 troglophiles, 1 stygobite, 10 unknown cavernicoles, and 3 "special case" species). Currently, only four cave-adapted taxa are known to occur in the Grand Canyon. The authors also provide an annotated checklist of all known invertebrates from caves over the Colorado Plateau. Because this information represents data on only about 5 percent of the known caves in Grand Canyon National Park, the authors suggest that more endemic cave-adapted invertebrates are expected to be discovered in the future.

ADDRESSING VEGETATION ISSUES

Vegetation studies are introduced into this book with Chapter 14, where Thomas et al. provide an analysis of plant community composition and structure at Petrified Forest National Park (PEFO). This is the first complete survey of all vegetation types to be published for PEFO. The vegetation at PEFO is complex and varied, containing many dominant plants of low stature, with a rich

mosaic of grasslands, steppe, and shrubland types that have different species dominating at different locations. In their description of PEFO vegetation associations, alliances, and park specials, the authors emphasize how the topography and soil types within the park are correlated with the vegetation distribution patterns that they documented. Other factors that influenced the expression of the park vegetation were drought and invasive plant species distributions. Precipitation in this area of the Colorado Plateau is biseasonal, with winter precipitation and a summer monsoon period. The authors found that grasses in the park responded particularly strongly to the seasonality of precipitation. Some of the PEFO grasses showed the most growth in the spring warmup (these were cool season grasses) and others showed the most growth in response to the summer monsoons (warm season grasses). Climatic events that reduce precipitation during the winter, summer, or both seasons inhibit plant growth and reproduction, and may ultimately kill plants. For example, the authors point out that the drought in the U.S. Southwest in the early 2000s greatly reduced vegetation cover at the park, as plants responded with reduced vegetative growth and dieback. Climate change, especially warmer temperatures and decreases in precipitation and/or changes in the monsoon pattern, can be expected to dramatically change the characteristics of plant distribution in the park.

Thomas et al. also provide a thorough inventory of invasive non-native plants within Headquarters Mesa, the Puerco River corridor, and portions of the southern park. The authors found more than 25 different invasive (non-native) plants, with the most prolific being Russian thistle, which occurred in more than 75 percent of the sampled area. Many of the earlier book chapters in the Colorado Plateau series (e.g., Floyd et al. 2001; Falzarano et al. 2005; Nabhan et al. 2005) have demonstrated that invasive plants are increasingly threatening ecosystems over the Colorado Plateau. The invasive plants not only interact with native plants and animals, but can also increase the frequency and magnitude of fires (Floyd-Hanna et al. 1999). In the event of a prolonged drought, invasive species can magnify the effects of reduced water on native species by sprouting earlier, thus removing soil moisture that would have been available for native plants.

Chapter 15 moves west, from the short-grass prairie at PEFO to the higher elevation ponderosa pine forests of northern Arizona. Speer and Bailey provide forest managers with information on under-story vegetation responses to tree harvesting and prescribed fire in overly dense ponderosa pine forest stands. Throughout the past century, ponderosa pine forests over the Colorado Plateau have become increasingly dense; this change was brought about by Euro-American settler land-use practices beginning in the late 1800s (Covington and Moore 1994; Covington et al. 1997; Moore et al. 1999). The current high density of ponderosa pine forests allows little sunlight for understory vegetation development (Naumburg and DeWald 1999). Continuous heavy grazing by domestic livestock and by recently expanding elk populations has further depleted the rich understory of grasses and forbs that once out-competed pine seedlings. The pre-historic understory once enabled frequent surface fires that further prohibited extensive pine regeneration (Korb and Springer 2003). In addition to supporting a natural fire regime (Laughlin et al. 2004), the earlier understory enhanced net primary productivity, nutrient cycling, and forage for wildlife communities, in addition to promoting a number of ecosystem functions such as hydrology and soil stabilization (Korb and Springer 2003). Of particular interest to the authors, in regards to harvesting, burning, and general soil disturbance, was the introduction and spread of introduced (alien) grass species (Crawford et al. 2001; Sieg et al. 2003; Korb et al. 2004). These alien species are of importance to ecosystem function and health because they alter successional pathways by out-competing native pioneer species, thereby altering the ecosystem functions normally performed by native species (Fornwalt et al. 2003). Speer and Bailey, as

also documented in the chapter by Hurteau et al., conducted their research on one of 13 sites in the national Fire and Fire Surrogate (FFS) Program. Their goal was to learn more about how perennial and annual understory plants respond to increasing intensities of management (burn only, harvest only, harvest and burn), with a focus on species richness and ground cover of native and exotic vegetation in ponderosa pine forests.

Speer and Bailey examined harvesting and burning, alone and in combination, focusing on any increases or decreases in native and alien species richness and abundance. They found that as management intensities increased (burn only, harvest only, harvest & burn), understory responses increased. In areas that were treated mechanically, understory showed significant but small increases in native species ground cover (2–3%) and native richness (~5 species, a 20% increase). Tree harvesting also resulted in smaller increases in alien species richness and ground cover that were significantly greater than that of their controls. Both native and alien species richness and cover responded most strongly to the combination of harvesting and burning treatments, yielding levels significantly higher than in the controls, where alien cover and native richness and cover stayed relatively consistent during the 4 years of their study. Burning alone stimulated insignificant increases in native species richness and cover, given only minor changes in overstory condition and relatively little site disturbance.

On their control plots, Speer and Bailey observed 9 of the 15 total alien species found in this study. Occurrences of field bindweed (*Convolvulus arvensis*), lambsquaters (*Chenopodium album*), prickly lettuce (*Lactuca serriola*), and the common dandelion (*Taraxacum officinale*) all declined across sites during the 4 study years. They felt that this decrease in alien species richness and frequency was due to species-specific natural germination cycles of annual and biennial plants relative to their limited 2-year sampling period. Alien species richness decreased (0.15) with time in their control areas. Several recent studies (see Chapter 16) clearly document that management disturbance provides a vector for alien species to colonize (Crawford et al. 2001; Sieg et al. 2003; Korb et al. 2004), but there is little documentation of a decrease in alien species richness when left undisturbed.

In response to harvesting and burning treatments in their study design, Speer and Bailey found that more intensive management regimes yield higher understory vegetation cover and richness for both native and alien species. This general trend has been documented repeatedly (e.g., Crawford et al. 2001; Laughlin et al. 2004), particularly in ponderosa pine forests. The authors suggest that if promoting a more robust understory is a desired management objective for enhancing grazing and foraging, promoting soil stabilization and nutrient cycling, influencing fire behavior, ecological restoration, or simply for aesthetics, then these goals can be achieved at different levels by changing intensities of management activities. However, the risk of invasive alien species colonizing after treatment should be weighed carefully, as any of the management activities presented here provide a vector for colonization in this Colorado Plateau landscape. They conclude this chapter by suggesting that managers continue to monitor these permanent plots to adequately document whether the trends that the authors found will continue over time, or will differ as time from disturbance passes.

The response of Colorado Plateau vegetation communities to fire is further explored in Chapter 16, where the 2000 Outlet fire in Grand Canyon National Park is examined. In this chapter, Julie Crawford brings to light the need to investigate high-severity fire and the effects of fire-fighting activity on vegetation and understory recovery in mixed conifer forests. The Outlet fire burned more than 13,000 acres (5261 ha) of mostly mixed conifer forest on the North Rim of Grand Canyon National Park and Kaibab National Forest. This chapter documents a study that examined post-fire vegetation change in relation to three types of disturbance: high-severity burned areas, fire-fighting staging areas, and fire-fighting

handlines. Crawford employed an indicator species analysis, nonmetric multidimensional scaling, and ANOSIM to determine indicator species and trends among disturbance types and across years. She found statistically significant differences in floristic composition, cover, and diversity over time and among disturbance types. Burned sites had the highest vegetation cover in all years through 2004. Diversity in the burned areas decreased following dieback of the initial invasion and by 2004 had become largely floristically homogeneous with high cover of two native rhizomatous species. Few exotic species were present in high-severity burn transects, although by 2004, cheatgrass (*Bromus tectorum* L.) had become an indicator species. Staging areas used in fire-fighting contained the greatest number of exotic species in all years of study, but this may be related to continued use of these roadside areas by park visitors. Areas of handlines showed no statistically significant differences between 2000 and 2004, indicating that no vegetation recovery had occurred.

Several studies have found that the damage associated with fire control activities is a legitimate concern that should be examined carefully (see also Chapter 15). Crawford found that following handline construction, the current methods of site rehabilitation do not improve vegetation recovery. In addition, she suggests that managers should (1) require mitigation for fire fighters and their equipment to eliminate the spread of exotic plants, (2) continue and expand this study to investigate vegetation responses at additional sites of fire and fire-fighting activity, (3) conduct experiments on the effectiveness and efficiency of restorative seeding using locally collected native species, and (4) encourage the local collection and storage of native seed for post-disturbance management. The author also states that continued monitoring is essential for understanding long-term changes in vegetation due to high-intensity fires and fire-suppression crews operating in high-elevation forests on the Colorado Plateau.

Chapter 17 by Crall et al. evaluated relationships between native and non-native plant species richness and cover within and across 15 vegetation types in the Grand Staircase–Escalante National Monument in Utah. This chapter extends the vegetation portion of this book into the Colorado Plateau region of southern Utah, focusing on a Bureau of Land Management area. The authors discuss how various theories have been proposed to explain patterns of species richness using measures of productivity, with the most widely accepted theory suggesting that this relationship results in a hump-shaped/unimodal curve, with species richness increasing and then decreasing as productivity increases (e.g., Grime 1973a, 1979; Huston 1979, 1994; Tilman 1982; Rosenzweig 1992; Grace 1999). However, some authors have suggested that surveys of species richness conducted over limited productivity ranges are less likely to detect a hump-shaped relationship than are studies conducted over a broad productivity range (Begon et al. 1990; Rosenzweig 1992, 1995; Huston 1994; Grace 1999). Therefore, data are clearly lacking to establish only one relationship between native species richness and productivity, and the authors examine this perceived need throughout the chapter.

The authors develop four hypotheses: (1) That the common unimodal relationship between species richness and total cover would be found for native and non-native species when looking across all vegetation types, and that this relationship should only show a monotonic increase for xeric and a monotonic decrease for mesic vegetation types; (2) that native and non-native species richness and cover would be greatest in the mesic vegetation types (when compared to the xeric vegetation) because of greater resource availability; (3) that non-native species richness and cover would be positively correlated with native species richness and cover within and across vegetation types at the plot scale, but that the reverse would be found at smaller spatial scales due to competitive interactions; and (4) that disturbance would increase non-native species richness and cover because disturbance is known to facilitate the establishment and potential dominance of non-native plant species. The

objectives of their study were to evaluate the relationships between native and non-native plant species richness and cover in the Grand Staircase–Escalante National Monument, and to provide some insight into how these relationships might be affected by productivity and disturbance across vegetation types at different spatial scales. In addition, to help guide and direct future BLM management efforts they determined where non-native species have successfully established and gained dominance in the monument.

The authors discuss the various mechanisms that can make species-rich vegetation types (e.g., riparian vegetation communities) more easily invaded than species-poor vegetation types. Species richness tends to be low in stressful environments as a result of few species being able to survive under harsh conditions (Grime 1973a, 1973b). If species-poor vegetation types are a result of limited resources, the authors argue that non-natives are also unlikely to establish and succeed in those areas. Stohlgren et al. (1998, 1999) also suggest that non-natives would more likely be found in areas of greater species richness and resource availability. Natural and anthropogenic disturbances are also correlated with the vulnerability of habitats to invasion (Fox and Fox 1986; Hobbs 1989; Hobbs and Huenneke 1992). As niche space in a vegetation type becomes available through disturbance, the establishment of a non-native species may be possible because of open space and increased nutrient availability (Robinson et al. 1995). However, establishment of non-native species into these areas may still be limited by dispersal or seed availability (Rosentreter 1994).

To add to this complexity, it is not known which factors make a vegetation type vulnerable to plant invasion. But a long-held theory of invasion asserts that disturbed, species-poor communities are more susceptible to invasion by non-natives due to a lack of biotic resistance from such factors as competition or predation (Elton 1958; Simberloff 1986). The authors point out that all of these theories are confounded by studies being conducted at multiple spatial and temporal scales (Levine and D'Antonio 1999; Stohlgren 2002). Several multi-scale observational studies have shown both a negative and a positive relationship between native and non-native species richness at small spatial scales (Brown and Peet 2003; Fridley et al. 2004), whereas in most cases a positive relationship was seen at larger spatial scales (Stohlgren et al. 1998, 1999). This may be a consequence of differences in primary controls on diversity. At smaller spatial scales (plant neighborhoods), native and non-native species richness may be negatively correlated because of competitive exclusion, while at larger spatial scales the effects of competition might be reduced or reversed because most competitors have similar habitat requirements (Levine and D'Antonio 1999). Nevertheless, differences at multiple scales have made it difficult for researchers to develop broad generalizations related to non-native species invasions.

In addition, the authors point out that research findings are dependent on the vegetation type's stage of invasion at a particular point in time (i.e., on a temporal scale). Positive relationships between native and non-native species richness may occur only in the early stages of invasion, while later in the invasion process certain non-native invaders might have the capability to drastically alter an ecosystem (e.g., Vitousek et al. 1987; D'Antonio 2000). In such cases, native species richness is likely to be reduced as a result of the non-native species' ability to gain dominance under these new conditions. Thus, it remains unclear as to what role productivity and disturbance may play in determining native and non-native species richness.

Crall et al. did find that, at all scales, regressions across all vegetation types showed an increase in species richness as total cover increased. They also demonstrated a monotonic increase in total cover for the xeric vegetation types, at both large and smaller scales on this BLM monument. Thus, they suggest that the hump-shaped model (see Grime 1973a, 1979; Huston 1979, 1994; Tilman 1982; Rosenzweig 1992; Grace 1999) may not be applicable to less-productive

landscapes such as occur in the monument. This may be an indication that productivity should be used in multivariate analyses, along with the other factors, in order to better explain patterns of species richness over the Colorado Plateau.

Chapter 18, which concludes the group of chapters that address vegetation issues, examines techniques of ecological restoration on forest roads. Across many landscapes, and especially on the Colorado Plateau, forest roads are a common component of the environment. Nearly half of all ponderosa pine forest lies within 0.25 miles of a road. The impacts of roads and trails in forests of the Colorado Plateau are of particular concern to people who deal with ecological restoration (Covington 2003). Forest-road removal is increasingly being used as a method of restoring pre-disturbance hydrology, ecosystem processes, and habitat continuity. The physical aspects of road rehabilitation are well studied (e.g. Luce 1997), but little research has been done to assess the effectiveness of these procedures in restoring critical ecosystem attributes and processes. When forest roads are constructed, the organic soil layers are removed, leaving a surface that is primarily mineral soil, which lacks symbiotic and other fungi that assist with essential processes in the soil food web, such as nutrient cycling and plant community support (Harvey et al. 1979).

The purpose of the study by Joseph Trudeau was to examine one possible method for increasing the effectiveness of road restoration through the utilization of fungal inoculum that would assist plant and microbial communities to achieve pre-disturbance conditions. He investigated the effects on plant establishment using ground waste-wood (mulch) and fungal inoculum, and then evaluated the effectiveness of inoculated saprophytic fungi in colonizing ponderosa pine mulch. This experiment was conducted on areas that had formerly been forest roads, with three experimental roads selected at Northern Arizona University's Centennial Forest near Flagstaff. Each road was divided into five experimental blocks containing four identical treatments. Treatments were (1) control, (2) mycorrhizal inoculum, (3) mulch, saprophytic fungal inoculum, and mycorrhizal inoculum, and (4) mulch only. All plots were seeded with the same mix of native plants. Trudeau collected data at 2 and 14 months after application of treatments, and found that mycorrhizal inoculum had no effect on grass seedling establishment, species richness, or abundance, while mulch was found to significantly suppress plant establishment. The author also found that mulched plots had lower species richness and abundance. However, he did discover that Gambel oak seedlings were frequent on mulched plots but not common on non-mulched plots. Saprophytic inoculum showed poor survivorship; after 14 months, only 34 percent of the inoculated sites were colonized, while most mulched plots were naturally colonized by resident soil fungi. Trudeau concludes this chapter by suggesting that inoculation is less effective than natural colonization, and that until sources of inoculum that are adapted to local conditions are developed, the methods that he examined are less effective than natural revegetation processes.

GAINING INSIGHTS FROM THE PAST

The final two chapters of this book focus on research that provides managers with insights from the past. In Chapter 19, Cole et al. provide readers with a compelling argument that climate change will have a dramatic effect on plant species distributions over the Colorado Plateau. They describe new techniques for paleo-botany modeling, using the widespread Southwest tree species Colorado pinyon pine (*Pinus edulis*). Their model requires knowledge of the plant's current distribution, climate tolerances, and migratory response to change, as well as the geography of future climates, and it incorporates all of the climate-modulated physical and biological variables occurring near the continental range of the species during the twentieth century. The authors developed models of future potential geographic ranges by applying this climatic envelope to future climate predictions from

general circulation model (GCM) results. Finally, to distinguish between this future potential climate range and the species' likely future range, they apply a spatial model of the species' observed migration rate in response to past and ongoing climate warming. Through the compilation of spatially detailed data for the twentieth century climate model, the GCM modeling, and current pinyon distribution data, their results are projected to a landscape grid scale of ~1 km^2.

The modeling results of Cole et al. for pinyon pine suggest that over the next 100 years, the range of pinyon pine will continue to profoundly contract throughout Arizona, Utah, and southern New Mexico, but will expand in Colorado and northernmost New Mexico. The results from this one GCM scenario imply a large magnitude of change for this species, and delineate useful areas in which managers can focus future monitoring efforts. This detailed projection allows their results to be easily applied by individual land managers as well as providing specific predictions of future distributions that would assist land-management agencies with future monitoring efforts.

In the final chapter of the book, Draut and Rubin examine the role of wind-blown (aeolian) sediment on the preservation of archaeological sites along the Colorado River corridor in Grand Canyon National Park. They document that aeolian deposits in the river corridor fall broadly into two categories: (1) modern fluvial sourced (MFS) deposits, which form as the wind transports sand inland from < 1270 m^3/s (45,000 ft^3/s)-stage sandbars, creating aeolian dunes directly downwind, and (2) relict fluvial sourced (RFS) deposits, which formed as wind eroded and redistributed sediment of extensive pre-dam fluvial terraces. Archaeological material is known to occur in aeolian deposits of both types. The authors then describe how Glen Canyon Dam operations have caused a reduction in sandbar size, thereby reducing the supply of sand available for transport from upwind sources to provide cover to some archaeological sites. They also show how past and present sedimentary processes can be evaluated, along with modern wind and sand-transport rates, to assess the sensitivity to dam operations of specific areas and associated cultural sites along the Colorado River corridor in the Grand Canyon.

The authors found that some archaeological sites in MFS dunes have been negatively affected by the loss of aeolian sand caused by decreased sand supply on upwind sandbars, a process attributable to dam operations. They suggest that these sites could benefit from aeolian redistribution of new sand deposited on fluvial sandbars by sediment-rich controlled floods. The November 2004 Colorado River high flow resulted in major deposition of new sand in many areas that are sediment sources for MFS aeolian deposits, and wind reworking of 2004 flood sand has also been observed to fill in small eroded gullies. The authors document that 3 months of high daily flow fluctuations in 2004 removed much of the new sand prior to the start of the first post-flood windy season in April 2005. Draut and Rubin conclude their chapter by suggesting that the restoration potential for cultural sites in aeolian deposits can be maximized by using dam operations (controlled floods and post-flood flows) that maximize the exposed sand area on fluvial sandbars from April through early June, when wind-borne sediment transport is greatest in the Grand Canyon.

SUMMARY

The 20 chapters of this book have brought together much of the current research on the Colorado Plateau, particularly that which is applicable to land managers. More and more we see people from diverse backgrounds coming together on the Colorado Plateau to achieve common conservation goals. The beginning portion of the book provides examples of collaborative processes that have worked; these chapters also provide recipes of the "ingredients" necessary to assure fruitful collaborations. If the public and private land stewards in Arizona, Utah, Colorado, and New Mexico—and in particular managers of our national parks, the U.S.

Forest Service, Fish and Wildlife Service, Bureau of Reclamation, tribal lands, and the many new BLM national monuments—utilize the ideas and concepts presented within this portion of the book, they will be better able to launch efforts toward enhanced management and stewardship of their lands. Along with the collaborative tools, these groups will also find useful some of the large-scale land-use tools that are presented in the second section of the book. GAP programs have now reached a level of development that makes them powerful tools for addressing large-scale questions and issues over the Colorado Plateau.

The chapters on assessing wildlife and vegetation issues, like many of the chapters in this series' previous books, provide species- and location-specific information that managers can use to better preserve their wildlife and vegetation resources. From looking at the history and movement patterns of pronghorn, and responses of that species to fenced transportation corridors, to relocation of bighorn sheep, wildlife managers have new information and tools that will better enable them to properly manage wildlife. Land managers who are concerned with the monitoring and preservation of birds will find current information on monitoring and the responses of avian communities to forest management; in particular, the power analysis provided by Spence in Chapter 9 should serve as an example that all managers should follow in the analyses of their monitoring information. For the first time, the Colorado Plateau manager is supplied with a complete inventory of all herpetofauna that they should expect to occur on their managed lands. There is also valuable information provided on the potential impacts of fire on native trout populations. Scientists and managers are also provided with insight into potential cave invertebrate resources over the Colorado Plateau.

As in previous books of this Colorado Plateau series, a number of chapters examine the impact of fire on vegetation communities. This is the first time that fire and restoration ecology are examined together in the same context, within the ponderosa pine ecosystem. Finally, there are sections in the book that provide the reader who is interested in natural and cultural resources, with a glimpse into the past and some predictions about the future state of the Colorado Plateau. It truly is our hope that the material in this volume will provide land managers with useful information and tools, and that this information can in some way act as a stimulus of future research support for cultural, natural, and physical resources over the Colorado Plateau.

LITERATURE CITED

Arrington, O. N. 1947. Predator control as a management factor in antelope reproduction. Project 22-R, August 20, 1947. Arizona Game and Fish Commission.

Arrington, O. N., and A. E. Edwards. 1951. Predator control as a factor in antelope management. Transactions North American Wildlife Resource Conference. 16: 179–193.

Bailey, J. D., M. R. Wagner, and J. J. Smith. 2001. Stand treatment impacts on forest health (STIFH): Structural responses associated with silvicultural treatment. In Proceedings of the Fifth Conference of Research on the Colorado Plateau, edited by C. van Riper III, K. A. Thomas, and M. A. Stuart, pp. 137–145. U.S. Geological Survey/Forest and Rangeland Ecosystem Science Center USGSFRESC/COPL/2001/21 Rep. Ser., Flagstaff, Arizona.

Begon, M., J. L. Harper, and C. R. Townsend. 1990. Ecology: Individuals, Populations and Communities. Blackwell Scientific Publications, Cambridge, Massachusetts.

Belnap, J., J. H. Kaltenecker, R. Rosentreter, J. Williams, S. Leonard, and D. Eldridge. 2001. Biological soil crusts: Ecology and management. In BLM Technical Reference 1730-2, edited by P. Peterson, p. 110. USDI Bureau of Land Management, USGS Forest and Rangeland Ecosystem Science Center, Denver.

Boykin, K. G., B. C. Thompson, R. A. Deitner, D. Schrupp, D. Bradford, L. O'Brien, C. Drost, S. Propeck-Gray, W. Rieth, K. Thomas, W. Kepner, J. Lowry, C. Cross, B. Jones, T. Hamer, C. Mettenbrink, K. J. Oakes, J. Prior-Magee, K. Schulz, J. J. Wynne, C. King, J. Puttere, S. Schrader, and Z. Schwenke. 2006. Predicted animal habitat distributions and species richness. In Southwest Regional Gap Analysis Final Report, edited by J. S. Prior-Magee. USGS Gap Analysis Program, Moscow, Idaho. Available at http: //fws-nmcfwru.nmsu.edu/swregap/ (see also http: //nhd.usgs.gov/).

Bright, J. L., and C. van Riper III. 2000. Pronghorn home ranges, habitat selection and distribution around water sources in northern Arizona. USGS, Forest and Rangeland Ecosystem Science Center, Colorado Plateau Field Station Technical Report USGSFRESC/COPL/2000/.

Brown, B. T., S. W. Carothers, and R. R. Johnson. 1987. Grand Canyon Birds. University of Arizona Press, Tucson.

Brown, R. L., and R. K. Peet. 2003. Diversity and invasibility of southern Appalachian plant communities. Ecology 84: 32–39.

Byers, C. R., and G. A. Bettas, editors. 1999. Records of North American big game, 11th ed. Boone and Crockett Club, Missoula, Montana.

Carothers, S. W., and B. T. Brown. 1991. The Colorado River through Grand Canyon. University of Arizona Press, Tucson.

Cody, M. L. 1985. An introduction to habitat selection in birds. In Habitat Selection in Birds, edited by M. L. Cody, pp. 3–56. Academic Press, San Diego, California.

Covington, W. W. 2003. The evolutionary and historical context. In Ecological Restoration of Ponderosa Pine Forests, edited by P. Friederici, pp. 26–47. Island Press, Washington, D.C.

Covington, W. W., and M. M. Moore. 1994. Southwestern ponderosa pine forest structure and resource conditions: Changes since Euro-American settlement. Journal of Forestry 92: 39–47.

Covington, W. W., P. Z. Fulé, M. M.Moore, S. C. Hart, T. E. Kolb, J. N. Mast, S. S. Sackett, and M. R. Wagner. 1997. Restoring ecosystem health in ponderosa pine forests of the Southwest. Journal of Forestry 95: 23–29.

Crawford, J. A., C. H. A. Wahren, S. Kyle, and W. H Moir. 2001. Responses of exotic plant species to fires in Pinus ponderosa forests in northern Arizona. Journal of Vegetation Science 12: 261–268.

Crist, P. J., and J. M. Scott. 1999. Identifying the gaps, locating the reserves: Some thoughts on getting gap analysis into conservation practice. Gap Analysis Bulletin 8: 14–16.

Crist, P. J., T. C. Edwards Jr., C. G. Homer, S. D. Bassett, and B. C. Thompson. 2000. Mapping and categorizing land stewardship. A Handbook for Gap Analysis, Version 2.1.0. Gap Analysis Program, Moscow, Idaho.

Cunningham, S. 1989. Evaluation of bighorn sheep habitat. In The Desert Bighorn Sheep in Arizona, edited by R. M. Lee, pp. 135–160. Arizona Game and Fish Department, Phoenix.

D'Antonio, C. M. 2000. Fire, plant invasions, and global changes. In Invasive Species in a Changing World, edited by H. A. Mooney and R. J. Hobbs, pp. 65–93. Island Press, Washington, D.C.

Edwards, A. C. 1947. Antelope airplane survey. Project 26-R-1, Job 1. Arizona Game and Fish Commission, Phoenix.

Edwards, T. C., C. Homer, and S. Bassett. 1994. Land management categorization: A user's guide. A Handbook for Gap Analysis, Version 1. Gap Analysis Program, Moscow, Idaho.

Elton, C. S. 1958. The Ecology of Invasions by Animals and Plants. Methuen, London.

Erling, H. G. 1956a. Report on a study of reproduction in antelope. Special Report, January 1956. Project W-53-R-5, WP2, J1.

Erling, H. G. 1956b. Report on a study of reproduction in antelope. Project W-53R-6,WP2, J1. Special Report, April 1956.

Falzarano, S., K. Thomas, and J. Lowry. 2005. Using decision tree modeling in gap analysis land cover mapping: Preliminary results from northeastern Arizona. In The Colorado Plateau II: Biophysical, Socioeconomic, and Cultural Resources, edited by C. van Riper III and D. J. Mattson, pp. 87–100. University of Arizona Press, Tucson.

Felley, D. L., and M. K. Sogge. 1997. Comparison of techniques for monitoring riparian birds in Grand Canyon National Park. In Proceedings of the Third Biennial Conference of Research on the Colorado Plateau, edited by C. van Riper III and E. Deshler, pp. 73–83. U.S. Department of the Interior National Park Service Transactions and Proceedings Series NPS/NRNAU/NRTP-97/12.

Fleming, B. 2006. Review of: The Colorado Plateau: Cultural, Biological, and Physical Research. New Mexico Historical Review 81(1): 115–117.

Floyd-Hanna M. L., A. DaVega, D. Hanna, and W. H. Romme. 1999. Fire vegetation monitorin and mitigation. In Proceedings of the Fourth Biennial Conference of Research on the Colorado Plateau, edited by C. van Riper III and M. A. Stuart, pp. 61–75. USGS Forest and Rangeland Ecosystem Science Center CPFS Rep. Ser., 99/16. Flagstaff, Arizona.

Floyd, M. L., D. D. Hanna, and G. Salamacha. 2001. Post-fire treatment of noxious weeds in Mesa Verde National Park, Colorado. In Proceedings of the Fifth Conference of Research on the Colorado Plateau, edited by C. van Riper III, K. A. Thomas, and M. A. Stuart, pp. 147–157. U.S. Geological Survey/ Forest and Rangeland Ecosystem Science Center USGSFRESC/COPL/2001/21 Rep. Ser., Flagstaff, Arizona.

Fornwalt, P. J., M. R. Kaufmann, L. S. Huckaby, J. M. Stoker, and T. J. Stohlgren. 2003. Non-native plant invasions in managed and protected ponderosa pine/Douglas-fir forests of the Colorado Front Range. Forest Ecology and Management 177: 515–527.

Fox, M. D., and B. J. Fox. 1986. The susceptibility of natural communities to invasion. In Ecology of Biological Invasions: An Australian Perspective, edited by R. H. Groves and J. J. Burdon, pp. 57–66. Australian Academy of Sciences, Canberra.

Fridley, J. D., R. L. Brown, and J. E. Bruno. 2004. Null models of exotic invasion and scale-dependent patterns of native and exotic species richness. Ecology 85: 3215–3222.

Garret L. D., and M. H. Soulen. 1999. Changes in character and structure of Apache/Sitgreaves forest ecology: 1850–1990. In Proceedings of the Fourth Biennial Conference of Research on the Colorado Plateau, edited by C. van Riper III and M. A. Stuart, pp. 25–29. USGS Forest and Rangeland Ecosystem Science Center CPFS Rep. Ser., 99/16, Flagstaff, Arizona.

Garret L. D., M. H. Soulen, and J. R. Ellenwood. 1997. After 100 years of forest management: "The north Kaibab." Proceedings of the Third Biennial Conference of Research on the Colorado Plateau, edited by C. van Riper III and E. Deshler, pp. 129–149. U.S. Department of the Interior National Park Service Transactions and Proceedings Series NPS/NRNAU/NRTP-97/12.

Gay, S. M. 1984. Winter range forage availability and utilization of range forage by pronghorn (Antilocapra americana) near Anderson Mesa. Master's thesis, Northern Arizona University, Flagstaff.

Grace, J. B. 1999. The factors controlling species density in herbaceous plant communities: An assessment. Perspectives in Plant Ecology, Evolution, and Systematics 2: 1–28.

Grime, J. P. 1973a. Competitive exclusion in herbaceous vegetation. Nature 242: 344–347.

Grime, J. P. 1973b. Control of species diversity in herbaceous vegetation. Journal of Environmental Management 1: 151–167.

Grime, J. P. 1979. Plant Strategies and Vegetation Processes. John Wiley and Sons, New York.

Harvey, A. E., M. J. Larsen, and M. F. Jurgensen. 1979. Comparative distribution of ectomycorrhizae in soils of three western Montana forest habitat types. Forest Science 25 (2): 350–358.

Hilden, O. 1965. Habitat selection in birds: A review. Ann. Zool. Fenn. 2: 53–75.

Hobbs, R. J. 1989. The nature and effects of disturbance relative to invasions. In Biological Invasions: A Global Perspective, edited by J. A. Drake, H. A. Mooney, F. di Castri, R. H. Groves, F. J. Kruger, M. Rejmanek, and M. Williamson, pp. 389–405. Wiley and Sons, New York.

Hobbs, R. J., and L. F. Huenneke. 1992. Disturbance, diversity, and invasion: Implications for conservation. Conservation Biology 6: 324–337.

Holmes, J., J. R. Spence, and M. K. Sogge. 2005. Birds of the Colorado River in Grand Canyon: A synthesis of status, trends and dam operation effects. In The State of the Colorado River Ecosystem, edited by S. P. Gloss, J. E. Lovich, and T. E. Melis, pp. 123–138. U.S. Geological Survey Circular 128.

Huston, M. A. 1979. A general hypothesis of species diversity. American Naturalist 113: 81–101.

Huston, M. A. 1994. Biological Diversity: The Coexistence of Species in Changing Landscapes. Cambridge University Press, Cambridge, UK.

Korb, J. E., and J. D. Springer. 2003. Understory vegetation. In Ecological Restoration of Southwestern Ponderosa Pine Forests, edited by P. Friederici, pp. 251–267. Island Press, Washington, D.C.

Korb, J. E., N. C. Johnson, and W. W. Covington. 2004. Slash pile burning effects on soil biotic and chemical properties and plant establishment: Recommendations for amelioration. Restoration Ecology 12 (1): 52–62.

Laughlin, D. C., J. D. Bakker, M. T. Stoddard, M. L. Daniels, J. D. Springer, C. N. Gildar, A. M. Green, and W. W. Covington. 2004. Toward reference conditions: Wildfire effects on flora in an old-growth ponderosa pine forest. Forest Ecology and Management 199: 137–152.

Levine, J. M., and C. M. D'Antonio. 1999. Elton revisited: A review of evidence linking diversity and invasibility. Oikos 87: 1–11.

Lewis, N. L. 2000. Arizona Wildlife Trophies. Arizona Wildlife Federation, Mesa.

Loeser, M. R., T. D. Sisk, T. E. Crews, K. Olsen, C. Moran, and C. Hudenko. 2001. Reframing the grazing debate: Evaluating ecological sustainability and bioregional food production. In Proceedings of the Fifth Conference of Research on the Colorado Plateau, edited by C. van Riper III, K. A. Thomas, and M. A. Stuart, pp. 3–18. U. S. Geological Survey/ Forest and Rangeland Ecosystem Science Center USGS FRESC/COPL/2001/21 Rep. Ser., Flagstaff, Arizona.

Luce, C. H. 1997. Effectiveness of road ripping in restoring infiltration capacity of forest roads. Restoration Ecology 5 (3): 265–270.

Manis, G., J. Lowry, and R. D. Ramsey. 2001. Preclassification: An ecologically predictive landform model. GAP Analysis Bulletin 10. U.S. Geological Survey, Biological Resources Division. Available at http: //www.gap.uidaho .edu/Bulletins/10/preclassification.htm.

Moore, M. M., W. W. Covington, and P. Z. Fulé. 1999. Reference conditions and ecological restoration: A Southwestern ponderosa pine perspective. Ecological Applications 9 (4): 1266.

Muñoz-Erickson, T. A., and B. J. Aguilar-Gonzalez. 2003. The use of ecosystem health indicators for evaluating ecological and social outcomes of the collaborative approach to management: The case study of the Diablo Trust. Prepared for the National Workshop on "Evaluating Methods and Environmental Outcomes of Community-based Collaborative Processes." Online Journal of the Community-based Collaborative Research Consortium (www.cbcrc).

Nabhan, G. P., S. Smith, M. Coder, and Z. Kovacs. 2005. Land-use history of three Colorado Plateau landscapes: Implications for restoration goal-setting. In The Colorado Plateau II: Biophysical, Socioeconomic, and Cultural Resources, edited by C. van Riper III and D. J. Mattson, pp. 101–119. University of Arizona Press, Tucson.

National Research Council. 1987. River and Dam Management: A Review of the Bureau of Reclamation's Glen Canyon Environmental Studies. National Academy Press, Washington, D.C.

National Research Council. 1996. River Resource Management in the Grand Canyon. National Academy Press, Washington, D.C.

National Research Council. 1999. Downstream. Adaptive Management of Glen Canyon Dam and the Colorado River Ecosystem. National Academy Press, Washington, D.C.

Naumburg, E., and L. E. DeWald. 1999. Relationships between Pinus ponderosa forest structure, light characteristics, and understory graminoid species presence and abundance. Forest Ecology and Management 124 (2-3): 205–215.

Neff, D. J., and N. G. Woolsey. 1980. Coyote predation on neonatal fawns on Anderson Mesa, Arizona. Proceedings Pronghorn Antelope Workshop 9: 80–93. Rio Rico, Arizona.

Neff, D. J., R. H. Smith, and N. G. Woolsey. 1985. Pronghorn antelope mortality study. Arizona Game and Fish Department Project W-78-R Final Report 1-22. Phoenix.

Ockenfels, R. A., C. van Riper III, and W. K. Carrel. 1997. Home ranges and movements of pronghorn in Northern Arizona. In Proceedings of the Third Biennial Conference of Research on the Colorado Plateau, edited by C. Van Riper III and E. T. Deshler, pp. 45–62. 97/12, 2 NPS Transactions and Proceedings Series NPS/NRNAU/NRTP-56.

Robinson, G. R., J. F. Quinn, and M. L. Stanton. 1995. Invasibility of experimental habitat islands in a California winter annual grassland. Ecology 76: 786–794.

Rosentreter, R. 1994. Displacement of rare plants by exotic grasses. In Proceedings, Ecology and Management of Annual Rangelands, edited by S. B. Monsen and S. G. Kitchen, pp. 170–175. General Technical Report INT-GTR-313. USDA-USFS, Intermountain Research Station, Ogden, Utah.

Rosenzweig, M. L. 1992. Species diversity gradients: We know more and less than we thought. Journal of Mammalogy 73: 715–730.

Rosenzweig, M. L. 1995. Species diversity in space and time. Cambridge University Press, Cambridge, UK.

Scott, J. M., B. Csuti, K. Smith, J. E. Estes, and S. Caicco. 1991. Gap analysis of species richness and vegetation cover: An integrated biodiversity conservation strategy. In Balancing on the Brink of Extinction: The Endangered Species Act and Lessons for the Future, edited by K. Kohm, pp. 282–297. Island Press, Washington D.C.

Scott, J. M., F. Davis, B. Csuti, R. Noss, B. Butterfield, C. Groves, H. Anderson, S. Caicco, F. D'Erchia, T. C. Edwards Jr., J. Ulliman, and R. G. Wright. 1993. Gap analysis: A geographic approach to protection of biological diversity. Wildlife Monographs 123: 1–41

Sieg, C. H., B. G. Phillips, and L. P. Moser. 2003. Exotic invasive plants. In Ecological Restoration of Southwestern Ponderosa Pine Forests, edited by P. Friederici, pp. 251–267. Island Press, Washington, D.C.

Simberloff, D. 1986. Introduced insects: A biogeographic and systematic perspective. In Ecology of Biological Invasions of North America and Hawaii, edited by H. A. Mooney and J. A. Drake, pp. 3–26. Springer-Verlag, New York.

Sisk, T. D., and J. Palumbo. 2005. Collaborative science: Making research a participatory endeavor for solving environmental challenges. The Quivira Coalition 7 (3).

Sisk, T. D., T. E. Crews, R. T. Eisfeldt, M. King, and E. Stanley. 1999. Assessing impacts of alternative livestock management practices: Raging debates and a role for science. In Proceedings of the Fourth Biennial Conference of Research on the Colorado Plateau, edited by C. van Riper III and M. A. Stuart, pp. 89–103. USGS Forest and Rangeland Ecosystem Science Center CPFS Rep. Ser., 99/16, Flagstaff, Arizona.

Spence, J. R. 2004. The riparian and aquatic bird communities along the Colorado River from Glen Canyon Dam to Lake Mead, 1996–2000. Final report to the U.S. Geological Survey

Grand Canyon Monitoring and Research Center, Flagstaff. Resource Management Division, Glen Canyon NRA.

Steidl, R. J., J. P. Hayes, and E. Schauber. 1997. Statistical power analysis in wildlife research. Journal of Wildlife Management 61: 270–279.

Stohlgren, T. J. 2002. Beyond theories of plant invasions: Lessons from natural landscapes. Comments on Theoretical Biology 7: 355–379.

Stohlgren, T. J., K. A. Bull, Y. Otsuki, C. A. Villa, and M. Lee. 1998. Riparian zones as havens for exotic plant species in the central grasslands. Plant Ecology 138: 113–125.

Stohlgren, T. J., D. Binkley, G. W. Chong, M. A. Kalkhan, L. D. Schell, K. A. Bull, Y. Otsuki, G. Newman, M. Bashkin, and Y. Son. 1999. Exotic plant species invade hot spots of native plant diversity. Ecological Monographs 69: 25–46.

Taylor, B. L., and T. Gerrodette. 1993. The uses of statistical power in conservation biology: The vaquita and northern spotted owl. Conservation Biology 7: 489–500.

Tilman, D. 1982. Resource competition and community structure. Princeton University Press, Princeton, New Jersey.

Tuhy, J. S., P. Comer, D. Dorfman, M. Lammert, J. Humke, B. Cholvin, G. Bell, B. Neely, S. Silbert, L. Whitham, and B. Baker. 2002. A Conservation Assessment of the Colorado Plateau Ecoregion. The Nature Conservancy, Moab Project Office, Moab, Utah.

Utah Division of Wildlife Resources. 2005. Utah's Watershed Restoration Initiative: Coming together to help rangelands. Available at http://www.wildlife.utah.gov/watersheds/upcd.php.

van Riper, C. III, editor. 1995. Proceedings of the Second Biennial Conference on Research in Colorado Plateau National Parks. NPS Transaction and Proceedings Series NPS/NRNAU/NRTP-95/11.

van Riper, C. III., and K. A. Cole, editors. 2004. The Colorado Plateau: Cultural, Biological, and Physical Research. University of Arizona Press, Tucson.

van Riper, C. III., and E. T. Deshler, editors. 1997. Proceedings of the Third Biennial Conference of Research on the Colorado Plateau. NPS Transactions and Proceedings Series NPS/NRNAU/NRTP-97/12. Denver, Colorado.

van Riper, C. III., and D. J. Mattson, editors. 2005. The Colorado Plateau II: Biophysical, Socioeconomic and Cultural Research. University of Arizona Press, Tucson.

van Riper, C. III, and R. A. Ockenfels. 1998. The influence of transportation corridors on the movement of pronghorn antelope over a fragmented landscape in northern Arizona. In Proceedings of the Second International Conference on Transportation and Wildlife Ecology, edited by D. Zeigler, pp. 241–248. Ft. Meyers, Florida.

van Riper, C. III, and M. K. Sogge. 2004. Bald eagle abundance and relationships to prey base and human activity along the Colorado River in Grand Canyon National Park. In The Colorado Plateau: Cultural, Biological, and Physical Research, edited by C. van Riper III and K. A. Cole, pp. 163–185. University of Arizona Press, Tucson.

van Riper, C. III, and M. A. Stuart, editors. 1999. Proceedings of the Fourth Biennial Conference of Research on the Colorado Plateau. USGS Forest and Rangeland Ecosystem Science Center CPFS Rep. Ser., 99/16, Flagstaff, Arizona.

van Riper, C. III, K. A. Thomas, and M. A. Stuart, editors. 2001. Proceedings of the Fifth Conference of Research on the Colorado Plateau. U.S. Geological Survey/ Forest and Rangeland Ecosystem Science Center USGSFRESC/COPL/2001/21 Rep. Ser. Flagstaff, Arizona.

Vitousek, P. M., L. R. Walker, L. D. Whiteaker, D. Mueller-Dombois, and P. A. Matson. 1987. Biological invasion by *Myrica faya* alters ecosystem development in Hawaii. Science 238: 802–804.

Wakeling, B. F., and W. H. Miller. 1990. A modified habitat suitability index for desert bighorn sheep. In Managing Wildlife in the Southwest, edited by P. R. Krausman and N. S. Smith, pp. 58–66. Arizona Chapter of the Wildlife Society, Phoenix.

Wallmo, O. C. 1951. Antelope range preference study. Completion Report, July 24, 1951. Project 46-R-2, J-5, Arizona Game and Fish Commission, Phoenix.

West, N. E., and J. A. Young. 2000. Intermountain Valleys and Lower Mountain Slopes. In North American Terrestrial Vegetation, 2nd ed., edited by M. G. Barbour and W. D. Billings. Cambridge University Press, New York.

Wilkins, A. S., and P. Welles. 1944. Antelope survey—1944. Project 9-R, Special Report. May 22 through June 1, 1944. Arizona Game and Fish Commission, Phoenix.

Yoakum, J. D. 2003. An assessment of pronghorn populations and habitat conditions on Anderson Mesa, Arizona: 2001–2002. Report prepared for Arizona Wildlife Federation, Mesa. Western Wildlife, Verdi, Nevada.

CONTRIBUTORS

Samantha Arundel
Department of Geography, Planning, and Recreation
Northern Arizona University
P.O. Box 15016
Flagstaff, AZ 86011
Samantha.Arundel@nau.edu

Terence Arundel
U.S. Geological Survey
Southwest Biological Science Center
Colorado Plateau Research Station
Northern Arizona University
P.O. Box 5614,
Flagstaff, AZ 86011
terry_arundel@usgs.gov

John Bailey
Oregon State University
Department of Forest Resources
Peavy Hall 235
Corvallis, OR 97331-5703
john.bailey@oregonstate.edu

William Block
USDA Forest Service
Rocky Mountain Research Station
2500 S. Pine Knoll Dr.
Flagstaff, AZ 86001
WBlock@fs.fed.us

Kenneth Boykin
New Mexico Cooperative Fish and Wildlife Research Unit
New Mexico State University
Box 30003, MSC 4901
Las Cruces, NM 88003
kboykin@nmsu.edu

David Brown
School of Life Sciences
Arizona State University
P.O. Box 874501
Tempe, AZ 85287-4501
david.e.brown@asu.edu

Jim Chew
USDA Forest Service
Rocky Mountain Research Station
Forestry Sciences Lab
800 E. Beckwith Ave.
Missoula, MT 59807
jchew@fs.fed.us

Neil Cobb
Merriam-Powell Center for Environmental Research
and Colorado Plateau Museum of Arthropod Biodiversity
Northern Arizona University
P.O. Box 6077
Flagstaff, AZ 86011
neil.cobb@nau.edu

Kenneth Cole
U.S. Geological Survey
Southwest Biological Science Center
P.O. Box 5614
Flagstaff, AZ 86011
ken_cole@usgs.gov

Craig Conley
The Quivira Coalition
1413 Second Street, Ste. 1
Santa Fe, NM 87505
bionomicssw@aol.com

Alycia Crall
Natural Resource Ecology Laboratory
Colorado State University
Fort Collins, CO 80523-1499
Current address: 868 Manchester Ct.
Geneva, IL 60134
mawaters@nrel.colostate.edu

Julie Crawford
Grand Canyon National Park
823 N. San Francisco St., Ste. C
Flagstaff, AZ 86001
Current address: 336 N. Main St.
Mancos, CO 81328
columbine_julie@yahoo.com

Brett Dickson
USDA Forest Service
Rocky Mountain Research Station
2500 S. Pine Knoll Dr.
Flagstaff, AZ 86001
dickson@cnr.colostate.edu

Amy Draut
U.S. Geological Survey
University of California-Santa Cruz
400 Natural Bridges Dr.
Santa Cruz, CA 95060
adraut@usgs.gov

Charles Drost
U.S. Geological Survey
Southwest Biological Science Center
2255 N. Gemini Dr.
Flagstaff, A 86001
Charles_Drost@usgs.gov

Philip Duffy
Lawrence Livermore National Laboratory
and University of California
7000 East Avenue
Livermore, CA 94550
duffy2@llnl.gov

Andrea Ernst
New Mexico Cooperative Fish and Wildlife
Research Unit
P.O. Box 3003, MSC 4901
Las Cruces, NM 88003
ernstae@nmsu.edu

Paul Evangelista
Natural Resource Ecology Laboratory
Colorado State University
Fort Collins, CO 80523-1499
paulevan@nrel.colostate.edu

Lisa Floyd-Hanna
Department of Environmental Studies
Prescott College
Prescott, AZ 86301
lfloyd-hanna@prescott.edu

Deb Guenther
Natural Resource Ecology Laboratory
Colorado State University
Fort Collins, CO 80523-1499
Current address: 2101 Salopek Rdl, #3
Las Cruces, NM 88005
debra.guenther@us.army.mil

Monica Hansen
U.S. Geological Survey
Southwest Biological Science Center
2255 N. Gemini Drive
Flagstaff, AZ 86001
mlhansen@usgs.gov

Jan Hart
U.S. Geological Survey
Southwest Biological Science Center
Colorado Plateau Research Station
Northern Arizona University
P.O. Box 5614
Flagstaff, AZ 86011
jan.hart@nau.edu

Sarah R. Hurteau
USDA Forest Service
Rocky Mountain Research Station
2500 South Pine Knoll Dr.
Flagstaff, AZ 86001
scr36@nau.edu

Kirsten Ironside
Merriam-Powell Center for Environmental
Research
Northern Arizona University
P.O. Box 6077
Flagstaff, AZ 86011
kirsten.ironside@nau.edu

Michele James
Ecological Monitoring and Assessment
Program and Foundation
Northern Arizona University
Flagstaff, AZ 86011
michele.james@nau.edu

Lisa Langs
Utah State University
College of Natural Resources - RS/GIS
Laboratory
5275 Old Main Hill
Logan,UT 84322
lisa.langs@usu.edu

George Leavesley
U.S. Geological Survey
M.S. 939, Federal Center
Box 25046
Denver, CO 80225
george@usgs.gov

Jonathan Long
Cooperative Extension
University of Arizona
P.O. Box 627
Peach Springs, AZ 86434
jwlong@ag.arizona.edu

John Lowry
Utah State University
College of Natural Resources - RS/GIS
Laboratory
5275 Old Main Hill
Logan, UT 84322
jlowry@gis.usu.edu

Janet Lynn
Ecological Monitoring and Assessment
Program and Foundation
Northern Arizona University
P.O. Box 5845
Flagstaff, AZ 86011
Janet.Lynn@nau.edu

David J. Mattson
U.S. Geological Survey
Southwest Biological Science Center
Colorado Plateau Research Station
Northern Arizona University
P.O. Box 5614, Flagstaff, AZ 86011
david_mattson@usgs.gov

David Mikesic
Navajo Natural Heritage Program
Navajo Nation Dept. of Fish and Wildlife
P.O. Box 1480
Window Rock, AZ 86515
dmikesic@hotmail.com

Mark Miller
U.S. Geological Survey
Southwest Biological Science Center
Canyonlands Research Station
2290 S. West Resource Blvd.
Moab, UT 84532
mark_miller@usgs.gov

Tischa Muñoz-Erickson
Arizona State University
School of Sustainability
P.O. Box 872511
Tempe, AZ 85287
Tischa.Munoz-Erickson@asu.edu

Erika Nowak
U.S. Geological Survey
Southwest Biological Science Center
Colorado Plateau Research Station
P.O. Box 5614
Northern Arizona University
Flagstaff, AZ 86011-5614
erika.nowak@nau.edu

Trevor Persons
USGS Southwest Biological Science Center
Colorado Plateau Research Station
Northern Arizona University
Box 5614
Flagstaff, AZ 86011-5614
Current address: 206 Bigelow Hill Road
Norridgewock, ME 04957
trevor.persons@nau.edu

Julie Prior-Magee
U.S. Geological Survey
Biological Resources Discipline
Gap Analysis Program
P.O. Box 3003, MSC 4901
Las Cruces, NM 88003
jpmagee@nmsu.edu

Erin Riddering
Arizona Game and Fish Department, Game
Branch
2221 W. Greenway Rd.
Phoenix, AZ 85023
eriddering@azgfd.gov

John Rihs
Grand Canyon National Park
P.O. Box 129
Grand Canyon, AZ 86023
John_Rihs@nps.gov

William Romme
Dept. of Forest, Rangeland and Watershed
Stewardship
Colorado State University
Ft. Collins, CO 80523
romme@warnercnr.colostate.edu

David Rubin
U.S. Geological Survey
400 Natural Bridges Dr.
Santa Cruz, CA 95060
drubin@usgs.gov

George San Miguel
Mesa Verde National Park
P.O. Box 8
Mesa Verde, CO 81330
george_san_miguel@nps.gov

Keith Schulz
NatureServe
4001 Discovery Drive, Ste. 2110
Boulder, CO 80303
Keith_Schulz@natureserve.org

John Shaw
USDA Forest Service
Rocky Mountain Research Station
507 25th St.
Ogden, UT 84401
jdshaw@fs.fed.us

Thomas Sisk
Center for Environmental Science and
Education
Northern Arizona University
P.O. Box 5694
Flagstaff, AZ 86001
Thomas.Sisk@nau.edu

Mark Sogge
U.S. Geological Survey
Southwest Biological Science Center
Colorado Plateau Research Station
Northern Arizona University
P.O. Box 5614
Flagstaff, AZ 86011
mark_sogge@usgs.gov

Robert Speer
City of Flagstaff
GIS Division
Flagstaff, AZ 86001-5018
rspeer@ci.flagstaff.az.us

John Spence
Resource Management & Interpretation Div.
Glen Canyon National Recreation Area
P.O. Box 1507
Page, AZ 86040
john_spence@nps.gov

Thomas Stohlgren
U.S. Geological Survey
Fort Collins Science Center
2150 Centre Avenue
Fort Collins, CO 80526
tom_stohlgren@usgs.gov

Kathryn Thomas
U.S. Geological Survey
Southwest Biological Science Center
Sonoran Desert Research Station
University of Arizona
125 Biological Sciences East
Tucson, AZ 85721
kathryn_a_thomas@usgs.gov

Whitney Tilt
Sonoran Institute
201 S. Wallace Ave., Ste. B3C
Bozeman, MT 59715
whitney@sonoran.org

Joseph Trudeau
Ecological Restoration Institute
Northern Arizona University
P.O. Box 15017
Flagstaff, AZ 86011-5017
Current address: Preserve Land Works
52 Kimball Rd., Hancock, NH 03449
jm_trudeau@msn.com

Christine Turner
U.S. Geological Survey
M.S. 939, Federal Center
Box 25046
Denver, CO 80225
cturner@usgs.gov

Charles van Riper III
U.S. Geological Survey
Southwest Biological Science Center
Sonoran Desert Research Station
University of Arizona
125 Biological Sciences East
Tucson, AZ 85721
charles_van_riper@usgs.gov

Roland Viger
U.S. Geological Survey
M.S. 939, Federal Center
Box 25046
Denver, CO 80225
rviger@usgs.gov

Brian Wakeling
Arizona Game and Fish Department, Game
Branch
2221 W. Greenway Rd.
Phoenix, AZ 85023
bwakeling@azgfd.gov

Peter Warren
The Nature Conservancy
Arizona Chapter
1510 E. Fort Lowell Rd.
Tucson, AZ 85719
pwarren@tnc.org

J. Judson Wynne
U.S. Geological Survey
Southwest Biological Science Center
2255 N. Gemini Dr.
Flagstaff, AZ 86001
jwynne@usgs.gov

Richard Zirbes
U.S. Geological Survey
M.S. 939, Federal Center
Box 25046
Denver, CO 80225
rzirbes@aol.com

INDEX